Overcoming Challenges in Software Engineering Education:

Delivering Non–Technical Knowledge and Skills

Liguo Yu
Indiana University South Bend, USA

A volume in the Advances in Higher
Education and Professional Development
(AHEPD) Book Series

An Imprint of IGI Global

Managing Director:	Lindsay Johnston
Production Editor:	Jennifer Yoder
Development Editor:	Vince D'Imperio
Acquisitions Editor:	Kayla Wolfe
Typesetter:	John Crodian
Cover Design:	Jason Mull

Published in the United States of America by
Engineering Science Reference (an imprint of IGI Global)
701 E. Chocolate Avenue
Hershey PA 17033
Tel: 717-533-8845
Fax: 717-533-8661
E-mail: cust@igi-global.com
Web site: http://www.igi-global.com

Library of Congress Cataloging-in-Publication Data

Overcoming challenges in software engineering education : delivering non-technical knowledge and skills / Liguo Yu, editor.
    pages cm
  Summary: "This book combines recent advances and best practices to improve the curriculum of software engineering education, bridging the gap between industry expectations and what academia can provide in software engineering education"-- Provided by publisher.
  Includes bibliographical references and index. ISBN 978-1-4666-5800-4 (hardcover) -- ISBN 978-1-4666-5801-1 (ebook) -- ISBN 978-1-4666-5803-5 (print & perpetual access) 1. Software engineering--Study and teaching (Higher) I. Yu, Liguo, 1969-
  QA76.758.O85 2014
  005.107--dc23
                                        2014000885

This book is published in the IGI Global book series Advances in Higher Education and Professional Development (AHEPD) (ISSN: 2327-6983; eISSN: 2327-6991)

British Cataloguing in Publication Data
A Cataloguing in Publication record for this book is available from the British Library.

For electronic access to this publication, please contact: eresources@igi-global.com.

Advances in Higher Education and Professional Development (AHEPD) Book Series

Jared Keengwe
University of North Dakota, USA

ISSN: 2327-6983
EISSN: 2327-6991

MISSION

As world economies continue to shift and change in response to global financial situations, job markets have begun to demand a more highly-skilled workforce. In many industries a college degree is the minimum requirement and further educational development is expected to advance. With these current trends in mind, the **Advances in Higher Education & Professional Development (AHEPD) Book Series** provides an outlet for researchers and academics to publish their research in these areas and to distribute these works to practitioners and other researchers.

AHEPD encompasses all research dealing with higher education pedagogy, development, and curriculum design, as well as all areas of professional development, regardless of focus.

COVERAGE

- Adult Education
- Assessment in Higher Education
- Career Training
- Coaching & Mentoring
- Continuing Professional Development
- Governance in Higher Education
- Higher Education Policy
- Pedagogy of Teaching Higher Education
- Vocational Education

IGI Global is currently accepting manuscripts for publication within this series. To submit a proposal for a volume in this series, please contact our Acquisition Editors at Acquisitions@igi-global.com or visit: http://www.igi-global.com/publish/.

Titles in this Series

For a list of additional titles in this series, please visit: www.igi-global.com

Cases on Teacher Identity, Diversity, and Cognition in Higher Education
Paul Breen (Greenwich School of Management, UK)
Information Science Reference • copyright 2014 • 346pp • H/C (ISBN: 9781466659902) • US $195.00 (our price)

Trends in European Higher Education Convergence
Alina Mihaela Dima (Bucharest University of Economic Studies, Romania)
Information Science Reference • copyright 2014 • 402pp • H/C (ISBN: 9781466659988) • US $215.00 (our price)

Overcoming Challenges in Software Engineering Education Delivering Non-Technical Knowledge and Skills
Liguo Yu (Indiana University South Bend, USA)
Engineering Science Reference • copyright 2014 • 343pp • H/C (ISBN: 9781466658004) • US $215.00 (our price)

Multicultural Awareness and Technology in Higher Education Global Perspectives
Tomayess Issa (Curtin University, Australia) Pedro Isaias (Universidade Aberta (Portuguese Open University), Portugal) and Piet Kommers (University of Twente, The Netherlands)
Information Science Reference • copyright 2014 • 322pp • H/C (ISBN: 9781466658769) • US $195.00 (our price)

Cutting-Edge Technologies and Social Media Use in Higher Education
Vladlena Benson (Kingston University, UK) and Stephanie Morgan (Kingston University, UK)
Information Science Reference • copyright 2014 • 462pp • H/C (ISBN: 9781466651746) • US $215.00 (our price)

Building Online Communities in Higher Education Institutions Creating Collaborative Experience
Carolyn N. Stevenson (Kaplan University, USA) and Joanna C. Bauer (Kaplan University, USA)
Information Science Reference • copyright 2014 • 473pp • H/C (ISBN: 9781466651784) • US $205.00 (our price)

Using Technology Tools to Innovate Assessment, Reporting, and Teaching Practices in Engineering Education
Firoz Alam (RMIT University, Australia)
Engineering Science Reference • copyright 2014 • 409pp • H/C (ISBN: 9781466650114) • US $215.00 (our price)

Cross-Cultural Online Learning in Higher Education and Corporate Training
Jared Keengwe (University of North Dakota, USA) Gary Schnellert (University of North Dakota, USA) and Kenneth Kungu (Tennessee State University, USA)
Information Science Reference • copyright 2014 • 337pp • H/C (ISBN: 9781466650237) • US $175.00 (our price)

Cases on Critical and Qualitative Perspectives in Online Higher Education
Myron Orleans (California State University at Fullerton, USA)
Information Science Reference • copyright 2014 • 604pp • H/C (ISBN: 9781466650510) • US $195.00 (our price)

www.igi-global.com

701 E. Chocolate Ave., Hershey, PA 17033
Order online at www.igi-global.com or call 717-533-8845 x100
To place a standing order for titles released in this series, contact: cust@igi-global.com
Mon-Fri 8:00 am - 5:00 pm (est) or fax 24 hours a day 717-533-8661

Table of Contents

Section 3
Supporting Communications

Section 4
Improving Soft Skills

Section 5
Promoting Project-Based Learning

Section 6
Engaging Classroom Games

Section 7
Experiencing Case-Based Teaching and Problem-Based Learning

Section 8
Meeting Industry Expectations

Section 9
Using Open-Source Tools

Detailed Table of Contents

Section 1
Developing Project Management Skills

Chapter 1

Kasi Periyasamy, University of Wisconsin – La Crosse, USA

Software project management is an inherent part of software engineering. While technical expertise is an important factor to complete a software product, knowledge and experience in project management are equally important. Teaching software project management is always a challenge. Most software engineering courses teach technical skills and knowledge on software development but lack project management guidance. On the other hand, project management courses taught by management faculty do not connect to technical activities. Therefore, a blend of technical and managerial skills must be taught together to train software engineers. This chapter describes the author's experience in teaching a graduate level software project management course with emphasis on blending technical and non-technical skills. The chapter includes the different modes/styles in which the course was taught, the challenges faced, the benefits gained, and the current status of the course.

Chapter 2

Esperanza Marcos Martínez, Rey Juan Carlos University, Spain
Juan M. Vara Mesa, Rey Juan Carlos University, Spain
Verónica A. Bollati, Rey Juan Carlos University, Spain
Marcos López-Sanz, Rey Juan Carlos University, Spain

This chapter summarizes the experience of applying coaching techniques to the teaching of leadership and team management, which is taught as part of the project management course in the Computer Science degree at a large public Spanish university. These skills have, until now, been taught through lectures. However, there are some key strengths or abilities that a good leader or team manager possesses, such as the ability to work well in a team, communication skills, etc., which a student cannot learn in a lecture hall. The authors have therefore decided to change the teaching method to one in which the student is converted into the protagonist and the professor takes on the role of the coach, thus becoming a facilitator for the student's learning process. The classes were organized in workshops, carried out in seminars outside the usual lecture room, and each workshop was dedicated to a specific skill or ability. In the first session, the students felt disconcerted and a little shy and were reluctant to participate in some of the professor's proposals. Nevertheless, during the five workshops of which the experience consisted, the students became more participative and were highly contented, and it will therefore be refined and repeated during the next academic year. This chapter provides details of the experience, highlighting the methodology used in each of the workshops, in addition to the conclusions eventually reached and possible improvements that could be made in the future.

Section 2
Encouraging Collaborations and Teamwork

The agile methodologies are part of a shift from predictive to adaptive approach towards software development. This change has had a notable impact on Software Engineering Education (SEE). In this chapter, a glimpse into the state-of-the-art of incorporating agile methodologies in software engineering courses is presented. In doing so, the reasons for including a project component in software engineering courses, and for committing to agile methodologies in software engineering courses, are given. To lend an understanding to the notion of collaboration in agile methodologies, a conceptual model for collaboration is proposed and elaborated. The pivotal role of collaboration in agile course projects is emphasized. The use of certain means for facilitating collaboration, including the Social Web, is discussed.

A major challenge to teaching software engineering is achieving functioning teams that enforce individual accountability while integrating software engineering principles, approaches, and techniques. The two-semester software engineering course at the University of Texas at El Paso, referred to as the Team-Oriented Software Engineering (TOSE) course, establishes communities of practice that are cultivated through cooperative group practices and an improvement process model that enables learning from past experiences. The experience of working with incomplete, ambiguous, and changing software requirements motivates the need for applying disciplined software engineering practices and approaches throughout project development. Over the course of the two-semester sequence, the nature of students' participation in project teams changes: they begin to influence others in software engineering practice, and their identities as software engineers begins to develop. The purpose of the chapter is to describe how to structure a software engineering course that results in establishing communities of practice in which learners become increasingly more knowledgeable team members who embody the skills needed to work effectively in a team- and project-based environment.

At the Department of Mathematics and Informatics, Faculty of Science, University of Novi Sad, elements of Web 2.0 have been used in teaching for several years. In particular, this is emphasized in encouragement of teamwork, through usage of Wiki technology within several courses. Initially, those courses were created as a part of a large international project that recommended the use of teamwork. Over the years, additional elements of Web 2.0 were introduced, while employment and utilization of teamwork was largely enhanced and suitably organized. In this chapter, the authors share their experiences with such work, starting from introductory methods of enhancing the chosen learning management system, Moodle, with the mentioned activities, up to looking beyond their simple application and extracting additional value for courses.

Section 3
Supporting Communications

Chapter 6

This chapter discusses the teaching of Requirements Engineering (RE) through a segmented approach. The idea is to teach this field, step by step, beginning with the requirements elicitation phase, which is the main focus of the chapter. The recommended linguistics-based method advocates the training of students in textual analysis techniques in order to develop their metacognitive and interpersonal skills, specifically, abstraction and comprehension. These skills are key soft skills for the practice of requirements elicitation.

Chapter 7

Teaching non-technical skills such as communication skills to Software Engineering (SE) students is relatively more difficult than teaching technical skills. This chapter explores teaching the peer feedback aspect of communication skills to SE students. It discusses several peer feedback mechanisms that can be used in an SE course, and for each mechanism, it discusses the potential challenges and the various approaches that can be used to overcome those challenges.

Section 4
Improving Soft Skills

This chapter explores the findings from an Action Research project that addressed the Professional Capability Framework (Scott & Wilson, 2002), and how aspects of this were embedded in an undergraduate Engineering (Software) degree. Longitudinal data identified the challenges both staff and students engaged with. The interventions that were developed to address these are described and discussed. The results of the project show that making soft skills attainment explicit as part of the learning objectives went a long way in assisting students to engage with the activities that exercised these skills.

Chapter 9

Software Engineering requires a specific profile of technical expertise combined with context-sensitive soft skills. Therefore, university education in software engineering should foster both technical knowledge and soft skills. Students should be enabled to cope with complex situations in real life by applying and combining their theoretical knowledge with team and communication competencies. In this chapter, the authors report findings from a software engineering project course. They argue that project work is a suitable approach to foster soft skills. To that end, the authors provide justification from a pedagogical point of view, setting project-based learning into relation to action-orientated didactics. As teaching goals, they focus on experiencing a complete development project from end to end, following a software process model that needs to be adapted to the specific situation, self-determined planning and acting, including the organization of the project, teamwork and team communication, and self-reflection on individual roles and contributions, and on the performance of the project team as a whole. In order to achieve these goals, the authors form teams of bachelor students, which are headed by one master student each. It turned out that a clear separation of roles is inevitable within the team, but also with respect to instructors. Self-reflection processes concerning the team roles and the individual competencies are explicitly stimulated and cumulate in individual self-reports and post-mortem analysis sessions. The authors share findings of how well the approaches have worked and outline some ideas to improve things.

Chapter 10

The job profile of a Software Engineer not only includes so-called "hard-skills" (e.g. specifying, programming, or building architectures) but also "soft skills" like awareness of team effects and similar human factors. These skills are typically hard to teach in classrooms, and current education, hence, mostly focuses on hard rather than soft skills. Yet, since software development is becoming more and more spread across different sites in a globally distributed manner, the importance of soft skills increases rapidly. However, there are only a few practical guides to teach such tacit knowledge to Software Engineering students. In this chapter, the authors describe an approach that combines theoretical lectures, practical experiments, and discussion sessions to fill this gap. They describe the processes of creating, planning, executing, and evaluating these sessions, so that soft skill topics can be taught in a university course. The authors present two example implementations of the approach. The first implementation lets students experience and reflect on group dynamics and team-internal effects in a project situation. The second implementation enables students to understand the challenges of a distributed software development setting. With this knowledge, the authors critically discuss the contribution of experimentation to university teaching.

Chapter 11
Lynette Johns-Boast, Australian National University, Australia

Although industry acknowledges university graduates possess strong technical knowledge, it continues to lament the lack of commensurately strong personal and professional skills that allow graduates to apply their technical knowledge and to become effective members of the workforce quickly. This chapter outlines a research-backed course design that blends experiential learning to create an industrial simulation, the rewards of which go well beyond the usual benefits of group-project capstone design courses. The simulated industrial context facilitates the graduation of software engineers who possess the requisite personal and professional attributes. Innovations include combining two cohorts of students into one, engaging industry partners through the provision and management of projects, and implementing proven education approaches that promote the development of personal and professional skills. Adoption of the suggested practices will help institutions produce "work-ready" graduates repeatedly, year after year, even by software engineering academics who may not have received teacher training and who may not possess significant industry experience themselves.

Section 5
Promoting Project-Based Learning

Chapter 12
Luís M. Alves, Instituto Politécnico de Bragança, Portugal
Pedro Ribeiro, Universidade do Minho, Portugal
Ricardo J. Machado, Universidade do Minho, Portugal

The lack of preparation of Software Engineering (SE) graduates for a professional career is a common complaint raised by industry practitioners. One approach to solving, or at least mitigating, this problem is the adoption of the Project-Based Learning (PBL) training methodology. Additionally, the involvement of students in real industrial projects, incorporated as a part of the formal curriculum, is a well-accepted means for preparing students for their professional careers. The authors involve students from BSc, MSc, and PhD degrees in Computing in developing a software project required by a real client. This chapter explains the educational approach to training students for industry by involving them with real clients within the development of software projects. The educational approach is mainly based on PBL principles. With the approach, the teaching staff is responsible for creating an environment that enhances communications, teamwork, management, and engineering skills in the students involved.

Ezequiel Scott, ISISTAN (UNICEN-CONICET) Research Institute, Argentina
Guillermo Rodríguez, ISISTAN (UNICEN-CONICET) Research Institute, Argentina
Álvaro Soria, ISISTAN (UNICEN-CONICET) Research Institute, Argentina
Marcelo Campo, ISISTAN (UNICEN-CONICET) Research Institute, Argentina

Software Engineering courses aim to train students to succeed in meeting the challenges within competitive and ever-changing professional contexts. Thus, undergraduate courses require continual revision and updating so as to cater for the demands of the software industry and guarantee academic quality. In this context, Scrum results in both a suitable and a flexible framework to train students in the implementation of professional software engineering practices. However, current approaches fail to provide guidance and assistance in applying Scrum, or a platform to address limitations in time, scope, and facilities within university premises. In this chapter, the authors present a software engineering training model based on the integration of the Agile Coach role and a virtual-reality platform called Virtual Scrum. The findings highlight the benefits of integrating this innovative model in a capstone course. Not only does this approach strengthen the acquisition of current software engineering practices but also opens new possibilities in the design of training courses.

Marc Lainez, Agilar, Belgium
Yves Deville, Université Catholique de Louvain, Belgium
Adrien Dessy, Université Catholique de Louvain, Belgium
Cyrille Dejemeppe, Université Catholique de Louvain, Belgium
Jean-Baptiste Mairy, Université Catholique de Louvain, Belgium
Sascha Van Cauwelaert, Université Catholique de Louvain, Belgium

This chapter shows how a lightweight Agile process has been used to introduce Agile project development to young computer science students. This experience has been conducted on a project aimed at developing Android applications. The context, the process, and the results of this experiment are described in this chapter.

Section 6
Engaging Classroom Games

Sakgasit Ramingwong, Chiang Mai University, Thailand
Lachana Ramingwong, Chiang Mai University, Thailand

Software development is uniquely different especially when compared to other engineering processes. The abstractness of software products has a major influence on the entire software development life cycle, which results in a number of uniquely important challenges. This chapter describes and discusses Engineering Construction for Software Engineers (ECSE), an effective workshop that helps software engineering students to understand some of these critical issues within a short period of time. In this workshop, the students are required to develop a pseudo-software product from scratch. They could learn about unique characteristics and risks of software development life cycle as well as other distinctive phenomenon through the activities. The workshop can still be easily followed by students who are not familiar with certain software development processes such as coding or testing.

Chapter 16

Elizabeth Suescún Monsalve, Pontifical Catholic University of Rio de Janeiro, Brazil
Allan Ximenes Pereira, Rio de Janeiro State University, Brazil
Vera Maria B. Werneck, Rio de Janeiro State University, Brazil

This chapter addresses the application of computer games and simulations in order to explore reality in many educational areas. The Games-Based Learning (GBL) can improve the teaching and learning experience by training future professionals in real life scenarios and activities that enable them to apply problem-solving strategies by putting into use the correct technique stemming from their own skills. For that reason, GBL has been used in software engineering teaching. At Pontifical Catholic University of Rio de Janeiro, the authors have developed SimulES-W (Simulation in Software Engineering), a tool for teaching software engineering. SimulES-W is a collaborative software board game that simulates a software engineering process in which the player performs different roles such as software engineer, technical coordinator, project manager, and quality controller. The players can deal with budget, software engineer employment and dismissal, and construction of different software artifacts. The objective of this chapter is to describe the approach to teaching software engineering using SimulES-W and demonstrate how pedagogical methodology is applied in this teaching approach to improve software engineering education. The teaching experience and future improvements are also discussed.

Section 7
Experiencing Case-Based Teaching and Problem-Based Learning

Chapter 17

Salamah Salamah, The University of Texas – El Paso, USA
Massood Towhidnejad, Embry Riddle Aeronautical University, USA
Thomas Hilburn, Embry Riddle Aeronautical University, USA

While many Software Engineering (SE) and Computer Science (CS) textbooks make use of case studies to introduce difference concepts and methods, the case studies introduced by these texts focus on a specific life-development phase or a particular topic within software engineering object-oriented design and implementation or requirements analysis and specification. Moreover, these case studies usually do not come with instructor guidelines on how to adopt the introduced material to the instructor's teaching style or to the particular level of the class or students in the class. The DigitalHome Case Study aims at addressing these shortcomings by providing a comprehensive set of artifacts associated with the full software development life-cycle. The project provides an extensive set of case study modules with exercises for teaching different topics in software engineering and computer science, as well as guidance for instructors on how to use these case modules. In this chapter, the authors motivate the use of the case study approach in teaching SE and CS concepts. They provide a description of the DigitalHome case study and the associated artifacts and case modules. The authors also report on the use of the developed material.

Oisín Cawley, The National College of Ireland, Ireland

Stephan Weibelzahl, The National College of Ireland, Ireland

Ita Richardson, University of Limerick, Ireland

Yvonne Delaney, University of Limerick, Ireland

With a focus on addressing the perceived skills gap in Software Engineering (SE) graduates, some educators have looked to employing alternative teaching and learning strategies in the classroom. One such pedagogy is Problem-Based Learning (PBL), an approach the authors have incorporated into the SE curriculum in two separate third-level institutions in Ireland, namely the University of Limerick (UL) and the National College of Ireland (NCI). PBL is an approach to teaching and learning which is quite different to the more typical "lecture" style found in most 3rd level institutions. PBL allows lecturers to meet educational and industry-specific objectives; however, while it has been used widely in Medical and Business schools, its use has not been so widespread with computing educators. PBL is not without its difficulties given that it requires significant changes in the role of the lecturer and the active participation of the students. Here, the authors present the approach taken to implement PBL into their respective programs. They present the pitfalls and obstacles that needed to be addressed, the levels of success that have been achieved so far, and briefly discuss some of the important aspects that Software Engineering lecturers should consider.

Section 8
Meeting Industry Expectations

Chapter 19

Bonnie K. MacKellar, St. John's University, USA

Mihaela Sabin, University of New Hampshire, USA

Allen B. Tucker, Bowdoin College, USA

Too often, computer science programs offer a software engineering course that emphasizes concepts, principles, and practical techniques, but fails to engage students in real-world software experiences. The authors have developed an approach to teaching undergraduate software engineering courses that integrates client-oriented project development and open source development practice. They call this approach the Client-Oriented Open Source Software (CO-FOSS) model. The advantages of this approach are that students are involved directly with a client, nonprofits gain a useful software application, and the project is available as open source for other students or organizations to extend and adapt. This chapter describes the motivation, elaborates the approach, and presents the results in substantial detail. The process is agile and the development framework is transferrable to other one-semester software engineering courses in a wide range of institutions.

Chapter 20

Paolo Ciancarini, University of Bologna, Italy
Stefano Russo, University of Naples Federico II, Italy

In this chapter, the authors describe their experiences in designing, developing, and teaching a course on Software Architecture that tested both in an academic context with their graduate Computer Science students and in an advanced context of professional updating and training with scores of system engineers in a number of different companies. The course has been taught in several editions in the last five years. The authors describe its rationale, the way in which they teach it differently in academia and in industry, and how they evaluate the students' learning in the different contexts. Finally, the authors discuss the lessons learnt and describe how this experience is inspiring for the future of this course.

Section 9
Using Open-Source Tools

Chapter 21

Jagadeesh Nandigam, Grand Valley State University, USA
Venkat N Gudivada, Marshall University, USA

This chapter describes a pragmatic approach to using open source and free software tools as valuable resources to affect learning of software industry practices using iterative and incremental development methods. The authors discuss how the above resources are used in teaching undergraduate Software Engineering (SE) courses. More specifically, they illustrate iterative and incremental development, documenting software requirements, version control and source code management, coding standards compliance, design visualization, software testing, software metrics, release deliverables, software engineering ethics, and professional practices. The authors also present how they positioned the activities of this course to qualify it for writing intensive designation. End of semester course evaluations and anecdotal evidence indicate that the proposed approach is effective in educating students in software industry practices.

Chapter 22

Liguo Yu, Indiana University South Bend, USA
David R. Surma, Indiana University South Bend, USA
Hossein Hakimzadeh, Indiana University South Bend, USA

Software development is a fast-changing area. New methods and new technologies emerge all the time. As a result, the education of software engineering is generally considered not to be keeping pace with the development of software engineering in industry. Given the limited resources in academia, it is unrealistic to purchase all the latest software tools for classroom usage. In this chapter, the authors describe how free/open-source data and free/open-source tools are used in an upper-level software engineering class at Indiana University South Bend. Depending on different learning objectives, different free/open-source tools and free/open-source data are incorporated into different team projects. The approach has been applied for two semesters, where instructor's experiences are assembled and analyzed. The study suggests (1) incorporating both free/open-source tools and free/open-source data in a software engineering course so that students can better understand both development methods and development processes and (2) updating software engineering course regularly in order to keep up with the advance of development tools and development methods in industry.

Section 10
Adopting Digital Learning

Collaborative learning methods have been widely applied in online learning environments to increase the effectiveness of the STEM programs. However, simply grouping students and assigning them projects and homework does not guarantee that they will get effective learning outcomes and improve their collaboration skills. This chapter shows that students can improve their learning outcomes and non-technical skills (e.g. collaboration and communication skills) through the cyber-enabled learning environment. The data was collected mainly from software engineering and object-oriented design classes of both graduates and undergraduates. The authors apply a blended version of education techniques by taking advantage of online environment and classroom teaching. Based on the study, the authors show that students can improve their collaboration and communication skills as well as other learning outcomes through the blended version of learning environment.

Software Engineering education involves two learning aspects: (1) teaching theoretical material and (2) conducting the practical labs. Currently, Software Engineering education faces a challenge, which comes from the new learning opportunities afforded by the Web technologies. Delivering a Software Engineering curriculum by online distance learning requires innovative and flexible approaches to present and manage the theoretical and practical learning materials. E-Learning could support Software Engineering education through utilizing special e-Learning concepts, techniques, and tools. E-Learning could also change the mode of teaching from knowledge-as-transmission to knowledge-as-construction. This is called "Software Engineering e-Learning." This chapter provides a review on Software Engineering education and e-Learning technology. It explores the need to adopt a Software Engineering e-Learning model to help the facilitators/instructors prepare and manage the online Software Engineering courses. This chapter also addresses how e-Learning environment could simplify the application of the constructivist learning model towards Software Engineering education.

Preface

We are living in an evolving society, where we can feel the changes in every aspect of our lives: communication, transportation, healthcare, education, career building, financial management, entertainment, social networking, and so on. Part of the driving forces of the current evolution of human society is information technology, where software products play an important role in shaping the way we live and how the society works. Accordingly, the functions and the qualities provided by software products could affect everyone in our society.

Software engineering is the discipline we follow to produce software products. The discipline itself and how well it is being followed could affect software products, which in turn could affect our lives. Software engineering education is the process we educate and train professional software engineers to deliver disciplined software engineering knowledge and skills. Therefore, software engineering education has an indirect impact on our evolving society.

Currently, software engineering education has two major challenges. On the one hand, software development is a fast-changing area. New methods and new technologies emerge every year. As a result, the education of software engineering is generally considered not to be keeping pace with the development of software engineering in industry. On the other hand, lecture-based software engineering education hardly engages and convinces students. Students often view software engineering principles as mere academic concepts and do not know how to use them in practice. In some cases, students even consider software engineering courses less interesting, less valuable, and less practical. However, the reality is computer science graduates often find that software engineering knowledge and skills are more in demand after they join the industry.

Ideally, most of the software engineering principles should be learned through real-world industry practices. However, given the limited resources in academia, it is not an easy task to deliver industry-standard knowledge and skills, especially non-technical knowledge and skills, in a software engineering classroom. For example, most software engineering educators are computer scientists instead of communication or management experts; they often have difficulties finding resources to support the delivery of communication skills, team-working skills, management skills, and so on. Accordingly, there is a gap between the industry expectations and what academia can provide in software engineering education.

This book presents the recent advances in software engineering education to overcome the aforementioned challenges. Through assembling the current best practices in software engineering education, this book intends to provide guidelines for software engineering educators to improve their curricula and instruct computer science students with ready-to-apply software engineering knowledge and skills.

This book aims to be useful for both research and teaching. The target audience will be software engineering researchers and educators in computer science programs, software engineering programs, or information technology programs of higher education institutions.

The book is organized into 10 sections, 24 chapters, which cover *developing project management skills, encouraging collaborations and teamwork, supporting communications, improving soft skills, promoting project-based learning, engaging classroom games, experiencing case-based teaching and problem-based learning, meeting industry expectations, using open-source tools,* and *adopting digital learning*. Below, we provide a brief summary of the 24 chapters.

BOOK LAYOUT

Sections

1. Developing Project Management Skills (Chapters 1-2)
2. Encouraging Collaborations and Teamwork (Chapters 3-5)
3. Supporting Communications (Chapters 6-7)
4. Improving Soft Skills (Chapters 8-11)
5. Promoting Project-Based Learning (Chapters 12-14)
6. Engaging Classroom Games (Chapters 15-16)
7. Experiencing Case-Based Teaching and Problem-Based Learning (Chapters 17-18)
8. Meeting Industry Expectations (Chapters 19-20)
9. Using Open-Source Tools (Chapters 21-22)
10. Adopting Digital Learning (Chapter 23-24)

Chapters

In chapter 1, Kasi Periyasamy describes his project management course, which has been used as a vehicle to deliver nontechnical skills to software engineering students. Various approaches, such as teamwork and conflict resolution, are used to improve students' managerial skills. Through team projects, domain knowledge and development skills are also delivered to students.

In chapter 2, Marcos Martínez et al. present their approach and experience of teaching leadership and team management skills in computer science program. Their coaching-based approach is realized through workshops, where students play the major roles and the instructor acts as a coach. Through participating in these workshops, students could also practice other nontechnical skills, such as team-working skills and communication skills.

In chapter 3, Pankaj Kamthan discusses collaborations in software engineering courses. Various interactions between students and other stakeholders are discussed. Specifically, the use of Social Web, such as wiki, is discussed to support collaborations and communications in an agile course project. In addition, the author proposes a conceptual collaboration model to represent human-human interactions.

In chapter 4, Ann Gates et al. describe their approach to teaching team working skills following a framework called TOSE (Team-Oriented Software Engineering). TOSE allows students to develop team-working skills through cooperative learning that encourages continuous improvement of team functions during the software development process. Trained through this approach, students could be increasingly more knowledgeable and become effective software engineering practitioners.

In chapter 5, Mirjana Ivanovic et al. present their experience of encouraging students to communicate and cooperate through Web 2.0 and social networking. Quantitative analysis is performed to evaluate their teaching approach. These experiences could be shared by other instructors who are going prepare their students with distributed working skills.

In chapter 6, Marcel Fouda Ndjodo and Virginie Blanche Ngah analyze their approach to improving students' communication skills in requirement elicitation. To communicate efficiently with the clients, the authors propose a five-step (know, comprehend, apply, analyze, and synthesize) approach to enriching students linguistic knowledge and improving their communication skills, interpretation skills, and representation skills.

In chapter 7, Damith Rajapakse explains how peer feedback is used as a communication mechanism in his software engineering course. Peer feedback has been used for team meetings, code reviews, and peer mentoring. Overall, peer feedback is demonstrated to be an effective mechanism to improve the communications and collaborations among developers.

In chapter 8, Jocelyn Armarego details a series of interventions used in a software engineering course aimed at better teaching employability skills to students. The purpose of putting the interventions in place is to better align the capabilities of graduates of a software engineering course with employer expectations in the area of soft skills. The three approaches tried in this research are the Cognitive Apprenticeship model, the Problem-Based Learning, and the Studio approach. This chapter also provides extensive reports on student reactions to the interventions.

In chapter 9, Yvonne Sedelmaier and Dieter Landes discuss practicing soft skills through a project-based didactical approach. One of the interesting parts about their approach is that each team contains both undergraduate students and graduate students, where graduates work as team leaders and are responsible for scheduling and project management, and undergraduate students are responsible for implementation. The project-based didactical approach is proven to be successful in delivering both technical skills and soft skills.

In chapter 10, Marco Kuhrmann et al. propose teaching soft skills in software engineering classes through controlled experiments, a new teaching approach that combines lectures, practical experiments, and discussions. Two experiments on group dynamics and global software development are implemented to demonstrate this approach. This teaching approach is proved to be an effective method to deliver soft skills.

In chapter 11, Lynette Johns-Boast discusses how large-scale group projects can help students develop skills that are necessary to work in the industry environment. After describing the motivation and background, the author presents her teaching approach and teaching experience, which include problem-based learning, teamwork, and assessment.

In chapter 12, Luis Alves et al. describe their project-based approach in teaching software engineering. In their classes, students work in teams to solve real-world problems. Through working on real-world projects, students interact with real-world clients and practice management skills and quality controls. One specific feature of their project is that the students' documents are assigned with ISBN, which makes the documents official and easy for future reference.

In chapter 13, Ezequiel Scott et al. describe integrating their team coaching method called Agile Coach with Scrum, an agile software development framework, in a capstone software engineering course. The development team uses Virtual Scrum, a prototype tool that aims to help students set up a virtual working environment. They conclude that their teaching strategy may facilitate students to integrate themselves in the software industry environment.

In chapter 14, Marc Lainez et al. provide their experience of teaching agile software development in project-based environment. Specifically, students are asked to follow lightweight agile process to implement a mobile app. Besides technical skills, this project also helps students develop cross-disciplinary skills such as modeling, teamwork, planning, management, and communication. This teaching approach is proven to be successful because agile-based development encourages communication and collaboration.

In chapter 15, Sakgasit Ramingwong and Lachana Ramingwong describe their one-day software development life cycle game, which is used to help students get an initial understanding of the complexity of software process. In this game, students are asked to build a pseudo-software product, a board house. Through playing this game, students understand the importance of communication, teamwork, and customer relations. They also learn basic concepts of project management.

In chapter 16, Elizabeth Monsalve et al. illustrate their teaching experience of using SimulES-W, a computer-based board and card game, in their software engineering course. Through playing the game, students could learn software engineering knowledge and practice software development skills. Statistical studies are also performed to evaluate the effectiveness of this game-based teaching approach, which is proved to be as effective as the regular teaching approach.

In chapter 17, Salamah Salamah et al. present their case-study-based approach to engaging students to learn software engineering concepts. Their case study can provide students with the development experience of a full software life cycle. In particular, they talk about one case study, DigitalHome, a Web-based system that allows home users to manage devices that control the environment of a home remotely. Through working on these case studies, students could learn software engineering principles that are not easily obtained through regular lectures.

In chapter 18, Oisin Cawley et al. review problem-based learning approach, especially its application in software engineering education. They describe two case studies. The students' experience, instructors' experience, and the assessment are also presented in this chapter. They conclude that problem-based learning is an important approach to educating computer science graduates to meet industry expectations.

In chapter 19, Bonnie MacKellar et al. introduce a Client-Oriented Open Source Software (CO-FOSS) model for undergraduate software engineering courses in order to bridge the academia-industry gap. The model has been applied to three universities, which are diverse in terms of size, demographics of students, and curricula. Four case studies are presented in this chapter, which demonstrate that CO-FOSS model can be flexibly adapted in software engineering education based on different needs.

In chapter 20, Paolo Ciancarini and Stefano Russo present the rationale and the experience of teaching software architecture in industrial and academic contexts. The authors compare two different environments and two different expectations and provide suggestions to reduce the gap between industry and academia. Future research directions in this area are also presented.

In chapter 21, Jagadeesh Nandigam and Venkat Gudivada describe their experience of using industry-standard open-source tools in their software engineering classes. The open-source tools and open-source data are the backbones of their class structure. Each learning objective is presented with the corresponding open-source data/tools, which make this approach easy to be replicated by other educators.

In chapter 22, Liguo Yu et al. provide their experience of applying open-source tools and open-source data in their graduate level software engineering course. The using of open-source tools and open-source data is incorporated into five team projects. One interesting project is to estimate the development cost of Linux kernel. Through working on these projects, students could develop industry-expected skills, such as measuring, configuration management, and estimation.

In chapter 23, Yujian Fu presents the experience of using cyber-enabled environment to improve student collaboration skills. The study was performed on software engineering courses of both undergraduates and graduates. The technology-based teaching is implemented on Blackboard, which supports Yahoo Messenger, Google Talk, video, and audio demo. Their study demonstrates that technology-based teaching could engage students, especially those students who are less motivated and less prepared for college.

In chapter 24, Zuhoor Abdullah Salim Al-Khanjari talks about distance learning in software engineering. The author provides a review on software engineering education and e-learning technology. It explores the need to adopt software engineering e-learning model to help facilitators/instructors prepare and manage online software engineering courses. This chapter also addresses how e-learning environment could simplify the application of the constructivist learning model towards software engineering education.

This book is an assembly of education research and classroom experiences collected from educators around the world. Some techniques and experiences have been proven successful and some are still under experiment and refinement. I hope the readers will find this book useful and inspiring in adapting software engineering education with new challenges.

Liguo Yu
Indiana University South Bend, USA

Acknowledgment

I am very much grateful to the authors of this book and to the reviewers for their tremendous service by critically reviewing the chapters. I would like to thank Ms. Monica Speca and Allyson Gard of IGI-Global for the editorial assistance and excellent cooperative collaboration to produce this important work.

Finally, I want to dedicate this book to my wonderful parents, Mr. Jiye Yu and Mrs. Shuying Wang for believing in and supporting me ever since I was a kid. They know little about software engineering education, but are ordinary people who are great role models for their children.

Liguo Yu
Indiana University South Bend, USA

Section 1

Developing Project Management Skills

Chapter 1
Teaching Software Project Management

Kasi Periyasamy
University of Wisconsin – La Crosse, USA

ABSTRACT

Software project management is an inherent part of software engineering. While technical expertise is an important factor to complete a software product, knowledge and experience in project management are equally important. Teaching software project management is always a challenge. Most software engineering courses teach technical skills and knowledge on software development but lack project management guidance. On the other hand, project management courses taught by management faculty do not connect to technical activities. Therefore, a blend of technical and managerial skills must be taught together to train software engineers. This chapter describes the author's experience in teaching a graduate level software project management course with emphasis on blending technical and non-technical skills. The chapter includes the different modes/styles in which the course was taught, the challenges faced, the benefits gained, and the current status of the course.

INTRODUCTION

Managing a software development project is a required skill for a good software engineer. A software project will fail (may not be implemented correctly, may be delayed or may overshoot the budget) if it is not managed properly right from the beginning. For large complex software systems, effective project management is mandatory because failures in such projects lead to major business loss. Often, software engineering courses focus on technical aspects of software development such as requirements engineering, design, coding and testing. Even though some instructors of software engineering courses discuss management issues briefly, the students in these courses do not get adequate exposure to apply and to practice management skills. It is therefore evident that a separate project management course is necessary

DOI: 10.4018/978-1-4666-5800-4.ch001

in a software engineering curriculum. This course is different from a project management course taught by the management faculty in the sense that the former focuses more on the management skills required through software development life cycle activities.

The author introduced the course titled "Management Issues in Software Engineering" as part of the Master of Software Engineering (MSE) program at the University of Wisconsin-La Crosse (UW-L). The aim of the course is to teach software project management but at the same time the students must apply the management skills to a software project. The students in this course are expected to complete a course on technical skills first. Therefore, it would be easier to teach only the managerial skills and ask the students to apply both technical and managerial skills at the same time. The contents of both the courses fit very well with the guidelines posted in Software Engineering Standard (GSwE, 2009). The students in the management course are divided into teams, usually with three students in each team. The set of deliverables includes technical documents such as requirements document, design diagrams, code and test cases, as well as managerial documents related to meeting logs, project plan, cost estimation and team members evaluation.

The management course was taught in different formats over the years by using different life cycle models, various tools and technologies, and by selecting appropriate projects that train the students towards the job market. The rest of the chapter describes the various techniques the author used in teaching the course.

BACKGROUND

The importance of project management skills for software engineers has been extensively discussed in many books and articles.

Software Engineering Management has been listed as one of the knowledge areas of software engineering in the Guidelines for Graduate Degree Programs in Software Engineering (GSwE, 2009), and in the Guide for Project Management Body of Knowledge (PMBOK, 2011). The following topics are included under Software Engineering Management (GSwE, 2009):

- Software Project Planning
- Risk Management
- Software Project Organization and Enactment
- Review and Evaluation
- Closure
- Software Engineering Measurement
- Engineering Economics

While the topics seem to cover all aspects of project management, teaching these topics in a software engineering course is always a challenge. A first-hand experience in this context was given by McDonald (2000) who taught project management courses both to industries and in academia. In this paper, McDonald focused on the differences between industrial workshops and academic courses and concluded that there is more freedom and time in academic environment.

Kruchton (2011) describes his experience in following the project management guidelines while teaching the project management course to industries and in academic settings. In particular, he indicates that the guidelines assume a waterfall-like life cycle model and needs to be revised to accommodate more advances in software development such as agile method. This is because models such as Agile method are quite different from the traditional models such as waterfall and incremental prototyping. The differences in life cycle models require significant changes in expectations from the students, deliverables and evaluation of students.

Koolmanojwang and Boehm (2011) describe how teaching software engineering should be elevated to system engineering by bringing in some of the topics shown in the guidelines. Their goal is to ensure that system engineers should be capable of performing cost estimation, business case analysis, team collaboration and project contracting. In teaching system engineering, their recommendation includes workshops, focus groups, and mentorships.

Teamwork is an essential part of software engineering. Humphrey describes the importance of team work for software development, how to form an effective team and how team members should work cooperatively (Humphrey, 2000). Of particular importance from his book is the quote on "jelled team" (first defined by DeMarco and Lister in their 1987 book). Humphrey's work served as the basis for the management course described in this chapter. In particular, many of his guidelines such as forming a jelled team, monitoring team progress and evaluation of teams were used in teaching this course. A few topics from his book on Personal Software Process were also included in the lectures (Humphrey, 1997). These include contribution of each team member towards project plan and schedules. Both books by Humphrey (Humphrey 1997, 2000) also include a lot of forms that were helpful in monitoring various project activities. The author simplified these forms to suit the projects at hand and the skills of the students which varied considerably over the years.

Some of the lectures on technical activities are taken from Christensen and Thayer's book (2001). Though titled as a project manager's guide, Christensen and Thayer's book focused mostly on integrating technical activities into group projects. Jalote (2002) wrote his personal experiences when observing project management at InfoSys, one of the leading software companies in the world. Though a lot of hints could be extracted from Jalote's book regarding project management skills, the book neither explains how to practically apply these skills to projects in another organization nor how these skills can be taught in a project management course. But the book serves as a good reference.

Our Project Management Course

Managing a software development project involves some critical activities such as project planning, cost estimation, project team formation, work allocation to team members, work schedule plan, evaluation of the progress and reporting the end results at each milestone. All these activities are important irrespective of which approach or life cycle model used (for example, waterfall model, incremental prototyping model or agile approach). In addition, a project manager may also contribute to technical activities such as requirements engineering, design, coding and testing. Quite often, project managers do not get involved in technical activities; similarly, not every software developer participates in management activities. With the result, the blending of technical and non-technical (managerial) activities suffers. As an example, the feasibility analysis of a software project requires information on the number and type of resources needed for the project, the expertise of the development team with the technology and tools, and so on. It clearly shows that a mere discussion between a project manager and software developers might not be sufficient to do the analysis.

Having emphasized the need for project management skills for software engineers, the next question is "How does software engineering education support project management skills?" Most undergraduate software engineering courses focus only on technical activities except for team work. The students in these courses are expected to work collaboratively to complete a project but most of the managerial work such as allocation of work to team members, work schedules, cost estimation and resource allocation are all done by project supervisor, in this case, the instructor for the course (Bavota et al., 2012). It is still acceptable at an undergraduate level because the emphasis in an undergraduate curriculum is on technical activities. However, in a graduate level software engineering course, the students are expected to learn more on project management skills along with technical activities. Some institutions teach a two-semester course on software engineering, the first semester focusing on management activities and the second one on technical activities. While it is true that the students in these courses learn both technical and managerial skills, there seems to be a discontinuity between the two semesters and blending of the two sets of skills is not smooth.

This chapter describes the author's experience in teaching a graduate level software engineering course with the aim to provide the students a blending of technical and managerial skills.

COURSE STRUCTURE

The management course in software engineering at the University of Wisconsin-La Crosse (UW-L) was first introduced in 2001. While some part of the course structure remained the same in all these years (for example, team size and format of lectures), the projects, life cycle models, deliverables and expectations of the course changed considerably over the years. Table 1 shows some of the significant changes in the topics covered in this course.

The introduction of new topics and refinement of previous topics arose because of the need and demands of software industries at the time of offering. At the beginning, some of the topics were only covered in the lectures providing additional

Table 1. Topics covered in the management course

2001 – 2007	2008-2012	2013
Group working, Work schedule, Hierarchy of an organization and its impact on software development, Cost and deadline estimation, Project planning	Previous topics PLUS Risk management, Software configuration management, Copyright and software licenses	Previous topics PLUS Agile development method (Scrum approach in particular)

information but the project requirements did not use these topics. For example, in the first three years, the students did not use any cost estimation model or project planning template. They were asked to do these tasks on their own. However, a majority of the students did not have any experience in managerial activities and needed more support from the instructor. Hence, in later years, the instructor provided specific cost estimation models (COCOMO I to start with and then switched to Function Point approach and finally settled down in Use Case Point approach). Further, some parts of the templates given in Watt Humphrey's books (Humphrey 1997, 2000) were used for project planning. Similarly, risk management was taught in-depth in later years (2008-2012) but the students only partially used risk management activities, especially risk mitigation, during these years. This is mainly because of lack of time in a one-semester course. However, the students are trained to identify risks in the given project and suggest how to handle them. Some topics such as copyright and license issues are still covered only in lectures.

Life Cycle Models

The significant changes in topics also resulted in changes to life cycle models used for project work. During 2001-2007, students used waterfall model with the result that there was only one final demonstration of the product. The outcome showed that more interaction needed from the customer (in this case, the instructor) and hence during 2008-2012, incremental prototyping approach was used. The students were asked to show at least one or two prototypes before the demonstration of the final product. There was a significant improvement in learning because of the feedback given during prototype demonstrations. Due to its increasing popularity, in 2013, the agile method was introduced into the course. The students were required to show a prototype every alternate week which resulted in six prototypes. The end product was quite stable and complete compared to the products in previous years as expected when using agile method. Within the spectrum of agile method, the Scrum technique was used. Other techniques such as Extreme programming and Dynamic System Development Method may be used in the near future.

Teams

During the 12 years the course was offered, various techniques were tried to form the teams. During 2001-2004, the students were asked to select their team members by themselves. This led to a biased view of technical skills. Several students, especially new students, complained about their inability to choose the right team members. So in 2005 the instructor decided to assign the team members based on the students' expertise. In this case, the instructor collected a brief resume of each team member before forming the team. The students were asked to list their technical background and their preferences. The instructor then assigned team members with a goal of having a balanced mix of skills. While this technique seems to be reasonable, it was quite difficult to choose the right set of students for each team because of the widely varied skill set. So from 2006 onwards the instructor finally settled down with a random selection of students. Though this may sometimes lead to an imbalanced technical expertise of team members, it was easier to handle. Moreover, the justification came from the real world scenario in industries; that is, quite often newcomers to industries do not know with whom they are going to work with and what skill set they possess. Interviews are supposed to help in this case but there is not enough time in a one semester course for such interviews. Hence, the random assignment of team members was considered to be the final solution for team formation. Sometimes this approach may lead to conflicts among team members. A subsequent section in this chapter discusses how these conflicts are resolved.

As mentioned earlier, the focus of this course is on teaching managerial skills but at the same time the students are required to apply both technical and managerial skills together. In order to achieve this goal, the students are asked to work in groups of three (sometimes two or four depending on the total number of students in the class) and to develop a software product from scratch. The students are expected to complete the course "Software Engineering Principles" before taking the management course and hence they are assumed to have sufficient technical experience for developing the product. Members of each team are required to meet quite frequently and also required to submit meeting logs to the instructor. The meeting logs are formal documents describing the topics discussed, issues resolved, contributions by team members and decisions taken. A sample meeting log is shown in the Appendix.

Selection of Projects

One of the criteria given in the guidelines for graduate software engineering education (GSwE, 2009) is the expectation of domain knowledge. According to the guidelines, the students must be exposed to developing software products in at least one application domain. To achieve this goal, a wide variety of projects from various application domains were given to the students. These include data-oriented systems, scheduling systems (bus schedules, course schedules, game schedules and work schedules), and control systems (train gate controller, traffic light controller, and robotic controller for a boiler system). In the first few years of offering this course, all teams were given the same project. This was easier for the instructor to compare and grade the projects. Sometimes the members across the teams talked with each other, and shared some of their team decisions which resulted in conflicts and frustration within team members. So in later years, each team was given a different project. Care was taken to ensure that each project has somewhat similar complexity to the others so that all projects in a given semester are equally weighed. Some of these projects are listed in the Appendix.

It is important to notice that a student taking the project management course will experience the domain knowledge in only one area since the student will be taking the course only once. However, the practice and training given to the students will help them to identify domain specific issues so that when they work on problems in other domains, they can use the training given in the project management course. This is evident from the feedback collected from the students (discussed shortly) which indicate how they applied the knowledge at their work place as well as when they do their graduate capstone project.

Outsourcing Experience

Outsourcing became a popular trend in software industries in the past decade. There are several reasons for a company to outsource a project and some of the dominating reasons are listed below:

- The company does not have adequate resources (human and computing resources). However, the company would like to take the project from its client in order to maintain the volume of projects and so it outsources the project to another company. The client who first gave this project to the company does not know anything about the outsourcing.
- The company does not have sufficient expertise in the application domain of the new project. Instead of taking the risk of handling the new project, the company first wants to outsource it to another company who has the expertise in the application domain and observes or learns from the product. Based on the observations, the company may want to try a similar project later.

- The cost of outsourcing a project outweighs the cost of in-house development.

It is therefore important for the students to gain experience in outsourcing projects.

In order to provide the training in outsourcing projects, every team is assigned a project that the team should outsource to another team. For example, if there are three teams A, B and C, team A will outsource the project to team B, team B in turn will outsource the project to team C and finally team C will outsource the project to team A. This way, every team is required to complete an in-house project that was outsourced by another team. At the same time, the same team must monitor the progress of the project that it outsourced. As a requirement by the instructor, when a team X outsources a project to team Y, team X must (i) participate in requirements elicitation process by team Y, (ii) evaluate the requirements document submitted by team Y, (iii) generate black-box test cases for team Y, and (iv) finally evaluate all prototype demonstrations by team Y. Team Y must use the test cases generated by team X during the final product demonstration.

Change in Requirements and Feasibility Analysis

One of the major concerns in software industries is to handle unexpected changes in requirements from customers. Even for a well-defined problem, customers may change their minds during the tenure of the project and may introduce major and minor changes in the requirements. For example, customers may ask for a different style of GUI, or they may want to integrate one of their existing products into the newly developed one.

To prepare the students for such real life experience, the students in this course are told at the very beginning that there will be changes in requirements at some point during the development process. The students are therefore expected to design the product with the goal that it should be easier to change the design when new or modified requirements are announced. In addition, the students are also given an opportunity to refuse any changes but it can be done only by giving adequate justification as a result of feasibility analysis on the new or modified requirements.

In more than 90% of the cases, the students successfully accommodated the new and modified requirements. They are also required to document the changes made to the requirements and design .

Evaluation of Students

Students in this course were evaluated for both technical and managerial skills. For technical skills, the following were used for evaluation:

- Technical contribution by each student in terms of writing requirements document, design document, code and test cases.
- Discussion and technical contribution during team meetings.
- Presentation by individual members during prototype and final product demonstration.
- Presentation of a peer-reviewed publication (journal paper, conference paper or book).

Managerial skills were evaluated using the following:

- Contribution as a team leader for one or more phases of the team.
- Contribution to report writing (meeting logs and deliverables).
- Contribution to discussions during project meeting and decision making. In particular, whether or not this member provided any suggestion or point during conflicts.
- Contribution to organizing team meetings.
- Contribution to work schedules.
- Contribution to project makeover when another team member does not fulfill the assigned task.

In addition to the instructor's evaluation of the students, each team member is required to evaluate the other team members and submit a report at the end of the semester. The instructor will use these individual team member evaluations and compare them with the meeting logs and other deliverables to ensure consistency.

As mentioned earlier, some of the topics are only taught in the lectures. There is a final exam for the course which testifies the knowledge on these topics. One portion of the final exam includes questions on definitions, concepts, and on core knowledge such as models, reviews and procedures. The other portion includes questions to evaluate the managerial skills gained by doing the project.

Deliverables

When using the waterfall model and incremental prototyping model, the students were asked to submit the following deliverables:

- Requirements document written using a simplified version of IEEE standard 830-1998. The students were given a template and also a tool developed in the department (Periyasamy & Garbers, 2006) that assists in formatting the document and validating some sections of the document. During 2011-12, the students were given a template of a use case document and they were asked to write the requirements as use cases.
- Design document with its structure similar to that of the requirements document. This document includes an architectural design using UML class diagram showing major classes, GUI design and Entity Relationship diagram for database design if the project requires a database.
- Black-box test cases that were used in testing the final product.

- Meeting logs that show the discussions and contributions by the team members in every meeting.
- Cost estimation report showing the estimated efforts for the project. The students were expected to use one of three models – COCOMO, Function Point, and Use Case Point.
- A project plan showing milestones, project leader and work schedule. This document would be revised depending on the progress of the project. Sometimes it changed significantly when conflicts occurred and eventually resolved.
- Team member evaluation report, submitted by each member.
- Outsourcing report indicating how the team managed the outsourcing project. Typically it includes the initial proposal given to the outsourced team, evaluation of the requirements and design documents from the outsourced team, and evaluation of the prototype and final product demonstrations by the outsourced team.
- Submitting a user manual was given as an option to the students. Though it will give a good opportunity for improving technical writing skills, there was not enough time to write and to evaluate user manuals in a one-semester course.

In 2013, the agile approach was used and the following deliverables were asked to be submitted:

- An initial set of requirements. This served as the product backlog which was revised consistently during the tenure of the project. The students were asked to submit the revised document prior to the demonstration of the prototype.
- An initial object-oriented design.
- Sprint reports showing detailed description of tasks selected and implemented during

each sprint. In addition, each sprint report also included the changes made to the previous functionalities based on the feedback from the instructor.
- Test cases used during each sprint cycle.
- Team member evaluation report, submitted by each member at the end of the semester.
- A final detailed object-oriented design. Though this document might not be used in the current project again, it may be used for maintenance purposes. It is the developers' responsibility to provide adequate documentation for maintenance.

Communication Skills

Most computer science courses focus on technical activities such as design and coding and do not provide adequate opportunities for the students to improve their communication skills especially in technical writing and presentation. On one hand, computer science is evolving continuously and hence there are a lot of new topics and updates in course work. Hence instructors find it difficult to spare more time on teaching communication skills. On the other hand, employers expect graduating students to have good communication skills which support customer interactions, discussion in project meetings, written and oral presentations of software systems and so on.

The project management course has been designed with all these goals in mind – improving communication skills but at the same time working closely with technical activities. In order to achieve the goals, the following requirements have been introduced:

- Every student must work in a team of at least three members and must actively participate in project discussions, review and presentations. The meeting logs indicate their contributions; the deliverables indicate their technical writing skills; and the communication with outsourcing team and the instructor indicates their customer interaction capabilities.

- Every student is required to give a presentation on a peer-reviewed publication on software project management. Selection of the paper for presentation is approved by the instructor prior to preparation for the presentation. This exercise gives an opportunity for the student to learn something new that is not presented during lectures and also helps him/her improve presentation skills.

Conflict Resolution

Conflicts occur when one or more team members do not agree on a decision. For example, one person may prefer to use a cascaded windows style of GUI while another person may prefer to use a menu or tab structure. It is quite natural that conflicts may arise when working in a group. In particular, when students are assigned randomly to teams, they may not know their team members very well. Hence, personal characteristics and egoism may lead to conflicts. Conflicts may also arise when one or more team members do not have the expected skill set.

It is an important requirement for a team member to understand every other team member, their expertise and their personal preferences to make a "jelled team." Humphrey (2000, pages 20-25) discusses the concept of "jelled team" and the requirements to form a jelled team. An important characteristic of a jelled team is to find solutions for conflicts. It is important to have some plans to tackle conflicts when they arise so that major damages to the project can be avoided.

In the project management course, three solutions are taken to resolve conflicts. They are listed below:

- The project leader (if the team has chosen one person as the project leader) will try to resolve conflicts by listening to the team members who are involved. Often, this will solve conflicts at the entrance and will not create any damage to the project. However, choosing a project leader is an option, not a mandatory requirement and hence there may not be a project leader for a team. Or, the project leader is unable to resolve the conflict.
- If there is no project leader or if the project leader is unable to resolve the conflict, the issue is brought to the instructor. At this time, the instructor will individually meet with the concerned team members and try to find a solution to the conflict. In most cases, this helped resolving the issues. Conflicts at this stage may delay the submission of one or more deliverables because of lack of coordination among team members. However, the instructor will give appropriate time extensions to adjust with the project schedule. It is not a major damage to the project but one or two team members may have to change their personal preferences in order not to deviate the progress of the project from its initial path.
- If the first two solutions fail, the instructor will split the team into two or three teams (depending on the severity of conflict). This causes a major damage to the project. Since the number of team members is reduced, the team may not be able to meet the complete set of requirements which the team started with. In addition, depending on the remaining time in the semester, some of the tasks need to be rushed in. For example, there may not be sufficient time to do adequate testing. The students will be penalized for not able to work in a group but they will be given consideration to complete the rest of the project.

Guest Lectures

In addition to the theoretical background taught in the lectures and practical knowledge gained through project work, the students in the project management course are also exposed to current industrial practices through guest lectures. Every semester, at least one guest lecturer from an industry is invited. This person explains how managerial skills taught in this course are practically used in industries. For example, guest lectures from IBM Rochester described their version of agile method. It gave a different perspective for the students and they were able to compare the techniques they used in the project with those used at IBM. Another guest from an insurance company indicated the need for a thoroughly reviewed requirements document that is used heavily in validation and verification. Former students of this course are often invited to give guest lectures on the correlation between what they learned in this course and how they applied the knowledge to their current work.

There are two types of students in the management course: (1) those who are already working in industries or have worked in the past and returning for full time study, and (2) those who do not have any work experience. For the first set of students, the guest lectures may provide information to compare their experience with those presented by the guest. It is for the second set of students the guest lectures are most useful. The interaction between the guest and the students also provide opportunities for the students to establish new contacts.

Feedback from Students

For every course taught at UW-L, the students submit course evaluations at the end of the semester. So far, feedback for the management course obtained through course evaluations was quite encouraging. There is ample evidence in it

to believe that the students have benefitted immensely from the lectures, the group projects, and the guidance provided by the instructor. Several students communicated with the instructor later, especially after graduating from the degree program, that this course was one of the most useful courses to their current work.

Delvin and Phillips (2010) indicated that assessment process used in software engineering courses generally assess the students' technical skills. In order to evaluate managerial skills, they suggest additional mechanisms which include the following: (1) Evaluate the skills on critical thinking, leadership and communication abilities through peer evaluation. That is, each student will evaluate their team members not only for technical skills but also for managerial skills. (2) Assess essential competencies of the students by evaluating how well they perform in a team, how much help they expect from others and how much help they themselves provide to others. The team evaluation form used in our course includes some of these criteria. The template for team member evaluation is given in the Appendix.

A lot of former students are willing to come and give guest lectures in the course citing their experience in applying what they learned. The department has received several positive comments from former students and also maintains contacts with them.

Current Status and Challenges

Based on the feedback from past students and employers who hired these students, the project management course offers very good training to the students to improve their managerial skills. At the same time, the students are also assessed of their knowledge through the final exam which includes a few questions on project management skills. For example, one of the questions includes the description of a small project and asks the students to write how they would approach the problem as a project manager, and what critical decisions they would make in developing that product.

The course spans one semester, typically for about four months. Within one semester, the students must get to know their team members, design, implement, test and demonstrate a software product and at the same time must also apply the managerial skills that they are learning. The latter includes cost estimation, feasibility analysis, project planning, and team coordination and evaluation. The author felt that one semester is too short to teach all the managerial skills and expect the students to practice most of them. Because of the short duration of the course, some skills such as software engineering measurement and legal issues (concerning copyright and licenses) are taught only in theory and are tested through exams. Many decisions such as choosing the appropriate technology and tools, review of technical documents and testing are done in a hurry. It is preferable therefore to teach a software project management course which spans at least two semesters in order to give more time for the students to practice the managerial skills. However, it has a huge impact on the curriculum especially in terms of university graduation requirements.

Since software technology evolves quite rapidly, teaching the state of the art technology to students is a big challenge. In particular, the management course involves integrating current technology into the curriculum so that the students will be able to start working immediately when they graduate without need for additional training. However, frequent changes in technology, tools, and life cycle models make it difficult to keep the course up to date. For example, the management course at UW-L started teaching the agile method from 2013. This transition required considerable changes to the team structure, deliverables expected from the students and intermediate reports. Though agile method is becoming increasingly popular in industries, only a few academic institutions started teaching agile method in their mainstream software engineering courses. Some discussions on this transition at other academic institutions are given next.

Extreme programming seems to be one of the earlier popular techniques for introducing agile principles. Hedin, Bendix, and Magnusson (2003) describe how they switched from waterfall model to some sort of incremental prototyping approach using agile principles. The main purpose of this switch is to closely monitor the progress of the projects and provide immediate feedback to the students. Tan, Tan, and Teo (2008) adopted a similar approach. Though the intention was clear in introducing agile method in teaching software engineering, their approach was more or less an incremental prototyping method with only fewer interactions between customers and developers and based on the requirement of documentation. Devedzic and Milenkovic (2011) has given some guidance in teaching agile method in software engineering courses. Their experience includes teaching this course both at the undergraduate level and at the graduate level; in the latter, it was taught as an elective course only. Besides, they also indicated that the success of the graduate course came out of the students who had prior industrial experience as opposed to undergraduate students who were still learning other computer science techniques. More recently, the Scrum technique is used in teaching agile method due to its increasing use in industries. Besides, the basic components of Scrum such as user stories, short sprints and daily meetings all enable the students learn faster and better (Mehnic, 2012).

Based on the observations on teaching agile methods, the author decided to use the Scrum approach for the management course. The first time when Scrum was used, the students enjoyed in doing the projects but it required considerable revisions because of lack of experience in teaching agile methods by the author. It is believed that this problem will be alleviated in the near future by continuous refinement of the method.

CONCLUSION

The objective of this chapter is to describe how managerial skills are taught with a blend of technical activities in a software project management course. The students in this course are exposed to a variety of managerial skills which are briefly summarized below:

- **Project planning and decision making:** The mandatory meetings, minutes of each meeting, submission of various deliverables, and presentation of several prototypes all contribute to learning planning and decision making skills.
- **Customer interaction skills:** Meeting with their own team members, meeting with outsourcing team members and presentation of the prototypes help learn and improve oral communication skills. The mandatory presentation of a paper on project management is a plus in this category. The mandatory reports and documents help improve written communication skills.
- **Knowledge on real-life projects:** The projects for this course are selected from real-world examples. Though not possible all the time, sometimes the sponsors of the real-world projects visit the classroom and interact with the students. The guest lectures from industries also provide valuable insights into real-world examples.
- **Choose and apply the state-of-the-art technologies:** The projects in this course are chosen to provide ample opportunities for the students choose the most recent and appropriate technology and tools. For example, the students used MonGo database, a no-SQL database for data storage, Github for source control and Twitter Bootstrap for developing a Web-based data-oriented application.

In summary, the management course at UW-L is designed to teach the most important managerial skills expected from a software engineer. Based on the feedback received from the students, employers and project sponsors, the author strongly believes that it is one of the most successful courses in the department.

REFERENCES

Bavota, G., De Lucia, A., Fasano, F., Oliveto, R., & Zottoli, C. (2012). Teaching software engineering and software project management: An integrated and practical approach. In *Proceedings of the 2012 International Conference on Software Engineering* (pp. 1155–1164). IEEE Press.

Christensen, M. J., & Thayer, R. H. (2001). *The project manager's guide to software engineering's best practices*. Los Alamitos, CA: IEEE Computer Society Press.

Delvin, M., & Phillips, C. (2010). Assessing competency in undergraduate software engineering teams. In *Proceedings of IEEE EDUCON Education Engineering 2010 – The Future of Global Learning Engineering Education* (pp. 271–278). IEEE Computer Society Press.

Devedzic, V., & Milenkovic, S. R. (2011). Teaching agile software development: A case study. *IEEE Transactions on Education, 54*(2), 273–278. doi:10.1109/TE.2010.2052104

GSwE. (2009). *Graduate software engineering 2009: Curriculum guidelines for graduate degree programs in software engineering, version 1.0.* Retrieved in July 2013 from http://www.gswe2009.org/fileadmin/files/GSwE2009_Curriculum_Docs/GSwE2009_version_1.0.pdf

Hedin, G., Bendix, L., & Magnusson, B. (2003). Introducing software engineering by extreme programming. In *Proceedings of the 25ᵗʰ International Conference on Software Engineering* (pp. 586–593). IEEE Computer Society Press.

Humphrey, W. S. (1997). *Introduction to the personal software process*. Reading, MA: Addison-Wesley Longman, Inc.

Humphrey, W. S. (2000). *Introduction to the team software process*. Reading, MA: Addison-Wesley Longman, Inc.

Jalote, P. (2002). *Software project management in practice*. Boston, MA: Addison-Wesley.

Koolmanojwang, S., & Boehm, B. (2011). Educating software engineers to become systems engineers. In *Proceedings of the 24ᵗʰ Conference on Software Engineering Education & Training* (pp. 209–218). IEEE Computer Society Press.

Kruchten, P. (2011). Experience teaching software project management in both industrial and academic settings. In *Proceedings of the 24ᵗʰ Conference on Software Engineering Education & Training* (pp. 199–208). IEEE Computer Society Press.

Mahnic, V. (2012). A capstone course on agile software development using Scrum. *IEEE Transactions on Education*, 55(1), 99–106. doi:10.1109/TE.2011.2142311

McDonald, J. (2000). Teaching software project management in industrial and academic environments. In *Proceedings of the 13ᵗʰ Conference on Software Engineering Education & Training* (pp. 151–160). IEEE Computer Society Press.

Periyasamy, K., & Garbers, B. (2006). A light weight tool for teaching the development and evaluation of requirements documents. In *Proceedings of the Annual Conference of American Society of Engineering Education*. American Society of Engineering Education.

PMBOK. (2011). *A guide to the project management body of knowledge* (4th ed.). IEEE Computer Society.

Tan, C., Tan, W., & Teo, H. (2008). Training students to be agile information systems developers: A pedagogical approach. In *Proceedings of the 2008 ACM SIGMIS CPR Conference on Computer Personnel Doctoral Consortium and Research* (pp. 88–96). ACM.

ADDITIONAL READING

Becker, P. R. (2009, August). Technology management degree programs: Meeting the needs of employers. In *Proceedings of Portland International Conference on Management of Engineering & Technology* (pp. 2171–2183). IEEE.

Boehm, B. et al. (2000). *Software cost estimation with COCOMO II*. Upper Saddle River, NJ: Prentice Hall, PTR.

Chao, J. (2005, July). Balancing hands-on and research activities: A graduate level agile software development course. In *Proceedings of 2005 Agile Conference* (pp. 306–311). IEEE.

Chatfield, C., & Johnson, T. (2013). *Microsoft project 2013: Step by step*. Redmond, WA: Microsoft Press.

Garcia, S., & Turner, R. (2007). *CMMI: Survival guide*. Boston, MA: Pearson Education, Inc.

Kerzner, H. R. (2013). *Project management: Case studies*. Hoboken, NJ: John Wiley & Sons, Inc.

Laplante, P. A. (2006). An agile, graduate, software studio course. *IEEE Transactions on Education*, 49(4), 417–419. doi:10.1109/TE.2006.879790

Paulish, D. J. (2002). *Architecture-centric software project management: A practical guide*. Boston, MA: Pearson Education, Inc.

Pyster, A., Lasfer, K., Turner, R., Bernstein, L., & Henry, D. (2009). Master's degrees in software engineering: An analysis of 28 university programs. *IEEE Software*, 26(5), 94–101. doi:10.1109/MS.2009.133

Shore, J., & Warden, S. (2008). *The art of agile development*. O'Reilly Media, Inc.

Tsui, F. (2004). *Managing software projects*. Sudbury, MA: Jones and Bartlett Learning.

Tsui, F., Karam, O., & Bernal, B. (2013). *Essentials of software engineering*. Burlington, MA: Jones and Bartlett Learning.

Wysocki, R. K. (2012). *Effective project management: Traditional, agile extreme*. Indianapolis, In John Wiley & Sons, Inc.

KEY TERMS AND DEFINITIONS

Conflict: With regard to software project management, this term refers to a situation when there is no agreement between the team members.

Conflict Resolution: It refers to a set of amicable solutions to resolve conflicts when they arise during software development.

Software Project Management: It refers to a set of technical and managerial activities for successfully developing and deploying a software product.

APPENDIX A: SAMPLE MEETING LOG

CS 744 – Management Issues in Software Engineering

Project Meeting

Project title: Re-engineering a type checker for Object-Z specification language
Group number: 2
Meeting #: 5
Held on: Feb 02, 2012 at 10:15 A.M.
Location: Wing Tech classroom
Members of the group: John Wiley, Kate Johnson and Mike Burge
Members attended: all
Topic(s) for discussion: Update on GUI
Report writer: Kate Johnson

The meeting started with a brief discussion on the previous (committed) work by each team member. Kate indicated that she had a problem in accessing some of the elements in the syntax table. It looked like a technology issue. John suggested a fix; Kate agreed to try it out next time. John's and Mike's work were accepted by the team unanimously.

As decided at the end of the previous meeting, this week's focus will be extending the GUI to include a section to introduce various function types to be used in the specification. John listed the functions to be included and explained how they should be accessible from the GUI. There are six different function types in the specification language. Mike and Kate agreed on the list and John's comments on acceptability. Mike agreed to take the responsibility of implementing the display elements. Kate suggested that Mike should look into Amsfonts to ensure that the font symbols are displayed correctly. Mike mentioned that he already looked into that and he knew how to do it. Kate agreed to help re-organizing the screen to accommodate the space for the new section. Mike and Kate will submit the prototype by Friday. John will test them.

The next meeting will focus on developing the functionalities behind the scene for each function type displayed. The meeting is planned to be on Feb 10, 2012 at 4:00 P.M. in Murphy Library, Room 127.

APPENDIX B: SOME OF THE PROJECTS GIVEN IN THIS COURSE

1. Electronic Voting System
2. Online Driving License Exam Conductor
3. Bus Transportation Services for a Big Company
4. Attendance and Billing Management in a Daycare
5. Patient Administration and Appointment Scheduler in a Hospital
6. Data Management in a Pharmacy

7. Data Management in a Medical Insurance Company
8. Data Management for Budget Office in a University
9. Data Management for Admissions Office in a University
10. Student Information System
11. Payroll System in a University
12. Investment Management for a Stock Broker
13. Inventory Management in a Manufacturing Industry
14. Intramural Game Scheduler in a University
15. Classroom Allocation Subsystem in a University
16. A Bug Tracking System in a Software Industry
17. A Content-based Indexing System for Document Retrieval

APPENDIX C: TEMPLATE FOR TEAM MEMBER EVALUATION

Team Members Evaluation Report

For each team member, use the following questionnaire for evaluation:

Project title:
Group name:
Your name:
Team member's name:

1. For every meeting you had, does this person contribute enough to discussions? You can enumerate each meeting, if you want. Otherwise, give a summary of contributions by this person in overall discussions during project meetings.
2. Did this person attend all meetings? If not, specify the meetings which this person did not attend and whether or not this person provided adequate contributions for those missed meeting(s).
3. How would you rate (in percentage) the communication skills of this person? Give your answer for all three types of communication (oral, reading and written).
4. Give overall contributions of this person (in percentage) for the following tasks:
    a. Project plan
    b. Requirements
    c. Cost estimation
    d. Design
    e. Implementation (coding)
    f. Testing
    g. Reviews
    h. Communication with other teams
5. Did this person finish the assigned work on time? If missed, indicate when and what task he/she did not finish on time?

6. If this person wrote a portion of the code for the project, did this person have sufficient knowledge to start coding? If not, did this person require additional training? If this person required additional training, who did provide the training? Did you help him/her?

7. Did this person act as a team leader? If yes, give an overall ranking of the person's management skills (in percentage). Also indicate in which phases of the development process the person acted as a team leader.

8. Additional comments.

Chapter 2
Applying Coaching Practices to Leadership and Team Management Learning in Computer Science:
A Practical Experience

Esperanza Marcos Martínez
Rey Juan Carlos University, Spain

Juan M. Vara Mesa
Rey Juan Carlos University, Spain

Verónica A. Bollati
Rey Juan Carlos University, Spain

Marcos López-Sanz
Rey Juan Carlos University, Spain

ABSTRACT

This chapter summarizes the experience of applying coaching techniques to the teaching of leadership and team management, which is taught as part of the project management course in the Computer Science degree at a large public Spanish university. These skills have, until now, been taught through lectures. However, there are some key strengths or abilities that a good leader or team manager possesses, such as the ability to work well in a team, communication skills, etc., which a student cannot learn in a lecture hall. The authors have therefore decided to change the teaching method to one in which the student is converted into the protagonist and the professor takes on the role of the coach, thus becoming a facilitator for the student´s learning process. The classes were organized in workshops, carried out in seminars outside the usual lecture room, and each workshop was dedicated to a specific skill or ability. In the first session, the students felt disconcerted and a little shy and were reluctant to participate in some of the professor´s proposals. Nevertheless, during the five workshops of which the experience consisted,

DOI: 10.4018/978-1-4666-5800-4.ch002

the students became more participative and were highly contented, and it will therefore be refined and repeated during the next academic year. This chapter provides details of the experience, highlighting the methodology used in each of the workshops, in addition to the conclusions eventually reached and possible improvements that could be made in the future.

INTRODUCTION

This chapter presents an experience involving the application of the principles of active learning (Stemp-Morlock, 2009) based on coaching (Cardón, 2003) to the teaching of the "Advanced Software Engineering" course (from the area of software project management) in the computer science degrees at a large public Spanish university. This experience was particularly focused on the teaching of personal skills, such as leadership and team management, which have traditionally been recognized as playing key roles in software development projects (Pressman, 2010).

The demand for not only technical but also personal skills as the object of teaching in any field has risen in recent years. For example, the area of psychology has, for decades, been advocating the importance of training people in skills related to what Goleman referred to as "Emotional Intelligence" (Goleman, 1996). This necessity is, to a great extent, linked with some of the principles of the new education framework that defines the implementation of the European Higher Education Area (EHEA[1]) which places emphasis on an education that is geared towards competency-based learning rather than the traditional transmission of knowledge (Voorhees, 2001).

The teaching of emotional intelligence is deeply rooted in preschools and primary schools (Petrides et al., 2006), and has been successfully applied in Business Schools for many years (Tucker et al., 2000). We believe that the application of new teaching methodologies of this type will also allow university student's to obtain their maximum potential. And why wait until they are professionals? As with other skills that can be trained (at least up to a certain point), such as languages or sports,

the sooner that individuals begin to develop these skills, the easier it will be for them to do so. In this respect it is worth highlighting the initiative of YPD[2], which aims to develop the potential and the talent of young people and which rests on four pillars: energy, creativity, communication and leadership.

If the training that is aimed at acquiring personal skills and developing emotional intelligence is important in general, then it is much more so for software professionals, who work in a field in which one of the main assets is people and in which emotional intelligence or abilities, such as leadership and team work, are key aspects, as can be demonstrated by the current trends in software engineering such as agile methodologies (Cockburn, 2006) or global software development (Herbsleb & Moitra, 2001).

Goleman himself cites the importance of these skills in IT professionals: "One of the fields in which emotional intelligence curiously has most bearing is in that of computer programming, a field in which the efficiency of the elite who occupy the highest 10% is 320% greater than that of average programmers, which in the case of the 'raraavis' who make up 1% of the total reaches 1,272%!" (Goleman, 1998, pp. 62). Some years ago ACM-IEEE Computing Curricula[3] therefore started to consider the teaching of skills and competences, although the majority of institutions continue to use lectures as the cornerstone of their teaching, supported by digital media at best (Wirth, 2003), even though this type of methodology is not the most adequate for competency-based learning.

Nevertheless, one of the main differences between the educative model represented by the traditional approach and the educative model proposed by the EHEA is the alteration in the roles of students and teachers. The former change from being passive subjects and mere receptors of information to being active subjects and entrepreneurs, while the latter are no longer protagonists but rather mediators and facilitators whose objective in this context revolves around teaching the student how to learn (King, 1993). That is to say, the teacher should guide the work that the students have to do in order for them to learn. The student will thus acquire the competences needed through activities which, while directed or supervised by the teacher, will be carried out by the student both inside and outside the classroom.

Our objective as regards the planning and development of this experience was therefore to make an important change in the teaching of the personal skills involved in direction and team management, moving from the traditional approach to a methodological proposal based on the principles of the EHEA. In particular, we propose the use of a teaching methodology based on coaching techniques.

The word coach comes from Hungary, where the term Koaching is used to refer to the transport service which is based on the utilization of a type of carriage called Kocs. So, while many definitions of the term coaching exist, that which is derived etymologically from this term is probably the most appropriate. This allows us to refer to coaching as the process of accompanying a person, or a team, from the point at which they find themselves to the point to which they wish to go. The use of this definition of coaching became popular in England in the area of sport, where good coaches managed to get the best from each sportsman they trained or coached. In view of the good sports results that this practice provided, its use became more widespread in other fields, fundamentally in that of business (Kampa-Kokesch & Anderson, 2001).The utility of coaching has more recently been defended for educational purposes (Gabriel, 2012). In this context, teacher-coaches leave their role as protagonist to one side in order to become a mediator and facilitator of the student's learning.

In other words, teacher-coaches help their students to get to "where they want to go," thus making the most of their potential.

Finally, it is worth highlighting that if the student's active participation is important in any learning activity, it is much more so as regards learning based on competences, which should be "acquired" rather than "learned." The nature of coaching, which is essentially an experience-based discipline, becomes particularly appropriate in providing a transformational education, such as that which one attempts to obtain by teaching personal skills.

The remainder of this chapter is structured as follows: Context Section offers a general view of the context in which this experience and its objectives were developed; Active Learning Through Coaching Techniques Section describes the experience in detail; Assessment of the experience Section summarizes the student's opinions, which were obtained via the surveys that were carried out afterwards, and finally the chapter concludes presents the main conclusions drawn from the development of the experience.

CONTEXT

This section provides a brief description of the educational framework in which this experience was developed. Details of our objectives, and their definition and development, are also shown.

Characteristics of the Course

The experience described in this chapter took place in the framework of the "Advanced Software Engineering" course, which is taught during the third year of the Computer Science degree at a large public Spanish university, in both classroom-based and online modes of study. This course is focused on software project management, which entails contents such as planning, cost estimation or risk management. In particular, as mentioned previously, the human factor plays a key role in project management (Pressman, 2010), since each software project manager is in essence a team manger.

The experience presented in this chapter is focused exclusively on this part of the course, which corresponds to approximately 20% of the subject. More concretely, owing to the limitation regarding the amount of time that could be dedicated to it, the experience was focused solely on the following skills: the ability to assimilate and abstract thought, creativity, oral communication and team work.

The intention of the experience was thus to teach the concepts of leadership and team management, which are essential for all Computer Science engineers, while the students develop some of the abilities needed to effectively manage teams, such as that of assimilate and abstract thought, or oral communication. This was achieved by doing away with almost all of the lectures and replacing them with an active learning method, in which coaching techniques (Cardón, 2003; Crane & Patrick, 1998; O'Connor & Lages, 2004) were applied through diverse activities and workshops that took place outside the traditional classroom environment.

Objectives

The main objective of this experience was to favor the student's autonomous learning through the use of coaching techniques which, by fostering their active participation, would introduce them to the concepts of leadership and team management. If this objective is to be achieved then the students must first understand and assimilate some basic concepts in order to later acquire and develop some of the key competences needed for leadership and team management, such as the ability to assimilate and abstract thought; creativity; oral communication and teamwork.

Likewise, the proposed methodology favors student's autonomous and self-directed learning, providing them with the resources and materials needed. It also uses innovative methods that encourage the student's participation and collaborative work.

Finally, a secondary objective was to promote continuous assessment since the final mark is computed from the student's participation in the different activities developed in the context of the experience.

ACTIVE LEARNING THROUGH COACHING TECHNIQUES

In this section we shall describe the teaching methodology that was defined and applied in the experience presented herein.

As mentioned previously, one of the objectives of the course, and in particular the objective of this experience, was for the students to acquire both the knowledge and the basic skills needed to be able to lead and effectively manage teams of people. In our opinion, if this knowledge and skills are to be acquired it is not sufficient for the student to listen to the teacher, study, or simply read the books on the booklist; the student must actively participate, gradually assimilating and interiorizing the necessary knowledge and skills. On the part of the course which is centered on leadership and team management we therefore proposed the elimination of lectures and their substitution with workshops.

Given that each workshop focused on a specific skill, it was not simply a case of the teaching methodology not always being the same, but that each workshop should follow a method adapted to the skill that was the object of learning for that workshop. The contents and methodology of each of the workshops are detailed in the following section. That said; please note that although each workshop was focused on a particular skill, the students also worked on the other skills, as they are all inter-related. Similarly, although each workshop followed its own dynamics, the methodological basis for all the workshops was the application of coaching techniques, which were applied horizontally in all of them.

Coaching is a discipline that drinks from many different sources. It is particularly based on psychology: cognitive, humanist, transformational, etc.; sociology, philosophy and linguistics. Furthermore, as in all disciplines, there are distinct approaches; coaching based on NLP (Neuro-Linguistic Programming) (O'Connor & Lages, 2004), transformational coaching (Crane & Patrick, 1998), etc. We focused on two of those approaches: ontological coaching (Echevarría, 2008), whose main base is the use and the potential of language as a generator of realities, and the systemic coaching of teams (Neenan & Dryden, 2002), whose key basis is the consideration of the teams as systems or groups of inter-related elements for a common objective. The coaching techniques that were considered to be the most appropriate for the teaching of the subject matter in the experience presented herein are summarized as follows:

1. All the workshops took place outside the classroom. This was done with the intention of breaking the typical classroom dynamic, in search of another means of teaching that would favor the student's collaboration and active participation, in addition to using spaces that had been adapted to the activities involved in each workshop. For example, the rooms used allowed us to change the layout of the tables in order to favor interaction between the team members.

2. Each group of students was considered as a team in itself, with its own rules, which were defined by the students themselves in the first session utilizing the alliance technique, commonly used in team coaching (Cardón, 2003). This basically involves the students making pacts or coming to a series of agreements concerning the rules that will govern the different sessions. The establishment of this alliance helps the students to think of the class as a team.

3. The teacher's role changes to that of the leader-coach, whose objective, rather than simply teaching the students, is to guide them and boost their own qualities. It is the students who should act, given that their skills or abilities are acquired in an experiential manner. Likewise, the teacher´s role as leader serves the students as a reference to what a good leader should be in contrast with the traditional concept of a boss or a superior. At the end of the experience, the students are capable of sensing that the attitude maintained by the teacher as a leader or director has been much more beneficial to their learning process than if the teacher had acted as a mere transmitter of knowledge.

4. As the process was being carried out, some of the basic distinctions of coaching (Guarnieri & Ortiz de Zarate, 2010) were used to facilitate learning: victim/person responsible, excellence/requirement, dream/vision, petitions/offers, complaint/claim, expectation/compromise, mistake/learning, problem/challenge, etc. For example, when a certain objective, such as meeting a deadline for a specific exercise, is not achieved, the student could adopt two attitudes: that of the victim of the events (The computer did not work on the last day …) or the person responsible for the events (If I had been more prepared, starting work earlier, I would have met the deadline …).As will be appreciated, these distinctions were used with the intention of transmitting to the students the values of responsibility, compromise, effort, etc. that are essential not only in the role of a leader but also in anyone who is part of a group, for the smooth running of that group.

5. All of the student's presentations were based on the principles of communication in coaching: verbal and nonverbal communication, active listening and empathy. In particular, every ICT worker should acquire the last two competences, although they are regularly undervalued. To be able to convince the

client, the first and most important step is to understand him. The professional should therefore put herself in the interlocutor's shoes, regardless of whether he is another colleague, a superior, a subordinate, or more frequently, an end user.

6. Finally, as proposed by different team coaching techniques (Cardón, 2003), both the composition of teams and their physical makeup has varied from one session to the other. This has given the students the opportunity to interact with the majority of the class, thus fostering the creation of the concept of "class as a team." This varied distribution of the class additionally allowed us to analyze the dynamics of the group generated by the various compositions.

7. Furthermore, these techniques have been combined with one of the oldest teaching methodologies: that of maieutics (Scraper, 2000), which is usually used in coaching activities. Maieutics is based on dialectic, and more concretely, on interrogation as a method with which to attain knowledge in an inductive manner. In the context of this experience, the teacher poses questions to the students but leaves them work out the answers through dialogue and induction in order to extract conclusions. This means of working has obviously also served to incentivize the student's participation and convert them into active players in the learning process, as opposed to the passive role played in traditional lectures.

The following section shows how the principals and techniques presented above were put in to practice during the course. The planning of the activities that have been followed during the development of the experience is therefore discussed in detail.

Description of Activities

As mentioned previously, the experience was structured in workshops which were distributed throughout different sessions. Table 1 describes the objectives and activities of each workshop, grouped by session. Note that according to the principles of the Bologna declaration, a large amount of the student's work should take place outside the classroom in those periods known as Inter-sessions.

It is important to stress that the division of the workshops into sessions took the class´s timetable into consideration. Nevertheless, we would recommend combining workshops 3 and 4 in a single session, and the exposition sessions 5 and 6 in another single session. In general, longer sessions have various advantages:

- Firstly, longer sessions allow the learning of the abilities, which are ultimately connected, to be combined. For example, the activities related to the design and promotion of a new product that are carried out in session 4 (team work) are closely related to oral communication (session 3). It would therefore be more appropriate to find a 4 hour slot in which these sessions could take place.

- On the other hand, the students might find the learning of certain skills more or less complicated than was expected. A longer session allows the teacher to balance the time dedicated to the different activities depending on whether or not the students are perceived to need more time to grasp the concepts.

- As the session advances the students will become more comfortable with the new way of working, the rest of the members of their team and/or class, etc. This causes communication to flow better and it benefits the class dynamics

Table 1. Descriptions of workshops

Session 1: Presentation and Motivation
Objectives: • To present and motivate the importance of leadership and team management for software or computer science engineers. • To start with the dynamics of the "class as a team." **Activities:** • An Alliance is defined, which we shall define the basic norms that will govern the workings of the rest of the workshops: establishing break times, allowing each person to speak, use, or not to use mobile phones in the classroom, etc. This alliance is defined in a collaborative manner by the students and the teacher working together. • The students are randomly divided into teams, with 5-7 students per team. The teams sit at tables placed together like a round table. • The teacher asks the students two questions: what professional profile would they like to have when they finish their degrees? And, what skills do they think are most important for a computer science or software engineer? • The students are left to discuss this in teams. A debate in which feedback is given to the class then takes place, and some conclusions are eventually drawn.

Session 2: Leadership Workshop
Objectives: • That the student understands that different types of leadership exist, each of which has its own advantages and inconveniences. • That the student understands the main characteristics that a good leader should possess. **Activities:** • Again, the students are randomly divided into teams of 5-7 people that are different from those in session 1. The teams sit at tables set out like a round table. • The students watch a short video that illustrates the different types of leadership. • Continuing on from this, the teams discuss what the main characteristics of a good leader are, in their opinion. The different teams then provide the rest of the class with feedback in a debate before some conclusions are drawn. • To finish, the teacher introduces the concepts of mission, vision and values. Note that this is a very short lecture (only 5-10 minutes long) whose aim is to establish a set of particular concepts that are needed.

Inter-Session 2.5
Objectives: • To encourage teamwork • To encourage creativity • To encourage the use of new technologies • To work on oral and written communication • To work on the ability to assimilate and abstract thought. **Activities:** (the students, working in their teams formed in session 2, should carry out the following activities) • The alliance of the team is defined (as opposed to the class alliance). • The mission, vision and values of the team are defined. • Team roles are defined (leader, ideas man, etc.). • The team chooses one, and only one, of the desirable skills a leader should have. • The team writes a brief description (not more than one page) of this ability. They should explain what it consists of and why they chose it. • The teams should prepare a presentation on the skill they have selected, which they will give in front of their fellow students. The presentation could be a video, a role-play or whatever the students feel is appropriate, bearing in mind that creativity is highly valued.

Session 3: Oral Communication Workshop
Objective: • In this workshop oral presentation skills are worked on, along with non-verbal communication and active listening, with the objective of the students improving their capacity to communicate effectively, something which is relative to public speaking, personal inter-communication, giving or receiving feedback, etc. **Activities:** • Once the students have been randomly distributed in teams of 5-7 people that are differing from the previously formed teams, the teams sit around tables set out in a circle. • The teacher shows a video concerning nonverbal communication. • Once the video has finished, the class moves on to a role-play related to nonverbal communication. • Active listening is demonstrated and the students, in pairs, then have to practice quality in a role-play. • The teacher presents some guidelines on how to give and receive feedback and a role-play is carried out to put this into practice. • The teacher starts by asking each team to choose a speaker (teacher, lecturer, businessman, etc.) whose presentations, discourses or speeches they like and another that they do not like. • With these people in mind as references or models, they should then draft a list of five positive characteristics that a speaker should have and five that they should not. • The teams provide the class with feedback, adding their reasons for choosing them. • The teacher finishes by providing some guidelines on presentations or public speaking, which ties in with the following session.

continued on following page

Table 1. Continued

Session 4: Workshop on Teamwork
Objectives: • That the students improve their ability to work as part of a team, learning the difference between a group and a team. • Work on enhancing the student's creativity. • Practicing oral communication and feedback. **Activities:** (this workshop took place in a large room in which the students had sufficient space to move about freely) • The workshop starts with some motivating images concerning the difference between a group and a team. The teacher discusses a team´s systemic focus and the students take part in an exercise that helps to illustrate this. • Once again, the students are randomly distributed in teams of 5-7 people that are different to previous sessions. • In these teams, the students should think up and design a new product, using LEGO blocks. • Each team should then "sell" their product to the rest of the class, using the guidelines on verbal communication skills the teacher had provided them with in the previous session. • Finally, each team receives feedback from the other teams. The teams should think about the guidelines that the teacher gave them in order to say something positive and something that can be improved on (note the different nuance between something "negative" and "an area that can be improved on") in their fellow students´ work.
Session 5 and 6: Expositions
Objectives: • Work on oral communication and feedback. • Work on enhancing the student's creativity. **Activities:** • These two sessions are dedicated to showcasing the teams´ work carried out in inter-sessions 2-5. As previously mentioned, this exposition can take the form of an oral presentation, a video, a roleplay, etc. • When each exposition ends, the team leaves the floor open to questions from the other students or the teacher. • Rather than on the content, value is placed on the ability to communicate a message to the rest of the students, in addition to creativity, the ability to motivate the listener and the means of responding to the questions at the end of their exposition. • Finally, the rest of the teams provide feedback on their impression of the exposition.

• This possibility will therefore be contemplated in the future, depending on the student's disposition and the availability of rooms.

Adapting the Course to Online Study

If the experience is to be carried out online then we believe that it is necessary to make some modifications to the methodology presented in the previous section. Fundamentally, on an online course you cannot force a student to attend a workshop. However, given that we believe that the teaching of these skills is much more effective if it is carried out in classroom based training, this possibility was raised with the students from the start. The idea was to organize the course by taking into consideration the student's availability to attend these workshops, but not forcing them to do so. As the student's response was positive, and with the intention of minimizing their disruption, the workshops were structured in two sessions of four hours each as follows:

• **Session 1:** Presentation, Motivation and Leadership workshop
• **Session 2:** Workshop on oral communication and team work

The corresponding adaptations were also carried out using new technology: the oral presentations were substituted for videos that the students uploaded to YouTube, and they gave and received feedback by using the forums created by the teacher on the university´s online learning platform (campusvirtual.urjc.es).

Additionally, given that the students on the online course had hardly interacted before the workshops, in the first workshop we included a few presentation games in which the objective was to break the ice and for the students to get to know each other and learn each other's names. The game was set up as follows:

• The students stand in a circle.
• The first student steps forward, states his name and then returns to his place.

- The student to the right of the first student then steps forward and says both her classmate's name and her own before returning to /her place.
- The next student then steps forward, repeats the names of the two previous classmates and says his own before returning to his position.
- The process is repeated until the last student steps forward and says the names of all the classmates along with her own.

This exercise allows the students to memorize the names of their classmates, and serves to facilitate the dynamics of the workshops, thus making teamwork more fluid.

Evaluation

In the adaptation of the EHEA educative system, the traditional system of evaluation based on exams should be replaced with procedures that permit continuous assessment, in which each of the instruments used in the assessment is deliberated upon (Voorhees, 2001). Furthermore, if we are concerned about how to evaluate personal skills, then the traditional exam space does not seem to be the right tool for this.

As part of this experience we proposed a system of evaluation based on different factors that were considered to be the concepts that the student should have assimilated as a result of the course, along with the personal skills or abilities that they should have acquired and had practiced in the different workshops. The course was therefore evaluated by carrying out two assessments in which both content and skills were combined:

1. **Assessment 1:** Teamwork development (inter-sessions 2-5)
2. **Assessment 2:** Exposition of the work carried out (sessions 5 and 6)

The factors that were considered in the evaluation system used and the weight of each assessment are specified below. Note that the weight of the assessment was not associated with the assessment itself, but rather with the different factors that were being assessed (see Table 2).

Note that the total mark is weighted in favor of the skills rather than the assimilation of content, given that the content can be evaluated in the course's final exam, in which 20% of the marks are dedicated to questions related to this part of the material.

ASSESSMENT OF THE EXPERIENCE

At the end of the course the students responded to an anonymous and voluntary survey to evaluate the usefulness of these types of workshops. Both the online course and the traditional classroom based course responded to the survey online. These surveys, which were composed of mainly three groups of questions, are shown in the Appendix. The answers provided by classroom based students are summarized below. This group has been chosen because it was the most numerous and we therefore believe it to be the most representative.

The first group of questions (A-N) asked the students to assess (1 - 7) the contribution of the workshops as regards to improving specific aspects of their education, such as whether they have served to incentivize their attendance or participation in class, or whether they have increased their teamwork abilities. The bar chart shown in Figure 1 summarizes the student's answers, showing the average mark for each question.

As will be observed, the students evaluated the workshops very positively:

- All of the questions received an average score of 4.7 or higher.
- When the students were asked how positively they valued the workshops, the average value was 5.46 (Question M).

Table 2. Assessment and evaluations

Contents	The Students Are Assessed to See Whether They Have Assimilated the Content They Should Have Acquired as a Result of Course	20%
C1: Alliance	Assessment of the assimilation of the concept of allies to the Alliance the students had defined and collected in the document as the answer to Test 1.	5%
C2: Mission, vision, values	Assessment of the assimilation of the concepts of mission, vision, and values in agreement with the solution they provided in Test 1.	5%
C3: Description of the chosen trait of a desirable leader.	Assessment of the description of the skill chosen, and the reason for choosing it. The document produced in Test 1 is used for evaluation purposes. The assessment focuses on the content and not on the way in which it was carried out.	5%
	Assessment of the description of the skill chosen, in addition to the reason for choosing it. The oral expression which makes up Test 2 is assessed. The assessment focuses on the content and not on the means. The overall mark will be an average of the mark given by the teacher combined with that given by the other teams.	5%
	Total C3	**10%**
Skills	**The Students Are Evaluated on Whether They Have Acquired the Skills Focused on in the Different Sessions.**	**80%**
D1: Ability to assimilate and abstract thought	Evaluation of the document concerning a desirable skill of a good leader (Test 1). Evaluation of the oral exposition of this skill (Test 2).	10%
D2: Creativity	Evaluation of the document concerning a desirable skill of a good leader (Test 1). Evaluation of the oral exposition of this skill (Test 2). The overall mark will be an average of the mark given by the teacher and also that given by the other teams.	15%
D3: Oral communication	Evaluation of the oral exposition of their chosen skill (Test 1). Evaluation of the ability to respond to their class-mates' questions (Test 2). The overall mark will be an average of the mark given by the teacher and also that given by the other teams.	20%
	Evaluation of the way in which they were able to give feedback to other teams (part of Test 2).	5%
	Total D3	**25%**
D4: Teamwork	Evaluation of how the roles were allocated in the team (Tests 1 and 2).	5%
	Evaluation of the coordination and collaboration in the oral exposition and in the answer to the questions in Test 2.	15%
	Each student evaluates the work of the others in their team. This allows for the evaluation of teamwork "from the inside," a view that the teacher does not have (Tests 1 and 2).	10%
	Total D4	**30%**

- When asked if they would like to participate in these types of workshops in the future, the students responded very positively (Question N).

Likewise, according to the student's opinions, the aspects that leave more space for improvement are the contribution of the workshops as regards to improving leadership skills and the relationship with the teacher (Questions G and K), while the most positively evaluated aspects were the division of the students into random teams for the duration of the different activities and the contribution of the workshop as regards to improving the relationship the students had with their classmates (Questions D and J).

Continuing on from this, another group of questions were asked (scored from 1-5) to discover whether the students thought it would be interesting for the workshops to cover other themes. Figure 2 shows the average interest shown in each workshop that was suggested.

Figure 1. Contribution of workshops to improvement of certain themes

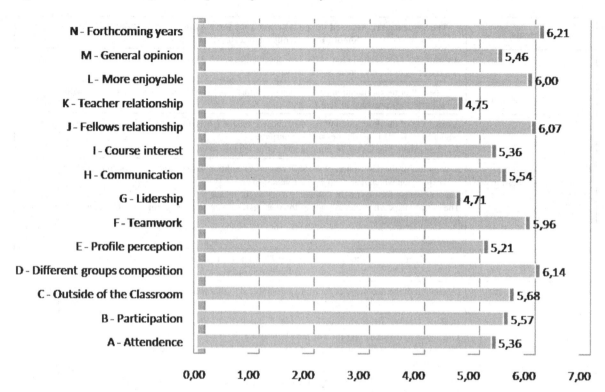

Figure 2. Interest in workshops that focused on other skills

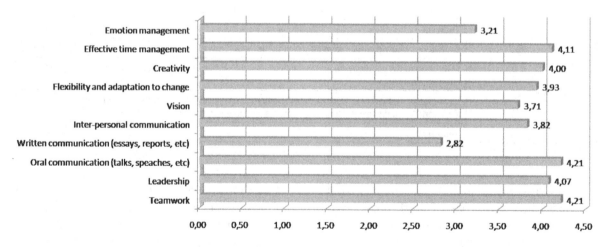

At a glance it will be noted that oral communication and teamwork are those skills that the students would like to work on in future workshops (both attained average scores of 4.21), while the students were less interested in workshops on written communication.

A final set of questions were used to determine how each student evaluated his/her own personal improvement. The students were asked what they thought their levels in the practiced skills were before and after the experience. These answers are summarized in Figure 3, which shows the average score.

Figure 3. Personal perception of the improvement made in the skills practiced in the workshops

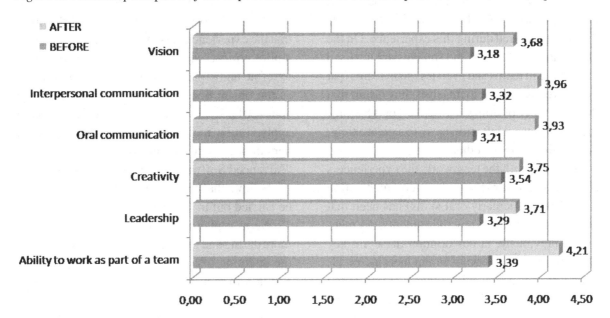

As will be observed, the students considered that the workshops had helped to improve their level in all of the skills but especially those of teamwork, oral communication and interpersonal communication.

CONCLUSION AND LESSONS LEARNED

In this chapter we have presented an experience of applying the principals of active learning, based on coaching techniques, to the teaching of certain human characteristics that are important for all engineers working in IT. The previous section summarizes the main results of the survey carried out with the students for the purpose of evaluating how much they had enjoyed the experience that was carried out.

Upon interpreting the results in order to extract conclusions, rather than a global positive evaluation, we would like to highlight that the student's initial reticence to work in groups, and particularly the random composition of the groups as chosen by the teacher, eventually became the most highly

valued aspect at the end of the experience, which they saw as the ideal way to improve their relationship with the rest of their classmates. This opinion was also reflected in the longer answers that the students wrote in the survey describing the positive aspects of the experience. On the other hand, the fact that the students perceived that although the experience contributed towards improving their relationship with the teacher, this was true to a lesser extent. This result is to be expected, since the teacher's role was, from the start, designed to be secondary, leaving the students to assume the role of the protagonist and making the class move forward by itself.

One aspect that could be improved on in the future is that although the students valued the contribution of the experience as regards to positively improving their leadership skills, they valued it less favorably than other aspects, as the answers to question G or the percentage of improvement that is shown in Figure 3 show. In our opinion, this perception is owing to two factors. On the one hand, part of the results of the applied teaching methodology are seen in the medium to long term, when the students are faced with a scenario in which they have to act as leaders and unconsciously apply the knowledge and skills they have acquired during this experience. On the other hand, the fact that the students are not used to working in teams and even less so in teams that they have not formed themselves, has meant that the possibility of doing so and the good feelings the students were left with, have, to a great extent, been eclipsed by the other results.

Together with the teamwork, the experience's contribution towards improving communication skills (and particularly oral skills) has been that most positively valued by the students. As mentioned previously, in our opinion these are key skills for anyone working in engineering or IT. Although they are usually included as part of the learning objectives for study plans, lack of time and resources are often used as a justification for minimizing these skills when education plans are being implemented. In fact, from the teacher's point of view, these types of methodologies help to implement a process of continuous assessment. In this experience the teacher was easily able to measure whether the students had acquired the skills at that moment and in situ in the class. The teachers only had to dedicate a short amount of time to correcting the short document drafted by the different teams, such as that in Test 1. To go even further, the introduction of the students marking their classmates' work during the evaluation process had not even been considered until now.

In our opinion, in which the spirit of the EHEA therefore emerges, these types of experiences should be a common practice in all computer science related degrees, if not in all the courses of the degree, at least in those in later years. The technical nature of IT teaching does not prevent new teaching methodologies from being applied to these degrees. On the contrary, its practical nature becomes ideal for active learning based proposals (King, 1993; Stemp-Morlock, 2009), which gives the students practical, "hands on" experience rather than them acting as mere receivers of knowledge.

To conclude, regarding the directions for future work that this work has opened, it is worth noting that the experience presented herein has served as a pilot project, whose lessons learned will serve to refine and broaden the number of skills acting as the object of learning for the next academic year (2012/2013). That said, our objective is to apply similar methodologies to those presented here in order to teach most of the contents of the "Advanced Software Engineering" course, since we believe that the contents of this subject can be carried out in this way.

We also plan to address the development of more daring and ground breaking workshops with the intention of fostering student's creativity, in addition to workshops focused on the assimilation of those competences in which, according to the surveys, the students have shown the most interest, such as the aforementioned creativity, oral communication skills or time management.

REFERENCES

Cardón, A. (2003). Coaching d'equipe. Ed.s d'Organisation.

Cockburn, A. (2006). *Agile software development: The cooperative game*. Reading, MA: Addison-Wesley.

Crane, T., & Patrick, L. N. (1998). *The heart of coaching: Using transformational coaching to create a high-performance culture*. San Diego, CA: FTA Press.

Echevarría, R. (2008). *Ontología del lenguaje*. Buenos Aires, Argentina: Granica.

Gabriel, G. (2012). *Coaching scolaire: Augmenter le potentiel des élèves en difficulté*. Brussels, Belgium: De Boeck.

Goleman, D. (1996). *Emotional intelligence: Why it can matter more than IQ*. Bantam Books. doi:10.1037/e538982004-001

Goleman, D. (1998). *Working with emotional intelligence*. Bantam Books.

Guarnieri, S., & Ortiz de Zarate, M. (2010). *No es lo mismo*. Madrid, Spain: LID Editorial Empresarial.

Herbsleb, J. D., & Moitra, D. (2001). Global software development. *IEEE Software, 18*(2), 16–20. doi:10.1109/52.914732

Kampa-Kokesch, S., & Anderson, M. Z. (2001). Executive coaching: A comprehensive review of the literature. *Consulting Psychology Journal: Practice and Research, 53*(4), 205–228. doi:10.1037/1061-4087.53.4.205

King, A. (1993). From sage on the stage to guide on the side. *College Teaching, 41*(1), 30–35. doi: 10.1080/87567555.1993.9926781

Neenan, M., & Dryden, W. (2002). *Life coaching: A cognitive-behavioural approach*. Brunner-Routledge. doi:10.4324/9780203362853

O'Connor, J., & Lages, A. (2004). *Coaching with NLP: How to be a master coach*. Element.

Peñalver, O. (2010). *Emociones colectivas: La inteligencia emocional de los equipos*. Barcelona, Spain: Alienta.

Petrides, K. V., Sangareau, Y., Furnham, A., & Frederickson, N. (2006). Trait emotional intelligence and children's peer relations at school. *Social Development, 15*(3), 537–547. doi:10.1111/j.1467-9507.2006.00355.x

Pressman, R. S. (2010). *Software engineering: A practitioner's approach*. New York: McGraw-Hill.

Scraper, R. L. (2000). The art and science of Maieutic questioning within the Socratic method. *International Forum for Logotherapy, 23*(1), 14–16.

Stemp-Morlock, G. (2009). Learning more about active learning. *Communications of the ACM, 52*(4), 11–13. doi:10.1145/1498765.1498771

Tucker, M. L., Sojka, J. Z., Barone, F. J., & McCarthy, A. M. (2000). Training tomorrow's leaders: Enhancing the emotional intelligence of business graduates. *Journal of Education for Business, 75*(6), 331–337. doi:10.1080/08832320009599036

Voorhees, R. A. (2001). Competency-based learning models: A necessary future. *New Directions for Institutional Research*, (110): 5–13. doi:10.1002/ir.7

Wirth, M. A. (2003). E-notes: Using electronic lecture notes to support active learning in computer science. *ACM SIGCSE Bulletin, 35*(2), 57–60. doi:10.1145/782941.782981

KEY TERMS AND DEFINITIONS

Active Learning: A particular kind of teaching which entails shifting the responsibility of learning to the students.

Coaching: The term coaching in this chapter is mainly related with two types of coaching: Ontological Coaching, based on the conception of languages as facilitator of realities and Systemic Coaching, which leans on the idea of considering teams as systems.

Competences-Based Learning: Kind of learning where the aim is at helping the students acquire a set of competences instead of just mastering a set of concepts.

Emotional Intelligence: The ability to identify, assess, and control own and others' emotions. In the context of this chapter, the term is used to refer to a set of (non-technical) skills, such as the ability to abstract thought, creativity, oral communication and team work.

Leadership: The personal ability to guide a group of people in order to achieve their objectives.

Teamwork: The ability of working collaboratively with a group of people.

Team Management: Set of techniques, processes and tools for organizing and coordinating a group of people working towards a common goal.

ENDNOTES

[1]   http://www.civiceducationproject.org/legacy/hesss/doc/bologna/Bologna%20Declaration.pdf

[2]   http://www.ypdgroup.com/

[3]   http://www.acm.org/education/curric_vols/cc2001.pdf

APPENDIX: END-OF-COURSE SURVEY

Advanced Software Engineering (ASE) Skills Workshop

1. Evaluate the following aspects from 1 to 7, bearing in mind that 7 is the highest level (the most positive) and 1 is the lowest (the most negative).

    a. To what extent do you think that the workshops carried out as a part of ASE helped to incentivize your class ATTENDANCE?

    b. To what extent do you consider that the workshops carried out as part of ESE have contributed towards incentivizing your PARTICIPATION in class?

    c. To what extent do you consider the carrying out of workshops in seminars OUTSIDE THE CLASS to be positive?

    d. To what extent do you consider that working in DIFFERENT GROUPS during the workshops was positive?

    e. To what extent do you consider that the workshops allowed you to acquire a vision of PROFILE required for a Computer Engineering?

    f. To what extent do you consider that the workshops carried out as part of this course contributed towards incentivizing and improving TEAMWORK?

    g. To what extent do you consider that the workshops carried out as part of this course contributed towards improving your LEADERSHIP skills?

    h. To what extent do you consider that the workshops carried out as part of this course have contributed towards improving your COMMUNICATION skills?

    i. To what extent do you think that the workshops carried out as part of this course have contributed towards MOTIVATING and/or increasing your INTEREST in the subject?

    j. To what extent do you consider that the workshops carried out as part of this course have contributed towards improving your RELATIONSHIP with the rest of your CLASSMATES?

    k. To what extent do you consider that the workshops carried out as part of this course have contributed towards improving your RELATIONSHIP with the TEACHER?

    l. To what extent do you consider that the workshops carried out as part of this course have contributed towards making it more ENJOYABLE?

    m. In general terms, to what extent do you consider that the workshops carried out as part of this course have been positive for your learning and personal maturity?

    n. If it was up to you, would you INCLUDE THESE TYPE OF WORKSHOPS in the course program, although it would leave you with less time to listen to the teacher or to carry out of lab work and exercises?

    Note: For students studying online, question (a) is reformulated and a new question (o) is added.

    a. Rate the convenience of having classroom based workshops (1-7)

    o. Indicate the number of classroom based workshops that you would be interesting in attending (those which you would be happy to attend): (1-4)

2. A number of workshops covering different subjects are listed below. Assign a value from 1 to 5 to each workshop to show the interest you would have in taking part in a workshop on them.
    a. Teamwork
    b. Leadership
    c. Oral communication (talks, speeches, etc)
    d. Written communication (essays, schemas, etc)
    e. Inter-personal communication (negotiation, feedback, active listening, etc)
    f. Vision
    g. Flexibility and adaptation to change
    h. Creativity
    i. Effective time management
    j. Management of emotions
3. Evaluate, from 1 to 5, your level in the following skills BEFORE and AFTER the workshops.
    a. Capacity to work in a team
    b. Leadership
    c. Creativity
    d. Oral communication
    e. Inter-personal communication
    f. Vision
4. Describe the positive aspects of taking part in these workshops:
5. Describe aspects of the workshops that could be improved:
6. Other comments or suggestions:

Section 2
Encouraging Collaborations and Teamwork

Chapter 3
Towards an Understanding of Collaborations in Agile Course Projects

Pankaj Kamthan
Concordia University, Canada

ABSTRACT

The agile methodologies are part of a shift from predictive to adaptive approach towards software development. This change has had a notable impact on Software Engineering Education (SEE). In this chapter, a glimpse into the state-of-the-art of incorporating agile methodologies in software engineering courses is presented. In doing so, the reasons for including a project component in software engineering courses, and for committing to agile methodologies in software engineering courses, are given. To lend an understanding to the notion of collaboration in agile methodologies, a conceptual model for collaboration is proposed and elaborated. The pivotal role of collaboration in agile course projects is emphasized. The use of certain means for facilitating collaboration, including the Social Web, is discussed.

INTRODUCTION

The increasingly significant role of software in society, and that of software development in industry, has led to attention by educational institutions and professional organizations towards software engineering education (SEE). There are a number of Universities around the world that offer courses, as well as programs, related to software engineering. There are also a number of initiatives by professional organizations towards 'standardization' of SEE-related bodies of knowledge and curricula.

The context of SEE comprises of a number of elements (Shaw, 2000), including the external, constantly evolving, industrial environment. In the past decade, there have been a number of notable changes in industrial software engineering,

DOI: 10.4018/978-1-4666-5800-4.ch003

including the movement towards *agility* (Highsmith, 2009). The prospects offered by agile methodologies are also associated with unique challenges towards software development, and effective collaboration (Whitehead, 2007) among stakeholders is one of those challenges.

The interest in this chapter is in exploring the manifestations of collaboration in agile course projects. In many disciplines, including investigative journalism, the Five Ws (and one H) are regarded as basic questions (or dimensions) in information gathering. In the context of this chapter, these questions can be posed as follows: *What* (is collaboration), *Why* (is collaboration necessary), *Who* (is involved in collaboration), *Where* (does collaboration occur), *When* (does collaboration occur), and *How* (does collaboration occur). In this chapter, the answers to these questions are pursued to varying extent.

The rest of the chapter is organized as follows. First, background and previous work relating course projects, agile methodologies, and SEE is considered. This is followed by a discussion aimed towards understanding the essential role of collaborations in agile course projects. Next, directions for future research are highlighted. Finally, concluding remarks are given.

BACKGROUND

In this section, arguments supporting the inclusion of projects in SEE are given, and the current state of commitment to agile methodologies in SEE is analyzed.

Motivation for Projects in Software Engineering Education

In software engineering courses, it is customary to have a project component. In general, a project could be carried out individually or collectively. However, for a number of reasons, the course projects are often carried out in a team (Hayes, Lethbridge, & Port, 2003; Devedžić & Milenković, 2011; Mahnic, 2012).

Realization of Active Learning

There are a number of theories of learning, of which *constructivism* (Hadjerrouit, 2005) and, based on it, *active learning* (Hazzan, Lapidot, & Ragonis, 2011), are applicable to SEE. The premise of active learning is that repetitive, rote memorization should be discouraged, and that opportunities for creativity (Paulus & Nijstad, 2003) and collaboration should be encouraged. This evidently requires that the students are engaged in practical knowledge that they can apply in the 'real-world'. In SEE, one way to realize active learning is through team projects.

Improvement of Scale

There are inevitable limits on the mental and physical abilities of a single person that prevent that individual to sole-develop software systems with certain characteristics in a given duration. The team projects allow the development of software systems with large size and/or complex domains, much like those that the students may come across in professional settings.

Instillation of Soft Skills

In recent years, soft professional skills have been recognized as being necessary in engineering (Surakka, 2007; Soundararajan, Chigani, & Arthur, 2012; Sedelmaier & Landes, 2013), but often addressed inadequately in standard curriculums (Kovitz, 2003; Mohan et al., 2010). The team projects act as a vehicle for students to rely upon and learn from each other. They also instill the habit among students of working with others over long periods of time. In particular, collaboration is viewed as a soft professional skill (Tabaka, 2006;

Mohan et al., 2010; Coleman & Lang, 2012), and the role of collaborative work in organizations is being seen as increasingly important (Davis, 2009; Sanker, 2011).

Practice of Software Engineering

In software engineering, there is interplay between theory and practice. There are certain concepts in software engineering, such as understanding the problem domain (Putnik & Cunha, 2008) or developing a glossary, that are best carried out in a collective, and certain concepts, such as software process (Parnas, 1999), that are best learned through practice. The team projects provide an avenue for students to experience (and thereby learn) a software development methodology, including the execution of activities and preparation of artifacts inherent to it.

Parity with Professional Environment

The onus of preparation of students for possible future careers rests, in part, on the administrators of SEE. The team projects provide the students with an opportunity to gain experience in working as part of a team, thereby preparing them for professional settings, including industry.

Conformance to Establishment

The presence of course projects is part of requirements to be satisfied by certain accreditation boards for software engineering programs, such as the *Accreditation Board for Engineering and Technology* (ABET) and the *Canadian Engineering Accreditation Board* (CEAB), and part of curriculum guidelines of certain professional organizations, such as the *Association for Computing Machinery* (ACM) and the *Institute of Electrical and Electronics Engineers* (IEEE).

A Brief Overview of Agile Methodologies

In the 1990s, a number of limitations of rigidity in approaches for the development of certain types of software systems were realized. The drive to cope with these limitations led to the inception of agility. The *Agile Manifesto* characterizes the term 'agile' and provides a vision for agile software development. For the sake of this chapter, an *agile methodology* is a software development methodology based on the Agile Manifesto, and an *agile course project* is a software project following an agile methodology in the auspices of a course offered by an educational institution. The other terms can be derived similarly.

The Agile Manifesto constitutes the basis for a number of agile methodologies, including *Extreme Programming* (XP), *OpenUP*, and *Scrum* (Highsmith, 2009; Williams, 2010). In the past number of years, there have been several surveys, including those conducted by *VersionOne* (http://www.versionone.com/) to assess the state of agile software development. The results of these surveys have regularly shown XP and Scrum, or their mutations, to be currently the most widely-used agile methodologies.

There are a number of identifying characteristics of an agile project, including a relatively small, highly experienced, team; incremental and iterative time-boxed development; and strong emphasis on testing. Most activities underlying an agile process are inherently collaborative, where collaborativeness is only accentuated and necessitated by small team size.

Evolution of Agile Methodologies

The agile methodologies continue to evolve in different directions, for different reasons. This evolution is initiated by the accumulated experience garnered from the use a single methodology, as well as the use multiple methodologies, in industrial software projects. In other words, agile methodologies have 'learned' from their own experiences, as well as from each other's experiences.

To better align with the development of interactive software systems, there is a movement towards making agile methodologies more *user-centric* (Beyer, 2010; Ratcliffe & McNeill, 2011), and to be more 'environment-friendly' (or 'green'), there is an assimilation in agile methodologies of the principles and practices of *lean development* (Gothelf & Seiden, 2013).

The evolution of agile methodologies has resulted in the refinement of old elements, as well as addition of new elements, as necessary. For example, in Scrum, the notion of user requirement has evolved to that of a user story (that is similar to, but not the same as, XP), the support for visibility has improved to accommodate software projects with teams that are geographically dispersed, the role of documentation has become more prominent than stated initially, and so on.

There have also been proposals for new agile methodologies, such as the *Discipline Agile Delivery* (DAD) (Ambler & Lines, 2012). For example, DAD relies on multiple agile methodologies and incorporates elements that have proven to be 'successful'.

Motivation for Agile Methodologies in Software Engineering Education

There are a number of technical as well as non-technical reasons for introducing agile methodologies in SEE.

Instructional Suitability

It is important for the curriculum design of any program, including software engineering, to be based upon broadly-accepted, essential, and mature knowledge. The agile methodologies have been recognized by national and international standards, included in maturity models, and described in textbooks related to software engineering. It has been shown (Séguin, Tremblay, & Bagane, 2012) that practices inherent to a number of agile methodologies, including Scrum, are aligned with 'conventional' software engineering principles (Ghezzi, Jazayeri, & Mandrioli, 2002).

Professional Relevancy

It has been shown in a number of studies that agile methodologies are being increasingly deployed in many organizations, of different sizes, for a variety of domains, and for software projects with teams that are geographically dispersed. This reality needs to be acknowledged and addressed by educational institutions, given that a significant proportion of students enrolled in software engineering programs seek careers in industry after graduation.

Operational Feasibility

The management of its resources is crucial for any organization. In particular, an appropriate distribution of budget is important for a not-for-profit organization with a fixed budget, as is the case with many public educational institutions and its units. There are a number of items competing for the same resources, in general, and for the same budget, in particular, at any given time. The accommodation of all these items may be desirable, but, at times, is infeasible. The infrastructure requirements of agile methodologies are modest, largely due to the nature of agile processes, types of agile artifacts, and the increasing availability of agile tools as open source software (Koranne, 2011).

Related Work on Agile Methodologies in Software Engineering Education

In this section, salient efforts towards the deployment of agile methodologies in SEE are discussed chronologically. The coverage is necessarily limited to those studies that are available publicly, appear in peer-reviewed contexts, describe the experience of one or more courses at an educational institution, and have nontrivial emphasis on and notable outcomes from the use of agile methodologies in course projects.

In (Bunse, Feldmann, & Dörr, 2004), a study of undergraduate students at the University of Kaiserslautern, involved in a course project using XP as the agile methodology, has been presented. It is concluded that it is easier for students to learn the practices of XP than to apply them, the quality of the system depends intrinsically on the background and experience of the students in software engineering, and that SEE should focus on rigid methodologies before committing to agile methodologies.

In (Dubinsky & Hazzan, 2005), a framework for teaching software engineering has been proposed. This framework has a number of principles and practices, and is based on the experience derived from undergraduate students using XP in their capstone-project-based courses at the Technion Israel Institute of Technology.

In (Bruegge, Reiss, & Schiller, 2009), a case study of undergraduate students at the Technische Universität München, involved in a course project using Scrum as the agile methodology, is given. The purpose of the project was to mimic a real-world scenario, and build a prototype for managing logistical information of certain items at the Munich Airport. The authors point out that the project was a success in many ways, including that the students had fun doing it.

In (Rico & Sayani, 2009), the experience of three teams comprised of graduate students at the University of Maryland University College, involved in a capstone course project using Scrum as the agile methodology, is reported. The purpose of the project was to build an e-commerce application, and the results were mixed. In their analysis, the authors point out a number of issues, including lack of training in agile methodologies, improper engagement with the customer, inadequate atten-

tion to usability, and preference of individualism over collaborativeness in teams.

In (Devedžić & Milenković, 2011), the authors outline their experience with teaching different agile methodologies to students at different levels, of different cultural backgrounds, at different institutions, in different countries. The authors make a number of recommendations for teachers, such as emphasizing to the students that the agile methodologies are not 'silver bullet', treating undergraduate and graduate students differently, having teams that are self-organized and of small size, having a leadership role in each team, and having short iterations in the agile process.

In (Venkatagiri, 2011), the experience of teaching an agile project management course to graduate students at the Indian Institute of Management Bangalore, involved in a course project using Scrum as the agile methodology, is reported. The author observes that the success of the project and the experience of the team are strongly correlated.

Finally, in (Mahnic, 2012; Mahnic & Rozanc, 2012), the experience of teaching undergraduate students at the University of Ljubljana, involved in a capstone course project using Scrum as the agile methodology, is reported. The purpose of the project was to build a student record system. The authors make a number of recommendations for teachers, such as having a leadership role in each team, integrating rigid practices in an agile process, and emphasizing to the students of the significance of communication in eliciting user stories.

Observations

There are a number of conclusions that can be drawn from previous work on the use of agile methodologies in course projects:

- **Cautious Optimism.** The deployment of agile methodologies in SEE is a new and relatively sporadic phenomenon. This could be attributed to the fact that there is scarcity of teachers that are trained in (and/or therefore believe in) agile methodologies, educational institutions tend to be rather cautious and conservative in making changes that are perceived as radical, and that there is an inevitable cascade of costs, monetary or otherwise, as a result of introducing and managing methodology-level changes.

- **Discerning Commitment.** There are several agile methodologies, although most have never been a choice for SEE. In particular, there is no evidence to support the deployment of newer developments in agile methodologies, such as those related to *user experience design* (Ratcliffe & McNeill, 2011; Gothelf & Seiden, 2013). In fact, the choice of agile methodologies for course projects has been limited to a few, and is contingent on the use of those agile methodologies in the industry. For example, initially, XP was the agile methodology of choice for SEE, but increasingly Scrum is being selected for course projects. In some cases, a *blend* of agile and rigid over a 'pure' agile approach in course projects is preferred. This 'striking a balance' is also among the recommendations for introducing agile methodologies in industry (Boehm & Turner, 2003).

- **Equivocal Outcome.** The results of introducing agile methodologies in SEE, as shown by some of the experiences, are mixed. The students, regardless of whether they were enrolled in an undergraduate or graduate program, performed well more on technical aspects and less on non-technical aspects of their course projects. Many of the successes could be attributed to the motivation of students and/or small team sizes, while many of the difficulties can be attributed to the inexperience and/or inability of students to effectively communicate, collaborate, estimate, and/or negotiate.

COLLABORATIVE WORK IN THE USE OF AGILE METHODOLOGIES FOR SOFTWARE ENGINEERING COURSE PROJECTS

In this section, 'collaboration' is situated in other, similar, concepts to provide an understanding of it through dissociation and disambiguation.

Understanding Collaboration

It has been observed that collaboration is one of the many different types of possible *human-to-human interactions* (HHIs) (Denning & Yaholkovsky, 2008). Figure 1 illustrates a conceptual model for collaboration. The purpose of this 'onion' model is to provide a methodology- and technology-independent understanding of collaboration, aligned with 'conventional' definitions of collaboration (Tabaka, 2006). This model has a number of layers. Each layer corresponds to a type of HHI and depends intrinsically (functionally) on the layer beneath it. The degree of HHI increases as lower to higher layers are traversed.

In this model, connection is the simplest and collaboration is the most complex form of HHI; connection allows maximum individual freedom (and is least socially constraining) and collaboration allows minimum individual freedom (and is most socially constraining); and connection is necessary, but not sufficient, for collaboration.

There are a number of concerns in software development, and *separation of concerns* (Ghezzi, Jazayeri, & Mandrioli, 2002) is one of the established software engineering principles to deal with those concerns. The identification of tasks, and allocation of those tasks to different members of a team, is a realization of separation of concerns.

Let A and B be two persons in some collective, and let T be some task. In the rest of the section, the layers depicted in Figure 1, and their interrelationships, are discussed in detail.

Figure 1. A conceptual model for collaboration from the viewpoint of space and time

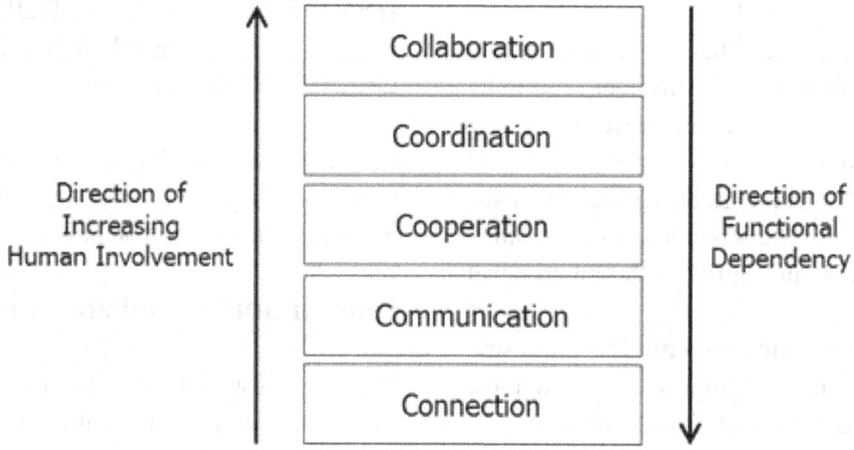

Connection

A connection is a means by which A and B share space and/or time. A and B can connect through some means, such as a room, telephone, or the Internet. For example, both A and B can be present in the same meeting room, be part of a conference call, have profiles on a social network, and so on. However, A and B may not interact with each other.

Communication

In communication, A and B interact with each other for some purpose related to T. However, A and B need not to agree with each other on anything. For example, A may solicit feedback from B on some aspect of T, but choose to ignore any suggestions made by B. The communication between A and B may be synchronous or asynchronous. For example, chat is synchronous and electronic mail is asynchronous.

Cooperation

In a cooperation, A and B work together on not necessarily the same task, and, in doing so, aid each other. For example, A and B may agree upon using the same presentation format, data schema, or international standard for quality for T. However, A and B need not to agree upon common space or common time constraints.

Coordination

In coordination, A and B work together on the same task T and, in doing so, aid each other. To do that, A and B may need to agree upon common space and/or time, and are therefore bound by such an arrangement. Thus, the requirements of coordination are more stringent than those of cooperation. For example, A and B may decide to meet at a specific place or on specific date to discuss specifics of T, or may decide to look at different aspects of T. There are a number of activities in software engineering, in general, and agile software development, in particular, such as interviews, brainstorming, user story authoring, interaction design, and usability testing, that require coordination.

Collaboration

In a collaboration, A and B work together on the same task T. It is possible that T cannot be carried out by any one individual, for some reason, such as feasibility constraints or lack of expertise. For example, A or B could complete T, but not before the deadline, or neither A, nor B, alone has the expertise to complete T, but together they complement each other. In a collaboration, the aim is to satisfy (or satisfice) the goal(s) of the collective, rather than the goal(s) of any of the individuals in that collective. In this sense, collaboration is *holistic*.

Types of Collaborations in Software Engineering

For the sake of this chapter, a *stakeholder* is a person, a group, or an organization that has interest in a course project for some reason, and a *team* is a non-decomposable unit defined as "a small number of people with complementary skills who are committed to a common purpose, performance goals, and approach for which they hold themselves mutually accountable" (Katzenbach & Smith, 1992).

There can be different types of stakeholders of an agile project, including student, team, teaching assistant, teacher, and representative. Therefore, there are different types of collaborations in software engineering course projects, initiated by and depending on the types of stakeholders involved (Robillard & Robillard, 2000; Goldberg, 2002):

- **Type 1 Collaboration. Student–Student:** This type of collaboration occurs among students in the same team. For example, in deploying Scrum, one student could play the role of a programmer, while another could play the role of *Scrum Master*.
- **Type 2 Collaboration. Team–Teaching Assistant:** This type of collaboration occurs between a team and a teaching assistant (if there is one). It is more common to have a teaching assistant in undergraduate courses than in graduate courses. For example, a teaching assistant could play the role of a representative user or a reviewer.

- **Type 3 Collaboration. Team–Teacher:** This type of collaboration occurs between a team and a teacher. For example, the teacher could play the roles of a customer or a manager. In certain agile methodologies, such as Scrum, the role of a manager is equivalent to a *Product Owner*.
- **Type 4 Collaboration. Team–Internal Representative:** In certain capstone projects, there are regular evaluations to assess team progress, and to provide feedback and suggestions for improvement. The evaluations usually involve people in the role of examiners. These examiners are external to the course, but internal to the educational institution offering the course, such as from the same department or faculty. This type of collaboration occurs between a team and one or more examiners, say, during project-related presentations.
- **Type 5 Collaboration. Team–External Representative:** It is common in capstone projects to have involvements of an external, for-profit or not-for-profit, organization. This type of collaboration occurs between a team and one or more representatives from an external organization. The representatives can play the role of a customer or a representative user. These representatives are usually involved in negotiating the requirements for the product, monitoring the status of the project, and acceptance testing of the product.

The scope of collaborations can vary across types. It is evident that collaborations of Type 1 are restricted to a team, collaborations of Type 2 and Type 3 are restricted to a course, collaborations of Type 4 are restricted to the educational institution, and collaborations of Type 5 are restricted by scope of the course project.

The Necessity of Collaboration in Agile Course Projects

There are a number of activities in a software project, in general, and an agile course project, in particular, that necessitate collaboration.

Understanding the Problem Domain

Every software system has a problem domain. For example, health care is a problem domain for a hospital information system.

Usually, the software engineering students, or even professional software engineers, do not have an in-depth knowledge of the problem domain at a level necessary for successfully building a software system. The knowledge of underlying problem domain is often acquired by software engineers, over the duration of the software project, through collaboration with *domain experts* (also known as *subject matter experts*). For example, while building income tax software, collaboration with accountants can be useful, even indispensable. In general, in mission- or safety-critical systems a proper acquisition of domain knowledge is vital.

Understanding Users

Every interactive software system has users. It is important to have an understanding of and empathy towards users at the beginning and during the software process so as to increase the likelihood that a completed system will be acceptable to its users at the end of the software process. Usually, such an understanding is created by means of interviews, user modeling, and feedback during the user interface design process. These activities, evidently, require close collaboration among all stakeholders involved.

Estimating

Planning is an essential component of many agile methodologies, and usually involves a number of things including an estimate of the time it will take, or the effort that will be required, to complete a project. *Planning Poker* (Cohn, 2005) and *Silent Grouping* (Power, 2011) are among the commonly used techniques to estimate an agile project, and rely strongly on collaboration to be effective (Tabaka, 2006).

Quality of Requirements

There are certain desirable properties, often called quality attributes, of requirements in an agile project (Leffingwell, 2010). The scope of a quality attribute can vary. A quality attribute may apply to a single requirement or to multiple requirements; may be relevant to some, but not all, stakeholders; and may be a concern of a single stakeholder or (collectively and simultaneously) that of multiple stakeholders.

For example, uniguity is related to a single requirement, and consistency is related to multiple requirements; priority of a requirement is relevant to a programmer, but not to a user; testability of a requirement is a concern of a single stakeholder, and negotiability of a requirement is a concern of multiple stakeholders.

The quality attributes of requirements that are a concern of multiple stakeholders, evidently, require close collaboration among all stakeholders involved.

Understanding the Solution Domain

Every software system has a solution domain. For example, user interface for doctors is a part of the solution domain for a hospital information system. In an abstract to concrete realization of the user interface, intimate and regular collaboration among user interface designers and programmers is necessary.

Usability Evaluation

There are several approaches for usability evaluation (Nielsen, 1994). In a *question-asking protocol*, a usability engineer and a (representative) user need to collaborate to assure that the results of the evaluation are credible, as well as useful, for subsequent analysis and action.

Support for Collaboration in Software Engineering Education

The support for realizing collaboration in a SEE can be broadly classified into experiential knowledge and information technology. It can be expected that experiential knowledge is essential, while information technology is ephemeral.

Experiential Knowledge

In the past decade, there have been a number of SEE-related initiatives by professional organizations, including the *Software Engineering Education Knowledge* (SEEK), the *Guide to the Software Engineering Body of Knowledge* (SWEBOK), the *Curriculum Guidelines for Undergraduate Degree Programs in Software Engineering* (Software Engineering 2004), the *Computer Science Curricula 2013* (CS2013) of the ACM and IEEE Computer Society, and the *Curriculum Guidelines for Graduate Degree Programs in Software Engineering* (Graduate Software Engineering 2009). The support for collaborative work exists in all these initiatives, but the support for agile methodologies is present only in recent initiatives.

The reliance on documented experience can be useful in many endeavors, including SEE. There are *pedagogical principles* and *pedagogical patterns* related to collaborations in SEE (Dubinsky & Hazzan, 2005; Coplien & Harrison, 2004; Bergin, 2006; Hayes et al., 2006), and can serve, for example, as a guide for collaborations of Type 1 and Type 3.

Information Technology

The advent and advancement of information technology has had a strong influence on the practice of collaboration. There are a number of reasons

that motivate the need for information technology as means for facilitating collaboration, including inherent limitations on memory and retention of humans, physical separation of team members in space and/or time, scale of managing interactions as their numbers increase, and so on.

To facilitate computer-mediated collaboration, a number of tools with affordances for collaboration have been developed (Tabaka, 2006). Traditionally, collaboration tools have been domain-specific and desktop-based. However, the emergence of the Web, and, especially, its successor, the *Social Web* (O'Reilly, 2007), have brought a paradigmatic change to the medium of collaboration, and increasingly collaboration tools are distributed and available as open source software. Indeed, there are broad implications of the Social Web for SEE (Kamthan, 2009). For example, the Social Web can be used for deliberating and disseminating user stories (Cohn, 2004; Fancott, Kamthan, & Shahmir, 2011). *Go2Web20* (http://go2Web20.net/) lists a variety of the tools pertaining to the Social Web that can serve as viable candidates for collaborations of Types 1-5. It is important that these tools are viewed as a complement to, not a replacement of, proximal face-to-face interactions.

Table 1 presents common activities, supported by the use of *Wiki* (Leuf & Cunningham, 2001), in agile course projects, and their corresponding types of collaborations. It can be noted that, normally, certain activities, such as "Acceptance Testing," related to collaborations of Type 2 and Type 5, cannot be carried out by using only a Wiki.

Challenges in Realizing Collaboration and Executing Agile Processes in Course Projects

There are a number of logistical challenges specific to SEE that need to be overcome in an optimal realization of collaborative work in agile course projects.

Table 1. A mapping between activities using Wiki and the types of collaborations in an agile course project

Wiki-Based Activity	Collaboration Type
Creating, updating, or annotating a product artifact	Type 1
Posting bug report	Type 1
Soliciting, suggesting, or negotiating a functionality	Type 1, Type 3, and Type 5
Sharing a resource or link to a resource	Types 1-3
Asking a question or responding to a question	Types 1-3
Arranging or announcing a meeting schedule	Types 1-5
Syndicating a change	Types 1-5
Commenting on a product artifact	Types 1-5

Course Registration

The course registration process can determine the composition of a team and the type of project. In some educational institutions, there are deadlines for adding or dropping a course during a semester. These deadlines can sometimes be several days, even weeks, after the commencement of the course. Therefore, there is no a priori guarantee of the stability of a team at the beginning of a semester. The issue of students dropping a course is especially detrimental as a team can be reduced to a size untenable for a project.

The formation of teams could be postponed until the registration deadlines have passed. However, that would evidently reduce the time available for carrying out a project. In some cases, such as a full-semester course, the remaining time available can negatively influence the type of project that can be selected and pursued. In other cases, such as a half-semester course, the remaining time available may simply not be realistic for a comprehensive software engineering project.

Practice Transfer

The agile methodologies are designed for professional software development, not for teaching or learning software engineering. There are certain agile practices that cannot be replicated as-is in an educational setting.

For example, it is unrealistic to expect all students in all teams, or even a majority of them, to always participate in *Daily Stand-Up Meetings*. There are similar limitations of *Collective Ownership* and *Pair Programming*. Furthermore, certain practices, such as *Retrospectives*, are usually not carried out at all. Therefore, the agile methodologies that include these practices cannot be embraced in their entirety, and may require necessary adjustments if adopted.

Performance Evaluation

The prospects offered by team-based course projects come with their unique set of issues. For example, it has been acknowledged that in a team project exact contribution and assessment of individual students poses a challenge that is only exacerbated in a 'purely' collaborative work. Even though a number of means to address this challenge have been suggested over the years (Wilkins & Lawhead, 2000; Hayes, Lethbridge, & Port, 2003; Clark, Davies, & Skeers, 2005; Coleman & Lang, 2012), they are context-dependent, and may not be applicable to every type of project or be successful in every offering of a course.

Furthermore, striking a balance between individual work and collective work is important, but non-trivial: on one hand, preference to individual contribution could reduce team spirit and introduce deceitfulness; on the other hand, preference to collective contribution could discourage aptitude and suppress individual creativity. The practice of assigning grades based entirely on relative marking favors individualism over collaborativeness in teams.

FUTURE RESEARCH DIRECTIONS

The convergence of hardware and software technologies related to mobility, and applications pertaining to the Social Web, has opened new vistas for communication and collaboration in SEE. For example, using appropriate *Mobile Social Web Applications*, meetings can be readily scheduled, pictures of low-fidelity prototypes can be effortlessly captured, and feedback on documents can be easily shared with team members separated by space and/or time. However, such flexibility is also associated with a number of challenges, some of which are related directly to the HHI emanating from collaborations of Type 1.

For example, collaboration is not automatic, especially among inexperienced students or those students of certain types of dispositions. It also takes time and effort to build *rapport and trust*: necessary conditions for effective collaboration that are beyond the realm of technology and tools. In addition, the breadth and depth of collaborations vary in many different ways. The *degree* of collaboration among team members, and the contribution to an agile project by those team members, both within a single iteration and across multiple iterations of an agile process, is relevant. To understand that, analysis of the social network formed by an agile team can be useful, and doing so falls naturally within the purview of *social network analysis* (SNA) (Golbeck, 2013). In SNA, *tie strength* is a measure of the strength of a particular type of relationship between people. It is understandable that successful collaboration is contingent on strong tie strength. Therefore, a study of collaborations of Type 1 from the perspective of SNA is of research interest.

In SNA, *centrality* refers to a collection of distribution metrics that aim to quantify the 'importance' or 'influence' (in some sense) of a particular node (or a set of nodes) within a social network. The identification of the most central actors in an agile social network can help deter- mine possible relationships between prominent collaborators, nature of information flow, and the quality of agile course projects, and is therefore also of research interest.

CONCLUSION

It could be said that software engineering education occurs in an ever changing context. To remain relevant, it must strive to be adaptive and, to do that, it needs to be sensitive to the variability of its context, and needs to maintain a balance between the elements in its context.

The agile methodologies and distributed means for collaboration are integral to the current context of software engineering education. This chapter is a step towards creating an understanding of different types of collaborations in agile course projects, as well as examining some of the associated challenges in realizing them.

In software engineering, in general, and in agile software development, in particular, the need for collaboration usually arises due to necessity rather than choice. Furthermore, as with other types of human interactions, a successful collaboration is not realized automatically or immediately. To be effective, the educational institutions need to create and cultivate a culture of collaboration in their agile course projects. This may necessitate a revisitation of current norms of assessment, as well as the currently held value system of software engineering education.

ACKNOWLEDGMENT

The author is grateful to Nazlie Shahmir for useful discussions related to the use of agile methodologies in industry, and to the anonymous reviewers for careful reading and suggestions for improvement.

REFERENCES

Ambler, S., & Lines, M. (2012). *Disciplined agile delivery: A practitioner's guide to agile software delivery in the enterprise.* IBM Press.

Bergin, J. (2006). Active learning and feedback patterns. In Proceedings of PloP 2006: Pattern Languages of Programs. PloP.

Beyer, H. (2010). User-centered agile methods. *Synthesis Lectures on Human-Centered Informatics, 3*(1), 1–71. doi:10.2200/S00286ED1V01Y-201002HCI010

Boehm, B., & Turner, R. (2003). *Balancing agility and discipline: A guide for the perplexed.* Reading, MA: Addison-Wesley. doi:10.1109/ADC.2003.1231450

Bruegge, B., Reiss, M., & Schiller, J. (2009, April). Agile principles in academic education: A case study. In *Proceedings of the 6th International Conference Information on Technology: New Generations* (pp. 1684–1686). IEEE.

Bunse, C., Feldmann, R. L., & Dörr, J. (2004). Agile methods in software engineering education. In *Extreme programming and agile processes in software engineering* (pp. 284–293). Berlin: Springer. doi:10.1007/978-3-540-24853-8_43

Clark, N., Davies, P., & Skeers, R. (2005). Self and peer assessment in software engineering projects. In *Proceedings of the 7th Australasian Conference on Computing Education* (vol. 42, pp. 91–100). Australian Computer Society.

Cohn, M. (2004). *User stories applied: For agile software development.* Reading, MA: Addison-Wesley.

Cohn, M. (2005). *Agile estimating and planning.* Pearson Education.

Coleman, B., & Lang, M. (2012). Collaboration across the curriculum: A disciplined approach to developing team skills. In *Proceedings of the 43rd ACM Technical Symposium on Computer Science Education* (pp. 277–282). ACM.

Coplien, J. O., & Harrison, N. B. (2004). *Organizational patterns of agile software development.* Upper Saddle River, NJ: Prentice-Hall.

Davis, B. (2009). *97 things every project manager should know.* Sebastopol, CA: O'Reilly.

Denning, P. J., & Yaholkovsky, P. (2008). Getting to we. *Communications of the ACM, 51*(4), 19–24. doi:10.1145/1330311.1330316

Devedžić, V., & Milenković, S. A. R. (2011). Teaching agile software development: A case study. *IEEE Transactions on Education, 54*(2), 273–278. doi:10.1109/TE.2010.2052104

Dubinsky, Y., & Hazzan, O. (2005). A framework for teaching software development methods. *Computer Science Education, 15*(4), 275–296. doi:10.1080/08993400500298538

Fancott, T., Kamthan, P., & Shahmir, N. (2011). Using the social web for teaching and learning user stories. In *Proceedings of the 6th International Conference on e-Learning.* Academic Press.

Ghezzi, C., Jazayeri, M., & Mandrioli, D. (2002). *Fundamentals of software engineering.* Upper Saddle River, NJ: Prentice-Hall.

Golbeck, J. (2013). *Analyzing the social web.* London: Elsevier.

Goldberg, A. (2002). Collaborative software engineering. *Journal of Object Technology, 1*(1), 1–19. doi:10.5381/jot.2002.1.1.c1

Gothelf, J., & Seiden, J. (2013). *Lean UX: Applying lean principles to improve user experience.* Sebastopol, CA: O'Reilly.

Hadjerrouit, S. (2005). Constructivism as guiding philosophy for software engineering education. *ACM SIGCSE Bulletin*, *37*(4), 45–49. doi:10.1145/1113847.1113875

Hayes, D., Hill, J., Mannette-Wright, A., & Wong, H. (2006). Team project patterns for college students. In *Proceedings of the 2006 Conference on Pattern Languages of Programs*. ACM.

Hayes, J. H., Lethbridge, T. C., & Port, D. (2003). Evaluating individual contribution toward group software engineering projects. In *Proceedings of the 25th International Conference on Software Engineering* (pp. 622–627). IEEE.

Hazzan, O., Lapidot, T., & Ragonis, N. (2011). *Guide to teaching computer science: An activity-based approach*. Berlin: Springer. doi:10.1007/978-0-85729-443-2

Highsmith, J. (2009). *Agile project management: Creating innovative products*. Upper Saddle River, NJ: Pearson Education.

Kamthan, P. (2009). A methodology for integrating the social Web environment in software engineering education. *International Journal of Information and Communication Technology Education*, *5*(2), 21–35. doi:10.4018/jicte.2009040103

Katzenbach, J. R., & Smith, D. K. (1992). *The wisdom of teams: Creating the high-performance organization*. Boston: Harvard Business Press.

Koranne, S. (2011). *Handbook of open source tools*. Berlin: Springer. doi:10.1007/978-1-4419-7719-9

Kovitz, B. (2003). Hidden skills that support phased and agile requirements engineering. *Requirements Engineering*, *8*(2), 135–141. doi:10.1007/s00766-002-0162-9

Leffingwell, D. (2010). *Agile software requirements: Lean requirements practices for teams, programs, and the enterprise*. Reading, MA: Addison-Wesley.

Leuf, B., & Cunningham, W. (2001). *The wiki way: Quick collaboration on the web*. Reading, MA: Addison-Wesley.

Mahnic, V. (2012). A capstone course on agile software development using Scrum. *IEEE Transactions on Education*, *55*(1), 99–106. doi:10.1109/TE.2011.2142311

Mahnic, V., & Rozanc, I. (2012). Students' perceptions of scrum practices. In *Proceedings of the 35th International Convention - Microelectronics, Electronics and Electronic Technology* (pp. 1178–1183). IEEE.

Mohan, A., Merle, D., Jackson, C., Lannin, J., & Nair, S. S. (2010). Professional skills in the engineering curriculum. *IEEE Transactions on Education*, *53*(4), 562–571. doi:10.1109/TE.2009.2033041

Nielsen, J. (1994). *Usability engineering*. London: Elsevier.

O'Reilly, T. (2007). What is web 2.0: Design patterns and business models for the next generation of software. *Communications & Strategies*, (1), 17.

Parnas, D. L. (1999). Software engineering programs are not computer science programs. *IEEE Software*, *16*(6), 19–30. doi:10.1109/52.805469

Paulus, P. B., & Nijstad, B. A. (2003). *Group creativity: Innovation through collaboration*. Oxford, UK: Oxford University Press. doi:10.1093/acprof:oso/9780195147308.001.0001

Power, K. (2011). Using silent grouping to size user stories. In *Agile processes in software engineering and extreme programming* (pp. 60–72). Berlin: Springer. doi:10.1007/978-3-642-20677-1_5

Putnik, G., & Cunha, M. M. (2008). *Encyclopedia of networked and virtual organizations* (Vol. 2). Hershey, PA: Information Science Reference. doi:10.4018/978-1-59904-885-7

Ratcliffe, L., & McNeill, M. (2011). *Agile experience design: A digital designer's guide to agile, lean, and continuous*. New Riders.

Rico, D. F., & Sayani, H. H. (2009). Use of agile methods in software engineering education. In *Proceedings of the 2007 Agile Conference* (pp. 174–179). IEEE.

Robillard, P. N., & Robillard, M. P. (2000). Types of collaborative work in software engineering. *Journal of Systems and Software, 53*(3), 219–224. doi:10.1016/S0164-1212(00)00013-3

Sanker, D. (2011). *Collaborate: The art of we*. New York: Wiley.

Sedelmaier, Y., & Landes, D. (2013). A research agenda for identifying and developing competencies in software engineering. *International Journal of Engineering Pedagogy, 3*(2).

Séguin, N., Tremblay, G., & Bagane, H. (2012). Agile principles as software engineering principles: An analysis. In *Agile processes in software engineering and extreme programming* (pp. 1–15). Berlin: Springer. doi:10.1007/978-3-642-30350-0_1

Shaw, M. (2000, May). Software engineering education: A roadmap. In *Proceedings of the 2000 Conference on Future of Software Engineering* (pp. 371–380). ACM.

Soundararajan, S., Chigani, A., & Arthur, J. D. (2012, February). Understanding the tenets of agile software engineering: Lecturing, exploration and critical thinking. In *Proceedings of the 43rd ACM Technical Symposium on Computer Science Education* (pp. 313–318). ACM.

Surakka, S. (2007). What subjects and skills are important for software developers? *Communications of the ACM, 50*(1), 73–78. doi:10.1145/1188913.1188920

Tabaka, J. (2006). *Collaboration explained: Facilitation skills for software project leaders*. Upper Saddle River, NJ: Pearson Education.

Venkatagiri, S. (2011). Teach project management, pack an agile punch. In *Proceedings of the 24th Software Engineering Education and Training Conference* (pp. 351–360). IEEE.

Whitehead, J. (2007). Collaboration in software engineering: A roadmap. In *Proceedings of 2007 Future of Software Engineering Conference* (pp. 214–225). IEEE.

Wilkins, D. E., & Lawhead, P. B. (2000). Evaluating individuals in team projects. *ACM SIGCSE Bulletin, 32*(1), 172–175. doi:10.1145/331795.331849

Williams, L. (2010). Agile software development methodologies and practices. *Advances in Computers, 80*, 1–44. doi:10.1016/S0065-2458(10)80001-4

ADDITIONAL READING

Chamillard, A. T., & Braun, K. A. (2002, February). The software engineering capstone: Structure and tradeoffs. *ACM SIGCSE Bulletin, 34*(1), 227–231. doi:10.1145/563517.563428

Dawson, R. (2000, June). Twenty dirty tricks to train software engineers. In *Proceedings of the 22nd international conference on Software engineering* (pp. 209–218). ACM.

Ebersbach, A. (2008). *Wiki: Web collaboration*. Springer-Verlag.

Larman, C. (2003). *Agile and iterative development: A manager's guide*. Addison-Wesley.

Larman, C., & Vodde, B. (2010). *Practices for scaling lean and agile development: Large, multisite, and offshore product development with large-scale scrum*. Addison-Wesley.

Lee, M. J. W., & McLoughlin, C. (2011). *Web 2.0-based e-learning: Applying social informatics for tertiary teaching*. IGI Global.

Mohammadi, S., Nikkhahan, B., & Sohrabi, S. (2009). Challenges of user involvement in extreme programming projects. *International Journal of Software Engineering and Its Applications*, *3*(1), 19–32.

Phuwanartnurak, A. J. (2009, May). Interdisciplinary collaboration through wikis in software development. In *Proceedings of ICSE Workshop on Wikis for Software Engineering* (pp. 82–90). IEEE.

Reagle, J. M. Jr. (2010). *Good faith collaboration: The culture of Wikipedia*. The MIT Press.

Selwyn, N. (2003). Apart from technology: Understanding people's non-use of information and communication technologies in everyday life. *Technology in Society*, *25*(1), 99–116. doi:10.1016/S0160-791X(02)00062-3

Treude, C. (2012). *The role of social media artifacts in collaborative software development*. Ph.D. thesis, University of Victoria, Victoria, Canada.

Weller, K. (2010). *Knowledge representation in the social semantic Web*. Walter de Gruyter. doi:10.1515/9783598441585

Wenger, E. (1998). *Communities of practice: Learning, meaning and identity*. Cambridge University Press. doi:10.1017/CBO9780511803932

KEY TERMS AND DEFINITIONS

Affordance: A property, or multiple properties, of an object that provides some indication to a user of how to interact with that object or with a feature of that object.

Agile Methodology: A software development methodology based on the Agile Manifesto.

Artifact: A document or a model produced during software development.

Social Web: The perceived evolution of the Web in a direction that is driven by 'collective intelligence,' realized by information technology, and characterized by user participation, openness, and network effects.

Software Engineering: The application of a systematic, disciplined, quantifiable approach to the development, operation, and maintenance of software; that is, the application of engineering to software.

Use Case: A sequence of actions performed by a system, which yields an observable result of value to an actor of that system.

User Story: A high-level requirement statement that contains minimally sufficient information to produce a reasonable estimate of the effort to implement it.

Wiki: A Web application developed cooperatively by a community of users, allowing any user to add, delete, or modify information.

Chapter 4
Developing Communities of Practice to Prepare Software Engineers with Effective Team Skills

Ann Q. Gates
The University of Texas – El Paso, USA

Elsa Y. Villa
The University of Texas – El Paso, USA

Salamah Salamah
The University of Texas – El Paso, USA

ABSTRACT

A major challenge to teaching software engineering is achieving functioning teams that enforce individual accountability while integrating software engineering principles, approaches, and techniques. The two-semester software engineering course at the University of Texas at El Paso, referred to as the Team-Oriented Software Engineering (TOSE) course, establishes communities of practice that are cultivated through cooperative group practices and an improvement process model that enables learning from past experiences. The experience of working with incomplete, ambiguous, and changing software requirements motivates the need for applying disciplined software engineering practices and approaches throughout project development. Over the course of the two-semester sequence, the nature of students' participation in project teams changes; they begin to influence others in software engineering practice, and their identities as software engineers begins to develop. The purpose of the chapter is to describe how to structure a software engineering course that results in establishing communities of practice in which learners become increasingly more knowledgeable team members who embody the skills needed to work effectively in a team- and project-based environment.

DOI: 10.4018/978-1-4666-5800-4.ch004

INTRODUCTION

A long-standing problem when teaching software engineering is achieving functioning teams that enforce individual accountability. Working as teams, students complete a large project while going through the appropriate training in team skills. The human aspect of software development also makes teaching software engineering and managing student-run projects challenging because of the following:

- General lack of maturity in the students' team and communication skills,
- Difficulty in ensuring that all team members contribute to the project,
- Differences in students' experiences and understanding, and
- Difficulty in evaluating and ensuring individual and team progress and work quality.

A two-semester, software engineering course, referred to as the *Team-Oriented Software Engineering (TOSE) course*, at the University of Texas at El Paso (UTEP) addresses these challenges by incorporating cooperative-learning principles with an aim of establishing a community of practice. *Cooperative learning* as an instructional approach (Johnson, Johnson, & Holubec, 1992; Johnson, Johnson, & Smith, 1991) is an evidence-based practice that contributes to team building while increasing student achievement and self-esteem (Johnson & Johnson, 1989). Using cooperative learning principles to structure groups generates positive interdependence in which each member is committed to supporting others in reaching their goals while at the same time working together to meet the group goal. The emphasis on cooperative behavior cultivates an environment in the software engineering course where communities of practice can emerge and grow. Drawing from the work of Lave and Wenger (1991) and Wenger (1998), a *community of practice* is defined as a group of individuals who share a common purpose,

contribute to each other's success, and develop shared practices that identify them as members of that group.

The purpose of the chapter is to describe how structuring a software engineering course using co-operative learning principles results in establishing communities of practice in which learners become increasingly more knowledgeable team members who embody the skills needed to work effectively in a team- and project-based environment. The objectives of the chapter are to: (1) present the challenges in developing functional teams; (2) outline how to structure a software engineering course in which teams move toward becoming a community of practice; and (3) describe how a community of practice serves to support functioning and practicing software engineers.

BACKGROUND

Overview

In the perspective of this chapter, cooperative learning is at the core of building functional and effective teams for addressing the issues, challenges, and concerns of ineffective student teams typically resulting from ill-structured group work (rather than team work). In such group work, a task is given to the group with the hope, for example, that group members will resolve any conflicts on their own and allow for a "leader" to emerge who can take charge. When groups are structured in this manner, those who are "followers" minimally contribute to deliverables and may be marginalized by the others. Rather, a cooperative learning framework is one that is intentionally designed to promote positive social interaction among the group members through the use of techniques and tools, such as active listening, active participation, conflict management, and individual accountability. When purposefully structured in a sustained manner, team members eventually embody these interpersonal skills; the functionality of the team reflects one of cooperation where team members have mutual respect and trust; and members' contributions are valued. In this case, the cooperative team most resembles an authentic community of practice. Using the term "authentic" distinguishes it from what is now a ubiquitous way to describe communities in action as "communities of practice" when too often they may simply be groups of individuals working together with a common purpose. In this section, more detailed descriptions of cooperative learning and communities of practice are discussed.

Cooperative Learning

Human beings are fundamentally cooperative, yet most individuals will argue that humans are competitive. This myth prevails in spite of the often-used phrases of *only the strongest survive* and *survival of the fittest*. Interest in cooperation as a scholarly endeavor began in the early twentieth century with the work of social psychologist Kurt Lewin (see Lewin, 1935; 1948) who investigated group dynamics and interplay of its members. Morton Deutsch, a student of Lewin's, continued investigating group dynamics and extended it to formulate a theory of competition and cooperation (see Deutsch, 1949; 1962). By the mid-20th century, David Johnson, under the guidance of Deutsch, further investigated competition and cooperation, including perspective taking and conflict, during the 1970s (Johnson, 1974; 1975). These studies were further extended to identify the following outcomes of cooperative learning: higher achievement, increased retention, greater intrinsic motivation, increased perspective taking, more positive heterogeneous relationships, higher self-esteem, and greater collaborative skills (Johnson & Johnson, 1989). Further efforts resulted in the identification of five essential components of cooperative learning (Johnson, Johnson, & Holubec, 1990; Johnson, Johnson & Smith, 1998) described in the next section.

The five essential elements of cooperative learning in the Johnson and Johnson model are as follows:

1. **Positive Interdependence:** Students are linked to others in the team in such a way that they cannot succeed unless their teammates do and that they must coordinate their efforts with the efforts of other members to complete a task. It is important to set clear team goals and tasks so that members believe that they cannot succeed unless everyone succeeds.

2. **Promotive Interaction:** Members of the team promote each other's success by sharing needed resources and helping, supporting, encouraging, and praising each other's efforts. Promotive interaction is characterized by individuals providing each other with efficient and effective help and assistance, as well as providing each other with feedback in order to improve their subsequent performance.

3. **Individual Accountability:** The group must be accountable for achieving its goals. Each member must be accountable for contributing his or her share of the work. The group has to be clear about its goals and be able to measure its progress in achieving them and the individual efforts of each of its members. Individual accountability exists when the performance of each individual student is assessed, the results given back to the individuals and the group, and the student is held responsible by team members for contributing his or her fair share to the group's success.

4. **Professional and interpersonal skills:** Group members must know how to provide effective leadership, decision-making, trust-building, communication, and conflict management and be motivated to use these skills. Skills must be taught and practiced.

5. **Group processing:** Group processing refers to reflecting on a team activity to describe which team behaviors were helpful and not helpful and to make decisions about what actions to continue or change. Team members describe how well they are achieving their goals and maintaining effective working relationships, as well as identify members' actions that are productive or not. The reflection helps clarify the team's goals and effective working relationships. Indeed, continuous improvement results from careful analysis of how members are working together and determining how team effectiveness can be enhanced.

The model requires the instructor to structure activities incorporating these five elements. Later sections describe how this is done in the TOSE course.

Theoretical Underpinnings

As mentioned earlier, the scholarly work of David and Roger Johnson builds on the work of Lewin and Deutsch in investigating how social interaction supports learning. They posit that social interaction occurs in three forms: competitive, cooperative, and individual. In describing the effect of cooperation in comparison to competition, Johnson and Johnson (1989) describe this as the result of bidirectional relationships:

Deutsch's (1985) crude law of social relations states that the characteristic processes and effects elicited by a given type of social interdependence also tends to elicit that type of social interdependence. Thus, cooperation tends to induce and be induced by mutual help and assistance, exchange of needed resources, influence, and trust. People tend to trust their collaborators but also to seek out opportunities to collaborate with those they

trust. There is a benign spiral of cooperation in which cooperation promotes trust, trust promotes greater cooperation, which promotes greater trust, and so forth (p. 9).

Also contributing to this theory are other strands of psychology, such as social, cognitive, and behaviorist. Most notably among these is the contribution of social psychologist Lev Vygotsky (1978, 1986) who theorized learning happening in social contexts through the interplay among and between individuals, objects, and language. Thus, learning is cultural and occurs in interaction with others. Other contributions include educator and philosopher John Dewey who wrote extensively about the importance of social interaction with others.

[In] this intercommunication one learns much from others. They tell of their experiences and of the experiences which, in turn, have been told them. In so far as one is interested or concerned in these communications, their matter becomes a part of one's own experience. Active connections with others are such an intimate and vital part of our own concerns (Dewey, 2009, p. 9).

Communities of Practice

Coined by Lave and Wenger (1991), the term *communities of practice* describes a concept of how novices, such as apprentices, in a particular discipline or craft learn the work of those more expert. As Wenger (1998) noted, communities of practice draw from scholarly work in sociology, anthropology, and psychology. He further posited that, as one participates in the practices of a community, meaning is negotiated in engagement with others, ideas, language, and objects. This draws in part from Vygotsky's (1978, 1986) socio-cultural theory of learning whereby individuals acquire concepts, theories, and the discourse of the group as their participation in the community gradually increases.

[Learning then results from] practices that reflect both our pursuit of our enterprises and the attendant social relations. These practices are thus the property of a kind of community created over time by the sustained pursuit of a shared enterprise (Wenger, 1998, p. 45).

The practice of sharing expertise and resources with others in a community allows members to thrive in appropriating such practices. Communities of practice are formed over time as participants engage in activities with a common purpose that defines the existence of the group. Newcomers enter communities either by invitation, as is the case for established communities, or by situation, as is the case for a college classroom where access is open to those who qualify in terms of their academic qualifications. In this case, however, the group is merely a collection of individuals who have the propensity to evolve into a community of practice when the group reflects the following characteristics:

1. Sustained mutual relationships
2. Substantial overlap in participants' descriptions of who belongs
3. Knowing what others know, what they can do, and how they contribute to the enterprise
4. Displaying certain styles of membership
5. Mutually defining identities
6. Shared ways of engaging in doing things together
7. Quick setup of problem to be discussed
8. Shared discourse
9. Ability to assess appropriateness of actions and products
10. Specific tools, representations, and other artifacts
11. Local lore, shared stories, inside jokes
12. Jargon and other communication shortcuts

13. Rapid flow of information and propagation of innovation (Wenger, 1998, p. 125).

Using the categorization defined by Wenger (2013), the aforementioned characteristics could be classified as follows: *the domain*-the group has an identity defined by a shared domain of interest and goes beyond being a connection among members (characteristics 1-5); *the community*-members interact and learn from each other (characteristics 6-9); and *the practice*-members are practitioners and develop a repertoire of resources (characteristics 10-13).

Theoretical Underpinnings

Learning theories have historically been theorized by cognitive psychologists and scientists, and more recently by social scientists as the product of our social and cultural interaction with others in thought and language. One of the preeminent theories to recently emerge is *situated learning theory* (Lave & Wenger, 1991), i.e., we learn values, language, knowledge and skills situated in everyday practice with others. Learning is situated and happens when learners engage with others in authentic and meaningful (i.e., real world) practice. Having a common purpose motivates participation of group members. Wenger (1998) postulated that communities of practice evolve where groups of people gather for a common purpose, such as family, church, or formal classrooms. In this interaction, we interpret what is said and what is done to obtain meaning of a practice. Members of the community gradually adopt these practices through immersion in the practice with others in the community. In communities of practice, learning the practices of the group takes time as newcomers, or novices, learn from those more expert as they socially interact in the practices of the community.

TEAM SKILLS DEVELOPMENT: ISSUES AND PROBLEMS

A recent paper (Radermacher & Walia, 2013) presented the results of a systematic literature review identifying the areas of greatest divergence between skill sets required by the industry and the actual skills of graduates. Teamwork, oral communication, project communication, and problem solving were identified as the skills with the widest gaps. Surveying engineering competencies of over 4000 graduates in eleven different engineering disciplines, Passow's study (2012) solicited opinion about the relative value of the prescribed Accreditation Board for Engineering and Technology (ABET) prescribed student outcome areas. The ability to function on a team and written and oral communication skills were among the highest rated competencies.

Peter Denning (1992) further noted the lack of skills in recent graduates as follows:

Employers and business executives complain that graduates lack practical competence. Graduates, they say, cannot build useful systems, formulate or defend a proposal, write memos, draft a simple project budget, prepare an agenda for a meeting, work on teams, or bounce back from adversity; graduates lack a passion for learning. They say the current concepts-oriented curriculum is well suited for preparing research engineers, but not the practice-oriented engineer on which their competitive advantage increasingly depends (p. 5).

Although educators have made much progress in preparing graduates to meet the competencies as described by Denning, a significant gap still exists between computer science graduates' capabilities and the needs/expectations of the industry. Because of the importance of team, communication, and other professional skills in the future workforce, a number of universities offer dedicated courses for engineers to develop professional skills. Furthermore, many computer science and software engineering programs include professional skills development in multiple courses using a variety of techniques and exercises to teach, for example, communication and team skills (Coleman & Lang, 2012). Communication skills are typically taught through instructor lectures and practiced through student presentations and written reports (Liu, Sandell & Welch, 2005).

Students often complain that, although they are expected to work in teams on projects, they are seldom given any advice or guidance on how to work in a team (Hart & Stone, 2002). Or, if they are given guidance, it is often from a business perspective and students find it difficult to integrate into their software development practice. Techniques for building team skills include lectures on group process and diversity. Team skills can be built through engagement in activities aimed at requiring individuals to complete a given task critical to the success of the team project (Lingard & Berry, 2002).

In software engineering capstone courses, students are typically grouped into teams to work on a one- or two-semester development project. Such courses provide the ideal setting for teaching and enhancing professional skills since they provide an opportunity to mimic real-world industrial settings (Broman, Sandahl, & Baker, 2012; Hogan & Thomas, 2005) through the use of team projects to apply the technical skills acquired through the coursework. One feature of these team projects is the need for communicating ideas to other members of the team, yet explicit directions or exercises to help students enhance these skills are often lacking (Smith, 2000). As mentioned in the Background section, merely grouping students without specific structure is a risk in achieving a desired outcome, particularly for developing teaming and communication skills. Swan, Magleby, Sorensen, & Todd (1994) noted that being a member of a dysfunctional team may result in students developing negative views of the whole experience of teamwork. Finally, McGinnes (1995) emphasizes that skills development, such as teaming, communication, and project management, should be integrated throughout the undergraduate curriculum rather than isolated in capstone courses.

THE TOSE COURSE: SOLUTIONS AND RECOMMENDATIONS

Drawing on both cooperative learning and communities of practice concepts, this section addresses the aforementioned challenges of software engineering projects by offering guidelines on how to incorporate cooperative learning into teams to eventually move them toward exhibiting the attributes of a community of practice. The structure for the *Team-Oriented Software Engineering (TOSE) course* uses a two-prong strategy: (1) develop student teams using a cooperative learning framework and (2) cultivate an environment to immerse teams in the practice of software engineering. TOSE is modeled after the computing research teams where cooperative learning principles were integrated into group meetings with an aim of developing students' skills as researchers. Findings from a qualitative study by authors Gates and Villa (Villa, Kephart, Gates, Thiry, & Hug, 2013) revealed that these research teams had attributes resembling those of an authentic community of practice as previously defined.

The project deliverables associated with the first semester (requirements engineering) of a TOSE course include the following: feasibility study, interview report, prototype, and software requirements specification document. The use case diagrams, data flow diagrams, class diagrams, and state transition charts are integrated into the documents. The deliverables for the second semester (design and implementation) are as follows: architectural design, detailed software design document, software code, and test plan. As the project teams works on the deliverables,

they evolve into communities of practice in which team members become increasingly more knowledgeable and effective software engineering practitioners. The cooperative approach prepares teams that function effectively and encourages continuous improvement of teams together with the software development process. Figure 1 elucidates the synergy between cooperative learning elements and communities of practice.

In particular, the TOSE course provides a foundation for project teams to evolve into communities of practice in which team members become increasingly more knowledgeable software engineering practitioners. To produce effective and functioning software engineering teams, instructors must first invest their time in understanding and then skillfully applying cooperative learning principles, in particular the deliberate and intentional professional skills building, in facilitating the software engineering teams. Like any good design, extensive planning is needed to facilitate an environment where students are intrinsically motivated and engaged in their learning.

The following section is organized around the elements of a community of practice, as defined by Wenger (1998, 2013): domain, community, and practice. For each element, we describe how cooperative learning is structured in the course and provide example activities of the TOSE course.

Setting the Domain

A community of practice sets the *domain* or identity of shared interest (Wenger, 1998). While typical communities of practice are created around an affinity and expertise for a particular topic, a software engineering course brings together a diverse group of students, who may or may not know each other and who typically have various levels of skills and capabilities. Rather than allowing students to self-select into their teams, the TOSE course emulates an actual work environment—students submit résumés featuring

Figure 1. The TOSE strategy

their professional and educational experiences, including relevant coursework. They apply for positions in which they are interested in taking the lead role, rank the positions in order of preference, and justify their choices. The positions are requirements analyst, system designer, detailed designer, lead programmer, and verification and validation supervisor. The lead for each position is responsible for the quality of the assigned deliverable and accomplishes this by delegating work, integrating individual assignments, interfacing with course instructors, and ensuring that deliverables are submitted on schedule. The instructor evaluates each application and determines the membership of the project teams based on each student's expertise, grade point average, choice, and personality. Students self-identify their personality based on a simple personality exercise.

Establishing positive interdependence among the team members lays the foundation for setting the domain for the team to evolve into a community of practice. Positive interdependence can be structured in the following ways: create positive role interdependence (each member is assigned a role), resource interdependence (members rely on a common resource to accomplish a task), positive reward interdependence (celebrate team's successes), goal interdependence (mutually shared team goal), and identity interdependence (establish a mutual identity).

After students are assigned to teams early in the semester, the team members brainstorm on a name for their "company" and create a logo, thus, establishing *identity interdependence. Goal interdependence* is inherent in the course goal since students work throughout the course to complete the course milestones and deliverables by the assigned due dates. Assigning each member of the team with a specific job during phases of the project establishes *role interdependence*. Roles are also assigned on in-class assignments, and leaders are asked to assign roles during meetings. Examples of roles include recorder, summarizer, direction giver, time manager, devil's advocate, and questioner. Through role assignments, students learn behaviors that are essential for teamwork.

Resources, which could include such things as project materials, specialized equipment, or tools, are limited; thus, team members must determine how best to share these resources, which is an example of *resource interdependence*. Finally, because the students' final grade includes how well they function in their team, students coordinate their efforts to complete tasks and realize that their success is dependent on the success of the other team members.

As a way of modeling promotive interaction, the instructor can acknowledge each team member as an expert in a skill they identified in their résumé. In addition, integrating role assignment throughout the course encourages students to help, support, encourage, and acknowledge each other's efforts.

Building Community

A community of practice engages in joint activities and discussions, helps one another, and shares information (Wenger, 2013). In addition to the two-semester TOSE project, the instructor assigns readings, delivers a brief lecture to reiterate important concepts from the readings, engages the class in discussion, and seeks clarification as needed. In-class assignments are used to build expertise in software engineering techniques, approaches, and tools. Students work in small groups to complete the assignment, which are structured as cooperative groups in which members share information and help each other learn.

For example, consider an assignment in which students are asked to create a state transition diagram for a given problem. Students can work in teams to create a diagram. The instructor can assign the role of devil's advocate to a student where he or she must challenge another member's contribution and reasoning. Such an exercise may lead to higher quality decision-making and can provide greater insight into problem solving. In an extension of this assignment, the instructor could use one or more diagrams to have the students conduct an inspection. For such an activity, the instructor would create a group of four or five students and assign the roles of leader, scribe, and inspectors. To help the students learn how to run an inspection, the instructor could prepare a script for the inspection, which would include a checklist that the inspectors would use to determine if the authors are using correct notations, and the script could have the inspectors review particular aspects of the diagram to determine if the notations capture the intended meaning of the problem and if the model is complete. The scribe would be instructed to record the responses and findings. A grade would be assigned based on the effectiveness of the inspection in finding shortcomings.

An essential TOSE component that leads to building community is constructive critique of assignments. Time is allocated during the class session to define, apply, and reflect on constructive critique and its importance as a team skill. Students are expected to resubmit work based on instructor and peer feedback to improve their understanding and add to their expertise. The instructor schedules weekly half-hour meetings with the teams to provide constructive critique of work-in-progress, ensure that the team is on track with respect to their next deliverable, and to ensure individual accountability. In addition, the instructor encourages team leaders and members to acknowledge contributions of others and praise each other's efforts, an example of the cooperative learning element of promotive interaction. It is important for the professor to emphasize constructive critique and to model promotive interaction while interacting with the groups. In a sense, the teaching assistant and instructor are the experts, and the students are the apprentices. It is important to advocate for their individual effort in achieving mutual goals, achieving the team goal, and becoming motivated to strive for mutual benefit. Practicing promotive interaction and incorporating positive interdependence are important in a developing a community practice.

Becoming a Software Engineering Practitioner

The TOSE course aims to move students from novices to practitioners by the end of the second semester; however, becoming a practitioner develops over time due to the numerous skills and software engineering knowledge students need to more fully participate in a community of software engineering practitioners. This section describes the TOSE course structure and the activities for practicing cooperative team, professional, and communication skills.

As described earlier, to move students toward becoming software engineering practitioners, the TOSE course requires students to apply for one of five positions within each team. The instructor provides the teams with a requirements definition document that is the starting point for working with a customer to elicit requirements for a project that is too complex for any one person to complete. As a result, students must work together throughout the year to develop the project from inception to implementation. When the teams are formed, the student, who has been assigned the lead for a particular deliverable, will distribute the tasks needed to the work. The instructor provides activities in which teams learn software engineering concepts needed to complete the deliverable. In addition, the teams practice team skills and reflect on how to improve the essential elements of functioning teams, including holding students individually accountable for their contributions to deliverables.

Earlier sections describe how positive interdependence and promotive interaction are structured in the project and in-class teams. To structure individual accountability, another element of cooperative learning, the instructor must assess individual efforts in mastering software engineering knowledge. There are several avenues for assessing students' team skills: observation, reflection, and weekly meetings with the instructor or assistant as described earlier. Observation forms (Scholtes,

1995; Johnson, Johnson & Smith, 1998) are used by the instructor to record particular behaviors observed during a team meeting and provide the feedback to the team at the end of the meeting. For example, the instructor may mark how often team members seek input from other members and how a team member ensures that the discussions are on task. The feedback provides the team with information on how they interact and identifies areas of improvement.

In a community of practice where members share resources, it is important to query members about his or her contributions, observe the team to check the frequency with which each member contributes, and have members explain concepts to others. In a TOSE course, the team leader completes a Task Assignment Record in which he or she logs the date and task assigned to each team member, and the date that the assignment was submitted. Together with meeting records, correspondence, and memoranda or e-mails, this paper trail is maintained by the team leader in a team notebook, thus, providing a reliable means of determining individual contributions.

Another effective approach is to have each team turn in a document at the end of the semester that summarizes individual, subgroup, and group contributions; each member of the team is required to sign the document to acknowledge their agreement with its contents. If a member does not agree and cannot come to consensus with the group after discussion, then a meeting is called with the professor to discuss differences. It is imperative that, at the beginning of the semester, students understand that they will be held individually accountable for producing quality work towards the team's deliverable.

For students to learn and be motivated to apply effective leadership, decision-making, communication, and conflict management skills, the instructor must integrate activities where these skills are taught and practiced. First, the TOSE instructor needs to assess the teams' abilities and identify appropriate skills to be learned and practiced. Second, the instructor must design an in-class activity, as previously described, and assign roles to students related to that skill to serve as practice. Prior to practicing skills, the instructor involves the students in describing the verbal and visual cues of a person applying the skill. Finally, after the group activity, the instructor asks students to reflect on how well they were able to apply the skill, and provides opportunities for them to continue their practice of the skills until they become automatic. As an example, consider teaching the skill "asking a probing question," which prepares students for the interview with the customer. To teach this, the instructor would ask the class for the verbal cues that are indicators that the question is probing. The class might contribute the following responses: the question does not have a yes or no answer; the question seeks to extend knowledge; the question would begin with the words "how" and "what;" and the question would require a comparison or analysis. The visual cues are not as appropriate for this particular skill, but could include raised hand, eye contact with the speaker, and nodding head (to indicate listening and seeking understanding). Students could be asked to create a list of probing questions over a particular topic and the questions could be critiqued with feedback to the author.

Basic skills, e.g., active participation checker, active listening checker, and recorder, can be taught early in the first semester and practiced by assigning particular roles during in-class group discussions. The skill of active participation checker is one in which a person ensures that all group members participate by calling on them or asking them to share their opinion of what others have contributed. The instructor identifies other skills from feedback by team members of issues or concerns arising from their team meetings. For example, if a team provides feedback that members come away from a meeting with different impressions about their newly assigned tasks,

the instructor can teach the role of summarizer or paraphraser and describe the importance of assigning someone this role.

The instructor provides students guidelines for conducting productive meetings and a template for planning an agenda and documenting the meeting. The template includes a section for entering the attendees' names, assigning roles, entering past action items for status checking, and setting the agenda with time allotment. The leader is responsible for ensuring that the following occurs during the meeting:

- The discussion remains focused on the topic and moves along.
- No one dominates or is overlooked.
- Discussions come to a close.
- The group is notified when the time allotted for an agenda item has expired or is about to expire.

Each meeting should have a scribe who records the names of those in attendance, key subjects and main points raised, decisions made including action items, and items that the group has agreed to defer or raise again at a future meeting. Team members can refer to minutes to reconstruct discussions, find decisions made, and review assigned action items. This role is rotated among the team members to ensure all learn these skills.

Every team meeting should include actions to facilitate discussion. It is expected that team leaders use skills for effective discussion; however, the team will be even more successful if each team member learns and practices them. While the following techniques are discussed in the framework of team meetings, they are useful whenever effective discussion is important. The behavior and roles that are important for conducting an effective meeting, which are learned over time, are as follows:

- Ask for clarification when the topic being discussed or the logic in another person's arguments is unclear (Role: clarifier/paraphraser).
- Encourage equal participation among group members by directly asking a member to express his or her opinion or making a general request for input, especially in the case when a member dominates the conversation (Role: active participation checker).
- Actively explore a person's ideas rather than debating or defending each idea that comes up (Role: elaborator).
- Occasionally compile what has been said and restate it to the group in summary form followed by a check for agreement (Role: summarizer).
- Ensure that there are no overlong examples or irrelevant discussion (Role: time keeper).
- Manage time by reminding the team of deadlines and time allotments so work can be either accelerated or postponed, or time readjusted appropriately (Role: time keeper).
- Be prepared to end a discussion when there is nothing to be gained from further discussion and help close the discussion (Role: facilitator).
- Summarize the group's position on an issue, state the decision that seems to have been made, and check whether the team agrees with the summary (Role: integrator).

Finally, group processing, another cooperative learning element, provides an opportunity for a team to improve how it functions. To ensure that group processing takes place, instructors should allocate some time for team members to process how effectively members worked together. In industry, this is referred to as an "after-action report" or "post-mortem review." While after-action reports take place at the culmination of a project, group processing in the TOSE course occurs throughout the two semesters as a mean of providing continuous quality improvement. To ensure effective group processing, sufficient time must be allowed for sharing; a structure must be provided for processing (such as "list three things your group is doing well today and one thing you could improve"); positive feedback is emphasized; the processing is specific rather than general; students are actively involved; instructors remind students to use their cooperative skills while they process; and the instructor communicates clear expectations about the purpose of processing.

After a major deliverable, the instructor asks students to write answers to the following basic questions: What are your contributions? What is working? What is not working? What can be improved? While the first question targets individual accountability, the remaining questions address group functioning. Students are told to write comments using first person as a way of capturing the individual's perception of the group. In addition, students are asked to phrase criticisms using a constructive tone. After reviewing the responses, the professor or teaching assistant groups them by teams, strips the headers, deletes the contribution section, and edits the text if personal remarks have been included. The altered responses along with comments from the instructor are sent to the respective teams. The teams discuss the content and devise strategies for team improvement. It is imperative that the instructor meets with groups who are not functioning well. The processing identifies teams with problems allowing for early intervention. The approach aligns with improvement process models such as Plan-Do-Check-Act (Deming, 1986) that provides a feedback cycle to enable learning from past results.

In addition to practicing cooperative skills, students review through class discussion the stages of team growth (Tuckman, 1965) as given below, as well as the roles that group members can practice to assist in developing their ability to function well in the team. Roles prescribe what an individual is obliged to do and provides a mechanism for building positive interdependence and developing interpersonal skills.

Stage 1: Forming

In this stage, members explore the boundaries of acceptable group behavior and test the leader's guidance. This stage includes the feelings of excitement, optimism, tentative attachment to the team, suspicion, fear, and anxiety. It also includes the following behaviors: complaints about the organization and barriers to the task, discussion of problems not relevant to the task, impatience with discussions, and decisions on what information needs to be gathered. The roles that support how a group functions are as follows:

- **Explainer of Ideas or Procedures:** Shares one's ideas and opinions.
- **Recorder:** Writes down the group's decisions and edits the group's report.
- **Gatekeeper:** Ensures that all members are contributing and gives both verbal and nonverbal support and acceptance through seeking and praising others' ideas and conclusions.
- **Direction Giver:** Giver direction to the group's work by reviewing the instructions and restating the purpose of the assignment, calling attention to the time limits, and offering procedures on how to complete the assignment most effectively.
- **Clarifier /Paraphraser:** Restates what other members have said to understand or clarify a message.

Stage 2: Storming

The Storming Stage is the most difficult: Team members realize that the task is different and more difficult than realized; members become testy, blameful, or overzealous. This includes feelings such as resistance to task and quality improvement

approaches, and sharp fluctuations in attitude about team and project's chance of success; it also includes behaviors such as arguing among members even when they agree on real issues, defensiveness and competition, questioning of wisdom of guidance team, leaders, establishing unrealistic goals, concern about excessive work, and some disunity, tension and jealousy. Roles that can help a team ferment ideas and resolve conflict are as follows:

- **Criticizer of Ideas, Not People:** Intellectually challenges member by criticizing their ideas while communicating respect for them as individuals.
- **Asker for Justification:** Asks members to give the facts and reasoning that justify their conclusions and answers.
- **Differentiator:** Differentiates the ideas and reasoning of group members so that everyone understands the differences in members' conclusions and reasoning.
- **Integrator:** Integrates the ideas and reasoning of group members into a single position to which everyone can agree.

Stage 3: Norming

Members in the Norming Stage reconcile competing loyalties and responsibilities; they accept the team and the individuality of team members. This stage includes feelings such as a new ability to express criticism constructively, acceptance of membership in the team, and relief that everything is going to work out. It includes behaviors such as an attempt to achieve harmony by avoiding conflict, more friendliness, confiding in each other and sharing of personal problems; team cohesion, common spirit and goals, establishing and maintaining team ground rules and boundaries (norms). As team members get used to working together, their initial resistance fades away. Some roles that can help formulate information are as follows.

- **Summarizer:** Restates the group's major conclusions or answers or what has been read or discussed as completely and accurately as possible without referring to notes or to the original material.
- **Accuracy Coach:** Corrects any mistakes in another member's explanations or summaries and adding important information that was left out.
- **Perspective-taking roles:** Each member is responsible for contributing one's perspective or viewpoint to the group's final product.

Stage 4: Performing

The team has settled its relationships and expectations. They can begin performing, diagnosing and solving problems, and choosing and implementing changes. All members have discovered and accepted each other's strengths and weaknesses, and learned what their roles are. It includes feelings such as members having insights into personal and group processes, and better understanding of each other's strengths and weaknesses. It includes behaviors such as constructive self-change, ability to prevent or work through group problems and close attachment to team. As team members become more comfortable with each other, and better understand the project and what is expected of them, they become a more effective unit with everyone working in concert. The following roles can move a team toward higher levels of thinking:

- **Extender:** Extends the ideas and conclusions of other members by adding further information or implications.
- **Prober:** Asks in-depth questions that lead to analysis or deeper understanding.
- **Cognitive roles:** Each member is responsible for contributing one aspect of the critical-thinking process to the group's final product (e.g., analysis, synthesis, evaluation, elaboration, application).

Understanding these stages of teaming allows student teams to reflect on the characteristics they may be experiencing. By providing roles for them to assume, the teams are more likely to transition from forming to performing teams.

FUTURE RESEARCH DIRECTIONS

As aforementioned, results from a qualitative study by authors Gates and Villa (Villa, Kephart, Gates, Thiry, & Hug, 2013), which focused on alumni who were former members of research groups using TOSE instructional approaches, provides evidence of graduates' ability to effectively transfer communication and team skills when they enter the workforce. A future research effort is to conduct a comparative study of TOSE and non-TOSE graduates to assess their level of preparedness in effectively working in teams upon entry into and their advancement in the workforce over time.

Social media tools are an emerging technology; such tools are currently not a feature of TOSE. Thus, an interesting future research effort would be to investigate the effectiveness of these tools in supporting team communication, individual accountability, and group processing. Finally, another research direction would be to individually follow students through TOSE using ethnographic research methods. This would contribute to a better understanding of specific features of TOSE that contribute (or hinder) development of identities as software engineers.

CONCLUSION

This chapter describes the TOSE two-prong strategy: (1) develop student teams using a cooperative learning framework and (2) cultivate an environment to immerse teams in the practice of software engineering. As a result, project teams evolve into communities of practice in which team members become knowledgeable and effective software engineering practitioners. The cooperative approach provides a framework for teams to learn what is needed to function effectively and to incorporate continuous quality improvement into the software development process.

The two-semester TOSE course incorporates the deliberate practice of team skills while providing students with an introduction to approaches, techniques, and methods to structured and object-oriented software engineering methodologies. Students work in teams to develop a cross-disciplinary, large-scale software system. The course provides student teams with an opportunity to deal with the challenges of developing a real-world product. The experience of working with incomplete, ambiguous and changing requirements motivates the need for applying disciplined software engineering practices and approaches, as well as effective team work throughout project development.

Students are assessed on their individual knowledge of software engineering through exams and homework assignments, which constitute 55 percent of their grade, and they are assessed on their team project, which constitutes 45 percent of their grade. The team receives a team grade on each deliverable and a grade for the team and individual presentation at the end of the semester. The team grade may vary among members based on the team notebook that documents individual contributions and group contributions to deliverables; leadership abilities on deliverables; and ability to describe and apply cooperative team skills. The chapter describes several avenues for assessing team skills and providing the team with feedback that enables their ability to improve how they work in teams: observation, reflection, and weekly meetings with the instructor or assistant.

In every phase of software development from elicitation to modeling and from analysis to design and implementation, students are instructed in techniques and principles of software engineering, then asked to apply this knowledge to their project. For example, shortly after the discussion of techniques for elicitation of requirements, the client is brought to campus for an interview. Immediately after discussing team development, strategies for conducting effective meetings, and time management, students are assigned to project teams and asked to keep minutes of their meetings and task assignments, which are handed in and graded along with other project deliverables. During the project lifetime, the project guidance team (consisting of faculty and teaching assistants for the course) guide project teams through the software development process.

To produce effective and functioning software engineering teams, instructors must commit to investing their time in understanding and then applying cooperative learning principles, in particular the deliberate and intentional professional skills learning, in facilitating the software engineering teams. Once instructors gain an understanding of how to skillfully apply cooperative learning principles into their software engineering course, they must reflect on its effective use in supporting community building and adjust as needed. They must also build in features for students to actively reflect on how these communities develop into professional software engineers in order for them to build such teams when they join the workforce. Like any good design, extensive planning is needed to facilitate an environment where students are intrinsically motivated and engaged in their learning.

REFERENCES

Broman, D., Sandahl, K., & Abu Baker, M. (2012). The company approach to software engineering project courses. *IEEE Transactions on Education*, *55*(4). doi:10.1109/TE.2012.2187208

Coleman, B., & Lang, M. (2012). Collaboration across the curriculum: A disciplined approach to developing team skills. In *Proceedings of the 2012 Special Interest Group on Computer Science Education Technical Symposium*. Academic Press.

Deming, W. E. (1986). *Out of the crisis*. Cambridge, MA: MIT Center for Advanced Engineering Study.

Denning, P. J. (1992). Educating the new engineer. *Communications of the ACM, 35*(12), 83–97. doi:10.1145/138859.138870

Deutsch, M. (1949). A theory of cooperation and competition. *Human Relations, 2*, 129–152. doi:10.1177/001872674900200204

Deutsch, M. (1962). Cooperation and trust: Some theoretical notes. In M. R. Jones (Ed.), *Nebraska symposium on motivation* (pp. 275–319). Lincoln, NE: University of Nebraska Press.

Dewey, J. (2009). *Democracy and education: An introduction to the philosophy of education*. New York: WLC Books.

Hart, G., & Stone, T. (2002). Conversations with students: The outcomes of focus groups with QUT students. In *Proceedings of the 2002 Annual International Conference of the Higher Education Research and Development Society of Australasia (HERDSA)*. HERDSA.

Hogan, J. M., & Thomas, R. (2005). Developing the software engineering team. In *Proceedings of 2005 Australasia Computing Education Conference*. Newcastle, Australia: Academic Press.

Johnson, D., Johnson, R., & Holubec, E. (1992). *Advanced cooperative learning*. Edina, MN: Interaction Book Company.

Johnson, D., Johnson, R., & Smith, K. (1998). *Active learning: Cooperation in the college classroom*. Edina, MN: Interaction Book Company.

Johnson, D. W. (1974). Communication and the inducement of cooperative behavior in conflicts. *Speech Monographs, 41*, 64–78. doi:10.1080/03637757409384402

Johnson, D. W. (1975). Cooperativeness and social perspective taking. *Journal of Personality and Social Psychology, 31*, 241–244. doi:10.1037/h0076285

Johnson, D. W., & Johnson, R. (1975/1994). *Learning together and alone: Cooperative, competitive, and individualistic learning*. Englewood Cliffs, NJ: Prentice-Hall.

Johnson, D. W., & Johnson, R. T. (1989). *Cooperation and competition: Theory and research*. Edina, MN: Interaction Book Company.

Lave, J., & Wenger, E. (1991). *Situated learning: Legitimate peripheral participation*. New York: Cambridge University Press. doi:10.1017/CBO9780511815355

Lewin, K. (1935). *A dynamic theory of personality*. New York: McGraw-Hill.

Lewin, K. (1948). *Resolving social conflicts*. New York: Harper.

Lingard, R., & Berry, E. (2002). Teaching teamwork skills in software engineering based on understanding of factors affecting group performance. In *Proceedings of the 32nd Frontiers in Education Conference* (Vol. 3, pp. S3G-1). IEEE.

Lui, C., Sandell, K., & Welch, L. (2005). Teaching communication skills in software engineering courses. In *Proceedings of the 2005 American Society for Engineering Education Annual Conference & Exposition*. ASEE.

McGinnes, S. (1995). Communication and collaboration: Skills for the new IT professional. In *Proceedings of the 2nd All-Ireland Conference on the Teaching of Computing*. Academic Press.

Passow, H. (2012). Which ABET competencies do engineering graduates find most important in their work? *Journal of Engineering Education, 101*, 95–118. doi:10.1002/j.2168-9830.2012.tb00043.x

Radermacher, A., & Walia, G. (2013). Gaps between industry expectations and the abilities of graduates: Systematic literature review findings. In *Proceedings of the 2013 Special Interest Group on Computer Science Education Technical Symposium*. Academic Press.

Scholtes, P. R. (1995). *The team handbook: How to use teams to improve quality*. Madison, WI: Joiner Associates, Inc.

Smith, K. A. (2000). Strategies for developing engineering student's teamwork and project management skills. In *Proceedings of the 2000 American Society for Engineering Education Annual Conference*. ASEE.

Swan, B., Magleby, M., Sorensen, C., & Todd, R. (1994). A preliminary analysis of factors affecting engineering design team performance. In *Proceedings of the 1994 American Society for Engineering Education Annual Conference* (pp. 2572–2589). ASEE.

Villa, E. Q., Kephart, K., Gates, A. Q., Thiry, H., & Hug, S. (2013). Affinity research groups in practice: Apprenticing students in research. *Journal of Engineering Education, 102*(3), 444–466. doi:10.1002/jee.20016

Vygotsky, L. S. (1978). *Mind in society*. Cambridge, MA: Harvard University Press.

Vygotsky, L. S. (1986). *Thought and language*. Cambridge, MA: The MIT Press.

Wenger, E. (1998). *Communities of practice: Learning, meaning, and identity*. New York: Cambridge University Press. doi:10.1017/CBO9780511803932

Wenger, E. (2013). *Communities of practice*. Retrieved August 9, 2013, from http://www.ewenger.com/theory/

ADDITIONAL READING

Brown, J. S., Collins, A., & Duguid, P. (1989). Situated cognition and the culture of learning. *Educational Researcher, 18*(1), 32–42. doi:10.3102/0013189X018001032

Brumm, T., & Henneman, L. F., & Mickelson, S, K. (2005). The data are In Student workplace competencies in the experiential workplace. In *Proceedings of the 112ᵗʰ American Society for Engineering Education Annual Conference & Exposition*

Burns, R., Pollock, L., & Harvey, T. (2012). Integrating hard and soft skills: Software engineers serving middle school teachers. In *Proceedings of the 2012 Special Interest Group on Computer Science Education Technical Symposium*

Hazzan, O., & Har-Shai, G. (2013). Teaching computer science soft skills as soft concepts. In *Proceedings of the 2013 Special Interest Group on Computer Science Education Technical Symposium*

Herrenkohl, L. R., Palincsar, A. S., DeWater, L. S., & Kawasaki, K. (1999). Developing scientific communities of practice: A sociocognitive approach. *Journal of the Learning Sciences, 8*(3/4), 451–493.

Johnson, D., Johnson, R., & Holubec, E. (1990). *Circles of learning: Cooperation in the classroom*. Edina, MN: Interaction Book Company.

Johnson, D. W., Johnson, R. T., & Smith, K. (2007). The state of cooperative learning in post-secondary and professional settings. *Educational Psychology Review, 19*(1), 15–29. doi:10.1007/s10648-006-9038-8

Kephart, K., Villa, E., Gates, A. Q., & Roach, S. (2008, Summer). The Affinity Research Group Model: Creating and maintaining dynamic, productive, and inclusive research groups. *CUR Quarterly*, *28*(4), 13–24.

Lave, J. (1998). Situated learning in communities of practice. In L. Resnick, J. Levine, & S. Teasley (Eds.), *Perspectives on socially shared cognition* (pp. 63–82). Washington, DC: American Psychological Association.

Martin, R., Maytham, B., Case, J., & Fraser, D. (2005). Engineering graduates' perceptions of how well they were prepared for work in industry. *European Journal of Engineering Education*, *30*(2), 167–180. doi:10.1080/03043790500087571

Redish, E. F., & Smith, K. (2008). Looking beyond content: Skill development for engineers. *Journal of Engineering Education*, *97*(3), 295–307. doi:10.1002/j.2168-9830.2008.tb00980.x

Rogoff, B. (1995). Observing sociocultural activity on three planes: Participatory appropriation, guided participation, and apprenticeship. In J. V. Wertsch, P. del Rio, & A. Alvarez (Eds.), Sociocultural Studies of mind (pp. 139–164). Cambridge, UK: Cambridge University Press. Reprinted (2008) in K. Hall & P. Murphy (Eds.), Pedagogy and practice: Culture and identities (pp. 58–74). London: Sage.

Rover, D. T., Smith, K., Kramer, B., Streveler, R., & Froyd, J. (2003). Communities of practice in engineering education. In *Proceedings of the 2003 Frontiers in Education Conference* (Vol. 2, pp. F2G-F21). IEEE.

Smith, K. (2010). Social basis of learning: From small-group learning to learning communities. *New Directions for Teaching and Learning*, *123*, 11–22. doi:10.1002/tl.405

Smith, K. (2011). Cooperative learning: Lessons and insights from thirty years of championing a research-based innovative practice. In *Proceedings from the 2011 Frontiers in Education Conference* (pp. T3E-1). IEEE.

Smith, K. A. (2000). Strategies for developing engineering students' teamwork and project management skills. In *Proceedings of the American Society for Engineering Education Annual Conference*.

Smith, K. A., Morgan, J., Ledlow, S., Imbrie, P. K., & Froyd, J. (2003). Engaging faculty in active/cooperative learning. In *Proceedings of the 2003 Frontiers in Education Conference* (Vol. 3, pp. S1A–S13). IEEE.

Smith, K. A., Sheppard, S. D., Johnson, D. W., & Johnson, R. T. (2005). Pedagogies of engagement: Classroom-based practices. *Journal of Engineering Education*, *94*(1), 87–100. doi:10.1002/j.2168-9830.2005.tb00831.x

Stevens, R., O'Connor, K., Garrison, L., Jocuns, A., & Amos, D. M. (2008). Becoming an engineer: Toward a three-dimensional view of engineering learning. *Journal of Engineering Education*, *97*(3), 355–368. doi:10.1002/j.2168-9830.2008.tb00984.x

Tuckman, B. (1965). Developmental sequence in small groups. *Psychological Bulletin*, *63*(6), 384–399. doi:10.1037/h0022100 PMID:14314073

Wenger, E., McDermott, R., & Snyder, W. M. (2002). *Cultivating communities of practice: A guide to managing knowledge*. Boston: Harvard Business School Press.

KEY TERMS AND DEFINITIONS

Communities of Practice: A group of individuals who mutually engage in pursuing a common purpose through shared resources and repertoire of concepts, stories, discourse, and action.

Cooperative Learning: A teaching approach in which students work in groups and formally structured to include the following elements: positive interdependence, promotive interdependence, skills developments, individual accountability, and group processing.

Positive Interdependence: Team members share a common purpose with each contributing to the success of others and to the group goal(s).

Professional Skills: A set of interpersonal skills to energize and influence a team in meeting its goal(s), such as oral and written communication skills, team skills, and leadership skills.

Promotive Interaction: Team members interact in positive ways by acknowledging, for example, the contributions of others.

Situated Learning: Negotiated meaning of values, language, knowledge and skills in the practice of doing.

Team-Oriented Software Engineering (TOSE): A software engineering course using cooperative learning principles as a pedagogical approach to building effective teams and cultivating an environment with an aim of developing a community of practice.

Chapter 5
Encouraging Teamwork, Web 2.0, and Social Networking Elements in Distance Learning

Mirjana Ivanović
University of Novi Sad, Serbia

Zoran Budimac
University of Novi Sad, Serbia

Zoran Putnik
University of Novi Sad, Serbia

Živana Komlenov
University of Novi Sad, Serbia

ABSTRACT

At the Department of Mathematics and Informatics, Faculty of Science, University of Novi Sad, elements of Web 2.0 have been used in teaching for several years. In particular, this is emphasized in encouragement of teamwork, through usage of Wiki technology within several courses. Initially, those courses were created as a part of a large international project that recommended the use of teamwork. Over the years, additional elements of Web 2.0 were introduced, while employment and utilization of teamwork was largely enhanced and suitably organized. In this chapter, the authors share their experiences with such work, starting from introductory methods of enhancing the chosen learning management system, Moodle, with the mentioned activities, up to looking beyond their simple application and extracting additional value for courses.

DOI: 10.4018/978-1-4666-5800-4.ch005

INTRODUCTION

This chapter presents the authors', now about a decade long, experience in combining different open source educational tools as means of introducing collaborative activities in teaching at the Department of Mathematics and Informatics at the Faculty of Sciences in Novi Sad, Serbia. This endeavour was motivated by both our belief that collaboration helps promote the development of critical thinking skills and co-creation of knowledge and meaning, as well as the proofs that contemporary literature provides for such a claim. It seems that the inclusion of collaborative activities in a course, for instance to conduct small group projects or case study work, makes a lot of sense. Collaborative learning is especially appropriate for many of courses in the field of informatics, since they often focus on the application of new knowledge to complex and often unstructured tasks.

Teams also represent vital organizational structure for professional software development today, since the increasing complexity of projects has made them unachievable for individuals. Thus projects are performed by development teams, which commonly distribute the work among their members by following well-defined structures of interdependent responsibilities (Benarek, Zuser, & Grechenig, 2005).

Experiences that form the basis of this chapter have been gained mainly while working on eCourses created as a part of a large educational project (Budimac, Ivanović, Putnik, & Bothe, 2011) under the auspices of the "Pact for Stability of South-eastern Europe" and DAAD foundation. Institutions (15 universities) from 9 countries took part in this project, creating several common courses, initially for the most part in a classic face-to-face form. Later on, each participating institution worked further on those courses, extending them and adjusting them to their individual needs.

Our Department developed several eCourses suitable for use in a learning management systems (LMS), where the most mature are: "Software Engineering," "Introduction to eBusiness," "Object Oriented Programming 1 and 2," "Data Structures and Algorithms 1 and 2," and "Web Design" at the bachelor level; and in addition "Software Testing," "Privacy, Ethics, and Social Responsibility," "Architecture, Design, and Patterns," and "Software Engineering for Critical Systems" at master level. Students taking our elective course on distance learning helped us during the development of the draft versions of some of those courses which we later completed and polished.

As presented in (Ivanović et al., 2009), in the very beginning we decided to use and, if needed, extend one of the existing eLearning platforms for our eCourses instead of developing a new one from scratch, which proved to be a reasonable choice. After consulting a number of comparative studies and research papers (Graf & List, 2005; Di Domenico, Panizzi, Sterbini, & Temperini, 2005; Stewart et al., 2007; Al-Ajlan & Zedan, 2008), and testing several (freely) available systems, we drew a set of conclusions on the offered tools. We inclined towards an open source solution (Ahmed, 2005), so the system we chose was one of the established general purpose LMS systems – Moodle. A platform like that offered flexibility and significant initial cost savings, but also the potential for extensibility and customization according to one's particular needs. That proved to be all rather valuable for an educational institution like ours. Overview of the typical resources and activities that can be implemented and offered to students using standard modules in Moodle is presented in Figure 1.

The facilities we chose initially were those that could be used for creating simple repositories of learning resources recommended to our students. After noticing their satisfaction and further possibilities the platform offered to us for improving our teaching methodology, we decided to move on. According to the trends in eLearning practice, over the years we have progressed through several phases:

Figure 1. Standard Moodle modules categorization

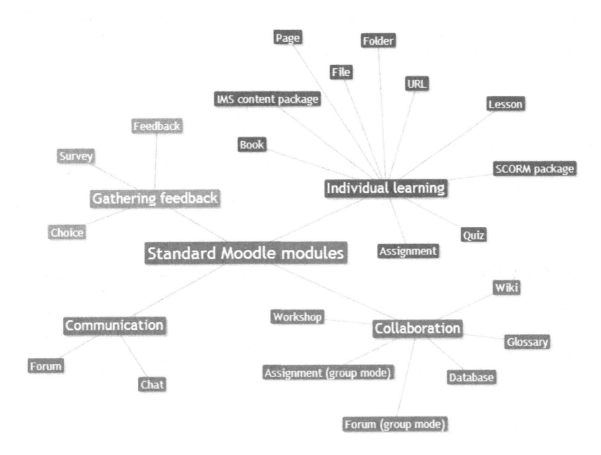

- Development of eLessons and presenting teaching resources in an active, multimedia form;
- Creation of quizzes and glossaries of important terms and notions;
- Management of assignments (their submission and assessment);
- Simulation of classroom activities through usage of chats and forums (in order to induce discussions, role-playing games, and similar) and finally
- Using Wikis for students' joint work on team assignments (acknowledging its ability to help us evaluate and grade those assignments in a more suitable way).

Our rising interest in eLearning in general and Moodle's features in particular led to our taking part in several educational and research projects in which we employed Moodle as the main experimental educational space: bilateral project of Serbia and Slovenia (Ivanović et al., 2009), multilateral project of Serbia, Czech Republic, and Greece (Ivanović, Xinogalos, & Komlenov, 2011), and four projects funded by international educational associations. As a result some additional eCourses were developed and two booklets were written on the key characteristics of Moodle and their possible applications.

In accordance with mentioned usage history, it is our belief that institutions having similar courses in their curricula, and particularly those that are currently struggling with incorporating more interactive Web 2.0 tools in their teaching practice could benefit from our experience.

The rest of the chapter is organized as follows. The second section presents current trends in the usage of collaborative learning and Web 2.0 technologies in teaching practice. In the third section, we present our own experiences, enlightening some difficulties and challenges we faced while introducing teamwork into our courses. Finally, in the last two sections we focus on declaring further research directions and stating valuable conclusions.

RELATED WORK

Collaborative work, one of the most important elements in different fields of education today is an integral part of our courses whenever it is possible and appropriate (Komlenov, Budimac, Putnik, & Ivanović, 2013). It is utilized not only with senior students, but it is courageously introduced to first-year students as well. Execution of such activities goes, as might be expected, through Web 2.0 tools. This is convenient since the freshmen are still unfamiliar with their colleagues and the course of studies, but at the same time majority of them are very accustomed to modern Web tools.

The main intention of applying collaborative learning in particular course is to allow students, grouped in appropriate teams (of 3-5 collaborators in average), to go together through phases of a certain project development, individually chosen as appropriate for each course by a lecturer. During that process they not only learn how to use Web 2.0 tools, but they also gain soft skills and valuable experience considering the team effort. Hopefully the social aspects of their future professional work will be easier to cope with, thanks to the collaborative/team work experiences.

Numerous reports testify to the successful usage of LMSs in everyday teaching practice at all levels of education and in different educational

contexts. However, the lack of interaction (the ability to share opinions, pose questions, or in some other way engage in dialogue), or presence (a sense of belonging to a group), and particularly of both, may result in students' negative observations on how well they may perform in an online class (Picciano, 2002; Song, Singleton, Hill, & Koh, 2004).

A strong positive correlation between the degree of social presence and perceived learning as well as perceived quality of the instructor is also reported on in literature (Richardson & Swan, 2003). The conclusions that students, who use opportunities in self-regulated and/or collaborative learning in general, are experiencing higher learning achievements (Paechter, Maier, & Macher, 2010) are also prevailing. According to all mentioned we did not hesitate to utilize the standalone communication and collaboration tools as well as those existent in Moodle to encourage collaborative work throughout the studies of informatics at our Department.

Wiki is the collaboration tool that has been most widely and deeply used at our Department, mainly the one integrated in Moodle as one of the standard activity modules. More than reasonable motivation for the use of this technology in education is given in (Educause, 2013), where the authors see Wiki as the easiest and most effective Web-based collaboration tool, using which students have the ability to easily perform collaborative group activities. This technology is thus, not surprisingly, becoming common in different educational landscapes (Parker & Chao, 2007; Hew & Cheung, 2009; Kane & Fichman, 2009) and has so far been applied with various causes in mind:

- To enhance social interaction amongst students (Augar, Raitman, & Zhou, 2004),
- For the dissemination of information to the student body,
- For building information repositories (Engstrom & Jewett, 2005),

- For collaborative production of documents that reflect the shared knowledge of the learning group (Cubric, 2007; Judd, Kennedy, & Cropper, 2010),
- For conducting peer reviews, etc.

Usage of this technology innovation prepares students to be not only readers and writers, but also editors, reviewers and collaborators, by developing their research, organizational, and negotiating skills (Richardson, 2010). Additional teachers' motivation for the use of Wikis is a possibility to gain regular insight into students' understanding of the learning material, in order to provide them with more targeted and prompt feedback (Cubric, 2007). We tried to give additional value to Wiki technology by using it to help prevent, or at least control cheating (Putnik, Ivanović, Budimac, & Samuelis, 2012), an everlasting issue when dealing with student assignments.

After further studying the research papers engaged in analysis of the influence of integration of social software into teaching (Stepanyan, Mather, & Payne, 2007), studies of results of empirical surveys (Bernsteiner, Ostermann, & Staudinger, 2008), and investigations of experiences in usage of Wikis and discussion forums (Wee & Abrizah, 2011), we employed discussion forums in some of our courses as well, particularly for "role-playing-game" type of assignments (Zdravkova, Ivanović, & Putnik, 2009; Zdravkova, Ivanović, & Putnik, 2012). We also tried to introduce chats, blogs, workshops, and similar, but those were not so gladly accepted by our students.

TEAMWORK IN PRACTICE

Within our courses students are offered a variety of Web resources and activities that enable them to exchange ideas and share newly gained insights with their colleagues. They are encouraged to use discussion boards, participate in chat sessions, form databases of useful links, and/or explore highly adaptive eLessons to test their advancement in gaining new knowledge and skills. However, it is true that most LMSs' and eLearning platforms focus more on distribution of learning material than social interaction or possibilities to construct shared knowledge.

Therefore we are aware of the fact that in order to support peer collaboration we as teachers need to offer some additional team projects where each team can be given its own learning environment, i.e. some sort of file exchange facility, discussion board, etc. to work in and achieve the required results. This is important to keep in mind and pursue in everyday teaching practice since various studies have reported that, regardless of the subject matter, students working in small teams tend to learn more of what is taught and retain it longer than when the same content is presented in any other instructional format (Davis, 1993). Introduction of elements of social networks/Web 2.0 proved to be particularly useful in this domain. Inclusion of team exercises, according to our experience, also helps students get diverse ideas, views, opinions and feedback, and thus makes improving their knowledge and grades easier.

Having all these potential benefits in mind, we prepared sets of team assignments for most of our courses. Types of assignments and the time designated to each of them (usually 2-4 weeks) differ depending on the topics covered during the particular courses, and other student obligations within them. Courses that we use for experimenting with collaborative activities were predominantly created within a more than a decade long international project, involving 15 universities from 9 countries, dealing with educational cooperation at the bachelor level (Budimac et al., 2011), and a subsequent project dealing with the design of joint master studies in the field of software engineering (Bothe, Budimac, Cortazar, Ivanović, & Zedan, 2009) that are now conducted at several universities in our region. The majority of the developed courses enforce teamwork, collaboration, home assignments, and usage of LMS facilities, yet allowing lecturers initiatives, and choice of tools and techniques. Figure 2 shows how a typical section of one of our Moodle courses ("Software engineering" to be precise) looks like and what sorts of resources and activities it offers to our students: lecture slides, an eLesson for self-studying, a glossary, an assignment with additional resources, discussion forum, etc.

Some of the team assignments we put in practice explore the application of gained knowledge in artificially produced "real-life situations," like the assignments in the "Software engineering" course in which students are required to perform a review of the given requirements specification for an existing software product, to go through the cost estimation process for the same product using the function point method, etc. Other assignments are concentrated on role-playing games, again situated around realistic problems, and/or practicing case-based learning approach (Voigt, 2008). Typical examples are two consecutive assignments in the "Introduction to eBusiness" course: developing a business model for an imaginary eCommerce company, and afterwards implementing it by building a Web shop. This

approach is in fact most commonly used in our practice as an instructional technique that, rather than presenting general concepts and theories, provides situations to analyse, open issues to deal with, and data based on which decisions must be made.

Students are generally intrigued by such assignments, yet at first some of them are unwilling to work in teams since they become aware of the fact that their grades will, to a certain extent, depend on their colleagues' effort or lack of it, which terrifies them even more. However, after an assignment or two, they realize that complex projects which should be completed in a reasonable amount of time have to be conducted in teams. Besides, since the results they achieved up to that moment are of acceptable quality, students start looking forward to further challenges. Of course there are always some lazy, too busy, or just not so ambitious individuals who delay taking part in team projects until the deadlines start knocking at their doors. The behaviour of such students influences the cohesion of the teams, the overall quality of the solutions they present at the end, and the satisfaction of the hard-working students

Figure 2. A typical section of one of the courses developed in Moodle to support the transition to blended learning and especially learner-centred activities

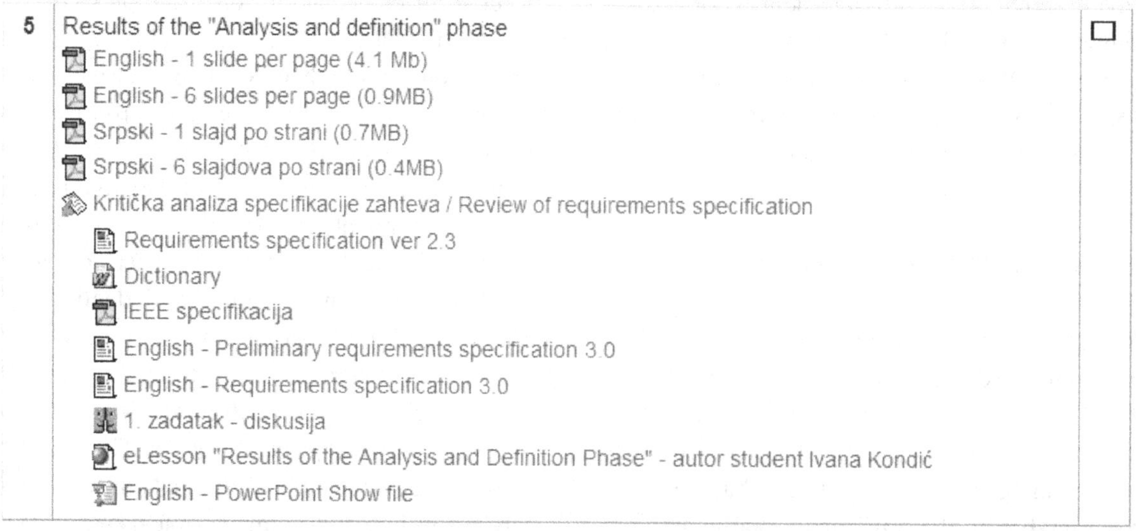

```
5 | Results of the "Analysis and definition" phase                              □
    English - 1 slide per page (4.1 Mb)
    English - 6 slides per page (0.9MB)
    Srpski - 1 slajd po strani (0.7MB)
    Srpski - 6 slajdova po strani (0.4MB)
    Kritička analiza specifikacije zahteva / Review of requirements specification
       Requirements specification ver 2.3
       Dictionary
       IEEE specifikacija
       English - Preliminary requirements specification 3.0
       English - Requirements specification 3.0
       1. zadatak - diskusija
       eLesson "Results of the Analysis and Definition Phase" - autor student Ivana Kondić
       English - PowerPoint Show file
```

with this approach to learning. But they also make these learning endeavours more realistic and challenging. We can also mention an option utilized at some courses, of "being able and allowed to fire a team member, for insufficient work." This option was exercised on several occasions over the years of collaborative work practice, more often with younger students who haven't yet formed too strong "friendly relationship" with the rest of the class.

What we experimented a lot with is the team creation process itself, switching between teams that were self-assigned and those that are created by lecturers a couple of years ago, all in order to create better, more functional teams that will produce high-quality results. The limit that we always keep the same is on the number of students per team: 3–5, since the size of working groups must represent a compromise between being large enough to have sufficient intellectual resources to complete the assignments and small enough to develop into true teams (Curtis & Lawson, 2001; Johnson & Johnson, 2002).

Issues, Controversies, Problems

In the first years of incorporating collaborative assignments into our courses we had some negative experiences with the way students handled those assignments, solution to which they, at that time, were required to submit by uploading the result in the assignment activity within the LMS. Self-formed teams, as a rule had friends as members, friends willing to cover for their other, non-working friends. This is not uncommon at other universities as well – overall participation can be high, but often a relatively small proportion of students perform the bulk of the work and many students' contributions can be regarded as superficial (Judd et al., 2010).

Even if we disregard this problem, the evaluation of the work performed by self-chosen teams was very difficult. A lot of less-ambitious students stopped participating in assignment solving after

fulfilling the obligatory part of the assignments, and achieving required minimal number of points. This problem was often brought up by students within regular surveys conducted at the ends of the courses, as an important negative aspect of this actual type of teamwork.

Nevertheless, this is a common issue that the colleagues using the same pool of assignments within the joint courses in Germany, FYR Macedonia, Croatia, etc. faced as well. Consequently, the rest of the team members had to work harder, without being fully rewarded for their efforts since teachers, being presented only with the final solutions, were not able to distinguish the individual contributions to those results that had been made by individual team members.

Solutions and Recommendations

Usage of Wiki Technology

The employment of Wiki technology for the collaborative development of the solutions to various types of assignments was seen as the way of introducing order in students' team efforts. Additionally, this tool provides teachers with the mechanisms that facilitate tracking the extent of the individual contributions to team solutions, which allows for a somewhat fairer process of teamwork evaluation.

Moodle offers Wiki implemented in a straightforward way, resulting in a stable working environment that students learn to use rather quickly, so we opted for such an integrated solution instead of introducing another standalone application into our teaching practice. What is even more important, it presents us with both the final result – document containing concluding solution, and the detailed history of changes, enabling us to appropriately assess and grade individual achievements of team members.

Analysis of the results of the teamwork (average grades on individual assignments, etc.) show us that most of our students managed to solve all the tasks successfully, at each of the courses. This is relieving since it tells us that they could deal both with new technologies and gaining soft skills needed for effective working in teams. Our principal goals are satisfied – to introduce students to teamwork and get them used to collaboration in problem solving, and to determine whether Wiki is in fact effective as a tool for collaborative eLearning.

Assessment of Students' Work

The assessment of each individual student remains rather complicated, since after grading a solution to an assignment as a whole and assigning a certain number of points to it lecturer has to go over the solution again, trying to assess individual contributions of team members. Luckily Moodle's Wiki module shows us the complete history of development of every solution. Each access of each participant of a team is recorded – when it happened, from which IP address, and the most importantly what has been done by that person. All of the changes, additions, or deletions are recorded. Thanks to that Wiki feature, the development of the whole solution can be followed step-by-step, which gives us a chance to determine the appropriate grade for each of the team members, according to their individual contribution to the final result their team achieved. Our experiences with this way of teamwork assessment have been considered and published in several papers (Komlenov et al., 2013; Putnik et al., 2012).

Team Formation Strategies

After gaining positive results by introducing new collaborative tool, we started paying more attention to the effect of the applied team formation mechanisms on the final results of team projects. For some time we worked with both self-formed and teacher-formed teams and tried to seize the differences in students' behaviour and the outcomes at the end of the course runs.

Differences in age, expert pre-knowledge and pre-knowledge in usage of the LMS between the students of the various years of studies induced our beliefs that they might act differently concerning Wikis and collaborative work in general. For instance, most of the freshmen had never used Moodle LMS before in their studies, while the final year students had been using it for several years before getting their first Wiki assignments. Another fact is that the first year students were usually unfamiliar with most of their colleagues, and the course of their studies, while the final year students had already shaped some friendships with the colleagues that formed their teams for assignment solving.

As a consequence, for several years teams consisting of the first year students were formed by the lecturers' choice, while the higher year students were allowed to choose their team members on their own. After getting some pleas from our students and having negative experiences in cheating/avoiding work by some of the students, we decided to disallow team self-creation altogether. Students of course prefer self-selected teams, but research shows that such teams under-perform when compared to the instructor-selected ones (Brickell, Porter, Reynolds, & Cosgrove, 1994) and that this method always runs the risk of further isolating some students or creating cliques within the class as a whole. Our experiences confirm these claims.

At first, we decided to select team members randomly, using an existing feature of our LMS. This assured complete fairness and, since random composition of a team is what our students will be faced with in a real life, seemed like a good and early introduction to that fact. Still, we were uncertain of the quality of the teams and their results. Therefore we decided to apply some specific techniques for team creation knowing that, ideally, teams should be diverse enough to include students with a range of intellectual abilities, academic interests, and cognitive styles.

Therefore, online surveys have been conducted with students within our courses at the beginning of the semesters since 2009/2010 school year. We collect answers about students' academic background, the study programme they enrolled, their gender, hometown, etc. In addition, students complete the Index of Learning Styles (ILS) questionnaire. This instrument is used to assess preferences on the four dimensions (active/reflective, sensing/intuitive, visual/verbal, and sequential/global) of the Felder-Silverman Learning Styles Model (FSLSM), which is possibly the most frequently used model for determining students' learning styles in eLearning environments (Felder & Silverman, 1988).

Additionally, in 2010/2011 we introduced another instrument in our team forming practice – the Belbin test. This test is based on the method acquired from the theory of management, the Belbin Team Role Inventory. This assessment strategy is used to gain insight into an individual's behavioural tendency in a team environment (Belbin, 2010). Our goal stayed the same – to balance the strengths and weaknesses of students, and provide them with team structures that would foster the development of their already existent capacities.

In summer semester 2011/2012 we started the development of a specific additional tool (i.e. plug-in) for Moodle in order to automate the team formation process using the described mechanisms. The plug-in called *Teammine* allows course instructors to define a set of criteria based on which the questionnaires that students are required to complete are automatically generated. It is possible and easy for teachers to choose whether the team formation process within a specific course will be based only on students' academic background, or it will include the analysis of the preferred learning styles and/or potentially best team roles in the process. Upon filling the appropriate questionnaire, students' answers are automatically analyzed and the system suggests a rough distribution into teams. The lecturer controls the process by defining the preferred team size

or the preferred number of teams, and since the plug-in is still in the testing phase, polishes the distribution manually. Such partially automated process saves a lot of time to teachers by giving them a better overview of the students' characteristics in just a few seconds (provided that they have already completed the required questionnaires) and some suggestions regarding the possible structure of the teams.

After applying this plug-in, we analysed the changes that occurred and the results students achieved. For example, when the final grades of students in the four consecutive course runs of the course "Introduction to eBusiness" are compared (Figure 3) it becomes obvious that we managed to approach the normal distribution of grades in the course run during which our Moodle plug-in was applied. At the same time, the number of students that failed the course was significantly reduced.

Students' Reactions and Behaviour

To show some of the obstacles we encountered during our attempts to implement collaborative activities into our eCourses, let us present some numerical results on the efforts and obligations this kind of work brings. We will give only some extracts of the data here, while additional numerical data and more discussions can be found in (Komlenov et al., 2013). For example, within our "Software Engineering" course in the last course run we had 86 students divided into 17 teams. For the first two assignments solved within the course students were required to use Wiki as the authoring tool for the production of their final solutions. Table 1 records the behaviour of students while solving those assignments.

The differences between the values achieved are striking and obviously suggest that it took some time for our students to adapt to the new environment and the methodology it proposes. To prove that, we collected the similar data for our freshmen students attending the "Introduction to eBusiness" course, over the two consecutive

Figure 3. Comparison of final grades of students during the four consecutive runs of the same course

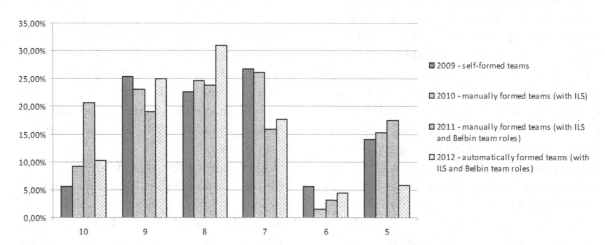

- 2009 - self-formed teams
- 2010 - manually formed teams (with ILS)
- 2011 - manually formed teams (with ILS and Belbin team roles)
- 2012 - automatically formed teams (with ILS and Belbin team roles)

Table 1. Student behaviour while solving two consecutive assignments within the "Software Engineering" course

Parameter	Assignment 1	Assignment 2
Total number of accesses	8812	2464
Total number of changes	1410	862
Number of accesses per student	2 – 1125	0 – 168
Average number of accesses per student	103	26

Table 2. Analysis of student behaviour while solving the Wiki assignment within the "Introduction to eBusiness" course in two consecutive school years

Parameter	Year 1	Year 2
Total number of accesses	19229	7691
Total number of changes	1398	1217
Number of accesses per student	0 – 574	0 – 821
Average number of accesses per student	154	118

school years. One of the assignments in this course, typically the first one, is solved using a Wiki (Table 2). In those school years 66 and 65 students consecutively participated in the course, forming 16 teams. For most of the students the Wiki assignment was not only the first contact with this technology, but the first team assignment they ever solved. Hence it does not surprise that as such it presented a challenge for many of them.

The need of students to adjust to the new tool was evident in each of the mentioned courses. It took 103 accesses to Wiki per student to properly formulate the opinions in the first assignment in "Software Engineering" course, and even worse – 154 accesses for the freshmen in the course "Introduction to eBusiness" in the first year of

our tracking student behaviour within these assignments, which led to the remarkable total of over 19000 accesses for the whole class. In the following year the difference was slightly smaller. Figure 4 shows the report on the work of one of those teams, which produces a rather good final result for which the four team members needed 475 actions in the Wiki during the 3-week period. Also, although perhaps 1125 accesses might seem too much for one person, there were several students with more than 1000. Even after subtracting the extreme ones, those with more than a thousand, and those with only a few accesses (≤ 5), the average number of accesses is still around 80.

Figure 4. Report on the actions performed in Wiki by one of the teams solving an introductory assignment within the "Introduction to eBusiness" course

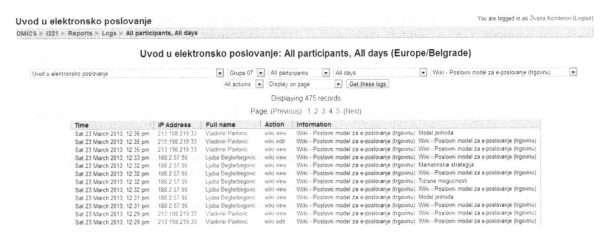

We also, with the help of Wiki technology, became aware that we need to reconsider our opinions about our students' attitudes and work ethics in some points.

As our team assignments are usually quite complex, we feel obliged to give our students at least two weeks for their solving, and usually they have 3–4 weeks to complete the current assignment. Before we introduced Wiki technology to our courses, we had no possibility to gain insight into the working schedule of any of the teams. However, after implementing Wiki as a tool for the assignments solving, and after taking some time to analyze students' post/update/edit activities one can come to a disappointing, although not too unexpected result. Almost all of the work is performed in the final few days before the deadline. Figure 5 supports that claim by presenting the distribution of workload for six assignments during the last run of the "Software Engineering" course.

It can be easily seen that students avoided the work for as long as possible and performed more and more activities as they were approaching the deadline. This is certainly not too commendable for them. The other possibility is that they simply

Figure 5. Percentage of work accomplished on the last 3 days while solving each of the six assignments within the "Software Engineering" course

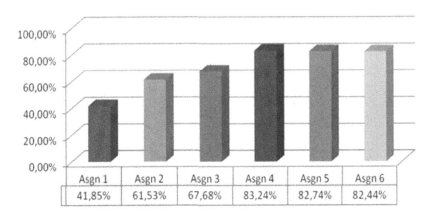

needed some rest from solving difficult assignments, i.e. needed to gather their strength after the previous assignment to carry out the next one easily and promptly. Let us comfort ourselves with that excuse.

Similar observation can be illustrated in yet another way, by showing the rising number of posts over time for the first (Figure 6), and the final, sixth assignment (Figure 7).

Even visually, it can be noticed that during the first assignment there was some limited activity in the beginning, yet nothing worth mentioning happened for the first two weeks in the last assignment. It is more than unsatisfactory that only

Figure 6. The difference in assignment solving approach over time – posts during the first assignment

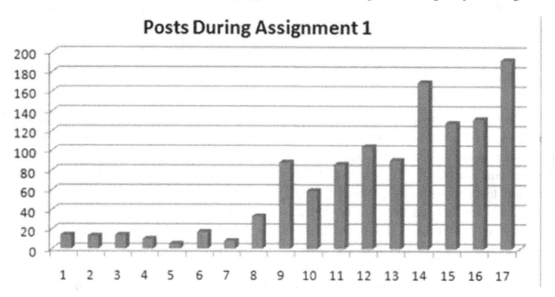

Figure 7. The difference in assignment solving approach over time – posts during the last assignment

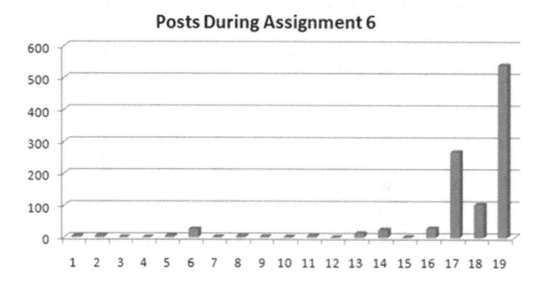

one group arranged their work over the whole given period for solving each of the assignments.

No matter how satisfied or not we are with certain aspects of the work performed by our students' when working in teams, we always wish to find out whether they are supporting our usage of collaborative activities and to get insights into their satisfaction with the courses containing them. For this purpose we conduct anonymous surveys towards the end of each semester (Ivanović et al., 2013). Over 52% of the surveyed students have so far done some collaborative assignments using Moodle's modules suitable for such efforts (Wiki, Workshop, etc.).

They generally found these activities both challenging and valuable as learning experiences, and responded very well to the team-building practice promoted through them.

Students who did not know so well other participants of the courses they took, joined some of the courses late, had no time for long live meetings, or are simply timid, etc., saw the benefits of online collaboration mechanisms and Wiki in particular. They liked the idea of fulfilling and correcting other team members' ideas, errors and solutions that eventually led to the creation of the best possible, joint one. What they also liked was the fact that teachers could have insights in the whole process of their collaboration and that grades were formed according to their individual contributions to the final solution. Most of the students were satisfied with their results, but even those less content admitted that we "took one great burden off their chests" by detecting the students who did not participate in the assignment solving process at all more precisely and awarding them with the deserved number of points: 0.

It is the general opinion that team assignments they solved are not difficult – around 43% of students think that their level of difficulty is appropriate, additional 36% of students find the assignments slightly more complex than the usual

ones, and only 7% of students think they are too complicated. Moreover, 86% of students find the assignments encouraging (motivating), and 79% of students think that working in a team was valuable for gaining realistic experience.

On the other hand, there were some students who wished they could change a member of a team with some other member, or just "fire" the unsuitable member, or that have actually done that in practice, but those make in total only 14% of all students that participated in the surveys. Approximately 50% of students said they never though of such an idea, others were undecided. Of course, a considerable number of students (29% of all that took part in the surveys) think that assignments solving would have been easier, better, and more successful if they had done it alone.

When asked whether it is fair of us to mark the team assignments based on how much effort each of the teams invested more than on anything else, satisfactory 60% of students think this is fine, only 14% of them think that this is not fair way of grading, while the rest of the students do not mind that strategy.

When it comes to the usage of communication tools within our courses, forums used for discussions, role-playing games, and similar (Putnik et al., 2012) were the ones that were best received by our students. At first we tried using live chats, yet problems with synchronicity of access and other obligations of students forced us to give up on them (a significant number of our students are already employed, especially those studying at the master level). Thus we switched to asynchronous method of communication, i.e. discussion forums.

Possibly the most successful application of those was within the course "Privacy, Ethics, and Social Responsibility" (Zdravkova et al., 2012) in which assignments were given to students together with the appropriate time-frame to participate in discussion on given topic(s). Each of their posts was graded. Special attention was paid to the transparency of the grading process, which was provided by always visible number of points per post. A well balanced combination of several graded forums and individual assignments contributed to the students' final grades.

FUTURE RESEARCH DIRECTIONS

Although the results collected and accomplished so far using collaborative technologies are more than satisfactory, there is still room for some further advancements. Onc of the most important aspects that need further interventions is the enormous workload, in particular the workload teachers are facing when incorporating collaborative activities into their classes. In our courses teachers on average perform almost 90 actions (edits/posts/views) in Wiki activities per day. Number of actions among students climbed up to 1327 while actively working on the assignments, which means that the person with the highest number of actions had on the average of around 13 actions per day during the whole semester. Even the person with the lowest number of actions, approached our LMS about 2.5 times per day while working on the assignments, and teachers need to carefully track all these activities to ensure the fairness of the grades. Of course, lecturers start working on their courses even before they are launched (preparing the study material and assignments), and after they end (final clean-ups, preparation for the next school year, and similar). Since the workload is enormous, some solutions that would relax and speed-up analyzing and grading the large number of assignment solutions is a necessity.

Another aspect worth further research regards the improvement of the team formation process. This process should not depend so much on the instructors but the automatic analysis of student data. We are trying to advance this process by applying some more precise discrete optimization techniques in the plug-in we are developing for Moodle. Still, even in this form, it already turned out to be of great use to teachers in terms of time saving, making the whole procedure unbiased and providing more balanced teams. For the courses held in the second or higher years of studies it would also be meaningful to incorporate knowledge from students' past results and experiences. For example, results of already filled questionnaires, detected behaviour in teams while working on previous assignments in any course they passed, etc. could be included in the team formation process for future tasks, so we plan to incorporate those possibilities into some next version of our plug-in as well.

CONCLUSION

Collaborative work performed by students certainly requires careful planning on the part of the instructor, and is not without its challenges and difficulties for students. Still, we find that the benefits can be substantial, including increased participation by students in all parts of the course, better understanding and retention of material, mastery of various soft skills, and increased enthusiasm for self-directed learning.

Students are generally satisfied with their newly developed competences for successful working in teams, valuable for both their future work in practice and advanced subjects awaiting them in the rest of their studies. The use of Web 2.0 tools is particularly convenient in the development of collaborative environment for a teacher having a lot of students, like in our courses. Student might not always be familiar with the course of studies and their colleagues, but they are in our experience almost surely very accustomed to modern tools they use online on daily basis so their application in the learning process is generally welcomed.

On the other hand, human factor, like in any other area where technology is used in order to improve the way we live and work, is sometimes the main obstacle for the ultimate success. Attempts of cheating, starting the work on the assignments too late (almost when it should be completed), burdening teammates with too much work, complaining about the nature of assignments and team work in general while doing more or less nothing and expecting good grades, all that is unfortunately not rare among our students. Still, we feel that those troubles have more to do with the human nature itself than the technology and it's a burden for teachers to try to use the available software tools to enhance the working habits of their students, to challenge them, and of course to track and evaluate their work promptly and fairly. Contemporary tools for collaboration we described in this chapter in our opinion are of great help in all these efforts.

REFERENCES

Ahmed, O. (2005). *Migrating from proprietary to open source learning content management systems*. (PhD thesis). Carleton University, Ottawa, Canada.

Al-Ajlan, A., & Zedan, H. (2008). Why Moodle. In *Proceedings of 12th International Workshop on Future Trends of Distributed Computing Systems* (pp. 58–64). IEEE.

Augar, N., Raitman, R., & Zhou, W. (2004). Teaching and learning online with wikis. In R. Atkinson, C. McBeath, D. Jonas-Dwyer, & R. Phillips (Eds.), *Beyond the comfort zone: Proceedings of the 21st Ascilite conference* (pp. 95–104). Perth, Australia: Australasian Society for Computers in Learning in Tertiary Education.

Belbin, M. (2010). *Management teams: Why they succeed or fail*. Oxford, UK: Butterworth Heinemann.

Benarek, G., Zuser, W., & Grechenig, T. (2005). Functional group roles in software engineering teams. In *Proceedings of the ACM Workshop on Human and Social Factors of Software Engineering* (pp. 1–7). ACM.

Bernsteiner, R., Ostermann, H., & Staudinger, R. (2008). Facilitating e-learning with social software: Attitudes and usage from the student's point of view. *International Journal of Web-Based Learning and Teaching Technologies*, *3*(3), 16–33. doi:10.4018/jwltt.2008070102

Bothe, K., Budimac, Z., Cortazar, R., Ivanović, M., & Zedan, H. (2009). Development of a modern curriculum in software engineering at master level across countries. *Computer Science and Information Systems*, *6*(1), 1–21. doi:10.2298/CSIS0901001B

Brickell, J. L., Porter, D. B., Reynolds, M. F., & Cosgrove, R. D. (1994). Assigning students to groups for engineering design projects: A comparison of five methods. *Journal of Engineering Education*, *83*(3), 259–262. doi:10.1002/j.2168-9830.1994.tb01113.x

Budimac, Z., Ivanović, M., Putnik, Z., & Bothe, K. (2011). Studies in wonderland – Sharing of courses, lectures, tasks, assignments, tests and pleasure. In *Proceedings of the 22nd EAEEIE Annual Conference* (pp. 213–219). EAEEIE.

Cubric, M. (2007). Using wikis for summative and formative assessment. In *Reap international online conference on assessment design for learner responsibility*. Academic Press.

Curtis, D. D., & Lawson, M. J. (2001). Exploring collaborative online learning. *Journal of Asynchronous Learning Networks*, *5*(1).

Davis, B. (1993). *Tools for teaching*. San Francisco: Jossey-Bass.

Di Domenico, F., Panizzi, E., Sterbini, A., & Temperini, M. (2005). *Analysis of commercial and experimental e-learning systems. Quality, Interoperability and Standards in e-Learning Team*. TISIP Research Foundation.

Educause. (2013). *7 things you should know about... Wikis*. Retrieved March 5, 2013, from http://Net.Educause.Edu/Ir/Library/Pdf/Eli7004.pdf

Engstrom, M. E., & Jewett, D. (2005). Collaborative learning the wiki way. *TechTrends, 49*(6), 12–15. doi:10.1007/BF02763725

Felder, R. M., & Silverman, L. K. (1988). Learning and teaching styles in engineering education. *English Education, 78*(7), 674–681.

Graf, S., & List, B. (2005). An evaluation of open source e-learning platforms stressing adaptation issues. In *Proceedings of the 5th IEEE International Conference On Advanced Learning Technologies* (pp. 163–165). IEEE.

Hew, K., & Cheung, W. (2009). Use of wikis in K-12 and higher education: A review of the research. *International Journal of Continuing Engineering Education and Lifelong Learning, 19*(2/3), 141–165. doi:10.1504/IJCEELL.2009.025024

Ivanović, M., Putnik, Z., Komlenov, Ž., Welzer, T., Hölbl, M., & Schweighofer, T. (2013). *Usability and privacy aspects of Moodle – Students' and teachers' perspective*. Informatica – An International Journal of Computing and Informatics.

Ivanović, M., Welzer, T., Putnik, Z., Hölbl, M., Komlenov, Ž., Pribela, I., & Schweighofer, T. (2009). Experiences and privacy issues – Usage of Moodle in Serbia and Slovenia. In *Proceedings of the Interactive Computer Aided Learning* (pp. 416–423). Academic Press.

Ivanović, M., Xinogalos, S., & Komlenov, Ž. (2011). Usage of technology enhanced educational tools for delivering programming courses. *International Journal of Emerging Technologies in Learning, 6*(4), 23–30.

Johnson, D. W., & Johnson, F. P. (2002). *Joining together: Group theory and group skills*. Boston: Allyn & Bacon.

Judd, T., Kennedy, G., & Cropper, S. (2010). Using wikis for collaborative learning: Assessing collaboration through contribution. *Australasian Journal of Educational Technology, 26*(3), 341–354.

Kane, G. C., & Fichman, R. G. (2009). The shoemaker's children: Using wikis for information systems teaching, research, and publication. *Management Information Systems Quarterly, 33*(1), 1–17.

Komlenov, Ž., Budimac, Z., Putnik, Z., & Ivanović, M. (2013). Wiki as a tool of choice for students' team assignments. *International Journal of Information Systems and Social Change, 4*(3), 1–16. doi:10.4018/jissc.2013070101

Paechter, M., Maier, B., & Macher, D. (2010). Students' expectations of, and experiences in e-learning: Their relation to learning achievements and course satisfaction. *Computers & Education, 54*(1), 222–229. doi:10.1016/j.compedu.2009.08.005

Parker, K. R., & Chao, J. T. (2007). Wiki as a teaching tool. *Interdisciplinary Journal of Knowledge and Learning Objects, 3*, 358–372.

Picciano, A. (2002). Beyond student perceptions: Issues of interaction, presence, and performance in an online course. *Journal of Asynchronous Learning Networks, 6*(1), 21–40.

Putnik, Z., Ivanović, M., Budimac, Z., & Samuelis, L. (2012). Wiki – A useful tool to fight classroom cheating? In *Advances in web-based learning - ICWL 2012 (LNCS)* (Vol. 7558, pp. 31–40). Berlin: Springer. doi:10.1007/978-3-642-33642-3_4

Richardson, J. C., & Swan, K. (2003). Examining social presence in online courses in relation to students' perceived learning and satisfaction. *Journal of Asynchronous Learning Networks*, 7(1), 68–88.

Richardson, W. (2010). *Blogs, wikis, podcasts and other powerful web tools for the classroom*. Thousand Oaks, CA: Corwin Press.

Song, L., Singleton, E., Hill, J., & Koh, M. (2004). Improving online learning: Student perceptions and challenging characteristics. *The Internet and Higher Education*, 7, 59–70. doi:10.1016/j.iheduc.2003.11.003

Stepanyan, K., Mather, R., & Payne, J. (2007). Integrating social software into course design and tracking student engagement: Early results and research perspectives. In T. Bastiaens & S. Carliner (Eds.), *Proceedings of the World Conference on E-Learning in Corporate, Government, Healthcare, and Higher Education* (pp. 7386–7395). Academic Press.

Stewart, B., Briton, D., Gismondi, M., Heller, B., Kennepohl, D., McGreal, R., & Nelson, C. (2007). Choosing Moodle: An evaluation of learning management systems at Athabasca University. *International Journal of Distance Education Technologies*, 5(3), 1–7. doi:10.4018/jdet.2007070101

Voigt, C. (2008). *Educational design and media choice for collaborative, electronic case-based learning (E-Cbl)*. (PhD thesis). School of Computer and Information Science, Division of Information Technology, Engineering and the Environment, University of South Australia.

Wee, M., & Abrizah, A. (2011). An analysis of an assessment model for participation in online forums. *Computer Science and Information Systems*, 8(1), 121–140. doi:10.2298/CSIS100113036C

Zdravkova, K., Ivanović, M., & Putnik, Z. (2009). Evolution of professional ethics courses from web supported learning towards e-learning 2.0. In *Learning in the synergy of multiple disciplines* (pp. 657–663). Berlin: Springer. doi:10.1007/978-3-642-04636-0_64

Zdravkova, K., Ivanović, M., & Putnik, Z. (2012). Experience of integrating web 2.0 technologies. *Educational Technology Research and Development*, 60(2), 361–381. doi:10.1007/s11423-011-9228-z

ADDITIONAL READING

Bielikova, M., & Navr, P. (2005). Experiences with designing a team project module for teaching teamwork to students. *Journal of Computing and Information Technology*, 13(1), 1–10. doi:10.2498/cit.2005.01.01

Brindley, J. E., Walti, C., & Blaschke, L. M. (2009). Creating effective collaborative learning Groups in an online environment. *International Review of Research in Open and Distance Learning*, 10(3), 1–18.

Bruns, A., & Humphreys, S. (2007). Building collaborative capacities in learners: The M/cyclopedia project revisited. In *Proceedings of the International Symposium on Wikis* (pp. 1–10).

Cannon, R., Hilburn, T. B., & Diaz-Herrera, J. (2002). Teaching a software project course using the team software process. In J. L. Gersting, H. MacKay Walker, & S. Grissom (Eds.), *33rd SIGCSE Technical Symposium on Computer Science Education* (pp. 369–370). New York: ACM.

Carr, N. (2008). Wikis, knowledge building communities and authentic pedagogies in pre-service teacher education. In *Hello! Where are you in the landscape of educational technology? – Proceedings of the Australasian Society for Computers in Learning in Tertiary Education* (pp. 147–151). Melbourne: Deakin University.

Chang, C. K., Chen, G. D., & Li, L. Y. (2008). Constructing a community of practice to improve coursework activity. *Computers & Education*, *50*(1), 235–247. doi:10.1016/j.compedu.2006.05.003

Drášil, P., & Pitner, T. (2006). E-learning 2.0: Methodology, technology, and solutions. In E. Mechlova (Ed.) *International Conference on Information and Communication*.

Figl, K. (2008). *Developing team competence of computer science students in person centered technology-enhanced courses*. PhD thesis, Vienna: Faculty of Informatics, University of Vienna.

Guzman, M., & Larkin, H. (2008). A Wiki as an intercultural learning environment for students of computer science in Australia and Spain. In Proceedings of the *EDEN Annual Conference*.

Hayes, J. H., Lethbridge, T. C., & Port, D. (2003). Evaluating individual contribution toward group software engineering projects. In Proceedings of the *25th International Conference on Software Engineering* (pp. 622–627). Washington: IEEE Computer Society.

Higgs, M., Plewnia, U., & Ploch, J. (2005). Influence of team composition and task complexity on team performance. *Team Performance Management*, *11*(7/8), 227–250. doi:10.1108/13527590510635134

Hogan, J. M., & Thomas, R. (2005). Developing the software engineering team. In *Proceedings of the 7th Australasian Conference on Computing Education* (pp. 203–210). Sydney: Australian Computer Society.

Ivanova, M., & Popova, A. (2011). Formal and informal learning flows cohesion in Web 2.0 environment. *International Journal of Information Systems and Social Change*, *2*(1), 1–15. doi:10.4018/jissc.2011010101

LeJeune, N. (2003). Critical components for successful collaborative learning in CS1. *Journal of Computing Sciences in Colleges*, *19*(1), 275–285.

Lund, A., & Smordal, O. (2006), Is there space for the teacher in a Wiki? In D. Riehle, & J. Noble (Eds.), *International Symposium on Wikis* (pp. 37–46). New York: ACM.

McKinney, D., & Denton, L. F. (2006). Developing collaborative skills early in the CS curriculum in a laboratory environment. In D. Baldwin, P. T. Tymann, S. M. Haller, & I. Russell (Eds.), *39th SIGCSE Technical Symposium on Computer Science Education* (pp. 138–142). New York: ACM.

Nance, W. D. (2000). Improving information systems students' teamwork and project management capabilities: Experiences from an innovative classroom. *Information Technology Management*, *1*(4), 293–306. doi:10.1023/A:1019137428045

Poindexter, S., Basu, C., & Kurncz, S. (2001). Technology, teamwork, and teaching meet in the classroom. *EDUCAUSE Quarterly*, *24*(3), 32–41.

Polack-Wahl, J. A. (2001). Enhancing group projects in software engineering. *Journal of Computing Sciences in Colleges*, *16*(4), 111–121.

Redecker, C., Ala-Mutka, K., Bacigalupo, M., Ferrari, A., & Punie, Y. (2009). Learning 2.0: The impact of Web 2.0 innovations on education and training in Europe, final report. Seville: Joint Research Centre, Institute for Prospective Technological Studies, European Commission. Retrieved June 20, 2013, from http://ftp.jrc.es/EURdoc/JRC55629.pdf

Smarkusky, D. L., Dempsey, R., Ludka, J., & Quillettes, F. D. (2005). Enhancing team knowledge: Instruction vs. experience. In W. Dann, T. L. Naps, P. T. Tymann, & D. Baldwin (Eds.), *36th SIGCSE Technical Symposium on Computer Science Education* (pp. 460–464). New York: ACM.

Stephens, C. S., & Myers, M. E. (2001). Developing a robust system for effective teamwork on lengthy, complex tasks: An empirical exploration of interventions to increase team effectiveness. In *16th International Academy for Information Management (IAIM) Annual Conference: International Conference on Informatics Education & Research* (pp. 351–361). New Orleans: International Academy for Information Management.

Tadayon, N. (2004). Software engineering based on the team software process with a real world project. *Journal of Computing Sciences in Colleges*, *19*(4), 133–142.

Trytten, D. A. (2001). Progressing from small group work to cooperative learning: A case study from computer science. *Journal of Engineering Education*, *90*(1), 85–91. doi:10.1002/j.2168-9830.2001.tb00572.x

Turhan, B., & Bener, A. (2007). A template for real world team projects for highly populated software engineering classes. In *Proceedings of the 29th International Conference on Software Engineering* (pp. 748–753). Washington: IEEE Computer Society.

Walker, E. L., & Slotterbeck, O. A. (2002). Incorporating realistic teamwork into a small college software engineering curriculum. *Journal of Computing Sciences in Colleges*, *17*(6), 115–123.

Wells, C. E. (2002). Teaching teamwork in information systems. In E. B. Cohen (Ed.), *Challenges of Information Technology Education in the 21st Century* (pp. 1–24). Hershey: Idea Group Publishing.

Wilson, J. D., Hoskin, N., & Nosek, J. T. (1993). The benefits of collaboration for student programmers. In B. J. Klein, C. Laxer, & F. H. Young (Eds.), *24th SIGCSE Technical Symposium on Computer Science Education* (pp. 160–164). New York: ACM.

KEY TERMS AND DEFINITIONS

Collaborative Learning: Learning that takes place in a peer-oriented environment. The development of collaborative tools such as Web conferencing, instant messaging, email, Weblogs etc. allow collaborative learning to take place between individuals/groups that are geographically dispersed.

Evaluation: Any systematic method for gathering information about the impact and effectiveness of a learning offering. Results of the measurements can be used to improve the offering, determine whether the learning objectives have been achieved, and assess the value of the offering to the organization.

Learning Management System (LMS): A Web based system that allows for the addition, deployment and tracking of learning content used for training purposes. Typically an LMS includes functionality for course catalogues (search/browse functionality), launching courses, registering new students, tracking current/completed student progress and assessments. Most of the learning management systems are developed to be independent of any content development/authoring packages. In addition, an LMS usually does not incorporate any authoring functionalities, but rather focuses on managing learning content.

Soft Skills: Personal attributes that enhance an individual's interactions, job performance and career prospects. Unlike hard skills, which are about a person's skill set and ability to perform a certain type of task or activity, soft skills are interpersonal and broadly applicable. They are often described by using terms associated with personality traits, such as: common sense, responsibility, integrity, etc. and abilities that can be practiced like teamwork, leadership, communication, negotiation, sociability, the ability to teach.

Teamwork: The combined action of a group of people, especially when effective and efficient.

Web 2.0: The second generation of the World Wide Web focused on the ability for people to collaborate and share information online. It basically refers to the transition from static HTML Web pages to a more dynamic Web that is more organized and is based on serving Web applications to users. Other improved functionality of Web 2.0 includes open communication with an emphasis on Web-based communities of users, and more open sharing of information.

Wiki: A set of Web pages developed collaboratively by a community of users, allowing any user to add and edit content.

Section 3
Supporting Communications

Chapter 6
From Textual Analysis to Requirements Elicitation

Marcel Fouda Ndjodo
University of Yaounde I, Cameroon

Virginie Blanche Ngah
University of Yaounde I, Cameroon

ABSTRACT

This chapter discusses the teaching of Requirements Engineering (RE) through a segmented approach. The idea is to teach this field, step by step, beginning with the requirements elicitation phase, which is the main focus of the chapter. The recommended linguistics-based method advocates the training of students in textual analysis techniques in order to develop their metacognitive and interpersonal skills, specifically, abstraction and comprehension. These skills are key soft skills for the practice of requirements elicitation.

INTRODUCTION

Most often the teaching of RE is reduced to its technical aspects and is therefore inefficient (Hanisch & Corbitt, 2004; Calelle & Makaroff, 2006; Danielsen, 2010). One of the reasons is that many soft skills which are important prerequisites in the RE process, and in the field of software engineering in general, are left aside (Chester 2011; Riemer 2007; Machanick 1998).Concerning the software engineering field, Jazayeri (2004) noted that:

A successful software engineer must possess a wide range of skills and talents.[...]; [he] must combine formal knowledge, good judgment and taste, experience, and ability to interact with and understand the needs of clients.

This is why many researchers (Machanick, 1998; Hanks, Knight, & Strunk, 2001; Hanisch & Corbitt, 2004; Hazzan & Kramer, 2007; Riemer, 2007; Burge & Wallace, 2008; Moràles, 2011; Reddy & Gopi, 2013) put emphasis on some non-technical skills, including metacognitive and interpersonal skills. These skills, which are considered essential by industry, should therefore be necessarily taken into account by Software

DOI: 10.4018/978-1-4666-5800-4.ch006

Engineering trainers. It is necessary to point out that the teaching of non-technical aspects of RE, and more generally of soft skills in engineering, is difficult because there are no established methods that allow efficient development of all expected skills in a formal education framework. In this regard, Macaulay & Mylopolous (1995) have examined issues in RE education and they concluded that the teaching RE is inherently challenging:

Requirements are variously described by practitioners as 'intangible', 'moving targets, 'inherently inconsistent', 'ever-changing' and host of other adjectives which fill the average university lecturer with horror... In contrast to this, university courses normally have a prescribed syllabus and strive to provide students with a solid foundation of knowledge, which guide practice and will direct future learning. […] The educational dilemma in teaching RE is to provide the student with the solid foundation in the subject matter while at the same time exposing the student to the inherent uncertainties, inconsistencies and idiosyncrasies associated with real requirements problems.

In order to improve the training of students in RE, and taking into account the fact that RE is made up of a set of activities, we suggest to address the teaching of RE in a segmented manner, i.e. going step by step. This chapter focuses on the first step of RE, namely Requirements Elicitation and Analysis. We propose a Linguistics-based teaching method. It puts at our disposal a number of fundamental concepts and techniques for clients' needs analysis and for the production of requirements specification documents. We think that some relevant aspects of linguistics can provide students with a solid foundation of knowledge in the subject matter while at the same time exposing them to the inherent and unavoidable uncertainties, inconsistencies and idiosyncrasies associated with real requirements problems.

The first section of the chapter gives a synopsis of the RE teaching. It presents the educational challenges facing this field and identifies some of the teaching approaches already underway. The second section focuses on the structure and the content used for the training of students in non-technical skills useful for the practice of RE in general, and Requirements Elicitation in particular. By clarifying the terms "text" and "analysis," this section explains what textual analysis is, and presents its different approaches and the reasons why we think that it is an appropriate pedagogical tool for Requirements elicitation and analysis. The third section establishes a parallel between linguistics concepts and the skills needed for requirements elicitation and analysis and identifies seven key themes of linguistics which should be taught as theoretical foundation for the discipline. Section 4 presents an experimental syllabus which is currently used at University of Yaounde (Cameroon). Section 5 suggests future research directions within the domain of the topic.

SYNOPSIS OF THE TEACHING OF REQUIREMENTS ENGINEERING

Of the many definitions proposed to describe RE, this chapter adopts Zave's view (1997), echoed by Nuseibeh & Easterbrook (2003) and Shams-Ul-Arif, Khan, & Gahyyur (2010):

Requirements engineering is the branch of software engineering concerned with the real-world goals, for functions of, and constraints on the software systems. It is also concerned with the relationship of these factors to precise specifications of software behavior, and to their evolution over time and across software families.

According to Shams-Ul-Arif, Khan, & Gahyyur (2010):

Requirements engineering is the most important activity in software project development as the other activities in the life cycle of software project development depend on this important activity. As the name implies, requirements engineering is a dig field responsible to cover all the activities involved in discovering, documenting, and maintaining a set of requirements for a computer based systems. A numbers of consequences may arise due to wrong requirements such as the system may be delivered late, more costly than the original estimation, customer and end-user will not be satisfied, the system may be unreliable and there may be regular defects.

Training students in RE suppose that they understand first what a requirement is. IEEE (1990) has defined a requirement as (1) a condition or capability needed by a user to solve a problem or achieve an objective; (2) a condition or capability that must be met or possessed by a system or a system component to satisfy a contract, standard, specification, or other formally imposed documents; (3) a documented representation of a condition or capability as in (1) or (2). Agreeing with the same direction, Christel & Kang (1992) think that:

A requirement is a "function or characteristic of a system that is necessary...The quantifiable and verifiable behaviors that a system must possess and constraints that a system must work within to satisfy an organization's objectives and solve a set of problems [...]. Requirements do not only consist of functions, a misconception introduced in part because the currently popular structured analysis techniques focus on articulating functional requirements. Different authors will present different definitions, but there are clearly nonfunctional requirements as well as functional requirements.

Several approaches are used for the RE teaching. We have identified some of them:

- **The interdisciplinary project approach (Suri & Durant, 2008):** The objective of this approach is to provide the students with a set of skills and knowledge enabling them to address real world problems. It is a learning-by-doing and collaborative work based approach. Students are incorporated in interdisciplinary working groups in order to allow everyone to put his experience and skills in contribution for the construction of the desired skills and their acquisition by the members of the team.

- **The role playing approach (Callele & Markoff, 2006):** This is one of the most widely used methods in the RE teaching. In this approach, emphasis is placed on real professional situations and students play successively the different roles attached to these situations: client, analyst, developer, user, etc. These different roles enable them to understand the difficulties that each of these actors can face depending on the angle in which it is situated.

- **The peer learning approach (Connor, Buchan, & Petrova, 2009):** This approach is based on the Research-Practice Gap concept, i.e., the study of the gap between research and actual practice in the field of RE. In order to identify the distance between theory and practice, emphasis is placed on collaboration among students and their ability to share their experiences and knowledge.

- **The holistic approach (Pérez-Martinez & Sierra - Alonso, 2003):** In this method which is not as widely used as the previous methods, it is assumed that students have a broad knowledge of all the activities of the discipline, and particularly that they have already made many practical works. The teaching covers almost all the subjects related to professional practices in the software industry in general and in the specific field of RE. The emphasis of the method is placed on the reproduction of real professional situations in order to provide students with a learning environment which is not very far from real-world professional environments.

These approaches that are far from being exhaustive, consider the RE activity in its entirety and do not develop the specific skills needed for each process of the activity. Indeed, it is a consensus today that RE is not a monolithic activity (Rahman & Sahibuddin, 2011), but consists of several processes that are not restricted to the first part of a software development life cycle (requirements elicitation and discovery, requirements analysis and reconciliation, requirements representation and modeling, requirements verification and validation and requirements management). The mastering and the efficient implementation of each process require specific skills that should be taught. This is why we propose to teach this activity in a segmented basis and we therefore focus here on its first stage which is Requirements Elicitation.

Our choice is suggested by the principle of modularization (Monteiro, 2011) according to which, in order to understand and solve a problem, it is better to break it down into subproblems. And in doing so in the case of the teaching of RE, one discovers that the teaching of Requirements Elicitation is itself complex. Indeed, this activity is not just a simple collection of data, but it goes further. Rahman & Sahibiddun (2011), quoting Zowghi & Coulin (2005), think that:

Requirements elicitation is the process of seeking, uncovering, acquiring, and elaborating requirements for computer based system. It is not just a gathering process but it is a process to understand the requirements that have been collected by going through activities with appropriate tools, techniques and approaches.

Although some researchers separate the requirements elicitation/discovery activity from the requirements analysis/reconciliation activity, we believe that the two activities constitute a continuum of linked activities. Consequently, when we talk about Requirements Elicitation, we rely on the Requirements Elicitation and Analysis model proposed by Welland (2006).

Even if it is still very little developed, the teaching of Requirement Elicitation and Analysis is however the subject of a few researches which aim is certainly to understand, explain and teach it, but mainly to practice it (Garcia & Moreno, 2003; Rahman & Sahibiddun, 2011). What we can say is that, more often, the interest on this activity comes more from its importance in professional practice and less from the place it occupies in the software engineering education system. One conclusion that can be drawn from the review of the teaching methods of RE in general, and Requirements Elicitation and Analysis in particular, is that most of them assume that students already have some knowledge and even a certain experience of the discipline. Conversely, the logic of our educational approach requires that some basic notions of Linguistics are given to students as prerequisites. These Linguistics fundamentals enable to teach a relevant body of knowledge and develop required skills to students having neither prior knowledge nor experience of Requirements Elicitation. We believe that the practice of Requirements Elicitation requires polyvalence and its teaching should therefore be multidisciplinary. But, since a major part of RE's work requires communication skills rather than technical abilities, we rely on linguistics to provide a theoretical basis to this discipline. In the same line, the argument of Nuseibeh and Easterbrook (2003) and Suri and Durant (2008) is the following:

Linguistics is important because RE is largely about communication. Linguistics analysis have changed the way in which the English language is used in specifications, for instance to avoid ambiguity and to improve understandability. Tools from Linguistics can also be used in requirements elicitation, for instance to analyze communication patterns within an organization.

For Hanks, Knight, and Strunk (2001),

The challenges provided by the domain-knowledge communication problem derive from far more than the goal of intact elicitation of domain knowledge. Elicitation is just the beginning. The challenges provided by the domain-knowledge communication problem derive at least much from the goal of intact propagation of domain knowledge. [...] Linguistics analysis demonstrates the way that a message can be degraded, and our knowledge of industrial processes indicates that the problem is recreated over and over in what are essentially new elicitation rounds every time message must be passed.

JUSTIFICATION FOR THE DIDACTIC AND METHODOLOGICAL CHOICE

Our motivation relies on our personal experience and on certain arguments in the literature. Narayan (2010) argues that the level of language proficiency always influence positively or negatively the level of understanding of texts given in natural language or in mathematical language. In the same vein, Bornat, Denhadi, and Simon (2008), following several other researchers, shows that high skills in natural language can actually have a positive impact on the students' capacity to understand and solve problems in the field of software engineering. We therefore strongly believe that the choice of Linguistics as a theoretical foundation for the teaching of Requirements Elicitation and Analysis can leads to better pedagogical results.

From a purely technical perspective, this didactic choice can be justified on several levels. First, it should be noted that developing the students' ability to analyze texts and documents in natural language strengthens their abstraction and communication skills. And among those skills, one is very decisive for the mastery of Requirements Elicitation, namely the comprehension skill. Indeed, an optimized identification of the needs of a software system requires the engineer to understand and analyze properly the desires of the various actors of the system. Secondly, based on the process of Shams-Ul-Arif, Khan, & Gahyyur (2009) and specifically on the detailed process of the REA of Welland (2006), we realize that the raw material that is transformed by this activity is made essentially of texts. In fact, REA is basically the extraction of information (i.e. data which are assigned a certain sense) from a set of data contained in different documents which are put at the engineer's disposal by the actors of the process (Figure 1). It should therefore be understood that the relevant information must be identified and retrieved from the mass of raw data available, and that the engineer should analyze that information in order to identify problems, i.e. the needs which express what the client wants (Danielsen, 2010). Those needs, gathered in a requirements catalog, are generally presented or given by the engineer in the form of texts written in natural language which the client finds easier to understand (McTavish & Pirro, 1990). Thirdly, an appropriate technical mastery of the client's language is essential for a requirements engineer. Obviously, the communication problems (oral or written) between a client and a non-native requirements engineer having a little knowledge of the client's language constitute an important obstacle for the engineer.

The above analysis reveals that one of the core competencies of a professional software engineer is the ability to extract relevant information from a huge set of raw data, to analyze this information in order to highlight problems and to properly

Figure 1. The requirements elicitation process

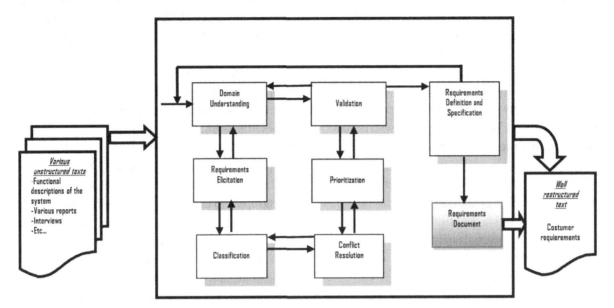

formulate needs in a natural language (Calelle & Makroff, 2006).

For decades, it has been established by consensus and the practice that training in computer science requires, as a prerequisite, an important knowledge and skills in mathematics (Levitin & Papalaskari, 2002; Liu, Huang & Brown, 2007). However, several studies (McCracken et al., 2001; Ford & Venema, 2010; Ma et al., 2011) paradoxically show that being successful in mathematics or programming is not a guarantee to solve software engineering problems. This is confirmed by Danielsen (2010) who found that, in the case of RE education, students with experience in programming or in the object-oriented paradigm have more difficulties to master the context and the problematic of RE than those students with no experience in these areas. Callele and Makaroff (2006) justify this situation by the fact that, most often, computer science's students are conditioned to expect that this portion of the project is already complete:

One of a software engineer's most skills is the ability to define the scope of the problem and ascertain the requirements engineering from general and vague precision. Teaching this skill is known to be difficult and is made more complex because (computer science) students are conditioned to expect that this portion of the projects is already complete.

We therefore believe that many problems facing the teaching of Requirements Elicitation are linked to the insufficient consideration of the importance of natural language and the very limited knowledge of linguistics by the trainers (Garcia & Moreno, 2003; Reddy & Gopi, 2013). This is expected in the actual context where most computer science teachers have a mathematical academic background. Increasingly, the development of metacognitive and communication skills - which are intimately linked to the level of student language proficiency - is increasingly proposed as one of the possible solutions for a better education in Requirements

Elicitation and Analysis (Teles & de Oliviera, 2003; Riemer, 2007; Rahman & Sahibiddun, 2011). We therefore think that it is essential to formally teach the science and the techniques of textual analysis, instead of mathematics, or less excessively in addition to mathematics, to software engineer students as fundamental courses of their training program. We remind, to those who remain skeptical, that the study of texts in the teaching of computer science and software engineering is not new. For instance, the object oriented (OO) analysis methodology actually highlights the role of texts and narrative descriptions, and therefore the importance of textual analysis, for the mastery of this methodology. Delisle, Barker, and Biskri (1993) indicated in this respect that:

The use of nouns, pronouns and noun phrases to identify objects and classes during the early stages of OO analysis is not a new idea. [...] It is fair to say that a majority of modern OO methodologies include equivalent notion in one form or another. [...] Some OO methodologies have generalized this idea to consider phrases as well. If the software engineer is going to consider noun or verb phrases, the first interesting question is where to look for them. The answer varies according to each OO methodology but one some typical are: in a concise summary of the subject matter; in a description of the problem space; in a description of the user's needs; in use cases; or in a summary paragraph describing the problem. The fundamental idea here is that these texts and narrative descriptions contain the basic elements to identify the building blocks of OO analysis of the problem. The second interesting question is how one should go about the analysis of relevant textual material.

Textual analysis deals with the meaning of words and phrases in a text. As a discipline, it refers to the scientific field which gives purist methodological approaches for the study of texts (Carr, 2009). From a methodological perspective,

the main considerations in the teaching of textual analysis are related to the selection of the types of texts to study, the acquisition of the appropriate texts and the choice of the analysis approach for studying these texts (Frey, Botan, & Kreps, 2000). Among the different textual analysis approaches (semantic, narrative, rhetorical, and stylistic), we believe that semantic analysis is more appropriate for Requirements Elicitation. Semantic analysis is the process of relating syntactic structures (from the level of words, phrases, clauses, sentences and paragraphs, to the level of the writing as a whole) to their language-independent meanings. It also involves the reformulation of elements specific to a particular linguistic or cultural context, to the extent that such a project is possible. Those specific elements of language, figurative speech or cultural being, are converted, as can be done, to elements of relatively invariant meanings in semantic analysis.

THE CONCEPTUAL FRAMEWORK

The proposed pedagogical approach implements the first principle of "abstract breakdown" proposed by Machanick (1998), namely, the "low-level-cognition-first." The idea behind this principle is summarized as follows: "Know, Comprehend and Apply, and finally, Analyze and Synthetize." In this perspective, our method which uses semantic textual analysis as a methodological framework, aims specifically to enable students to:

- **Know**: Equip students with linguistic concepts, notions and techniques that will serve as prerequisites for a better understanding of Requirements Elicitation.
- **Comprehend**: Conduct students to use their language skills and the acquired knowledge in textual analysis to better understand the principles, the rules and the techniques that govern Requirements Elicitation.

- **Apply**: Train students to apply their knowledge in order to extract information relevant to requirements elicitation real world problems presented to them in the form of texts containing raw data.
- **Analyze**: Train the students to the meticulous study of information gathered in texts presenting requirements elicitation problems with the objective to discern the different parts of a whole, or to determine and explain the relationship that they have each other.
- **Synthesize:** Equip students with skills enabling them to reformulate and organize information according to certain logic depending on the problem to be solved.

Taking as a starting point the relations between notions of linguistics and those of software engineering highlighted by the object-oriented analysis, we have identified a set of basic concepts and linguistic paradigms that allow introducing students and facilitating their understanding of requirements elicitation techniques. The teaching of these concepts and paradigms can require knowledge of other concepts that can be known or not by learners. Depending on the situation, the teacher will ensure that students have mastered these concepts that may be empirical for some of them and not for others. One can object the fact that most software engineering students are not linguistics majors, but it can be noted that linguistic concepts and paradigms are not new for them. Most often, they are studied less extensively or implicitly in primary and secondary education. Furthermore, research reveals today that linguistics is one of the theoretical foundations of requirements elicitation and analysis.

The expected difficulty of teaching linguistic to traditional software engineering students must not be an obstacle to the introduction of this teaching. The real concern, which is a didactic problem, is to find and experiment a syllabus relevant to those software engineering students, and which can be taught from the first year of university. The contribution of this chapter to this important issue is that it proposes a set of basic linguistic themes that should be taught to software engineering students. Other important issues such as the pedagogical progression, the syllabus, the selection of the types of texts to study and the acquisition of the appropriate texts depending on the level of the students are not fully addressed here. These issues are challenging questions, not only for the authors, but for the entire software engineering teachers' community.

The parallel we have established between the linguistic concepts and the skills needed for requirements elicitation and analysis (Figure 2) helped us to identify seven key themes of Linguistics which should be taught as a theoretical foundation for the discipline. These concepts are the following:

1. Lexical classes,
2. Lexical and semantic fields,
3. Taxonomic relations,
4. Textual coherence and cohesion,
5. Theme, rhyme and thematic progression,
6. Verbs and verbal tense value.
7. Summary of documents (synthesis).

Lexical Classes

In grammar, a part of speech (also referred as a word class, a lexical class, or a lexical category) is a set of words (or more precisely lexical items) that display the same formal properties, especially their inflections and distribution. Common English lexical classes include:

- **Noun**: A part of speech inflected for case, signifying a concrete or abstract entity,
- **Verb**: A part of speech without case inflection, but inflected for tense, person and number, signifying an activity or process performed or undergone,

Figure 2. Relations between linguistic notions and requirements elicitation skills

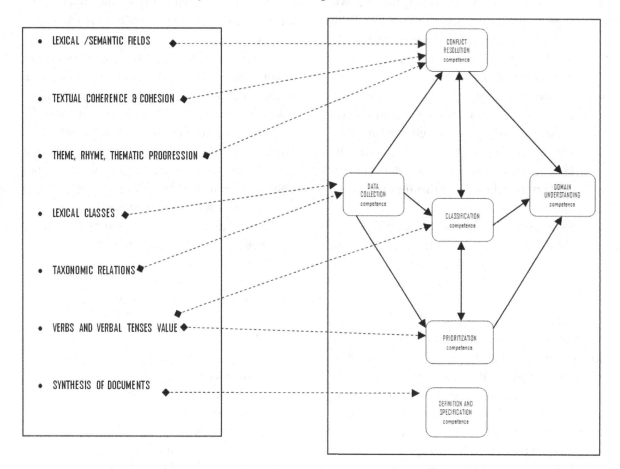

- **Participle**: A part of speech sharing the features of the verb and the noun,
- **Interjection**: A part of speech expressing emotion alone,
- **Pronoun**: A part of speech substitutable for a noun and marked for a person,
- **Preposition**: A part of speech placed before other words in composition and in syntax,
- **Adverb**: A part of speech without inflection, in modification of or in addition to a verb, adjective, clause, sentence, or other adverb,
- **Conjunction**: A part of speech binding together the discourse and filling gaps in its interpretation.

Although those classes are the traditional eight English parts of speech, modern linguists have been able to classify English words into even more specific categories and sub-categories based on function. Almost all languages have the lexical categories noun and verb, but beyond these there are significant variations in different languages. For example, Japanese has as many as three classes of adjectives where English has one; Chinese, Korean and Japanese have nominal classifiers whereas European languages do not. Many languages do not have a distinction between adjectives and adverbs, adjectives or adjectives and nouns, etc. This variation in the number of categories and their identifying properties entails that analysis be done for each individual language. Given the number of languages of the world, one can legitimately questions the universality of this approach. But beside the fact that requirements elicitation and analysis is necessarily done in a language the engineer must fully master (oral and written), what is more important and universal in a software engineering education's perspective is that the study of lexical analysis, irrespective of the language, develops categorization skills.

Categorization is a natural cognitive process for humans (Sutton & Rouvellou, 2005), that consists in grouping objects or non-identical events in cognitive categories considered as a set of equivalent objects (Mervis & Rosch, 1981). Many researchers think that categorization is the basis of abstraction. Going in the same direction, Hanks, Knight, & Strunk (2001) define the cognitive categories as follows:

Cognitive categories are collections of mental representations of entities encountered or imagined by an individual that are judged by that individual to be sufficiently similar to each other to count in some partitioning of reality as being the same.

The importance of lexical analysis for requirements engineers is that the mastery of the categorization helps them to determine intuitively, during the *classification* activity, the important elements in a written or an oral text describing a software system: components, objects, functions and processes, states, constraints, rules of operation, relations, etc.

Lexical/Semantic Fields

A semantic or lexical field is a set of words grouped by meaning referring to a specific subject. It denotes a segment of reality symbolized by a set of related words. The words in a semantic field share a common semantic property. A general and intuitive description is that words in a semantic field are not synonymous, but are all used to talk about the same general phenomenon. In fact, a meaning of a word is dependent partly on its relation to other words in the same conceptual area. Most often, fields are defined by subject matter, such as body parts, landforms, diseases, colors, foods, or kinship relations, etc. As a simple illustration, if a passage of writing included the words "weapon," "military," "dead," "destruction," "negotiations," the semantic field that would most likely be considered is 'war'.

In the context of software engineering training, the ability to identify the semantic fields in a spoken or written text allows the requirements engineer to easily organize the information gathered from raw data.

Taxonomic Relations

The analysis of a text often requires a meticulous study of the internal links that may exist between the different signs (words) of this text. This study is essential in the particular case of a semantic analysis (Ifversen, 2003). In the following statement, Carr (2009) also highlights this importance:

Within a semantic analysis, the description of internal relations between the different segments of meaning (lexemes, words), for instance the syntagmatic and paradigmatic relations, are important. Roughly speaking, syntagmatic relations are relating linguistic entities as well syntactically as semantically ('horizontal' relation), whereas paradigmatic relations designate those entities that are only related semantically ('vertical' relation). The latter relation concerns the existence of synonyms, antonyms, homonyms (the same sound/ signifier, but different meanings).

The study of semantic relations is very important for the activity of *"Domain Understanding."* Indeed, textual analysis seeks a thorough decomposition of the lexicon in which words, according to situations, contexts or particular domains can have several meanings which shall be circumscribed. In this perspective, the concepts of *signifier*, *referent* and *signified*, are often used. The signifier is the term used to refer to a thing, while the signified is the object or the thing designated by the signifier. Thus, a signifier can have many signified, meaning that the same term can be used to designate several realities or different objects. This is why, the different taxonomic relations between words, relate more to the meaning that these words have each other. Therefore, the existing lexical relations can be of several types.

We have formal relations that are interested in the identity of the signifiers (homonymy) or their similarity of form (paronymy). There are also semantic relations that bear on the hierarchies and taxonomies (hyperonymy, hyponymy, meronymy and holonymy), and equivalencies between the signified (synonymy, antonymy). Speaking of semantic relations, some are internal to the words, such as homonymy and polysemy, which are interested in the different signified of one word at a time. This relationship between a word and a concept is a semasiological relation. We also have relations, not only between words, but between groups of words and phrases. These include periphrastic and paraphrastic relations. The first relation is a semantic and syntactic equivalence relation (same meaning and same syntactic category) between a word and a group of words, while the second is a semantic and syntactic equivalence relation between two sentences.

The knowledge of these relations, the semantic relations in particular, permits to teach and understand more easily the various internal relational considerations of the classification of the information during the process of data collection. To clarify this argument, we highlight the importance of two of these paradigms:

- **Hyperonymy/Hyponymy**: Hyperonymy and the hyponymy are two asymmetric relations. Hyperonymy refers to the relation species/kinds, while the hyponymy establishes the relation kinds/species. By transposing these relations in software engineering, the hyperonym would correspond

more or less to the class of objects to which belong the referents of the hyponyms. In other words, it is the "generic term" used to represent a class and its subclasses, a set of objects with common characteristics. Hyperonymy is therefore equivalent to the logical relation of inclusion of a class in another.

- **Holonymy/Meronymy**: Holonymy is a relation between a whole and one of its parts, while conversely, the meronymy relation is established between a part and its whole. Thus, holonymy and meronymy are semantic relations between lexical units resulting from the description of a whole consisting of some elements (the parts of a machine, the steps in a process, etc.). Therefore, any description of a system includes relations of holonymy and meronymy, since in this case, the description designates the elements that make up the system.

In summary, the knowledge of lexical classes, lexical fields and taxonomic relations enables students to develop their ability to classify the various elements to be considered in a given problem, by identifying and categorizing them.

Theme, Rhyme, Thematic Progression and Textual Cohesion

The theme in a text refers to the general idea or the main subject discussed therein, and it allows to answer the question of what (who) speaks the text? The rhyme is precisely the discourse given on this subject, i.e. what is said about the theme. The rhyme answers the questions for what? for who? why? etc. The thematic progression is the specific manner in which the general theme is presented or evolves. There exist three types of thematic progression (TP): the simple linear TP (each rhyme becomes the theme of the next utterance), the constant TP (theme stays constant over sentences) and the derived TP (theme of successive sentences are derived from a single over-riding theme). The simple linear TP and the constant TP are used for the decomposition of a system in subsystems. The thematic progression of a text also reflects its cohesion. Cohesion, which verifies the overall coherence of a text, may be understood as the binding or the close harmony (syntactic and semantic) between the various elements of the text. So, the principle of cohesion requires the presence of a logical and non-contradictory relationship between the words or sentences of a text. Therefore, for a text to be consistent, its development should be accompanied by a constantly renewed semantic contribution. Each word or phrase should bring something new, which may be an additional information, justification, consequence, goal, condition, precision of time or place, etc.

The knowledge of these concepts helps students to clearly identify during the data collection process of Requirements Elicitation the system which is studied or the problem which should be solved. And, through the study of rhymes and the thematic progression, they can identify the relevant information describing the system.

Verbs and Verbal Tenses Value

A verb is a word with many morphological variations. It allows describing or situating in time, sequences of actions or events. Textual analysis is greatly interested with the nature, the tense and maybe the mode of conjugation of verbs. These features bring important implicit information for a good understanding of a text. In general, verbs are classified into two groups according to their nature: state verbs and action verbs. Action verbs are those that express or describe a movement, an operation or a process. State verbs express, describe or present the situation in which an object is in a moment of time; they present the states of the objects. In order to master the variety of tenses and modes of conjugation, as well as their values, it is useful for learners as part of their

training program in Requirements Elicitation, to study the different values of tense and modes of conjugation since they are used to situate in time an action, a sequence of events or even a text.

Synthesis of Documents

A synthesis is an abstract, condensed, well organized and restructured reformulation of documents of different nature as texts, books, articles, iconographic documents, etc. The synthesis skill requires the student to have some basic competencies which derive from the above mentioned knowledge. Synthesis which is an important activity in requirements elicitation and in software engineering in general, must therefore be taught to students. Machanik (1998) argues that:

Synthesis involves constructing complete solutions out of components. While there are aspects of this skill in lower levels of the hierarchy, synthesis requires more complete understanding of the overall process, and the ability to arrive at a more complete solution. The aspects of synthesis most relevant to Computer Science and Software Engineering include producing a plan to meet requirements of task (including proposing how to test an hypothesis and integrating results of research into a solution plan, ability to produce a complete design from a given specification and the ability to use theory to define a new process), ability to drive a new set of abstract relations (formulate hypotheses, ability to convert specific instances to a conceptual structure, and ability to make generalizations).

Apply Text Analysis Knowledge to Requirements Elicitation Case Studies

As indicated above, requirements elicitation consists in extracting, analyzing, and organizing information from a material made up essentially of written or oral texts. This information is then reformulated in a precise manner, without ambiguity and omissions in a customer requirements document. Since the purpose of the teaching of textual analysis is to give students the necessary scientific tools for the practice of requirements elicitation, it is desirable that texts used throughout the course be documents describing technological systems. As we have mentioned earlier, the pedagogical progression, the selection of the types of texts to study and the acquisition of the appropriate texts depending on the level of the students are still open questions.

EMPIRICAL FINDINGS

The ideas presented in this chapter are the result of an experiment that is underway at the University of Yaounde since 2007. The Faculty of science has a 3-2-3 based traditional computer science program leading to Bachelor's, Master's and PhD. The first author teaches software engineering in this program since 20 years in a traditional way. In 2007, an alternative and experimental access path to this program was opened at the College of Education (Higher Teachers' Training College). The same type of students are recruited from secondary schools (Mathematics as major), but they complete the first two years of the program, with alternative experimental methods, at the college of education, and then are merged at the Faculty of science with their faculty fellow from the third year. The objective is to identify and experiment innovative training methods that suit best to the development nontechnical skills (21th century skills) from the first year of university. Indeed, most of the time, the development of soft skills indispensable for engineers' professional practice is pushed to the highest level of university education.

The main differences between the experimental path (at the College of Education) and the traditional one (at the Faculty of science) is that

(software engineering) Applied linguistics (French and English) has been introduced as a fundamental and compulsory course in the experimental path in place of some mathematical courses. The first results show that, at the end of the third year, the performances of the students of the experimental path are far better than those of their faculty fellow. Even if we strongly believe that language skills play an important role in these performances, a scientific study has not been made to confirm this empirical observation.

The Experimental Syllabus

Our two experimental program, reviewed and delivered in collaboration with language teachers, is composed of two courses 30 hours each; one for the first year and the second for the second year. The guidelines of these courses are presented below.

Roughly speaking, the first year develops students' competencies on structural analysis of texts, while the emphasis of the second year is on semantic analysis (Table 1). More concretely, at the end of the first year, the student must be able to a) structurally decompose a text and, b) identify the main idea of a text and its supporting idea. At the end of the second year, the student must be able to a) logically reorganize a text, b) eliminate inconsistencies and ambiguities in a text, c) reformulate a text in a clear and precise manner, and c) synthesized a text. The course is given in one of the two official languages of Cameroon namely French or English.

The concepts which are taught during lectures are reinforced during tutorial classes through case studies based on the description of functional systems or systems to be developed. These case studies are given in the form of descriptive or explanative texts, and conducted in small projects groups, therefore allowing the evaluation of the knowledge and competencies targeted by trainers. The particularity in this evaluation lies in the fact that solutions to these case studies are found in an iterative manner and are presented in versions

Table 1. The experimental syllabus

First Year	Second Year

1) Basic lexical classes
2) Introduction to Semantic fields
3) Basic taxonomic relations
  - Homonymy
  - Synonymy
  - Paronymy
  - Antonymy
4) Types of texts
  - Descriptive
  - Narrative
  - Explanative
  - Informative
  - Injunctive
5) Textual coherence and cohesion
  - Explicit and implicit coherence
  - Theme
  - Rhyme
	1) Values of plain words
	  - Nouns
	  - Verbs
	  - Adjectives
	  - Adverbs
	
	2) Logical connectors
	
	3) Types of thematic progression
	  - Simple linear
	  - Constant
	  - Derived
	
	6) Text summary
	  - Case studies of level B3

which differ and evolves through the training. The most difficult part in designing the lectures is the selection of the appropriate texts depending on the level of the students. The suggested pedagogical progression consists in using short texts that are well structured (strong coherence and cohesion, less than one page) for beginners who need limited competencies in text analysis. Then move on less and less structured texts and documents whose study demands more elaborated competencies in textual analysis for advanced students.

We use two criteria to classify documents in 12 levels of difficulty for students. The first criterion is the length of the text which measures the difficulty to reduce that text. Four modalities are used for this criterion: A (very short text: less than one page), B (short text: less than 5 pages), C (long text: more than 5 pages), D: (more than one long text). The second criterion is the cohesion of the text which measures the difficulty to logically reorganize that document. Three modalities are for this criterion: 1 (strong cohesion), 2 (weak cohesion), and 3 (very weak cohesion).

The texts of length A are used for first year students and the texts of length B are used for second year students. The lengths C and D are reserved for advanced learners. At each level, the suggested pedagogical progression is 1 -> 2 -> 3, i.e we start with texts having a strong internal and ends with texts having weak internal cohesions. The main challenge for the trainer is to find enough case studies for each level of difficulty from the easiest level A1 (Table 2) to the most difficult level D3.

The experiment with students showed us that texts of level A1 are not so easy that it seems at first view. Indeed, if we limit the exercise on the comprehension aspect of the given texts, there is no apparent difficulty requiring the usage of sophisticated textual analysis techniques. But, the most important and difficult aspect of the requirements elicitation activity is the reformulation (in an active verbal tense) of the input documents in such a way that inaccuracies, omissions and ambiguities are eliminated. The reader can test this assumption by skipping questions a) to k) and answering directly the final question of the case study given in Table 2. This simple exercise is used at the beginning of the course in order to show the importance of textual analysis for the requirements elicitation activity.

FUTURE RESEARCH DIRECTIONS

Part of problems encountered by the teaching of Requirements Engineering, and specifically the discipline of Requirements Elicitation, comes from students' communication skills deficiency. The communication skill is a transverse skill which is essential to the software engineering domain. Chester (2011) emphasizes to this effect that despite the importance taken by technical com-

Table 2. A case study of level A1

A telephone exchange's commutator manages telephone lines and communications between subscribers. We consider a system with a single commutator. Each subscriber is identified by a unique telephone number. Some numbers are not assigned. During a call, the subscriber picks up the receiver. If the line is available (not in trouble), he composes a number and waits. If that number is assigned, the commutator establishes a linking with the target phone, otherwise an error message is sent to the caller. Several cases arise: a) if the phone of the callee is busy, a signal warns the caller, (b) if the callee's phone is not connected or if the line is in trouble, the behavior is not clearly defined (sounds normally or busy), (c) finally, if the phone is free, an alarm is activated and if he picks up the receiver, the communication starts and an alarm is activated on the callee's phone. The communication is maintained until a subscriber rings off or the line is busy or disconnected for any reason. The communication's time is counted by the commutator during the linking.

Questions:

Part I: Text analysis questions
   a) Theme?
   b) Rhymes?
   c) Thematic progression?
   d) Lexical and semantic fields?
   e) Taxonomic relations?

Part II: Requirements elicitation questions
   f) Based on the above analysis, identify inaccuracies, omissions and possible ambiguities in the description.
   g) Give a decomposition of the system.
   h) Identify all the types of objects manipulated by the system.
   i) Give the states of the system.
   j) Identify all the operations performed by the system.
   k) Give an input-output description of each operation.
   l) Reformulate the given description, using only the active tense, in such a way that inaccuracies, omission and ambiguities are eliminated.

petences, they are not sufficient to succeed in this domain. This is why many studies agree on the fact that the learning of Requirements Elicitation cannot be reduced to the mere acquisition of technical knowledge and competencies (Teles & de Oliveira, 2002). It is rather dependent mostly on the aspects which are considered to be the fundamentals of Requirements Engineering. The majority of these fundamentals rally with aptitudes unanimously considered as cognitive and interpersonal skills. This is why Gonzàlez-Morales (2011), in *Teaching "Soft" skills in Software Engineering*, regrets the fact that nontechnical competencies are not sufficiently developed during the very first years of training in higher education.

An important consideration for the research community is therefore, in our opinion, the approach and methods of teaching soft skills in software engineering and in computer science in general from the first year of university. In general, the first two years of undergraduate studies are devoted to the study of the theoretical foundations of the discipline. In the teaching of computer science and related disciplines, the important place of mathematics at this level is justified by the fact that the theoretical basis of several notions of computing lies in mathematics. But concerning Software Engineering, and Requirements Engineering in particular, mathematics does not seem to be the appropriate discipline from which the relevant theoretical fundamentals can be taught to students. It is therefore important to investigate other scientific disciplines, like linguistics which is proposed in this chapter, for developing effective fundamental and theoretical teachings and relevant training programs that can be taught from the first year of university.

On the other hand, the important role of computer science in all areas of life today pushes several countries to consider seriously the teaching of, not only ICT, but also computer science, in secondary education. Knowing the difficulties that a majority of students of this level have in mathematics, the teaching of computer science

and related disciplines, particularly software engineering, cannot be confidently envisaged if the theoretical basis of computer science is not extended to other scientific disciplines.

CONCLUSION

Our choice to base our method of teaching on textual analysis comes from the empirical use of textual analysis in object-oriented engineering. However, from our point of view, a good practice of textual analysis, as required by requirements elicitation, necessarily involves the learning of theoretical concepts of this discipline. We therefore believe that, like mathematics, linguistics should be introduced as a fundamental theoretical subject in training programs of software engineering. In doing so, some essential skills of this discipline, which for the moment are not sufficiently developed from the first year of training, will be taught in a gradual manner to students throughout their training. Many important issues such as the pedagogical progression, the syllabus, the selection of the types of texts to study and the acquisition of the appropriate texts depending on the level of the students are challenging questions, not only for the authors, but for the entire software engineering teachers' community.

REFERENCES

Bornat, R., & Dehnadi, S. (2008). Mental models, consistency and programming aptitude. In *Proceedings of the 10th Conference on Australasian Computing Education* (vol. 78, pp. 53–61). Australian Computer Society, Inc.

Burge, J., & Wallace, C. (2008). Teaching communication skills in the software engineering curriculum. In *Proceedings of the 21st Conference on Software Engineering Education and Training* (pp. 265–266). IEEE.

Callele, D., & Makaroff, D. (2006). Teaching requirements engineering to an unsuspecting audience. *ACM SIGCSE Bulletin, 38*(1), 433–437. doi:10.1145/1124706.1121475

Carr, D. (2009). Textual analysis, digital games, zombies. In *Proceedings of DiGRA 2009 Conference: Breaking New Ground: Innovation in Games, Play, Practice and Theory*. DiGRA.

Christel, M. G., & Kang, K. C. (1992). *Issues in requirements elicitation (No. CMU/SEI-92-TR-12)*. Pittsburgh, PA: Carnegie Mellon University.

Connor, A. M., Buchan, J., & Petrova, K. (2009). Bridging the research-practice gap in requirements engineering through effective teaching and peer learning. In *Proceedings of the 6th International Conference on Information Technology: New Generations* (pp. 678–683). IEEE.

Damian, D. E., & Zowghi, D. (2003). RE challenges in multi-site software development organisations. *Requirements Engineering, 8*(3), 149–160. doi:10.1007/s00766-003-0173-1

Danielsen, A. (2010). *Teaching requirements engineering: An experimental approach. Norsk Informatikkonferanse*. NIK.

Delisle, S., Barker, K., & Biskri, I. (1999). Object-oriented analysis: Getting help from robust computational linguistic tools. *Application of Natural Language to Information Systems, Oesterreichische Computer Gesellschaft*, 167–172.

Ford, M., & Venema, S. (2010). Assessing the success of an introductory programming course. *Journal of Information Technology Education, 9*, 133–145.

Frey, L. R., Botan, C. H., & Kreps, G. L. (2000). *Investigating communication*. New York: Allyn & Bacon.

González-Morales, D., de Antonio, L. M. M., & García, J. L. R. (2011). Teaching soft skills in software engineering. In *Proceedings of 2011 Global Engineering Education Conference* (pp. 630–637). IEEE.

Hanisch, J., & Corbitt, B. J. (2004). Requirements engineering during global software development: Some impediments to the requirements engineering process – A case study. In *Proceedings of the 13th European Conference on Information Systems* (pp. 628–640). Academic Press.

Hanks, K. S., Knight, J. C., & Strunk, E. A. (2001). *A linguistic analysis of requirements errors and its application* (Technical Report CS-2001-30). University of Virginia Department of Computer Science.

Hazzan, O., & Kramer, J. (2007). Abstraction in computer science & software engineering: A pedagogical perspective. *Frontier Journal, 4*(1), 6–14.

Ifversen, J. (2003). Text, discourse, concept: Approaches to textual analysis. *Kontur, 7*, 60–69.

Jazayeri, M. (2004). The education of a software engineer. In *Proceedings of the 19th IEEE International Conference on Automated Software Engineering* (pp. 18-27). IEEE Computer Society.

Levitin, A., & Papalaskari, M. A. (2002). Using puzzles in teaching algorithms. *ACM SIGCSE Bulletin, 34*(1), 292–296. doi:10.1145/563517.563456

Liu, D. K., Huang, S. D., & Brown, T. A. (2007). Supporting teaching and learning of optimisation algorithms with visualisation techniques. In *Proceedings of 2007 Australian Association of Engineering Education Conference*. Academic Press.

Ma, L., Ferguson, J., Roper, M., & Wood, M. (2011). Investigating and improving the models of programming concepts held by novice programmers. *Computer Science Education, 21*(1), 57–80. doi:10.1080/08993408.2011.554722

Macaulay, L., & Mylopoulos, J. (1995). Requirements engineering: An educational dilemma. *Automated Software Engineering, 2*(4), 343–351. doi:10.1007/BF00871804

Machanick, P. (1998). The skills hierarchy and curriculum. In *Proceedings of 1998 Conference of the South African Institute of Computer Scientists and Information Technologists* (pp. 54–62). Academic Press.

McCracken, M., Almstrum, V., Diaz, D., Guzdial, M., Hagan, D., Kolikant, Y. B. D., & Wilusz, T. (2001). A multi-national, multi-institutional study of assessment of programming skills of first-year CS students. *ACM SIGCSE Bulletin, 33*(4), 125–180. doi:10.1145/572139.572181

McTavish, D. G., & Pirro, E. B. (1990). Contextual content analysis. *Quality & Quantity, 24*(3), 245–265. doi:10.1007/BF00139259

Mervis, C. B., & Rosch, E. (1981). Categorization of natural objects. *Annual Review of Psychology, 32*(1), 89–115. doi:10.1146/annurev.ps.32.020181.000513

Monteiro, M. P. (2011). On the cognitive foundations of modularity. In *Proceedings of 2011 Psychology of Programming Interest Group Conference.* Academic Press.

Narayan, A. (2010). Rote and algorithmic techniques in primary level mathematics teaching in the light of Gagne's hierarchy. In *Proceeding of the 3rd International Conference to Review Research on Science, Technology and Mathematics Education.* Academic Press.

Nuseibeh, B., & Easterbrook, S. (2000). Requirements engineering: A roadmap. In *Proceedings of the Conference on the Future of Software Engineering* (pp. 35–46). ACM.

Pérez-Martínez, J. E., & Sierra-Alonso, A. (2003). A coordinated plan for teaching software engineering in the Rey Juan Carlos University. In *Proceedings of the 16th Conference on Software Engineering Education and Training* (pp. 107–118). IEEE.

Rahman, A. N., & Sahibuddin, S. (2011). Extracting soft issues during requirements elicitation: A preliminary study. *International Journal of Information and Electronics Engineering, 1*(2), 126–132.

Reddy, B. B., & Gopi, M. M. (2013). The role of English language teacher in developing communication skills among the students of engineering and technology. *International Journal of Humanities and Social Science Invention, 2*(4), 29–31.

Shams-Ul-Arif, M. R., Khan, M. Q., & Gahyyur, S. (2009). Requirements engineering processes, tools/technologies, & methodologies. *International Journal of Reviews in Computing.*

Sutton, S. M., Jr., & Rouvellou, I. (2001). Applicability of categorization theory to multidimensional separation of concerns. In *Proceedings of the Workshop on Advanced Separation of Concerns.* OOPSLA.

Teles, V. M., & de Oliveira, C. E. T. (2003). Reviewing the curriculum of software engineering undergraduate courses to incorporate communication and interpersonal skills teaching. In *Proceedings of the 16th Conference on Software Engineering Education and Training* (pp. 158–165). IEEE.

Welland, R. (2006). *Requirements engineering.* Simula Research Laboratory.

Westfall, L. (2005). Software requirements engineering: What, why, who, when, and how. *Software Quality Professional, 7*(4), 17.

Zave, P. (1997). Classification of research efforts in requirements engineering. *ACM Computing Surveys, 29*(4), 315–321. doi:10.1145/267580.267581

Zowghi, D., & Coulin, C. (2005). Requirements elicitation: A survey of techniques, approaches, and tools. In *Engineering and managing software requirements* (pp. 19–46). Berlin: Springer. doi:10.1007/3-540-28244-0_2

KEY TERMS AND DEFINITIONS

Linguistics-Based Teaching Approach for Requirements Elicitation: An educational approach that explains and teaches the fundamentals of requirements elicitation relying on linguistics. For us, this approach seems to be an effective alternative to the traditional teaching approach based on mathematics.

Metacogntive Skills: Refers to learners' automatic awareness of their own knowledge and their ability to understand, control, and manipulate their own cognitive processes.

Non-Technical Skills: Natural and generic skills used in everyday life that underpin and enhance technical tasks, improving safety by helping people to anticipate, identify and mitigate against errors.

Requirements Elicitation: Consists in extracting, analyzing, and organizing information from a material made up essentially of written or oral texts. This information is then reformulated in a precise manner, without ambiguity and omissions.

Requirements Elicitation Teaching: The activity that develops among the learners, in a formal educational setting, using the classical teaching processes which are theoretical and practical courses, essential skills for effective practice of requirements elicitation.

Technical Skill: A basic competence necessary to realize a procedural or highly structured activity. Technical skills are those skills learnt for a particular job.

Textual Analysis: A methodology for studying the meaning of a written text relying essentially on the morpho-syntactic elements that form this text.

Chapter 7
Peer Feedback in Software Engineering Courses

Damith C. Rajapakse
National University of Singapore, Singapore

ABSTRACT

Teaching non-technical skills such as communication skills to Software Engineering (SE) students is relatively more difficult than teaching technical skills. This chapter explores teaching the peer feedback aspect of communication skills to SE students. It discusses several peer feedback mechanisms that can be used in an SE course, and for each mechanism, it discusses the potential challenges and the various approaches that can be used to overcome those challenges.

INTRODUCTION

While programming in the small is often considered a solo activity, Software Engineering (SE) in the large is a team activity. When training SE students, we should also equip them with tools to communicate with team members in an effective way. Peer feedback is one such tool that is indispensable to a Software Engineer.

Teaching SE students how to use peer feedback effectively is not easy to do in the school environ-

ment. Here are some of the reasons why students lack the intuition or the motivation to give good peer feedback in school projects.

- The school environment is more flexible than the industry environment. In school, students are often allowed to pick their team members. If a picked member did not turn out to be a good fit, the other team members simply can bear with it for the semester and not team up with same person in the future. However, in the industry, one rarely has the option to choose team members.

DOI: 10.4018/978-1-4666-5800-4.ch007

- There is less at stake in the school environment. In school, only the course grade is at stake. If the teamwork is not going well, the student has many avenues to compensate for the grade, such as scoring more in individual components, complaining to the instructor in the hope of obtaining sympathy marks, or doing extra work to make up for the shortcomings of a team member. Therefore, students might consider the cost of frank peer feedback (e.g. unpleasantness created by giving negative feedback to team members) as not worthwhile compared to the potential benefits. In contrast, in the industry, the success of a project can usually be linked to tangible benefits, such as the career advancement opportunities, team members' job security, bonuses, and even the very viability of the company's future.

- Students are used to relying on academic staff to give feedback to others but not used to taking constructive actions to rectify the behavior of a team member.

In this chapter, we explore four mechanisms of peer feedback that we can use in an SE course.

1. Peer feedback during team meetings.
2. Code reviews as peer feedback.
3. Peer mentoring as peer feedback.
4. Peer feedback using online tools.

While this is not an exhaustive list, we believe these four can be good starting points to facilitate and guide effective peer feedback practices in an SE course. For each of the four, we discuss the potential challenges it poses, some of the practical tactics that can be used to overcome those challenges, and our experience in applying those techniques. The chapter content is based on the author's experiences in teaching SE for over a decade (since 2002) in various capacities, and in particular, building peer feedback tools in the recent years (since 2009).

BACKGROUND

While there is not much published work on using peer feedback in SE courses specifically, there are many prior publications about various aspects of student peer input (i.e. feedback, peer reviews, and peer assessments) in various other subject area courses. In this section, we describe a representative sample of such work.

Much of the prior work focuses on the benefits of peer input. For example, Morrow (2006) reports an experiment involving Psychology students. That study indicated that students felt they benefited from the opportunity to engage in peer feedback. Xie (2013) did a study that examined the relationships between motivation, peer feedback and students' posting and non-posting behaviors in online discussions in a distance learning class involving 57 college students. The study found significant correlations between students' posting and non-posting behaviors, suggesting that if learning occurs in online discussion activities, it happens in both posting and non-posting behaviors. Smith, May, & Burke (2007) did a study on Peer Assisted Learning (PAL) among first-year undergraduates of a School of Surveying. They found that while some students used PAL as a means of managing a comprehension problem (reactive) that had arisen, others used it as a means of preventing problems (proactive). Draper & Cutts (2006) studied peer mentoring as a form of intervention to help students weak in Computer Science. The work reported that the scheme generated some strongly positive qualitative feedback from the students. Wen & Tsai (2006) studied students' perceptions of and attitudes toward (online) peer assessment by collecting data from 280 university students in Taiwan. Their results revealed that participating students held positive attitudes toward the use of peer assessment activities.

Other work focused on the viability of and motivation for using student peer input in education. Liu & Lin (2007) reported that students are capable of using advanced-level cognitive and metacognitive strategies in providing feedback although students also displayed the tendency to provide less elaborate feedback with lower cognitive and metacognitive strategies. The study found that if students are requested to provide evaluative comments or modification suggestions, they are more likely to show a higher level of cognition. Liu and Carless (2006) argued that the literature on peer assessment has focused too much on the accuracy of peer assessments, thereby underplaying other features of peer learning processes. They make a case for peer feedback as an end in itself, as well as a precursor to peer assessment. In particular, they argue that peer feedback processes help to develop skills such as critical reflection, listening to and acting on feedback, sensitively assessing and providing feedback on the work of others.

Some prior work reports on novel approaches to using peer input. A recent example of such work is Kao (2013) which describes an innovative approach called "Peer Assessment with Positive Interdependence" (PAPI), intended to counter the threats of carelessness and favoritism in peer assessments.

Another segment of prior work proposes some guiding principles in using student peer input. For example, Nicol and Macfarlane-Dick (2006) identified some principles of good feedback practice that address the cognitive, behavioral and motivational aspects of self-regulation. They propose that good feedback (quoted from the original work):

1. Helps clarify what good performance is (goals, criteria, expected standards);
2. Facilitates the development of self-assessment (reflection) in learning;
3. Delivers high quality information to students about their learning;
4. Encourages teacher and peer dialogue around learning;
5. Encourages positive motivational beliefs and self-esteem;
6. Provides opportunities to close the gap between current and desired performance;
7. Provides information to teachers that can be used to help shape teaching.

Michaelsen and Schultheiss (1988) argue that even when the intent is positive, the outcome is likely to be negative unless the process is handled skillfully. They propose that helpful feedback is (quoted from the original work):

1. Descriptive, not evaluative, and is "owned" by the sender.
2. Specific, not general.
3. Expressed in terms relevant to the self-perceived needs of the receiver.
4. Timely and in context.
5. Desired by the receiver, not imposed on him or her.
6. Usable; concerned with behavior over which the receiver has control.

The same work also recommends that the person giving feedback should check, explicitly, with the receiver, to make sure that the receiver heard and understood what the feedback giver was trying to communicate.

PEER FEEDBACK DURING TEAM MEETINGS

Team meetings provide a 'natural' setting for peer feedback. Most Software Engineering courses have team projects that require students to have regular team meetings. While it may not be a formal requirement for students to give peer feedback during team meetings, it is certainly one of the avenues through which we can encourage peer feedback.

Here are some of the challenges of using team meetings as a means of training students to give proper peer feedback:

- **Student team meetings are too informal to be productive or constructive**: Team meetings conducted by students are often very informal. This does not imply that peer feedback given in a very informal setting is not useful or inferior to formal peer feedback. However, such informality can open the door for distractions such as gossip and can render the meeting unproductive in most aspects, including giving good peer feedback. Furthermore, as peer feedback provided in informal team meetings is not recorded, this makes it difficult for instructors to gauge the quality of peer feedback and guide students toward better peer feedback, if necessary.
- **Team meetings are not consistent/frequent enough**: As the semester progresses, team meetings often give way to the workload from other courses. It is often the case that students do not have enough time for team meetings during later stages of a semester.
- **Not enough manpower/expertise to guide team meetings**: In some cases, there are simply not enough instructors to supervise student team meetings. Even if team meetings are conducted under instructor supervision, because the instructors of SE courses are rarely communication experts themselves, they may lack the expertise to equip students with better peer feedback techniques.

Here are some approaches we can use to overcome the above challenges:

- **Include team meetings in the official course timetable**: By including team meetings as part of the official course timetable (e.g. schedule meetings during weekly tutorial time slot), we can ensure that students hold team meetings consistently and regularly. It will also help them

to keep the time slot free because this slot is now officially allocated to a particular course. If the team meeting did not use up the entire timeslot, students can use the remaining time to do other project-related work. One hidden cost of making project meetings 'official' is that instructors are now obliged to provide venues for those meetings. However, if there are not enough suitable meeting rooms for separate team meetings, one can also fall back on using large lecture halls for 'mass' team meetings as the latter can often accommodate several team meetings at the same time. Having regular 'official' team meetings is also a good deterrent to the usual tendency of students to do all the project work near the deadline instead of spreading it evenly over the semester.

We used this strategy in our courses *CS2103: Software Engineering* course, *CS3215: Software Engineering Project*, and the *CS4217: Software Development Technologies*. In all cases, the strategy helps to ensure that each team meets at least once a week. One downside of this strategy is that, because the number of "official" hours for a course is usually limited by the curriculum constraints, some tasks that used to occupy those official hours may have to be eliminated or redistributed to make room for team meetings in the course timetable.

- **Oversee team meetings**: Including team meetings in the official course schedule will force students to have team meetings but it does not necessarily yield good peer feedback. That is why some supervision during the meeting can be helpful. If there is a shortage of manpower for meeting supervision, we can use the technique of scheduling multiple team meetings in parallel where the instructor drops in on each meeting for a short period of time. For example, consider a scenario of scheduling eight team meetings in the same two-hour time slot in a large lecture hall. Two instructors can go around the lecture hall, taking turns to spend some time with each team. This setup gives students the freedom to run their own meetings, puts pressure on them to run the meeting properly (because the instructor can observe their conduct any time) and facilitates guidance by the instructor should students need it.

In the aforementioned three courses, we have used the described strategies in allocating instructor time to oversee team meetings, depending on the class size and manpower availability. While having such oversight clearly motivated students to try harder to make the meetings more productive, we were able to make it even more effective by defining some goals to be achieved in each meeting. For example, we could say to students, "you should finalize the architecture for the product in this week's meeting."

- **Twinning with Communications courses**: Most SE degrees include courses that teach communication skills taught by instructors with expertise in communication skills. Twinning such 'Communications' courses with SE courses can leverage the expertise of those instructors to help SE students learn how to give effective peer feedback.

For example, our *CS2103: Software Engineering* course (a first course in SE taught by the department of Computer Science) mentioned previously is twinned with the *CS2101: Effective Communication for Computing Professionals* (a course taught by our *Centre for English Language & Communication*). Each project team takes both courses together with the team project they do in the SE course used as the context for some assignments they do in the Communications course. For example, students are required to write progress reports and present their project proposals to a mixed audience. They are also required to provide oral feedback on the technical documentation written by other teams for their SE project. As a result, students get to learn peer feedback skills as well as other communication skills in the context of a real SE project. This also addresses the challenge posed by the lack of expertise/manpower among SE instructors to teach students how to give better peer feedback.

In our experience of twining CS2103 and CS2101, many of the teething problems such as misalignments between the deadlines of the two courses have been resolved with better coordination between the instructors of the two courses and iterative refinements of the course structure over several semesters. However, one lingering problem we have not yet managed to solve is the problem of forming teams in a way that maximizes learning in both courses. If the two courses were run independently, CS2101 instructor could form teams to ensure a good mix of communication skills in each team and the CS2103 instructor can similarly ensure a good mix of technical skills in each team. However, the twinned approach requires the same team of students to take both courses in parallel and the flexibility in forming teams to maximize learning could potentially be reduced. Fortunately, our experience in twinning CS2103 and CS2101 did not indicate this to be a big problem. Similarly, the flexibility in choosing assignments to maximize the learning is reduced because both courses use the same project as the backbone of the assignments. On the positive side, students get to use the same project for two courses, improving their workload to learning ratio.

CODE REVIEWS AS PEER FEEDBACK

Code reviews can play a role in training students to give peer feedback. Most SE projects involve a substantial coding component. In fact, students often become so preoccupied with the coding aspect that they tend to neglect spending enough time to hone their communication skills. One way we can leverage this student mindset to their own advantage is to use code as a platform to train students on how to give good peer feedback — specifically, to use code reviews to train students in giving peer feedback.

Doing proper code reviews is a time consuming task requiring pre- and post- session work and multiple roles such as author, moderator, and note keeper. However, there are often too few instructors to carry out sufficient instructor-led code reviews. On the other hand, getting students to do peer code reviews is challenging because students are not trained in doing peer reviews and are not confident/competent enough to review someone else's code. This is why code reviews are rare in SE courses. Here are some tactics that can increase the use of peer code reviews in SE courses:

- **Use code review meetings to train students**: A proper code review meeting is time consuming and requires multiple persons to come together for significant lengths of time. Therefore, it is not practical to enforce code review meetings for the code of an entire project. However, a few early code review meetings can be used as training exercises that show by example how to give peer feedback during a code review meeting. In an extreme case, this can be a done in front of the entire class where a piece of representative code is reviewed by the instructor while a tutor plays the role of the author.

- **Use pair programming to train students**: *Pair programming*, a practice popularized by agile methodologists, can be considered a mechanism of peer code review done during the coding process itself. While there are many prior instances of using pair programming in classroom settings (McDowell et al., 2002), particularly for beginner programming courses, it may not suitable for most SE courses in wherein students cannot 'pair' for most of their coding periods due to time table clashes and other logistical hurdles. However, a few well-planned pair programming sessions can be used in the early part of the course as a means to train students on programing and giving coding-related peer feedback.

- **Use offline code review tools**: While the two points above can be used for training students on how to give peer feedback in the form of code reviews during the early phases of a project, enforcing more frequent code reviews is only practical if done in a tool-assisted manner. For example, online code repositories such as Google code (n.d.), GitHub (n.d.), SourceForge (n.d.), and Bit Bucket (n.d.) have well-developed tools for reviewing code online. They are easy to use for students and convenient to oversee for instructors. In addition, most of these tools are free for educational use.

In our CS2103 (Software Engineering) course, we use all three strategies mentioned above, but in a very lightweight manner.

- First, we do a simple coding exercise where tutors review students' code and comment on areas that can be improved. These reviewed code submissions are published to the whole class so that students get to see many ways to program the same task and multiple ways to improve the code.

- Next, we get students to do coding-related work together as a team where they can observe a team member write code. For this, we use a work environment where each table has space for up to six students and any student at the table can connect their computer display to a big TV screen that the whole table can observe.
- Later in the semester, we use the GoogleCode online project management system where students can review team members' code if they wish to.

In spite of all these measures, we cannot claim that there is a lot of peer code reviewing going on in our course, which is understandable given that it is only a first course in SE. However, we have noticed a definite improvement in the code quality of student projects after implementing those measures. Furthermore, we also noticed that enforcing a coding style guide is a good way to increase peer feedback about code because it allows students to review peer code and point out superficial style guide violations even if they are not confident enough to comment on deeper aspects of the code.

PEER MENTORING AS PEER FEEDBACK

An underused approach to peer feedback is to get better (or senior) students to act as mentors. There are several challenges in using peers as mentors:

- **Peers are not experts**: Because peers are also students, their feedback is not going to be as useful as those received from an expert or the instructor.
- **Lack of peers**: In the case of using senior students as mentors, these students may be too busy to spend time mentoring their juniors.

- **Lack of consistency in mentoring**: Related to the above point, mentors might not meet their mentees consistently and the mentoring process could break down during the busy periods of the semester when the mentees need guidance most.

These are some tactics to help students tackle the above challenges:

- **Schedule, pace, and oversee**: As with the team meetings, mentor-mentee meetings can be scheduled as part of the official course schedule. Furthermore, this approach works better when the students are forced to use an incremental approach to their project. For example, students can be given a delivery schedule such as the example given below:
  - **Week 3:** First-cut high-level architecture is finalized.
  - **Week 5:** The first working skeleton of the code is ready.
  - **Week 7:** At least some unit tests have been written.
  - **Week 9:** The code is version controlled. At least one feature is available to the user.

The result of such an incremental approach is that we can predict roughly the type of work done and the kind of issues faced by students each week. This allows us to prepare the mentors to help students in that week.

- **Give course credits for mentoring**: If paying mentors a 'tutoring fee' is not viable, an alternative is to offer them credits under a course code. Mentoring a project team is a valuable learning experience for SE students and it is acceptable in most contexts to incorporate that learning experience officially into the curriculum as a course.

- **Create expertise by repeating projects**: While senior students used as peer mentors are technically 'more experienced' than their mentees (because they have taken the same course before), they may still not be experienced enough to mentor a team that is doing an entirely different project using different tools/technologies. One tactic that is useful here is to repeat the same or a similar project for multiple semesters so that mentors feel more confident in guiding their juniors because they have 'done it before'. For cxample, we can make the project requirements a slight variant of the previous semester's project, with some flexibility in interpretation so that each team produces a product slightly different from other teams and yet not drastically different from what the mentors did when they took the course. As for the concern of whether similar projects encourage plagiarism, it is considerably difficult for a team to copy an existing project and pretend to make weekly progress based on mentor feedback, especially if the code is expected to be version controlled.

In our CS2103 (Software Engineering) course, which has 200-250 students divided into 50-65 teams, we get students who took the course in previous semesters to act as mentors for the current batch. All students are given the same requirements outline for their course project with only some flexibility allowed in interpreting those requirements, which results in all student projects being 'variations of the same theme'. Students follow an incremental approach when doing the project. Project deliverables are small and spread out (as opposed to 'deliver everything at the end of the semester' approach), which forces all students to follow a mostly similar project schedule. This allows us to predict what students will be doing each week and brief mentors on issues likely to be raised by their mentees in their next meeting. Furthermore, we reuse the project requirements for multiple semesters with only minor changes,

which makes the current project similar to the project mentors did when they read the course. However, note that students prefer to define their own projects under the impression that such choice helps them learn more. Therefore, it is recommended that the rationale for any restriction on project requirements and the toolset should be explained to students beforehand.

PEER FEEDBACK USING ONLINE TOOLS

Online peer-feedback tools can play a vital part in helping to give peer feedback, especially for big classes. Given below are some of the points that can be considered in that regard:

- **Align with the purpose**: When choosing an online peer feedback tool, we have to consider if the tool is truly aligned with our purpose. Peer *feedback* tools are different in purpose compared to peer *evaluation* tools, although some tools can be used for both purposes. If our aim of peer feedback is formative (i.e. we want students to learn how to give peer feedback and learn from the peer feedback received), we should not choose a tool that is summative in nature (i.e. meant to be used for grading purposes). Furthermore, when using peer feedback for formative purposes, we recommend conducting peer feedback sessions early and frequently during the semester so that students get more practice in giving and responding to peer feedback. If peer feedback is also used for grading, only the peer feedback given during the final stages of the semester can be used for grading, thus giving more time for students to improve themselves based on peer feedback. We speculate that students are more likely to give constructive and genuine feedback to peers if they know that peer feedback is not used for grading.

In our courses CS2103 and CS4217, both of which have significant software-intensive team project components, we have three peer feedback and peer evaluation periods. The first two, scheduled roughly around third and two third into the team project duration, are mostly formative and not counted for grading. They are for students to understand peer expectations and adjust their conduct accordingly. The final peer feedback/evaluation submissions are considered in grading.

- **Know existing tools**: One of the easiest ways to collect peer-feedback online is to use a general purpose online survey tool such as *Google Forms* (n.d.) or *Survey Monkey* (n.d.). For more control and better results, one can opt for custom-built peer feedback tools. One such tool is *TEAMMATES*, built by a team lead by the author, and available to educators as a free service (http://teammatesonline.info). TEAMMATES is optimized for conducting frequent peer feedback sessions and it can collect, moderate, and publish quantitative estimates and qualitative comments about peers. Other similar tools include iPeer (n.d.), CATME (n.d.), and QuackBack (n.d.), to name a few.

We use TEAMMATES in our courses CS2103 and CS4217. In particular, systematic peer feedback in CS2103 would have been impractical without the help of a tool such as TEAMMATES because of the large class size (200-250). It is very satisfying to see the contribution level of certain students improve after realizing that their team expects more from them. However, there were cases where the team was consistently unhappy about the contribution level of a member. Such cases were rare and easy to spot with the help of TEAMMATES' reporting facilities.

- **Integrate peer feedback into other project work**: The overhead of using a peer feedback tool on the instructor and the students have to be factored in too. The more peer feedback is integrated into project work, the more it will be accepted and used by students. In that regard, online *project task trackers* can play a role. It is common practice for industry projects, open source projects in particular, to use task trackers (sometimes called *issue trackers* or *bug trackers*) to track task allocations and task statuses. Students can be asked to use such a task tracker in their projects too. This naturally leads students to record comments about expectations from the task owner and feedback about the task execution. These comments can be considered a form of peer feedback that is more 'natural' to give because students see it as an essential part of the project workflow. Online project collaboration tools such as GoogleCode, GitHub and Bit Bucket (mentioned previously in this chapter) come with well-developed and feature-rich task trackers that can be used in SE courses.

In CS2103 (Software Engineering), we require students to use GoogleCode issue tracker to record and track their project tasks. Note that it is quite hard to motivate students to use such a tool because they view it as an unnecessary overhead. In a similar vein, we have also noticed that recent batches of students are somewhat resistant to traditional group communication tools such as online forums and mailing lists, instead preferring to use tools such as group text-messaging (e.g. WhatsApp, n.d.) and social networks (e.g. Facebook).

FUTURE RESEARCH DIRECTIONS

One of the emerging trends in the higher education domain is Massively Open Online Courses (MOOC), spearheaded by initiatives such as Coursera (n.d.) and Udacity (n.d.). It would be interesting to study peer feedback in the context of MOOCs. As of now, MOOCs rely heavily on online forums, which can be considered as using peer feedback to help students clarify doubts by getting their questions answered by peers. Some MOOCs (e.g., Coursera) also use peer grading as an evaluation mechanism. However, the applicability of using peer feedback in SE team projects conducted in MOOC mode has yet to be explored in-depth at the point of this writing.

CONCLUSION

It is not easy to train SE students to give effective peer feedback in school projects. However, it is a valuable skill to have and there are several mechanisms we can use to facilitate and guide peer feedback practices in SE school projects. While each of those mechanisms poses multiple challenges, we have explained some practical tactics that can be used to overcome those challenges.

ACKNOWLEDGMENT

This chapter draws heavily from the author's experience in teaching the *CS2103 Software Engineering* course which has benefitted from the contribution from many colleagues over the years. In particular, the author wishes to express his gratitude to colleague Janet Chan-Wong Swee Moi and her team from the *Centre for English Language & Communication (National University of Singapore)* for their invaluable contribution to the twinning of *CS2101: Effective Communication for Computing Professionals* with CS2103, and to colleague Bimlesh Wadhwa for her con-

tribution as a co-lecturer of CS2103. We would also like to acknowledge the funding support for the TEAMMATES project from the *Centre for Development of Teaching and Learning (CDTL) of National University of Singapore*, under grants C-252-000-096-001 and C-252-000-105-001.

REFERENCES

Bitbucket. (n.d.). Retrieved from https://bitbucket.org

Catme. (n.d.). Retrieved from http://www.catme.org

Code. (n.d.). Retrieved from https://code.google.com

Coursera. (n.d.). Retrieved from https://www.coursera.org

Draper, S., & Cutts, C. (2006). Targeted remediation for a computer programming course using student facilitators. *Practice and Evidence of Scholarship of Teaching and Learning in Higher Education, 1*(2), 117–128.

Drive. (n.d.). Retrieved from http://www.google.com/drive

Github. (n.d.). Retrieved from https://github.com

iPeer. (n.d.). Retrieved from http://sourceforge.net/projects/ipeer/

Kao, G. Y. M. (2013). Enhancing the quality of peer review by reducing student free riding: Peer assessment with positive interdependence. *British Journal of Educational Technology, 44*(1), 112–124. doi:10.1111/j.1467-8535.2011.01278.x

Liu, E. Z. F., & Lin, S. S. (2007). Relationship between peer feedback, cognitive and metacognitive strategies and achievement in networked peer assessment. *British Journal of Educational Technology, 38*(6), 1122–1125. doi:10.1111/j.1467-8535.2007.00702.x

Liu, N. F., & Carless, D. (2006). Peer feedback: The learning element of peer assessment. *Teaching in Higher Education, 11*(3), 279–290. doi:10.1080/13562510600680582

McDowell, C., Werner, L., Bullock, H., & Fernald, J. (2002). The effects of pair-programming on performance in an introductory programming course. *ACM SIGCSE Bulletin, 34*(1), 38–42. doi:10.1145/563517.563353

Michaelsen, L. K., & Schultheiss, E. E. (1988). Making feedback helpful. *Organizational Behavior Teaching Review, 13*(1), 109–113.

Morrow, L. I. (2006). An application of peer feedback to undergraduates' writing of critical literature reviews. *Practice and Evidence of Scholarship of Teaching and Learning in Higher Education, 1*(2), 61–72.

Nicol, D. J., & Macfarlane-Dick, D. (2006). Formative assessment and self-regulated learning: A model and seven principles of good feedback practice. *Studies in Higher Education, 31*(2), 199–218. doi:10.1080/03075070600572090

Quackback Peer Review. (n.d.). Retrieved from http://goldapplesoftware.ca/portfolio/quackback-peer-review

Smith, J., May, S., & Burke, L. (2007). Peer assisted learning: A case study into the value to student mentors and mentees. *Practice and Evidence of the Scholarship of Teaching and Learning in Higher Education, 2*(2), 80–109.

Sourceforge. (n.d.). Retrieved from http://sourceforge.net

Surveymonkey. (n.d.). Retrieved from http://www.surveymonkey.com

Udacity. (n.d.). Retrieved from https://www.udacity.com

Wen, M. L., & Tsai, C. C. (2006). University students' perceptions of and attitudes toward (online) peer assessment. *Higher Education, 51*(1), 27–44. doi:10.1007/s10734-004-6375-8

Whatsapp. (n.d.). Retrieved from http://www.whatsapp.com/

Xie, K. (2013). What do the numbers say? The influence of motivation and peer feedback on students' behaviour in online discussions. *British Journal of Educational Technology, 44*(2), 288–301. doi:10.1111/j.1467-8535.2012.01291.x

KEY TERMS AND DEFINITIONS

Massive Open Online Course (MOOC): A distance education mechanism where an online course is opened up for access by the public, with the possibility of enrollment by a large number of students at a time.

Pair Programming: An agile software development technique in which two programmers work together at one Computer. One (called the *driver*) writes code while the other (called the *navigator*) reviews code as it is typed in. The two programmers switch roles.

Section 4
Improving Soft Skills

Chapter 8
Engaging Software Engineering Students with Employability Skills

Jocelyn Armarego
Murdoch University, Australia

ABSTRACT

This chapter explores the findings from an Action Research project that addressed the Professional Capability Framework (Scott & Wilson, 2002), and how aspects of this were embedded in an undergraduate Engineering (Software) degree. Longitudinal data identified the challenges both staff and students engaged with. The interventions that were developed to address these are described and discussed. The results of the project show that making soft skills attainment explicit as part of the learning objectives went a long way in assisting students to engage with the activities that exercised these skills.

INTRODUCTION

A number of recent studies have discussed the challenges facing 21st century graduates (e.g. Andrews & Higson, 2008; P. Brown & Hesketh, 2004; Cassidy, 2006). The fast pace of change, affecting the whole of society, has led to a knowledge-driven economy that asks its workforce to "flexibly adapt to a job market that places increasing expectation and demands on them" (Tomlinson, 2012). Being an employable graduate is no longer an automatic result of successful completion of formal study – tacitly acknowledged in the focus on identifying and developing graduate attributes in universities globally as a response to employer concerns.

However, an example definition of graduate attributes: "the qualities, skills and understandings that a university community expects its students to develop during their time at the institution and

DOI: 10.4018/978-1-4666-5800-4.ch008

consequently, shape the contribution they are able to make to their profession and as a citizen" (Bowden, Hart, King, Trigwell, & Watts, 2000) indicates that, although highlighting the required cognition and character development of students, graduate attributes do not have the focus expected by employers on *employability skills*. These have been identified as key transferable soft skills and competencies integral to graduate professional practice (Andrews & Higson, 2008). Employability skills are not job specific and "cut horizontally across all industries and vertically across all jobs from entry level to chief executive officer" (Sherer & Eadie, 1987).

As a result of trends in the sector, the nature of ICT employment has evolved and employability has become a major issue (Scholarios, van der Schoot, & van der Heijden, 2004). In the Australian context, a comprehensive study of employability skills conducted on behalf of the federal Department of Education highlighted the importance of abilities sought by employers in addition to technical knowledge and skills (ACCI, 2002). Studies undertaken since 2000 confirm earlier work (Doke & Williams, 1999; Lee, 1999; Snoke & Underwood, 1999; Turley, 1991) that, although the technical competency of ICT graduates can, in general, be assumed, other, softer, skills are considered by practitioners as lacking. In particular the demand and expectation of industry employing ICT graduates has changed: surveys show that they are dissatisfied with graduates in a number of areas (Hagan, 2004).

However, embedding soft skills into undergraduate programmes is not an easy task:

- From the academic's perspective, this is a tacit aspect of competency and therefore less easy to define and assess – for example, what soft skills should be included? Anecdotal evidence also suggests many academics with high technical expertise find themselves less comfortable in this space, and therefore show a preference for students taking an externally offered course (and therefore without discipline context (Smith, Belanger, Lewis, & Honaker, 2007)) rather than attempting to incorporate these within the discipline

- From the student perspective, technical skills have been viewed as the prime competency to be attained from their studies. Affective skills, cognitive skills, understanding of the 'context' in which the task being addressed, are undervalued, often until well after graduation (Lethbridge, 2000).

This chapter explores the findings from an Action Research project that identified one framework where soft skills appropriate to an engineering context were explicitly described - the *Professional Capability Framework* (Scott & Wilson, 2002) - and how aspects of this were embedded in an undergraduate Engineering (Software) degree. The interventions that were developed to address these are described and discussed. The results of the project showed that making soft skills attainment explicit as part of the learning objectives went a long way in assisting students to engage with the activities that exercised these skills. In effect, soft skills can be mastered in a learning environment that applies active and collaborative learning, acknowledges differences in learning styles, and contextualises the learning of engineering as a profession.

BACKGROUND

Employability

The concept of employability has emerged as a dominant issue in a number of contexts over the last decade (McQuaid & Lindsay, 2005). The quality of the graduate labour market, and graduate ability to meet employer needs, have been the focus of debate across many disciplines acknowledged as impacting on a knowledge-based society. Despite a focus on generic attributes, university graduates lack the ability to apply these in the work place (Precision_Consultancy, 2007). As a result, there is a widespread perception that employability skills are not effectively developed in an undergraduate programme.

Andrews and Higson (2008) indicate that serious concerns have been expressed about an increasingly wide 'gap' between the skills and capabilities of graduates, and the requirements and demands of the work environment they enter. Although expectations of graduates are similar across the globe (that graduates would be employment-ready; equipped with the necessary skills and competencies; and able to work with the minimum of supervision (Harvey & Bowers-Brown, 2004)) and, in fact across time, with only the priority changing (DEEWR, 2012), studies show that graduates focus on technical or 'hard' skills, despite, since the 1990s, these having being considered less important by employers (Hagan, 2004; Turner & Lowry, 2003).

From the work of McLarty (1998); Tucker, Sojka, Barone, and McCathy (2000) and others, Andrews and Higson (2008) identify: professionalism; reliability; the ability to cope with uncertainty; the ability to work under pressure; the ability to plan and think strategically; the capability to communicate and interact with others, either in teams or through networking; good written and verbal communication skills; ICT skills; creativity and self-confidence; good self-management and time-management skills; a willingness to learn and accept responsibility, as ideal generic skills and competencies required of graduates in the workplace. Although their work addresses graduates within the business discipline, other studies indicate similar requirements:

- More generally as higher order skills that assist individuals to be more flexible, adaptable, creative, innovative and produc-

tive. This requires skills and knowledge to work effectively as a member of a team, co-operate in ambiguous environments, solve problems, deal with non-routine processes, handle decisions and responsibilities, communicate effectively, and see 'the bigger picture' in which they are working. It also requires the skills and knowledge that allow individuals to understand the context in which they are working and to apply their existing skills in the new context (Curtin (2004) cited in DEEWR (2012))

- In specific disciplines, such as engineering and ICT, the focus of this chapter:
  - Professional skills for ABET accreditation for both engineering and computing programmes in the USA (namely: an ability to function in teams; an understanding of professional and ethical issues and responsibility; an ability to communicate effectively with a range of audiences; the broad education necessary to understand and analyse the impact of solutions in different contexts; a recognition of the need for, and an ability to engage in lifelong learning and continuing professional development and a knowledge of contemporary issues (ABET, 2012a, 2012b))
  - For Engineers Australia as professional and personal attributes (namely: ethical conduct and professional accountability; effective oral and written communication in professional and lay domains; creative, innovative and pro-active demeanour; professional use and management of information; orderly management of self, and professional conduct; effective team membership and team leadership (EA, 2011)).

Studies in the disciplines of ICT indicate technical abilities are covered in curricula, and that students and employers are well informed as to the nature of these (Lowry & Turner, 2005). However, a number of practitioner studies attest to a lag in other abilities. Examples across different ICT sub-disciplines include international studies (Lee, 1999, 2004; Lethbridge, 2000; Noll & Wilikens, 2002; Turley & Bieman, 1995; Zwieg et al., 2006) as well as studies in an Australian context (Kennan, Willard, Cecez-Kecmanovic, & Wilson, 2009; Scott & Wilson, 2002; Scott & Yates, 2002; Snoke & Underwood, 1999; Turner & Lowry, 2003)) and highlight the need for:

- An understanding of business functions and organisational knowledge
- The ability to teach themselves what they need to know to perform the task successfully
- Interpersonal skills and personal attributes;
- The adaptability and flexibility required to be ICT practitioners
- Generic attributes such as lifelong learning;
- As well as career resilience (Waterman, Waterman, & Collard, 1994).

Few studies address the skills and knowledge needed in Software Engineering specifically. Turley and Bieman (1995) examined professional Software Engineers in an attempt to identify the competencies and demographics that contribute to 'excellence' in performance. They identify four categories of competencies which are statistically significant in differentiating between exceptional (XP) and non-exceptional (NXP) performers. *Interpersonal Skills* (helping others); *Personal Attributes* (proactive role with management, exhibiting and articulating strong beliefs and convictions and maintaining 'big picture' view); *Task Accomplishment* (mastery of skills and techniques) were the top competencies displayed by XP performers. The final category - *Situational*

Skills (e.g. responding to schedule pressure by sacrificing parts of design) - was significant as a competency not exhibited by XPs. Lethbridge (1999)'s study highlighted the gap between formal learning and practitioners' perception of importance. Of the top ten gaps, soft skills figure at position 1 (negotiation – 84% gap between learning and importance in practice; position 3 (leadership – 73%); and position 7 (ethics and professionalism – 63%).

Australia, similar to many OECD countries, has a great deal of emphasis on the development of knowledge, skills and attitudes of its citizens, representing 'human capital' particularly in a knowledge-economy. This requires an understanding of the relevant skills and systematic development of these through effective education and training (OECD, 1996), as well as a commitment to lifelong learning. The comprehensive study of employability skills conducted on behalf of the Department of Education defined employability skills as "skills required not only to gain employment, but also to progress within an enterprise so as to achieve one's potential and contribute successfully to enterprise strategic directions" (ACCI, 2002, p. [3]). This definition highlights the importance of abilities sought by employers in addition to technical knowledge and skills.

Scott and Yates (2002) and Scott and Wilson (2002) undertook Australian studies tracking the experience of successful recent graduates from all 38 publicly funded Australian universities. Their work parallels an extensive government-initiated project analysing responses from recent graduates to the CEQ (Course Experience Questionnaire). Scott, Wilson and Yates operated from the assumption that graduates working in professional practice for between two and six years are well positioned to identify what is likely to be most relevant for those currently studying at university. Such people have sufficient experience to know what counts in the real world of the profession whilst not being too far away from their university courses to have forgotten what was covered.

In both the studies *Emotional Intelligence* ranked highest in importance, dominating the factors identified by graduates as important to their professional careers, closely followed by *Intellectual Capability*, addressing generic issues such as abstraction and contingency, while profession-specific knowledge ranks relatively low. The ability to work in teams, particularly cross-disciplinary teams that are common in the ICT workplace, is also considered vital.

Support for the results of global and Australian studies is provided by the surveys of ICT employers and ICT graduates in the workplace carried out in the ICT-DBI study (Koppi & Naghdy, 2009), commissioned by the Australian Learning and Teaching Council. The survey of ICT employers in this study reported that recent ICT graduate recruits met about half of the knowledge and skills required (but with the caveat that these were not at the bleeding edge of technological innovation employers and graduates wished for). However, the graduates only met some of employer needs concerned with understanding business processes, project management knowledge, and written communication ability. More than half of the employers indicated that graduates did not meet their needs for commercial awareness, and the interpersonal skills provided in the survey met less than half of the needs of employers. These findings indicate that the university ICT curriculum needs to be brought into line with industry requirements with respect to the development of employability-relevant soft skills.

So, the work of Litecky, Arnett, and Prabhakar (2004) may be a valid summary of the place of soft/professional/employability skills. They suggest technical skills are used a 'filtration' in the hiring process – technical screening by an employer is used to eliminate some candidates from the pool of potential candidates and to pass others on to the second stage. They argue that this second stage can reasonably be expected to consist of information on a different set of skills – this stage is based on the perceived soft skills of the candidate, obtained through face-to-face assessments. This second set of skill data is the more important for the final hiring of the candidate.

Thus, while the successful professional must possess a high level of profession-specific technical expertise, such skills have little value without the ability to handle uncertainty, deploy appropriate components of one's repertoire of generic and profession-specific expertise, accurately diagnose the unexpected and work productively with people from a wide variety of backgrounds.

Frameworks for Professional Competency

Scott and Yates (2002) and Scott and Wilson (2002) discuss the findings of their studies in relation to a *Professional Capability Framework* that identifies competency in five dimensions consisting of: *Emotional Intelligence* – personal and interpersonal; *Intellectual Capability*; *Profession-specific* skills and knowledge; and *Generic* skills and knowledge. An *Educational Quality* scale addresses issues of appropriateness, authenticity of tasks and assessment (Scott & Yates, 2002).

The *Professional Capability Framework* is further refined (Scott & Wilson, 2002): Emotional Intelligence (personal and interpersonal (now social)) becomes *Stance* and Intellectual Capability is now defined by two components: *Way of Thinking* (incorporating cognitive intelligence and creativity) and *Diagnostic Maps* (developed through reflection on experience). The characteristics of the revised framework are taken from Scott and Wilson (2002, pp. 7-9):

A. **Stance:** *Emotional Intelligence-social* includes an ability to empathise with the perspectives of others, to work constructively in a team, engage in reciprocal relationships, to be patient and to allow others 'room' to do things for themselves. *Emotional Intelligence- personal* includes a capacity and willingness to try new things, take informed risks, tolerate uncertainty, ambiguity and change, admit and learn from errors, defer judgement, pursue excellence and persevere, to work independently, withstand personal attacks, behave ethically, keep work in perspective, lead a balanced life and to 'pitch in' and do 'drudge work'

B. **Way-of-thinking:** Reflects an ability to think 'contingently' and creatively, to be able to trace out and assess the consequences of different options for action, to identify and accommodate conflicting interests and perspectives, to set priorities, identify the core issue in any situation and to think holistically, laterally and iteratively not rigidly, technically or in a linear manner. This way of thinking cannot operate well if the individual lacks the emotional intelligence to work with continuing complexity, uncertainty and ambiguity and in collaboration with a wide range of people

C. **Diagnostic Maps:** Refer to that which gives meaning to what an outsider or novice would find a complex and hard-to-fathom in a set of work-specific factors and relationships. These are generated through challenging practice, reflection, and transfer. These 'maps' are most effectively developed if the individual has been able to deploy a high level of social emotional intelligence to build networks

D/E. **Generic Skills and Technical Expertise**: Such knowledge actively assists with the process of problem interpretation and diagnosis. Generic skills such as the ability to effectively undertake team work, network, communicate and present ideas, manage project, learn one's organisation, mentor, carry out self-managed learning or to organise one's work and life, also have importance.

In summary, the framework supports their findings that, while technical and generic skills are necessary (D and E), they are not sufficient for competent professional practice: a high level of social and personal emotional intelligence (A); a contingent way of thinking (B); a set of 'diagnostic maps' (C) developed from handling previous practice problems in the unique work context (Scott & Wilson, 2002) are imperative. The framework has been applied or adapted to a number of disciplines, including higher education.

The Australian study of employability skills noted above (ACCI, 2002) proposed an *Employability Skills Framework*, originally published in 2006 (DEST, 2006) and in the process of being updated. This later version emphasises the "underpinning skills and knowledge that enable individuals to perform effectively in employment contexts and to apply their technical or discipline specific skills in different contexts" (DEEWR, 2012 p 6). The report highlights the fact that, while consensus exists on the types of skills sought, employers are more interested in the outcomes - that is, individuals demonstrating these skills, and the attitudes that are outcomes of proficiency in these skills. As one example provided: "adaptability is the result of an ability to take risks and respond to challenges and to adapt and apply prior learning" (DEEWR, 2012). The *Employability Skills Framework* will encompass both employability skills and aspects of the context which impact upon an individual's ability to develop and demonstrate these skills. Technical or discipline specific skills are detailed in educational curricula, while the core language, literacy and numeracy (LLN) skills of reading, writing, oral communication, numeracy and learning are addressed in the *Australian Core Skills Framework* (McLean, Perkins, Tout, Brewer, & Wyse, 2012).

Soft Skills in the ICT Curriculum

Bentley, Lowry, and Sandy (1999) suggest a developmental process in which personal attributes, which influence intellectual abilities and skills, are applied to the acquisition of knowledge in order to enable the development of higher cognitive activities. They note that, at the end of the educational process, students must be able to apply knowledge to new situations and problems. This ability to transfer requires certain generic intellectual abilities and skills, which, they suggest, although highly valued by employers of graduates, are sometimes given only 'lip service' in tertiary education curricula. The personal attributes identified as important in their model include attributes like curiosity, risk taking, personal discipline and persistence, which can influence in important ways the successful application of intellectual skills and abilities to knowledge and hence support the higher orders of thinking. Earlier Turley (1991) had suggested a significant area for research was to explore how competencies are reinforced. He concluded that education needed to support the development of differential skills (namely interpersonal skills and personal attributes) through the creation of learning situations which stress these.

The approach taken for examining education for ICT specialisations in general (and hence to some extent taking up the challenge in the discipline) has been the revision of the various model curricula (ACM/IEEE, 2008; GSwE, 2009; LeBlanc & Sobel, 2004; Topi et al., 2010) and the Bodies of Knowledge which underpin them (e.g. SWEBoK (Bourque & Dupuis, 2004; Fairly, 2013)) to identify guidelines that support soft skills learning. So, for example a curriculum guideline may state:

Guideline – Students should be trained in certain personal skills that transcend the subject matter: The skills below tend to be required for almost all activities that students will encounter in the workforce. These skills must be acquired primarily through practice:

- *Exercising critical judgment: Making a judgment among competing solutions is a key part of what it means to be an engineer. Curriculum design and delivery should therefore help students build the knowledge, analysis skills, and methods they need to make sound judgments. Of particular importance is a willingness to think critically. Students should also be taught to judge the reliability of various sources of information.*

- *Evaluating and challenging received wisdom: Students should be trained to not immediately accept everything they are taught or read. They should also gain an understanding of the limitations of current SE knowledge, and how SE knowledge seems to be developing.*

- *Recognizing their own limitations: Students should be taught that professionals consult other professionals and that there is great strength in teamwork.*

- *Communicating effectively: Students should learn to communicate well in all contexts: in writing, when giving presentations, when demonstrating (their own or others) software, and when conducting discussions with others. Students should also build listening, cooperation, and negotiation skills. (LeBlanc & Sobel, 2004, p. 40)*

How these and similar skills should be acquired and honed within the learning environment is less tangible (Voogt, Erstad, Dede, & Mishra, 2013).

In general, learning in higher education is complex (Claxton, 1998). In order to achieve complex learning, appropriate environments and opportunities should stimulate a learning which is a "constructive, cumulative, self-regulated, goal oriented, situated, collaborative and individually different process of knowledge building and meaning construction" (De Corte, 2000, p. 254). However, this raises an additional issue: Benson (2003) notes that within the emerging ICT discipline of the 1970s, academics were migrants to the discipline, with an overwhelming majority having qualifications in other areas, most often computer science. Practitioners also relied heavily on scientific, mathematic and engineering disciplines, with many migrating from engineering and manufacturing. Therefore the degree of alignment (or dissonance) between practice and education, in effect between the theory-in-practice (what practitioners do and what competencies they need to do what they do) and the espoused theory (what formal education says practitioners do, and how students are taught to do it) (Argyris & Schön, 1974) may be exacerbated by the learning/teaching models academics bring to the learning environment, usually heavily influenced by their own discipline learning.

Matching the gaps identified by practitioners, and learning models that purport to focus on these, shows that non-traditional approaches provide leverage for a graduate entering the profession

of Software Engineering. Such approaches are based on active learning models, addressing learning as more discursive and collaborative. Active learning approaches are seen to create situations which engage students in such higher order thinking tasks as analysis, synthesis and evaluation (Bloom, 1956). By contrast, non-active learning are directed to absorption and imprinting. Active learning methods also attempt to develop the cognitive (knowledge, understanding and thinking) and affective (emotive) dimensions of the learning process in such a way that learners' active involvement in the learning is improved (LTSN, 2002).

In developing a framework for advanced active learning Horvath, Wiersma, Duhovnik, and Stroud (2004) propose a topography based on the focus of the approaches and the nature of the applied methods (see Figure 1). These are described as a continuum from instructive through explorative

Figure 1. Topography of advanced approaches to active learning (adapted from Horvath et al. (2004))

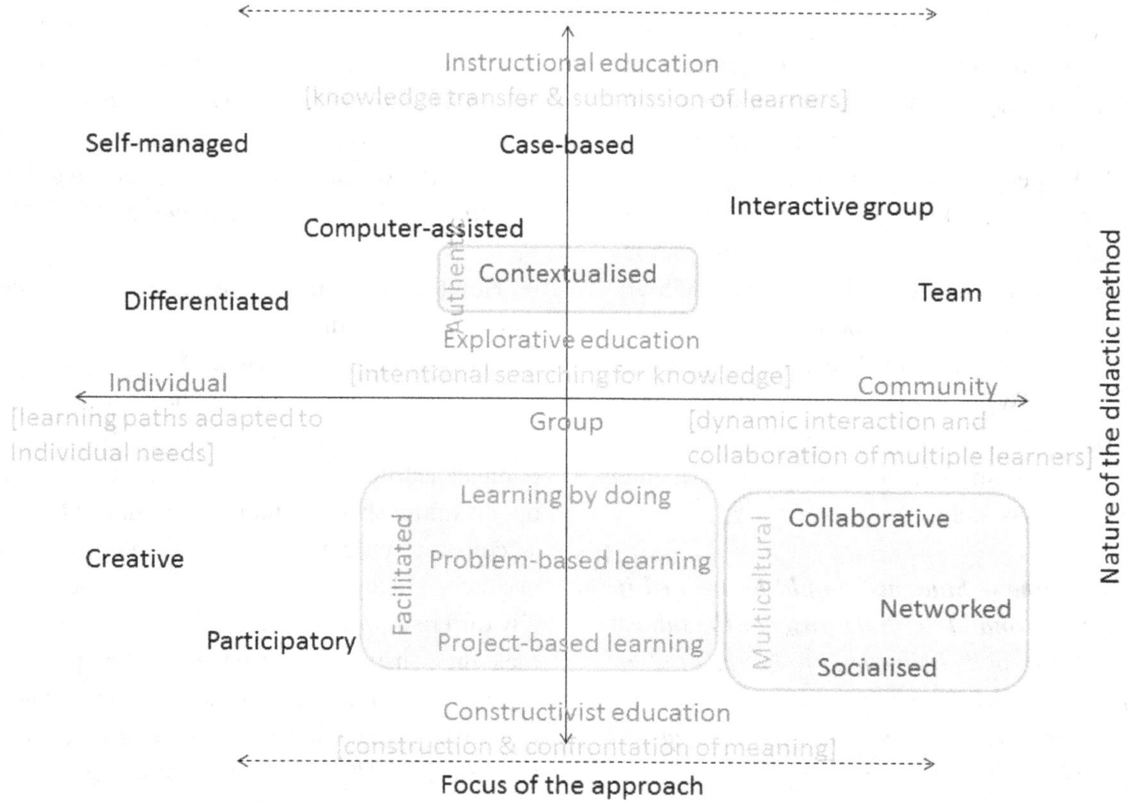

to constructive, with the latter assuming that learners construct knowledge for themselves by creative activities such as planning and design. The orthogonal axis describes the focus of the activities: from a priority given to sensory-motor and conceptual activities of individual students to the development of knowledge by a group or community of learners.

However, learning approaches that involve critical thinking and reflection are a challenge to students: in ICT they expect to be taught formulaic and recognised methods that will allow them to build successful systems. Banks (2003) even suggests an approach that requires reflective and active questioning that challenges previously held beliefs may be inappropriate for undergraduate students.

Development of Competency

McCracken (1997) suggests that formal education provides the entree into the discipline. A principled academic education enables the graduate to become, over a period of years after graduation, a member of the profession. McConnell and Tripp (1999) go so far as to suggest at least four years of apprenticeship necessary for Software Engineers.

Work on the transition from study to workplace (e.g. Lee (2004)) has shown that the onus of teaching themselves what they needed to know in order to perform the task successfully is one of the 'reality shock' involved in the socialisation of new graduates to work. He concludes "..educators should also help students to develop their initiatives and abilities to deal with ill-structured problems. This would require approaches which emphasize independent learning and collaborative teamwork" (Lee, 2004, p 135). Full integration into the organisation was seen to take up to two years, during which time they were not considered an 'insider' and 'working professional'. What the practitioner studies indicate is that this is too long (Benamati, Zafer, Ozdemir & Smith, 2010), and assumes too much 'on the job' training. The reality is that new graduates are expected to perform on

the same level as their experienced counterparts - the best that can be hoped for is a sympathetic, experienced mentor as coach. Higher education is required to reduce the amount of time taken to become a competent member of the profession – in effect being more than novices on graduation.

Within education competence or mastery are often linked to the concept of learning objectives or outcomes, and usually are seen as positions on a continuum which spans pre-novice to expert. Jonassen and Grabowski (1993) describe such a continuum as leading from ignorance to expertise. Dreyfus and Dreyfus (1986) and later Dreyfus (2001) developed a richer skills acquisition model that focusses on adult education. This model is based on five (then six) stages: *novices* and *advanced beginners* apply rules and maxims to solve problems; *competence* is process driven and problem-solving conscious; *proficiency* applies pattern recognition arising from extensive experience to identify the problem as intuitive reaction; intuitive situational response is automatic in the *expert*, based on abstract representations formed through reflection on experience; *mastery* provides a sense that there is no one right thing to do and that improving is always possible. Masters respond immediately to the whole meaningful context: reaching a new level of skilful coping beyond expertise and developing a 'style'. In relation to education, the Dreyfuses also suggest that the higher modes of functioning – intuitive expertise and mastery – require risk taking (and hence emotional engagement), direct experience and active involvement in the company of experts.

An additional concept is important here - competence is defined, contained and developed within communities of practice (Wenger, 2000). Since the discourse of a discipline is central to the way its knowledge is constructed and transmitted, the means by which learners acquire discursive knowledge is important. While expert members of a discipline share knowledge about discursive practices in their community, this is mostly tacit, and traditionally given little emphasis within the (formal) educational environment.

Therefore there is marked difference between the fragmentary knowledge structures of novice understanding and the integrated knowledge structures which underpin more robust knowledge and flexibility exhibited by the expert knower (Wood, 1999). The theories of learning highlight aspects of knowledge and skill development that are pertinent: advanced competence in a discipline is based on a rich framework of understanding and a commensurate reduction of effort in further learning achieved through:

- Problem-solving practice
- Correspondence with the learner's intuitive model of the phenomenon (and all that implies about previous experience, belief systems etc.)
- Ease of transfer (with its focus on strategic thinking and metacognitive skills)
- Facility with multiple representations – translation between different models facilitates the understanding of concepts, whilst the models themselves support differing insights, reasoning and problem-solving occurring within a social and cultural context (and all that implies about dialogue, self-explanation)
- Shared meaning where certain activities are seen as authentic and the discursive practices of a discipline are integrated.

Such characteristics of expertise are viewed as fundamental skills for Software Engineers.

This chapter argues that industry competency expectations of graduates assume cognitive skills related to higher order learning, the development of emotional intelligence, and problem-solving strategies, and therefore map to a higher level of competency on the novice-to-master continuum than is normally assumed for graduates. Practitioners look for graduates who are flexible, adaptable in the organisational environment and can continue learning.

ADDRESSING PRACTITIONER EXPECTATIONS

The purpose of the research reported in this chapter was to develop and implement learning strategies which address the issue of aligning the competency expectations of SE practitioners with formal education for SE, so that, by addressing the practitioner gaps, graduates gain leverage in becoming competent professional practitioners early in their career.

Constructivist learning, a theoretical framework based on the work of Papert (1980), provides the greatest opportunity to achieve this goal. Constructivism distinguishes itself from more traditional instruction, in part, by the degree of active learner engagement as well as the assumption that learners have the ability to create meaning, understanding, and knowledge - learners develop their own reasoned interpretations of their interactions with the world. Perhaps more important, constructivist learning environments allow learners to share and collaboratively reflect on these cognitive artefacts.

Friedman (2001) indicates that experiences of the pioneers show that problem-based learning, project-based learning and learning by doing are the methods that offer the largest potential for constructive learning, in particular, in groups and communities. The first two approaches can be differentiated by saying that problem-based learning concentrates on better understanding and the solving of recognised problems, while project-based learning focuses on the end product. Hence, project-based learning is more or less an artefact production process, while problem-based learning is a knowledge development process. They can appear in practice side-by-side or interwoven. Learning by doing places the learners in direct contact with the subject matter and facilitates finding information through social communication.

Other research on student learning of the discourse of a discipline draws on the metaphor of apprenticeship, described as some gradually mentored pathway to membership. Learners can be enculturated in the discipline through both being socialised to the 'forms of talk' (Berkenkotter & Huckin, 1995) in the community, and gaining conceptual knowledge. Students are seen to absorb the disciplinary practices rather than learn them through explicit teaching. Lave and Wenger (1991) considered the importance for newcomers of grasping knowledge and skills, achieved by means of 'legitimate peripheral participation'. In their view, the notion of participation deals with the process of situated learning, an integral and inseparable aspect of social practice which encourages newcomers to become part of a community of practice. Earlier, Lave, Smith, and Butler (1988) had suggested that learning is a process that involves becoming a different person with respect to possibilities for interacting with other people and the environment. The individual is no longer the same individual with new skills, but is a new person who has become more enculturated into the practice, negotiating meanings based on experiences as a student.

Students bring a complex assortment of beliefs, past experiences and expectations to a learning situation, which influence the approach to learning they take. In turn, this affects the quality of their learning outcomes, and reflects on their future learning intentions and behaviours (Prosser & Trigwell, 1999). While the achievement of high quality learning is important in all graduates, it has an extra relevance where the context of practice is continually changing and professions are continuously developing. There is a need, therefore, to identify aspects of the learners' conceptions of learning (Meyer & Shanahan, 2000) and approaches to it so that appropriate support can be provided. The approach to learning that a student takes is also sensitive to the context in which learning is done, with a demonstrated correlation between more advanced conceptions (e.g. abstraction of meaning and understanding of reality) and a deep approach to learning (see Wilson and Fowler (2005) for a discussion of the relevant studies). This conception to a large extent

determines the student's expectation of what the learning process and teaching entail. Thus, for some students, the ability to take a deep approach (and hence the quality of their learning) appears to be limited by the conception they hold of learning.

Context for the Project

Over a number of years, an action research project was undertaken at this university with the objective of graduating software engineers who could engage with practitioner requirements for soft skills. A variety of interventions, based on non-traditional learning environments, were designed and implemented to address these non-technical skills, and both qualitative and quantitative data collected and analysed. A practitioner perspective on how the 'work' of the discipline is undertaken also coloured the conceptual framework developed through the interventions.

The curriculum context for this project was an undergraduate 4-year degree in Software Engineering (BE (SE), accredited by both Engineers Australia (EA) and the Australian Computer Society. After a common first year students transitioned into the engineering discipline of their choice. The curriculum framework comprised elements of Computer Science (to cover fundamental aspects to form the basis of technical knowledge and skills in software and hardware), Software Engineering (a focus on SE theory and practice that forms the basis of core knowledge and skill in software development and evolution), and Engineering (knowledge and skills in engineering practice and principles including those elements of EA curriculum requirements not covered in the previous components. These are common to all Engineering students within the School).

It should be noted at this point that all content material was available online, so lectures and tutorials could be replaced by workshop-based discussion and exercises. Briefly, the *Software Factory* presented a coherent system and learning context. Topics were categorised mnemonically as sections. This allows for 'chunking big' and focusses on connections between topics in the same category for content- and context- dependent knowledge construction (Jonassen, 1992). Each course has a comprehensive study programme comprising a set of topics across relevant sections. Once authenticated as a member of the class, a student was able to access the unit material both through the *SENavigator* (see Figure 2) or the Production Line (see Figure 3). Within each course topics are sequenced and displayed on a map that provides alternative routes from commencement to completion. To a certain extent these provide choice in the order of topics studied and allow students to vary the sequencing of content, although support in the form of workshop schedules dictate the dates by which topics must be completed. Nevertheless, students had the flexibility to access any topic to assist them in solving problems encountered throughout the course.

While the learning environment has as its focus activities/real-world problem solving, on-line/interactive activities cease to be meaningful if the student hits a snag and is unable to progress from there. The purpose of scaffolding is to provide activity-sensitive help mechanisms. Examples include exercises; explanations and background information; monitoring tools (to help students keep track of their progress); modelling scratchpads (software tools to create and manipulate models); problem hints; process coordinators(which guide the students through the complete task cycle); and planning tools (Jong, 2006).

The *Software Factory* provided examples of all these scaffolds, as appropriate, both as purpose-built activity help and underlying manuals. The former was accessible through a 'help' icon on an activity screen, while the latter is best demonstrated through the underlying help in the FM (Formal Methods) topics, where help is activated through 'hot' spots in the notation itself. Both of these mechanisms are not imposed on the student, but are readily available. Links to the help are

Figure 2. SE navigator for RE course

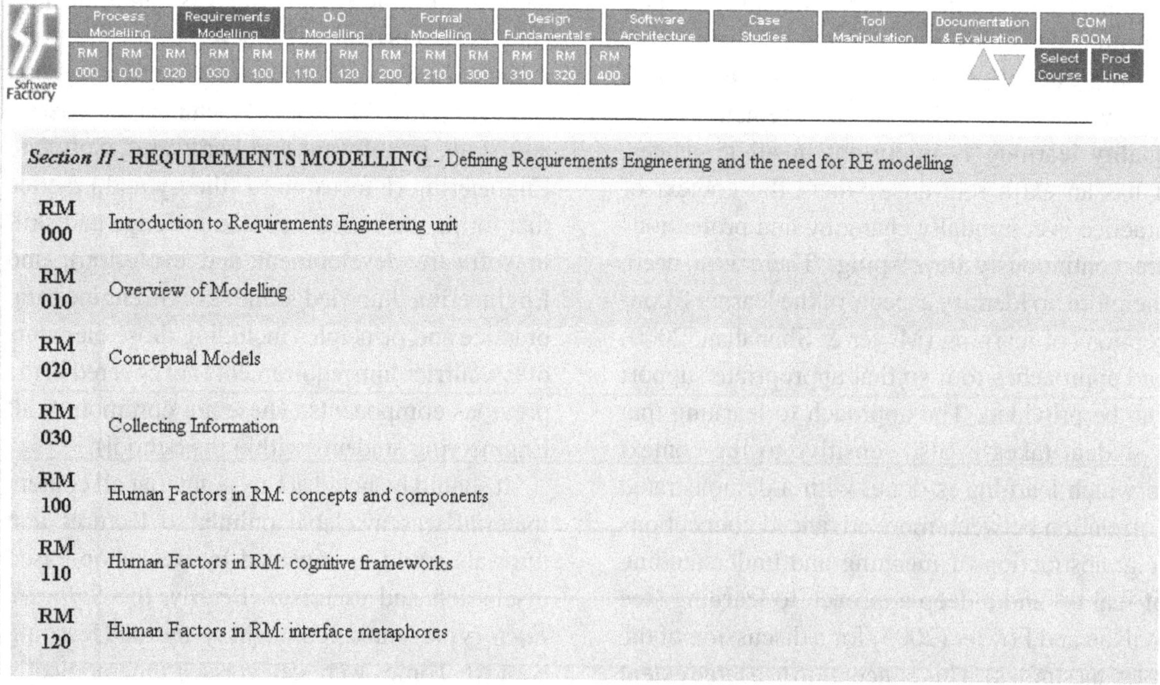

| Process Modelling | Requirements Modelling | O-O Modelling | Formal Modelling | Design Fundamentals | Software Architecture | Case Studies | Tool Manipulation | Documentation & Evaluation | COM ROOM |

| RM 000 | RM 010 | RM 020 | RM 030 | RM 100 | RM 110 | RM 120 | RM 200 | RM 210 | RM 300 | RM 310 | RM 320 | RM 400 |

Select Course | Prod Line

Section II - REQUIREMENTS MODELLING - Defining Requirements Engineering and the need for RE modelling

RM 000	Introduction to Requirements Engineering unit
RM 010	Overview of Modelling
RM 020	Conceptual Models
RM 030	Collecting Information
RM 100	Human Factors in RM: concepts and components
RM 110	Human Factors in RM: cognitive frameworks
RM 120	Human Factors in RM: interface metaphores

Figure 3. Production line for RE course

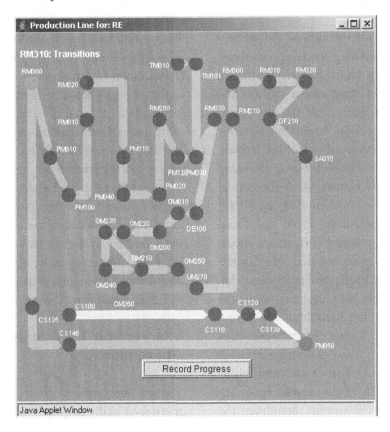

seamless, which enables the student to maintain focus on the learning activity, rather than on the task of retrieving aid.

Interventions 1 and 2 were situated within the first of the eight core SE courses (four courses is a full semester load). As noted above, all engineering students had undertaken a common first year: they had generally been immersed in a scientific/engineering paradigm where problem-solving through laboratory procedure, repeatability of experimentation and rigour in mathematics were key learning objectives. The result of this was strong preconceptions about both teaching and learning. Requirements Engineering (RE), offered in semester 1 of the second year of study addressed RE in an object-oriented environment. Students were exposed to O-O principles, tools

(Rational Rose at that time) and techniques based on UML in order to construct a Requirements Specification for a given problem. The course was taught in workshop mode (two sessions of 2-hours twice per week).

By the time Intervention 3 was to be operationalised, the curriculum had been modified from the eight SE courses to three Design Studios in each specialisation. Two Studios comprised a full semester load, with 20 hours 'student workload' allocated to each. Of this time, up to ten hours could be class contact. The specialisations would only kick in in year 3 (giving two common engineering years instead of the previous one). The original RE course became the foundational content of the first SE Studio with additional material absorbed from another original SE course.

INTERVENTION OVERVIEW

The three cycles of intervention of this project explored alternate learning models to evaluate their appropriateness for addressing a shift in focus from technical competency to the soft and metacognitive skills that enable the competent practice of SE. These were based on *Cognitive Apprenticeship*, as developed by Collins, Brown, and Newman (1989); *Problem-based Learning* (PBL), which draws from several theoretical traditions including pragmatism (Dewey, 1916), cognitive psychology (Piaget, 1968) and social constructivism (Vygotsky, 1978), and a modified *StudioLearning* approach that incorporates elements of PLB into a reflective practice model. Kuhn (2001) describes the characteristics of the Studio model: open-ended problems are solved, using a variety of media, through rapid iteration of design solutions, with frequent critique and contingent decision-making. To re-iterate from definitions quoted previously in this chapter, graduates from these learning environments should: see 'the bigger picture' in which they are working; be more flexible, adaptable, creative, innovative; cope with uncertainty; cooperate in ambiguous environments; deal with non-routine processes; plan and think strategically and be able to transfer; handle decisions and responsibilities; acquire good self- and time-management skills and self-confidence as professional software engineers (Andrews &

Higson, 2008; Curtin, 2004). The overarching aim was to expose students to a learning environment that would address components of the *Professional Capability Framework* as appropriate to achieve these outcomes. A mapping provides an indication of the synergies between them. However, it should be noted that dependencies that exist between soft skills (see the description of the components of the *Professional Capability Framework* for an example of this) meant that students were exposed to soft skills other than those explicitly noted in Table 1. As one example 'cooperate in ambiguous environments' assumes teamwork, communications, and management (self and task) skills. Each intervention strategy addressed specific concerns and, through evaluation of and reflection on each, strategies are refined for subsequent iterations.

Intervention 1: Cognitive Apprenticeship

The first intervention focussed on authenticity and transfer of skills acquired to other courses and, eventually, to the profession. The authenticity referred to providing students with an environment where they were treated (and acknowledged) as novice software engineers rather than generic engineering students. The aim of this strategy was to engender a strong sense of community. Transfer (the skilled application of understanding) has been acknowledged as a significant component of

Table 1. Synergies between framework, skills and interventions

Framework Dimension	Soft Competency	Intervention
Stance	cope with uncertainty cooperate in ambiguous environments handle decisions and responsibilities	*CreativePBL* *StudioLearning*
Way-of-thinking	see 'the bigger picture' in which they are working be more flexible, adaptable, creative, innovative plan and think strategically	*Cognitive Apprenticeship* *CreativePBL* *StudioLearning*
Diagnostic Maps	deal with non-routine processes	*Studio Learning*
Generic Skills and Technical Expertise	acquire good self- and time-management skills acquire self-confidence as professionals	*Cognitive Apprenticeship* *CreativePBL* *StudioLearning*

problem solving success – which in turn requires self-regulation metacognitive skills (directing, monitoring and evaluating learning) and engagement with multiple perspectives on the problem, its abstraction and its solution. Transfer is seen to be enhanced within a socio-cultural context (Brown, Collins, & Duguid, 1989) (in this case the community mentioned here) where the skill is de-coupled from the task (abstracted) and the problem-solving interpreted through mental modelling (Gott, Hall, Pokorny, Dibble, & Glaser, 1993; Patry, 1998). Peer-to-peer knowledge transfer is also required (Budgen, 2003), achievable through collaborative work.

In *Cognitive Apprenticeship* settings, learning is considered a process of active knowledge construction that is dependent on the activity, discourse, and social negotiations that are embedded within a particular community of practice (Collins et al., 1989). The curriculum for RE was addressed as a two-cycle spiral: the workshops during first part of the semester (8-9 weeks) focused on learning and practicing use of the tools, gaining an understanding of the conceptual framework (in this case object-orientation principles), and an appreciation for the context in which professionals practice (e.g., historical overview; issues in RE theory and practice; organisational involvement; group dynamics). The second part of the semester focused on issues of team work and knowledge transfer - students were involved in a group project

that required them to apply the tools to model a complex problem: the development of a 'complete' Requirements Specification. In broad terms, the phases (see Table 2) of the *Cognitive Apprenticeship* model were traversed throughout the semester, though without a clean break—the focus of the class sessions changed, but the ability to revisit any phase as required existed. The value of this model is its alignment with practitioner perception of learning in the discipline: they indicated an apprenticeship model of on-the-job learning for graduates was appropriate.

As applied in the SE course, the teacher initially modelled effective practices (e.g., given an informal system description, how do we create appropriate UML diagrams; which are appropriate; how are different elements of the notation applied, and when and why; etc.), demonstrating tool use, techniques and applying notation to address the problems presented. The students then undertook tasks they would encounter as practicing professionals, requiring proficiency with notations and tools, but also an appreciation of the context in which these must be applied (e.g., how/why different types of systems need different types of modelling – not all systems are necessarily best modelled as o-o, etc). This required an understanding of the underlying conceptual frameworks and paradigms used in the domain. Because these skills were all new to them, the students were closely coached by the teacher, both

Table 2. Phases of cognitive apprenticeship model as implemented

Phase	Component	Class Session#	Activities & Teacher Role
I	Modelling	1-6	Demonstration of the task as a process. Example approaches and sample solutions are provided as a basis for comparison and critique. Teacher explains strategies applied and use of modelling tools (e.g. notation) explicitly.
II	Coaching	7-16	Critique and whole class discussion of individual approaches applied. Focus is on exploration of multiple perspectives and the reasoning process.
III	Scaffolding	17-20	Teacher's role is to question, prompt and encourage students to stay on task.
IV	Fading	21-26	Student collaboration and peer discussion lead to a negotiated solution for submission.
#The course comprised of 2 * 2 hour workshop sessions for 13 weeks			

individually and in their groups, to apply a process for modelling each task as they reason about the issues being raised. At this point the learning becomes student-driven (though not yet student-centred). Students practice the techniques previously modelled by the teacher, with the step-by-step process acting as scaffolding. Coaching took the form of side-by-side work, protocol analysis (what is the goal and what steps are needed to get there) and reflective discussion of how the problem could be tackled and why specific models were a 'better' representation of the problem than others. Whenever the students reached an impasse, and were unable to continue or complete the task independently or with assistance from group peers, the teacher once again modelled the appropriate approach in a protocol analysis environment, for all students.

Gradually, students were able to complete tasks more independently, with the final class assessment item requiring the development of a complete model of a problem, including critique and justification of the approach taken, with minimal support from the teacher. Other assessment items (concept maps, portfolio) addressed the level of conceptual understanding, and student's willingness to explore outside the boundaries provided within the course.

The learning environment was one that some students found difficult to assimilate. Across a semester, ambivalence was quite noticeable: while some students appreciated it as co-operative and interactive (example student feedback included: *class discussions are very useful; casual workshop environment*), others felt it shifted the burden too heavily to their shoulders (*far too much content to read; too much workload leading to no marks*). Although collaboration strategies had been made explicit (with team roles and duties identified), in practice students were also unhappy about the amount of collaboration required: the need to negotiate with team members, explore alternatives and resolve conflict, co-ordinate tasks and produce a deliverable that had a unified look-

and-feel, taxed some students (*this leaves little time for other subjects; not clearly stated what we are required to do in terms of assessed work*) while others could relate the experience to the workplace (*can see industry advantages; shows how to analyse stuff*). In almost every case of a request for an extension beyond the due date, the comment was on the lines of *the work is done, just not put together*.

In effect, students saw software development as fundamentally scientific (where following a defined process will always lead to a quality product (Pfleeger, 1999)) and, although they accepted the collaborative nature of the process, were not comfortable with both the breadth of acceptable solutions to a problem and the amount of group work expected. This perspective is not unexpected – as Baxter-Magnola (2001) suggests, students at low intellectual (or epistemological) stages of development either believe that every intellectual and moral question has one correct answer and their (competent) teachers know what it is or are transitioning to believing that some knowledge is certain, and that making judgements following logical procedures prescribed by authority deserves full credit. Challenges to their belief systems within the courses they take and interactions with peers are necessary for them to gradually come to believe in the validity of multiple viewpoints. However, discussions during the class suggested these perceptions were very little changed.

Evaluation of the *Cognitive Apprenticeship* model in terms of the *Professional Capability Framework* indicated low levels of both component A and D. In relation to *Emotional Intelligence - social*, although students were required to work collaboratively, there was little demonstration of "ability to empathise with the perspectives of others, engage in reciprocal relationships, to be patient and to allow others 'room' to do things for themselves" (Scott & Wilson, 2002, p. 7). In terms of *Emotional Intelligence - personal*, students' capacity and willingness to try new things (in this case a different approach to learning), tolerate uncertainty, ambiguity and change were also not well demonstrated. Students exhibited some of the traits of surface learning - they focussed on learning the tools and techniques of SE (component E) at the expense of a more expansive view of the discipline: they did not see themselves as acquiring the more generic skills (component D) valued by practitioners, with the majority of students focussed on being able to apply the tools and techniques in order to achieve a pass. The low level of *Stance* affected other components as well, specifically components B (*Way of Thinking*) and D (development of *Diagnostic Maps* to deal with complexity at a non-novice level).

As a positive note, when the same student cohort undertook the follow-on course they showed an improvement in student success rate (in raw marks) on a similar task (creating a Requirement Specification), achieved closer to schedule. This suggests a measure of transfer (component C) had occurred – students were able to complete the task given in an appropriate time frame and at an appropriate level of competency. Also on the positive side, although attendance at workshops was not compulsory, participation was very high (generally 90 - 100%): active participation on the part of most students is seen as an indicator of engagement (*helpful class sessions; lab teaching is good; fun topic, encouraging; good workshops*).

In summary, the evaluation suggested the *Cognitive Apprenticeship* model could be applied reasonably successfully from an academic perspective. However, the majority of students were not comfortable with a 'master' who, towards the end of semester 'faded': they acknowledged that this placed the onus on them to do the learning (which they did not necessarily accept). A significant finding of this intervention also related to student emphasis on 'correct' answers to problem solving undertaken – the master should always be right. The results of this cycle therefore indicated the intervention was only achieving some part of the goal: while students appeared more confident, in subsequent courses, in applying the

knowledge they had gained, they still expected to be explicitly taught: that is, the master/apprentice relationship was assumed whether it remained appropriate or not.

In effect, elements of an 'incorrect' learning environment were identified in the evaluation of that intervention. Patel, Kinshuk, and Russell (2000) argue that learners in an ill-fitting setting focus on skills that will yield higher grades as an immediate objective. With the relevance of domain knowledge not fully understood, cognitive skills related to exam techniques acquire importance though they do not model real life situations. The learning, in many cases, is reduced to assignment hopping with 'just-in-time' and 'just-enough' learning to fulfil the assessment tasks. These are also characteristics of surface learning. The Apprenticeship cycle exhibited some of these traits – students focussed on learning the tools and techniques of SE at the expense of a broader (and more abstract) understanding within the discipline.

Apprenticeship models may be seen as limited in an academic context for other reasons which have heightened significance for the ICT discipline:

- They have difficulty accounting for the multiplicity of the work environment. Increasingly flexible response to the differing environments is required. Traditional apprentice learning is encapsulated in the social practices of a discrete community whose relations with other social practices are much more regularised and stabilised. Flexibility, reflectivity and critique are necessary in ICT, not simply the uncritical socialisation in a single social practice
- They have great difficulty accounting for the dialectical contradictions when apprentices have greater expertise in some areas that masters. The complex division of labour and cognition is not catered for by the apprenticeship model (Russell, 1998).

As the majority of student feedback was collected anonymously, it was not ethically acceptable to relate specific comments to particular student learning styles. An interesting insight may have been provided by the comparison: were particular learner types more/less comfortable with the Apprenticeship model (as the literature suggested), or were the comments across type? While this cohort's profile indicated the learning model would not inhibit their learning, raw course results suggested most students were not interested in exploring beyond the set requirements of the course – student comments regarding not being pressured to complete weekly tasks support this perception. This implied that, while the model enabled effective learning across the learning styles, the motivational element was missing.

Intervention 2: Adapting PBL

The conclusion reached after the first intervention was that the master/apprentice relation could be down-played so that students took early control of their own learning rather than relying on the teacher/master to direct them in what they needed to learn, and that a more open approach to describing the design of the course might be beneficial to students challenged by its non-traditional nature. The second intervention, therefore, focussed on student-centred learning; creativity and adaptability within an explicitly PBL environment.

Student-centred is a term used to refer to learning environments that pay careful attention to the knowledge, skills, attitudes, and beliefs that learner brings to the educational setting (Bransford, Brown, & Cocking, 1999). This implies a need for students to assume a high level of responsibility in the learning situation and be actively choosing their goals and managing their learning, and involves considerable delegation of power by the teacher. This addressed the reliance on the 'master' found after Intervention 1.

While there are many views about the nature of creativity, there is some agreement that the creative process involves an application of past experiences or ideas in novel ways (Hennessey & Amabile, 2010). Common characteristics of creative individuals include 'holistic thinking' (in the sense they look for an overall broad scope before moving into specific detail); proposing several candidate solutions early on (divergent thinking) in order to better examine the problem and help generate new concepts; a wide toolkit of domain relevant skills –and the ability to imagine/play out situations; creativity-relevant processes – including breaking perceptual and cognitive set and breaking out of performance 'scripts', suspending judgement; knowledge of heuristics (e.g. use of analogies, when all else fails try something counter-intuitive); adopting a creativity inducing work style (e.g. tolerance for ambiguity, high degree of autonomy, independence of judgement); and intrinsic task motivation (Amabile, 1996; Guindon, 1989, 1990). Csikszentmihalyi (1996)'s analysis emphasised the social nature of creativity. Focussing on creativity addressed the reliance on the 'correct' answer found in evaluating Intervention 1.

McCracken (1997) suggests that a principled academic education teaches how to place new knowledge in context and then use it in multiple contexts - providing the practitioner with adaptability. This addressed the "uncritical socialisation in a single social practice" Russell (1998) noted as an issue with apprenticeship models and also reinforced the importance of transfer.

The overarching aim in this intervention, therefore, was to focus on creativity and divergent thinking (as an indicator of adaptability), so that, instead of students aimed at finding the single, best, 'correct' answer to a standard problem in the shortest time (convergent thinking) they aimed at redefining or discovering problems and solving them by means of branching out, making unexpected associations, applying the known in unusual ways, or seeing unexpected implications (divergent thinking). Investigation of learning approaches described in the topography developed by

Horvath et al. (2004) suggested a problem-driven approach as a feasible course environment for achieving the aim identified for this intervention.

An extensive literature exists on the applicability of *Problem-based Learning* where an expanding knowledge base makes it impossible to include all the knowledge required for the beginning practitioner in the undergraduate curriculum. PBL emphasises 'learning to learn' by placing great responsibility for learning on the learner. It involves teaching both a method of approaching and an attitude towards problem-solving, and is characterised by its flexibility and diversity, since it can be implemented in a variety of ways in different subjects and disciplines (Savin-Baden, 2000). PBL is also designed to integrate the subject knowledge students require in order to solve a particular problem and therefore study issues at a deep rather than surface level. In this environment, the teacher becomes a facilitator for learners who are actively involved in the learning process.

Other characteristics of a PBL environment include (Boud, 1985; de Graaff & Kolmos, 2003; Jonassen, 2002):

- **Learning is collaborative with a focus on communication and interpersonal skills:** As PBL tends to take place in small groups, students have to work cooperatively to achieve their collective learning outcomes, with their level of independence measured by their ability to work with others. Consequently, communication skills, collaborative skills and reflective/self-evaluation skills can also be developed (thus addressing components A and D of the *Professional Capability Framework*)
- **Students are self-directed:** They are required to take responsibility for their own learning, with a leaning towards self- and peer-assessment (component D)
- **Its problem solving requires the mental representation of problematic situations:** The problem space must be con-

structed, either individually or socially through negotiation (component C)

- **Active, systematic manipulation of the problem space is required for PBL problem solving:** The 'problematic' nature of the situation enables learners to accommodate when their current experience cannot be assimilated into the existing mental models they possess (component B).

Its supporters claim PBL results in increased motivation for learning, better integration of knowledge across disciplines and greater commitment to continued professional learning (Boud, 1985). Without losing the benefits of a situated context, PBL is seen to offer the flexibility to cater for a variety of learning styles, integrating the learning of content and skills in a collaborative environment that reflect how learners might use them in real life (Oliver & McLoughlin, 1999). It also appeared to address aspects of expertise (that learning beyond the initial stages may best be achieved through situational case studies with rich contextual information (Dreyfus & Dreyfus, 1986)) and wider engineering education issues relating to generic skills and life-long learning.

The PBL environment applied as Intervention 2 was augmented by creativity-enhancing strategies drawn from Edmonds and Candy (2002), to enable innovative approaches to problem-solving in students, and to redress student perception of the discipline. An additional focus addressed metacognitive strategies and reflection as an aid to transfer of the skills and knowledge learnt.

The value of metacognition is confirmed in the recurring findings from Scott's work on applying the *Professional Capability Framework* (previously discussed). A focus on flexibility and productive thinking is also necessary, so that students learn to use past experience on a general level, while still being able to deal with each new problem situation in its own terms. Gott et al. (1993) posit that this adaptive/generative capability suggests the performer not only knows the procedural steps for problem solving but understands when to deploy them and why they work, in effect is 'expert' in the use of them.

The problem environment was set up to provide students with an opportunity to deal with a complex professional problem by 'living' it, with the Web providing a richly developed context for roleplaying in a blended mode (where online interaction and resource provision are supported by face-to-face activities). As McLaughlan and Kirkpatrick (2004) note, Internet-mediated roleplay-simulations can be designed to maintain effectively the interaction necessary for individuals and groups to work in a way that is truly collaborative rather than simply supporting distributed individual effort. The success of this approach has been reported in Engineering (McLaughlan & Kirkpatrick, 1999), while an analysis of student performance in such an environment suggests it supports student learning about alternative perspectives on problems and encourages transfer of learning to new contexts (McLaughlan & Kirkpatrick, 2004).

The development of the problem and its context was a considerable challenge: it had to be suitably complex and open-ended in order to instantiate the constructivist principles; it had to exhibit ill-structure so that students would engage with complexity, and present claims and rationale to negotiate understanding with their peers; it had to be too 'large' for individual students to complete successfully on their own; it had to address the learning outcomes identified, and provide a mechanism for students to demonstrate their competency in achieving these.

The scenario developed for the *CreativePBL* environment focuses on the secondment of the class to a (virtual) organisation – collaboration between a software house and the university. MurSoft required a team to work, on short-term placement, on a project to develop gaming software to be used as an educational resource within a tertiary institute. This provided an authentic context for learning: students were to have an opportunity, within their final year of study, to undertake an internship with a software-based organisation.

As well as providing a dedicated 'office' space – a laboratory specifically designated for use by SE students in third and fourth year exclusively, a collaborative on line work space was enabled for each group. Only group members and the teacher had access to this area – with students required to maintain all documentation on the site. This resource facilitated construction of the shared understanding necessary for collaborative learning, and enabled students to tune the accuracy and suitability of individual understanding through disentangling cognitive conflicts. It also introduced technical issues of version control and change management, also authentic in the domain. In addition, the representation tools provided (mindmapping, models through the CASE tool) acted as mediators for collaborative learning.

Students are expected to undertake the learning tasks collaboratively by means of the learning environment. All interaction with the 'client' was undertaken through Web-based material: a fictitious character, the Team Manager acted as go-between for the team and the client, while memos, minutes of meetings, telephone messages, 'talking heads', press releases etc. provided the triggers required by the PBL process. These acted as prompts to students to undertake some task identified in the problem. Support material was provided, as were expert consultants (either from industry or the School) as required. The lecturer was always available as a facilitator.

Course content was centred on the online teaching material (now viewed as a resource repository in the problem environment) within the *Software Factory* previously described, and a recommended text. These acted as a constraint: students initially explored this material in order to achieve the learning outcomes they have identified in a problem component, rather than having unlimited access to resources on the Internet and elsewhere. On the other hand, it was important that students became aware that other views existed. Again, providing environment constraints added to the authentic approach: as graduates, students would be expected to follow the operating procedures standardised within the employing organisation, rather than having the freedom to choose those which suited them individually. The Web material was expanded to include resources on the PBL methodology, including support documents for students on the PBL process, self-assessment items and trigger checklists (as triggers (the memos etc. previously mentioned) were released automatically).

The decision was taken to minimise technical complexity (e.g. in terms of advanced interactive media and 'appearance') so that students could easily access the resource off-site and with minimal software requirements (a Java runtime environment and Flash reader). Pragmatically, technological capacity could never keep up with student expectations (in general males in their late teens, very au fait with gaming environments). Rather, 'cognitive realism' (Herrington, Reeves, & Oliver, 2003) to the real-life task (as opposed to a technological-driven view) was seen as having (much) greater significance.

Students were expected to construct a problem representation based on the triggers provided, and manipulate the problem space so that an external representation (in this case the artefacts that were deliverables) could be created from the multiple individual representations at a fine enough grain to demonstrate group and individual achievement of the appropriate learning outcomes. The course still focussed on RE in an object-oriented context.

As with the *Cognitive Apprenticeship* environment, evaluating the success of the *CreativePBL* intervention was based on strategies in several dimensions: the success of the implementation of the PBL model and the effect of the intervention on the development of the student cohort, the latter in terms of both performance in assessable tasks and their perceptions of learning in this environment. Throughout the semester each problem required use of the products of the preceding problem. In

this way, students were able to activate (recent) prior knowledge in order to commence the PBL process for the next problem. Since rework was required to redress issues identified by their 'manager', any re-conceptualisation undertaken by the students (after feedback or when tackling a subsequent problem) was reflected in their increased understanding of the context (addressing components B and C of the *Professional Capability Framework*).

In order to facilitate learning in this environment three 'processes' were made explicit to the students:

- **The PBL process:** That advocated by Koschmann, Myers, Barrows, and Feltovich (1994) was used to anchor the student within the learning environment, and provides the discipline to assist in content learning. The first two workshop sessions were structured to introduce PBL as an explicit process with defined phases, provide an introduction to the role-playing environment and complete a small problem with the teacher, by 'visiting' each PBL phase. In effect what was offered was a mini-apprenticeship, with the process modelled and discussed. Support documentation for PBL was also made available through the online environment
- **Approaches to Learning:** Students were exposed to a variety of personality and learning style instruments (e.g. Keirsey and Bates (1984), Kolb (1995), Soloman and Felder (1999), Entwistle and Ramsden (1983)). The purpose was to provide strategies they could apply to help overcome any dissonance between their preferred method of learning and the learning environment, and to assist in dealing with the team dynamics encountered
- **Expertise:** Students were introduced to the 'process' for becoming expert in a dis-

cipline. This involved exposure to learning strategies that model expert learning (see Donald (2002) for a discussion on how people think and students learn in specific disciplines), including the ability to: manage knowledge structures (to plan) and exploit opportunistic and creative cognitive behaviour; collect, manipulate and analyse many different forms of data and then present them in meaningful ways; manipulate multiple representations of problem components and to effectively utilise the notations and symbol systems that are shared knowledge. In effect, exposure to cognitive flexibility (Spiro, Feltovich, Jacobson, & Coulson, 1991).

The results of this intervention indicated the PBL environment was also achieving some of the goals hoped for: students perceived themselves as more confident of their skills, and more willing to be innovative. However, they also raised issues about the relevance of individual components of the course (in particular non-problem-related items such as those included in the portfolio, which required reflection on individual and group achievements during the session), suggesting generic skills and reflection were still not considered important. Nevertheless, over 65% of the students expended effort to pass this component (up from 50% the previous year), with 41% of students (up from 12%) submitting well beyond the requirements of the course content. Despite some negative comments about the learning model (*don't really like how it's structured; does not have proper course structure; no lecture or tutorial*) other students acknowledged its value (*makes you think; helps with thinking about all areas of a problem (good for other units)*), and considered the course *well structured; interesting; practical; well presented; good for software industry; it's really good*.

Students perceived they learnt more in the areas of research, communications (*confidence to speak up; need to be heard & get ideas across*)

and team skills. They added: *concepts easier to grasp; forced to learn more for project relevant components* and, finally, they had to grapple with *various perspectives from others*. These student comments would indicate many components of the *Professional Capability Framework* had been successfully addressed.

Of major concern, when evaluating this intervention, was the realisation that the students were very product oriented – they saw the artefact (generally the code they developed) as the primary goal of the activities they undertook. Being made to focus on process to (in their perception) the detriment of the product was very frustrating, and had some negative effect. The PLB focus on process was ultimately in conflict with the aim of the intervention – to model professionals in practice. Schön (1983) notes that in the ordinary form of practical knowledge practitioners do not think about what they are doing, except when puzzled or surprised. He named this reflecting-in-action, and argued that it is central to the ability to act effectively in unique, ambiguous, or divergent situations.

Intervention 3: Studio Learning

The third intervention was developed to gain leverage from the positive elements of the both the apprenticeship model and the *CreativePBL* environment, while addressing the negatives. The PBL environment did not allow for a 'master' role, and therefore 'lost' the expertise the teacher provided: there the teacher acted as facilitator instead. The apprenticeship model that featured 'fading' as students gained competency, in contrast, allowed the students to depend on the 'master' extensively. Therefore, in the same way that the master/apprentice model addressed some aspects of discipline practitioner action and inhibited others, the facilitation aspect of the PBL model also exhibited elements of an 'incorrect' learning environment. A mentor/protégé relationship allows teacher and learner to seek to understand each other's position with the aim of agreement and/or defensible deviations. However, this requires a confidence on the part of the learner that is not often present at novice stage, and therefore needs to be fostered. The work of Laurillard (1993) develops this concept of learning as a dialogue.

Some learning models for adaptive and flexible learning (and, by implication, supporting soft skill development) are based on reflection: Kolb (1984)'s work presents learning as a process in which knowledge is created by the learner through some transformation of experience; Schön (1985) refers to reflection-in-action as the responses that skilful practitioners bring to their practice. This reflection consists of strategies of action (ways of framing the situations encountered in day-to-day experience), and may take the form of problem solving, theory building, or re-appreciation of the situation. The success of this learning is based on factors such as the degree of learner control, degree of correspondence of learning environment to real environment and degree of involvement of self and the adoption of strategies which have come to be identified as contributing to reflection. If students are being prepared to become reflective practitioners (Schön, 1983), opportunities for students to develop reflective skills and sensibilities should be embedded as a normal part of all professional courses (Boud, Keough, & Walker, 1985; Schön, 1987).

As noted previously, the approach to learning that a student takes is sensitive to the context in which learning is done, with a deep approach to learning required to develop advanced conceptions. Learning diagnostics (e.g. ASI) undertaken as part of Intervention 2 showed there was as strong a bias to surface learning as there was to deep learning in the student cohort. This related (and had impact on) the critical concept of lifelong learning. Therefore metacognitive strategies needed to be made explicit – students required a rationale for including reflection as an element of their learning (demonstrated through portfolios

and activity logs). Edwards (2004) reports on approaches to explicitly provide the opportunity for students to adopt expert strategies. These enable the teacher to guide students in the nature of expert processes and help them to reflect critically on their effectiveness. He notes that the best means of facilitating expertise is to provide the opportunity for practice. However, only through encouraging students to challenge their own effectiveness can they learn what this implies.

What was needed, then, was a model of education that added, to the positive aspects of the studios as an immersive environment, a focus on metacognition, so that learning integrated evaluation of the 'practical' outcomes of the problem with the creative process. To Laurillard (1993)'s learning model (where attention on key relationships between forms of discourse, academic knowledge, interactions with the world and reflection are emphasised) aspects of teachback and self-explanation were incorporated, forcing a focus on key aspects of the domain, deeper processing of the topic, and allowing failures and conflicts to emerge (Gobet & Wood, 1999).

These ideas formed the basis for the *StudioLearning* model developed and applied in this final intervention. The *StudioLearning* model allowed for approaches from previous interventions to be integrated. The teacher acted as expert consultant, and could be 'engaged' by the students to provide modelling demonstration and domain expertise. However, the students were still required to direct their own learning: developing the learning objectives to address each problem presented within the PBL environment previously developed. Here the strategy was to reach all types of learners by 'teaching around the cycle' (Kolb, 1984), thus enabling students to develop the mental dexterity required in professional practice, and emphasising the importance of contingency measures, opportunism, reflective practice and metalearning. This *StudioLearning* environment also provided the opportunity for students to adopt expert strategies

– the teacher acted as guide/mentor or 'consultant' in these processes, helping students to reflect critically on their effectiveness in specific contexts.

Kuhn (2001) describes characteristics of the studio environment:

- Student work is organised primarily into complex and open ended problems
- Students' design solutions undergo multiple and rapid iterations
- Students are exposed to relevant precedent and to rapid iteration of design solutions. This combination is seen as essential to the development of true expertise (Dreyfus & Dreyfus, 1986)
- Critique is frequent, and occurs in both formal and informal ways, from teachers, peers, and visiting experts. One of the hallmarks of studio education is the creation of a 'culture of critique', in which students, who spend long hours working side by side at their projects, give each other frequent feedback, and also get both formal and informal feedback from the academic in charge of the studio in order to reflect on their learning
- Students study precedents (past designs) and are encouraged to think about the big picture
- Teachers mentor students to impose appropriate constraints on their design process in order to navigate a complex and open ended problem and find a satisfactory design solution
- The appropriate use of a variety of design media over the course of the project significantly supports and improves students insight and designs.

A studio environment has been applied to SE education, most famously at Carnegie Mellon (Tomayko, 1996) as part of their Masters' programme. In undergraduate education, studios are more often applied to capstone courses rather than foundational learning (Ramakrishan, 2003), although recent success in this area has been reported (Narayanan, Hundhausen, Hendrix, & Crosby, 2012).

At this time, a decision was made that all 3rd and 4th year learning within the School (in effect the final two years of undergraduate engineering degrees) would apply a studio model, with two studios considered a full-time load. Later evaluation indicated the appropriateness of this approach. Students noted that with all their studies undertaken within studios, they felt they were much more in control of their efforts:

- Academic staff were perceived as more tolerant of the needs of other studios
- With a full-time load of only two studios student time was not as fragmented across different areas
- Except for the (negotiated) compulsory attendance, students could vary the time they spent on each studio in response to their total learning context. It was the team's role to ensure tasks were on schedule. They concluded that this flexibility reduced stress and allowed them to focus on the learning they needed to achieve for the task.

In order to encourage deeper learning, a portfolio (worth 20% of the final mark) comprising, as its major element, the concept maps for the topics explored throughout the course, was included as one of the artefacts developed throughout the semester. The maps were evaluated on a regular basis by team members. This peer review was seen as a mechanism to assist the development of a common understanding of the knowledge being developed within the course – simply, as a view into the minds of the students (Freeman and Urbaczewski, 2001). The portfolio was now both a group and individual effort and included a reading log compiled by the group (but initialled by whoever had written the particular summary), minutes of group meetings throughout the semester (with rotating minute taker) to indicate task allocation, and individual activity logs for 'in- and out-of-class' time spent on the course. For this component, students were required to indicate time spent on tasks such as: discussion (time spent discussing issues internally within group); consultation (time spent discussing issues with teacher); research (time spent researching new areas of knowledge to assist with engineering of project); design work (time spent on formulating and describing design solution); review & testing (time spent on reviewing the design solution and/ or testing it); personal management (time spent on individual management, documentation and related tasks); group management (time spent on project management related tasks for the group as a whole). Students also answered several questions on a weekly basis: *what did I achieve this week? what issues did I have this week? what can I do next week to address these issues?*

The evaluations of this intervention focussed on the level of success of *StudioLearning* in relation to its implementation, and on the effect of the intervention on the development of the student cohort - examined in terms of short term impact (performance of the students in assessable tasks during the intervention) and longer term impact (performance of the students in assessable tasks dependant on the learning objectives of the intervention, in the subsequent course).

For the discipline of SE the *StudioLearning* model appeared to hold the most promise – students had been provided with a process during an orientation week at the commencement of semester, and were able to apply it in the course, but also adapt or discard it as they perceived necessary. In this way, the environment supported both discipline and opportunism. The student data indicated they were taking control of their own learning and acknowledging the importance of skills such as time management: *(while I kept up with the readings and assignments I ended*

up getting behind in creating the mind maps and logging my personal journal. While I acknowledge that is my responsibility to manage my own time it would have been useful if this task was checked more frequently). Robillard (2005) noted that, in a typical opportunistic problem-solving activity, individuals spontaneously rely on teammates to provide missing information. Evaluation of the *StudioLearning* model in relation to the *Professional Capability Framework* shows this was occurring, and that other components of the Framework were being addressed:

Some of the knowledge I have learnt has resulted from the interaction with my team members i.e. I don't believe the level of understanding I now have, would have been achieved by working on the assignments by myself (component A)

we have to apply learning to a realistic problem which means it moves as out in the real world e.g. the lecturer pointed out errors in thinking and this resulted in us having to revise what we had completed previously in order to move to the next step. I found this gave me a greater depth of knowledge than the usual do an assignment get some of it wrong and move on to the next usually non related assignment; this method of teaching has provided me with a frame work that I can use to identify future problems and develop solutions (component B)

each of us shares the ability or outcomes from the ability e.g. ability to research (find information/ knowledge), understanding a problem that others don't and interpreting it into a context the other can understand so they can solve the problem (component C)

the personal journals can be a useful tool to ensure the unit learning's are integrated into the students existing knowledge/experience and that the students behaviours are modified to align with those required to be a successful engineering graduate. [...] weekly journals would be useful in ensuring they are being used as a tool to guide student development into the higher levels of Bloom's taxonomy and the non-technical aspects of their learning; this unit teaches a process that is built on knowledge but more importantly that knowledge is converted to a skill via practice on the problem. I don't believe this is achieved by the other style of teaching e.g. lectures and exercise type assignments (components D/E).

The portfolio entries showed that students were highly motivated to complete the tasks assigned – the suggestion is that general student attitude towards the controllability of the learning outcome (ie externally dictated and beyond student control, or within the internal control of the student through effort and personal interest) influences their motivation and the level of achievement in the learning process. In addition, motivation is heightened in situations which offer opportunities for increasing competence or intelligence. More interestingly, as students rotated into the role of Project Manager, they (individually) applied what they had previously learnt with regards to learning strategies and approaches to study in order to motivate their group members.

Observation and analysis of learning of some of the cohort in the subsequent course showed strong indications of willingness to transfer knowledge gained, to take control of their learning, and indicated motivation to deeper learning. Examination of student reflective comments, in conjunction with data regarding student learning, added another dimension to the issue of education for competent practice. This examination indicated a relationship existing between the learner and the learning model, so that students whose approach favoured deep learning for understanding were advantaged by a learning environment which challenged them (*I have noticed that the design studios require a lot more work from me than if I was working alone. For example I have to spend*

more time working on problems because of the extra overhead of working in a team (meetings and social interaction). There is also the need to do extra research to gain information that is normally just handed out in a lecture. However I don't mind putting in the extra effort because I feel the extra effort is worth it because I feel more confident that I do know the material (not an impostor) and can apply it to future situations). Use of the term 'imposter' is perhaps telling.

STUDENT DEVELOPMENT ACROSS INTERVENTIONS

Grow (1996) describes an approach to modelling learning from the learner's 'growth' towards life-long learning. This model reflects the principles advocated in student-centred learning environments: the learner determines the need for some education, decides on a preferred approach to learning, identifies and accesses learning resources and draws on the assistance of educators as a part of that overall strategy rather than as a central element:

- **Stage 1:** Learners need an authority figure to give them explicit directions on what to do, how to do it, and when. They either treat teachers as experts who know what the student needs to do, or they passively slide through the educational system, responding mainly to teachers who 'make' them learn. The teacher acts as *authority, coach*

- **Stage 2:** Learners are 'available'. They are interested or able to be interested. They respond to motivational techniques. They are willing to do those assignments they can see the purpose of. They are confident but may be largely ignorant of the subject of instruction. The teacher acts as *motivator, guide*

- **Stage 3**: Learners have skill and knowledge, and they see themselves as participants in their own education. They are ready to explore a subject with a good guide. They will even explore some of it on their own. But they may need to develop a deeper self-concept, more confidence, more sense of direction, and a greater ability to work with (and learn from) others. These learners benefit from learning more about how they learn, such as making conscious use of learning strategies. The teacher acts as *facilitator*

- **Stage 4:** Learners set their own goals and standards – with or without help from experts. They use experts, institutions, and other resources to pursue these goals. Learners at this stage are both able and willing to take responsibility for their learning, direction, and productivity. They exercise skills in time management, project management, goal-setting, self-evaluation, peer critique, information gathering, and use of educational resources. The teacher acts as *consultant, delegator*.

In summary, Intervention 1 showed that, even within a constructivist framework, the relationship between teacher and learner (or master and apprentice) can remain unidirectional – the former modelling behaviour for the latter and guiding attainment of the learning outcomes (Grow's Stage 1). Intervention 2 provided an environment for students to address Stages 2-3 – the PBL environment required a more explorative and collaborative approach to learning. Finally, Intervention 3 supported a transition to Stages 3- 4. Students were able to demonstrate reflective-practitioner behaviour, and were comfortable (and successful) in an environment where the teacher was not *the* authority but one of the resources they had access to.

SUPPORT FOR SOFT-SKILLS LEARNING

The results of this project show that a critical aspect of the challenge is integrating non-technical skills across the whole curriculum, so that students do not receive conflicting messages about their importance. Making acquisition of these skills explicit (by, for example, addressing specific soft skills learning outcomes, activities and assessment of them) was valuable. Finally, requiring students to incorporate metacognitive strategies and reflection in the evaluation of their own learning enabled students to track and acknowledge their increasing competence in these skills.

Implementing environments such as those described in this chapter requires a cultural change to occur within the educational environment for the discipline. Not only do students need to adapt their learning behaviour so that a reliance on lecture/tutorial/laboratory is minimised (if not removed totally), but academics also require orientation to less traditional learning models. In this context, although the interventions were 'public' within the School, and reported on within School seminars, in general it could be said that academic staff were not deeply conversant with the approaches described. Therefore a series of staff development workshops was initiated for academics in order to provide training and education on problem- and studio-based learning.

The staff development workshops were structured to address two separate concerns:

- **Provide a background in PBL and studio approaches by undertaking a PBL session:** The outcome of this was an understanding of what changes would be needed to enable studios across all courses, and a set of tasks to be completed by each academic staff member for each of the courses they coordinated. Tasks included curriculum mapping and problem development, amongst others. Concerns were also raised – the most critical being discipline content coverage. In addition, several staff members were not convinced that basic (ie foundational) knowledge could be learnt this way

- **Develop an appreciation of the issues raised by such active learning:** The School invited an academic who had a great deal of experience in applying PBL in an engineering context. The result of this workshop was clearer commitment on the part of academic staff to make studios work.

Students were also provided with exposure to the learning environment prior to semester start. A one week orientation programme, in which students were placed into multi-(engineering) discipline teams, provided the opportunity to model *StudioLearning* and establish the roles and responsibilities of students and academics within this model. These objectives were achieved through a small-scale design task as a means of identifying and exposing the *StudioLearning* approach. The Design Week provided an introduction to generic tools, techniques, methods and processes as support services made available with the learning environment. These might otherwise have needed to be duplicated in each studio. This orientation also provided an opportunity to pre-test student perceptions of the model.

The adoption of *StudioLearning* across the School may be considered 'transformational', a second order change that required a fundamental shift in behaviour and resources (Levy, 1986). A correlation appeared to exist between attendance at staff development sessions and the 'success' of the specific studio. The implication of this result is that no staff should teach within a studio environment without adequate and appropriate training (both in the learning theory behind the model and

in facilitation techniques). Therefore, while it is clear from the literature that an understanding of the underpinning pedagogical basics is necessary, as are special facilitation skills, it has resource implications within the tightening finance environment prevailing at many universities.

CONCLUSION

Learning profession-specific content provides the 'scaffold' for the important task of career-long professional learning: the skills to undertake this are of great importance, with the ability to know when and when not to deploy technical expertise, and how to continuously update it, the keys to successful professional practice. The interventions described in this chapter addressed the need to:

- **Provide students with authentic experiences which address competencies additional to specific discipline knowledge:** Students were exposed to learning both as a 'generic' metacognitive activity, and as a skill to be continually adapted and utilised within a discipline context. Flexibility in thinking - addressing creativity, opportunism and divergency/convergency – was made explicit and strategies to exploit it developed
- **Provide learners with a deep understanding of self and others in complex human activity systems in a collaborative environment:** Students became aware of and learnt to utilise each other's strengths and weaknesses in achieving the learning outcomes. They learnt how to 'jell' in their team, what to do if they did not, and to be empathetic to the contexts of other students; they learnt to value and exploit alternate perspectives brought to a problem by different stakeholders (client, teacher/consultant, other team members)

to enrich their learning; they became aware of the need to be self-motivated and learn independently - students were confident in questioning their own and others' assumptions within the learning environment

- **Allow time to explore new ideas and to reflect on possible processes and outcomes:** Students were open to discussion and feedback and willing to retrace their steps/redo the work in order to advance to a solution; they were willing to 'trust' each other's knowledge (implicit or not, technical or not), accepting the multi-disciplinary nature of the skills and knowledge required to achieve the learning objectives, within an environment that enabled the advantages of 'flow time' to be exploited

- **Provide challenges:** Students were motivated by the (increasing) complexity of the task, and were able to focus on cognitive and interpersonal skills to adapt to the changes required.

While technical know-how acts as the gatekeeper to professional practice, the non-technical, 'soft' or 'employability' skills are those which provide career resilience (Waterman et al., 1994). Ensuring graduates are ready to apply these as practitioners remains a challenge in higher education.

REFERENCES

ABET. (2012a). *Criteria for accrediting computing programs, 2013 - 2014*. Retrieved from http://www.abet.org/DisplayTemplates/DocsHandbook.aspx?id=3148

ABET. (2012b). *Criteria for accrediting engineering programs, 2013-2014*. Retrieved from http://www.abet.org/DisplayTemplates/DocsHandbook.aspx?id=3149

ACCI. (2002) Employability skills - An employer perspective: Getting what employers want out of the too hard basket. *ACCI Review, 88.*

ACM/IEEE. (2008). *Computer science curriculum 2008: An interim revision of CS 2001*. ACM/IEEE Computer Society.

Amabile, T. M. (1996). *Creativity in context*. Boulder, CO: Westview Press.

Andrews, J., & Higson, H. (2008). Graduate employability, 'soft skills' versus 'hard' business knowledge: A European study. *Higher Education in Europe, 33*(4), 411–422. doi:10.1080/03797720802522627

Argyris, C., & Schön, D. A. (1974). *Theory in practice: Increasing professional effectiveness*. San Francisco, CA: Jossey Bass.

Banks, D. A. (2003). Belief, inquiry, argument and reflection as significant issues in learning about information systems development methodologies. In T. McGill (Ed.), *Current issues in IT education* (pp. 1–10). Hershey, PA: IRM Press.

Baxter-Magnola, M. B. (2001). A constructivist revision of the measure of epistemological reflection. *Journal of College Student Development, 42*(6), 520–534.

Benamati, J. H., Zafer, D., Ozdemir, Z. D., & Smith, H. J. (2010). Aligning undergraduate IS curricula with industry needs. *Communications of the ACM, 53*(3), 152–156. doi:10.1145/1666420.1666458

Benson, S. (2003). Metacognition in information systems education. In T. McGill (Ed.), *Current issues in IT education* (pp. 213–222). Hershey, PA: IRM Press.

Bentley, J. F., Lowry, G. R., & Sandy, G. A. (1999). Towards the compleat information systems graduate: A problem based learning approach. In *Proceedings of the 10th Australasian Conference on Information Systems*.

Berkenkotter, C., & Huckin, T. N. (1995). *Genre knowledge in disciplinary communication: Cognition/culture/power*. Hillsdale, NJ: Lawrence Erlbaum Associates.

Bloom, B. S. (1956). *Taxonomy of educational objectives: The classification of educational goals Handbook 1: Cognitive domain*. New York: David Mackay.

Boud, D. (1985). Problem-based learning in perspective. In D. Boud (Ed.), *Problem-based learning in education for the professions* (pp. 13–18). Sydney: Higher Education Research Society of Australasia.

Boud, D., Keough, R., & Walker, D. (1985). *Reflection: Turning experience into learning*. London: Kogan Page.

Bourque, P., & Dupuis, R. (Eds.). (2004). *Guide to the software engineering body of knowledge - SWEBoK*. Los Alamitos, CA: IEEE Computer Society.

Bowden, J., Hart, G., King, B., Trigwell, K., & Watts, O. (2000). Generic capabilities of ATN university graduates. Sydney Teaching and Learning Committee, Australian Technology Network.

Bransford, J. D., Brown, A. L., & Cocking, R. R. (Eds.). (1999). *How people learn: Brain, mind, experience and school*. Washington, DC: National Academy Press.

Brown, J. S., Collins, A., & Duguid, P. (1989). Situated cognition and the culture of learning. *Educational Researcher, 18*, 32–42. doi:10.3102/0013189X018001032

Brown, P., & Hesketh, A. J. (2004). *The mismangement of talent: Employability and jobs in the knowledge-based economy*. Oxford, UK: Oxford University Press. doi:10.1093/acprof:oso/9780199269532.001.0001

Budgen, D. (2003). *Software design*. Harlow, UK: Pearson Education Ltd.

Cassidy, S. (2006). Developing employability skills: Peer assessment in higher education. *Education + training, 48*(7), 508–517.

Claxton, G. (1998). *Hare brain, tortoise mind*. London: Fourth Estate.

Collins, A., Brown, J. S., & Newman, S. E. (1989). Cognitive apprenticeship: Teaching the crafts of reading, writing and mathematics. In L. Resnick (Ed.), *Knowing, learning and instruction: Essays in honour of Robert Glaser* (pp. 453–494). Hillsdale, NJ: Erlbaum.

Csikszentmihalyi, M. (1996). *Creativity: Flow and the psychology of discovery and invention*. New York: HarperPerennial.

Curtin, P. (2004). Employability skills for the future. In J. Gibb (Ed.), *Generic skills in vocational education and training: Research readings* (pp. 40–41). Adelaide, Australia: National Centre for Vocational Education Research.

De Corte, E. (2000). Marrying theory building and the improvement of school practice: A permanent challenge for instructional psychology. *Learning and Instruction, 10*, 249–266. doi:10.1016/S0959-4752(99)00029-8

de Graaff, E., & Kolmos, A. (2003). Characteristics of problem-based learning. *International Journal of Engineering Education, 19*(5), 657–662.

DEEWR. (2012). *Employability skills framework stage 1 – Final report*. Canberra, Australia: Australian Government. DEEWR Retrieved from http://foi.deewr.gov.au/system/files/doc/other/employability_skills_framework_stage_1_final_report.pdf

DEST. (2006). Employability skills - From framework to practice. Canberra, Australia: Commonwealth of Australia. Department of Education, Science and Training.

Dewey, J. (1916). *Democracy and education*. New York: Macmillan.

Doke, E. R., & Williams, S. R. (1999). Knowledge and skill requirements for information systems professionals: An exploratory study. *Journal of Information Systems Education, 10*(1), 10–18.

Donald, J. G. (2002). *Learning to think.* San Francisco: Jossey-Bass.

Dreyfus, H. L. (2001). *On the internet.* London: Routledge.

Dreyfus, H. L., & Dreyfus, S. E. (1986). *Mind over machine.* New York: Free Press.

EA. (2011). *Stage 1 competency standard for professional engineer.* Retrieved from http://www.engineersaustralia.org.au/sites/default/files/shado/Education/Program%20Accreditation/110318%20Stage%201%20Professional%20Engineer.pdf

Edmonds, E., & Candy, L. (2002). Creativity, art practice and knowledge. *Communications of the ACM, 45*(10), 91–95. doi:10.1145/570907.570939

Entwistle, N. J., & Ramsden, P. (1983). *Understanding student learning.* London: Croom Helm.

Fairly, D. (Ed.). (2013). Guide to the software engineering body of knowledge (SWEBOK V3). ACM/IEEE Computer Society.

Friedman, K. (2001). *Creating design knowledge: From research into practice.* Paper presented at the IDATER 2000 Conference. Loughborough, UK. https://dspace.lboro.ac.uk/2134/1360

Gobet, F., & Wood, D. (1999). Expertise, models of learning and computer-based tutoring. *Computers & Education, 33*, 189–207. doi:10.1016/S0360-1315(99)00032-9

Gott, S. P., Hall, E. P., Pokorny, R. A., Dibble, E., & Glaser, R. (1993). A naturalistic study of transfer: Adaptive expertise in technical domains. In D. K. Detterman, & R. J. Sternberg (Eds.), *Transfer on trial: Intelligence, cognition and instruction* (pp. 258–288). Norwood, NJ: Ablex.

Grow, G. O. (1996). Teaching learners to be self-directed. *Adult Education Quarterly, 41*(3), 125–149. doi:10.1177/0001848191041003001

GSwE. (2009). *Graduate software engineering 2009 (GSwE2009), curriculum guidelines for graduate degree programs in software engineering.* Hoboken, NJ: Stevens Institute of Technology.

Guindon, R. (1989). The process of knowledge discovery in system design. In G. Salvendy, & M. J. Smith (Eds.), *Designing and using human-computer interfaces and knowledge based systems* (pp. 727–734). Amsterdam: Elsevier.

Guindon, R. (1990). Knowledge exploited by experts during software systems design. *International Journal of Man-Machine Studies, 33*, 279–304. doi:10.1016/S0020-7373(05)80120-8

Hagan, D. (2004). *Employer satisfaction with ICT graduates.* Paper presented at the 6th Conference on Australasian Computing Education. Canberra, Australia.

Harvey, L., & Bowers-Brown, T. (2004, Winter). Employability cross-country comparisons. *Graduate Market Trends.*

Hennessey, B. A., & Amabile, T. M. (2010). Creativity. *Annual Review of Psychology, 61*, 561–598. doi:10.1146/annurev.psych.093008.100416

Herrington, J., Reeves, T., & Oliver, R. (2003). Patterns of engagement in authentic online learning environments. *Australian Journal of Educational Technology, 19*(1), 59–71.

Horvath, I., Wiersma, M., Duhovnik, J., & Stroud, I. (2004). Navigated active learning in an international academic virtual enterprise. *European Journal of Engineering Education, 29*(4), 505–519. doi:10.1080/03043790410001716275

Jonassen, D. H. (1992). Semantic networking as cognitive tools. In P. A. M. Kommers, D. H. Jonassen, & J. T. Mayes (Eds.), *Cognitive tools for learning* (pp. 19–21). Heidelberg, Germany: Springer-Verlag.

Jonassen, D. H. (2002). Learning to solve problems online. In C. Vrasidas, & V. Glass (Eds.), *Distance education and distance learning* (pp. 75–98). Greenwich, CT: Information Age Publishing.

Jonassen, D. H., & Grabowski, B. L. (1993). *Handbook of individual differences, learning and instruction*. New York: Allyn & Bacon.

Jong, T. (2006). Technological advances in inquiry learning. *Science, 312*(5773), 532–533. doi:10.1126/science.1127750 PMID:16645080

Keirsey, D., & Bates, M. (1984). *Please understand me*. Prometheus Nemesis Book Company.

Kennan, M. A., Willard, P., Cecez-Kecmanovic, D., & Wilson, C. S. (2009). IS knowledge and skills sought by employers: A content analysis of Australian IS early career online job advertisements. *The Australasian Journal of Information Systems, 15*(2), 169–190.

Kolb, D. A. (1984). *Experiential learning experience as the source of learning and development*. Upper Saddle River, NJ: Prentice-Hall.

Kolb, D. A. (1995). *Learning style inventory: Technical specifications*. Boston: McBer & Company.

Koppi, T., & Naghdy, F. (2009). *Discipline-based initiative: Managing educational change in the ICT discipline at the tertiary education level*. Wollongong, Australia: University of Wollongong.

Koschmann, T. D., Myers, A. C., Barrows, H. S., & Feltovich, P. J. (1994). Using technology to assist in realising effective learning and instruction: A principled approach to the use of computers in collaborative learning. *Journal of the Learning Sciences, 3*(3), 227–264. doi:10.1207/s15327809jls0303_2

Kuhn, S. (2001). Learning from the architecture studio: Implications for project-based pedagogy. *International Journal of Engineering Education, 17*(4-5), 349–352.

Laurillard, D. (1993). *Rethinking university teaching: A framework for the effective use of educational technology*. London: Routledge.

Lave, J., Smith, S., & Butler, M. (1988). Problem solving as an everyday practice. In *Learning mathematical problem solving*. Palo Alto, CA: Institute for Research on Learning.

Lave, J., & Wenger, E. (1991). *Situated learning: Legitimate peripheral participation*. Cambridge, MA: Cambridge University Press. doi:10.1017/CBO9780511815355

LeBlanc, R., & Sobel, A. E. K. (Eds.). (2004). *Software engineering 2004: Curriculum guidelines for undergraduate degree programs in software engineering*. Los Alamitos, CA: IEEE Computer Society Press.

Lee, D. M. S. (1999). *Information seeking and knowledge acquisition behaviors of young information systems workers: Preliminary analysis*. Paper presented at the 1999 Americas Conference on Information Systems. New York, NY.

Lee, D. M. S. (2004). Organizational entry and transition from academic study: Examining a critical step in the professional development of young IS workers. In M. Igbaria, & C. Shayo (Eds.), *Strategies for managing IS/IT personnel* (pp. 113–141). Hershey, PA: Idea Group.

Lethbridge, T. C. (1999). *The relevance of education to software practitioners: Data from the 1998 survey*. Ottowa, Canada: School of Information Technology and Engineering, University of Ottowa.

Lethbridge, T. C. (2000). What knowledge is important to a software professional? *IEEE Computer, 33*(5), 44–50. doi:10.1109/2.841783

Levy, A. (1986). Second order planned change: Definition and conceptualisation. *Organizational Dynamics, 15*, 5–20. doi:10.1016/0090-2616(86)90022-7

Litecky, C. R., Arnett, K. P., & Prabhakar, B. (2004). The paradox of soft skills versus technical skills in IS hiring. *Journal of Computer Information Systems, 45*(1), 69–76.

Lowry, G., & Turner, R. (2005). Information systems education for the 21st century: Aligning curriculum content & delivery with the professional workplace. In D. Carbonara (Ed.), *Technology literacy applications in learning environments* (pp. 171–202). Hershey, PA: IRM Press. doi:10.4018/978-1-59140-479-8.ch013

LTSN. (2002). *Constructive alignment and why it is important to the learner.* Retrieved from http://www.ltsneng.ac.uk/er/theory/constructivealignment.asp

McConnell, S., & Tripp, L. (1999). Professional software engineering: Fact or fiction? *IEEE Software, 16*(6), 13–17. doi:10.1109/MS.1999.805468

McCracken, W. M. (1997). SE education: What academia can do. *IEEE Software, 14*(6), 27, 29.

McLarty, R. (1998). *Using graduate skills in small and medium sized enterprises.* Ipswich, UK: University College Suffolk Press.

McLaughlan, R. G., & Kirkpatrick, D. (1999). A decision making simulation using computer mediated communication. *Australian Journal of Educational Technology, 15*, 242–256.

McLaughlan, R. G., & Kirkpatrick, D. (2004). Online roleplay: Design for active learning. *European Journal of Engineering Education, 29*(4), 477–490. doi:10.1080/03043790410001716293

McLean, P., Perkins, K., Tout, D., Brewer, K., & Wyse, L. (2012). *Australian core skills framework.* Canberra, Australia: Government of Australia.

McQuaid, R. C., & Lindsay, C. (2005). The concept of employability. *Urban Studies (Edinburgh, Scotland), 42*, 197–219. doi:10.1080/0042098042000316100

Meyer, J. H. F., & Shanahan, M. P. (2000). Making teaching responsive to variation in student learning. In *Proceedings of the 7th Improving Student Learning Symposium.* Manchester, UK: Academic Press.

Narayanan, N. H., Hundhausen, C., Hendrix, D., & Crosby, M. (2012). *Transforming the CS classroom with studio-based learning.* Paper presented at the SIGCSE'12. Raleigh, NC.

Noll, C. L., & Wilikens, M. (2002). Critical skills of IS professionals: A model for curriculum development. *Journal of Information Technology Education, 1*(3), 143–154.

OECD. (1996). *Lifelong learning for all.* Paris: OECD.

Oliver, R., & McLoughlin, C. (1999). *Using web and problem-based learning environments to support the development of key skills.* Paper presented at the Responding to Diversity. Brisbane, Australia.

Patel, A., Kinshuk, R., & Russell, D. (2000). Intelligent tutoring tools for cognitive skill acquisition in life long learning. *Journal of Educational Technology & Society, 3*(1), 32–40.

Patry, J. L. (1998). *Transfer evaluation in educational processes: Models, results and problems - A theoretical approach.*

Pfleeger, S. L. (1999). Albert Einstein and empirical software engineering. *IEEE Computer, 32*(10), 32–37. doi:10.1109/2.796106

Piaget, J. (1968). *La structuralisme* (T. I. B. C. Maschler, Trans.). London: Routledge & Kegan Paul.

Precision_Consultancy. (2007). *Graduate employability skills*. Canberra, Australia: Canberra Commonwealth of Australia.

Prosser, M., & Trigwell, K. (1999). *Understanding learning and teaching: The experience in higher education*. Buckingham, UK: Oxford University Press.

Ramakrishan, S. (2003). MUSE studio lab and innovative software engineering capstone project experience. *SIGCSE Bulletin, 35*(3), 21–25. doi:10.1145/961290.961521

Robillard, P. N. (2005). Opportunistic problem solving in software engineering. *IEEE Software, 22*(6), 60–67. doi:10.1109/MS.2005.161

Russell, D. R. (1998). *The limits of the apprenticeship models in WAC/WID research*. Paper presented at the Conference of College Composition and Communication. New York, NY.

Savin-Baden, M. (2000). *Problem-based learning in higher education: Untold stories*. Buckingham, UK: Society for Research into Higher Education and Open University Press.

Scholarios, D., van der Schoot, E., & van der Heijden, B. (2004). *The employability of ICT professional: A study of European SMEs*. Paper presented at the e-Challenges Conference. Vienna, Austria.

Schön, D. A. (1983). *The reflective practitioner: How professionals think in action*. New York: Basic Books.

Schön, D. A. (1985). *The design studio*. London: RIBA Publications.

Schön, D. A. (1987). *Educating the reflective practitioner: Towards a new design for teaching in the professions. San Fransisco*. Jossey-Bass Inc.

Scott, G., & Wilson, D. (2002). *Tracking and profiling successful IT graduates: An exploratory study*. Paper presented at the 13th Australasian Conference on Information Systems. Canberra, Australia.

Scott, G., & Yates, W. (2002). Using successful graduates to improve the quality of undergraduate engineering programs. *European Journal of Engineering Education, 27*(4), 60–67. doi:10.1080/03043790210166666

Sherer, M., & Eadie, R. (1987). Employability skills: Key to success. *Thrust, 17*, 16–17.

Smith, W. J., Belanger, F., Lewis, T. L., & Honaker, K. (2007). Training to persist in computing careers. *Inroads -. SIGCSE Bulletin, 39*(4), 119–120. doi:10.1145/1345375.1345429

Snoke, R., & Underwood, A. (1999). *Generic attributes of IS graduates - A Queensland study*. Paper presented at the 10th Australasian Conference on Information Systems. Wellington, NZ.

Soloman, B., & Felder, R. (1999). *Index of learning styles (ILS)*. Retrieved from http://www2.ncsu.edu/unity/lockers/users/f/felder/public/ILSpage.html

Spiro, R. J., Feltovich, P. J., Jacobson, M., & Coulson, R. (1991). Cognitive flexibility, constructivism and hypertext: Random access instruction for advanced knowledge acquisition in ill-structured domains. *Educational Technology, 31*, 24–33.

Tomlinson, M. (2012). Graduate employability: A review of conceptual and empirical themes. *Higher Education Policy, 25*, 407–431. doi:10.1057/hep.2011.26

Topi, H., Valacich, J. S., Wright, R. T., Kaiser, K. M., Nunamaker, F. J., Sipior, J. C., & de Vreede, G. J. (2010). *IS 2010: Curriculum guidelines for undergraduate degree programs in information systems*. Association for Computing Machinery and Association for Information Systems.

Tucker, M. L., Sojka, S., Barone, F., & McCathy, A. (2000). Training tomorrow's leaders: Enhancing the emotional intelligence of business graduates. *Journal of Education for Business*, *75*(6), 331–338. doi:10.1080/08832320009599036

Turley, R. T. (1991). *Essential competencies of exceptional professional software engineers*. Fort Collins, CO: Colorado State University.

Turley, R. T., & Bieman, J. M. (1995). Competencies of exceptional and non-exceptional software engineers. *Journal of Systems and Software*, *28*(1), 19–38. doi:10.1016/0164-1212(94)00078-2

Turner, R., & Lowry, G. (2003). Education for a technology-based profession: Softening the information systems curriculum. In T. McGill (Ed.), *Current issues in IT education* (pp. 153–172). Hershey, PA: IRM Press.

Voogt, J., Erstad, O., Dede, C., & Mishra, P. (2013). Challenges to learning and schooling in the digital networked world of the 21st century. *Journal of Computer Assisted Learning*, *29*, 403–413. doi:10.1111/jcal.12029

Vygotsky, L. S. (1978). *Mind and society: The development of higher psychological processes*. Cambridge, MA: Harvard University Press.

Waterman, R. H., Waterman, J. A., & Collard, B. A. (1994). Toward a career resilient workforce. *Harvard Business Review*, *69*, 87–95.

Wenger, E. (2000). Communities of practice and social learning systems. *Organization Science*, *7*(2), 225–246. doi:10.1177/135050840072002

Wilson, K., & Fowler, J. (2005). Assessing the impact of learning environments on students' approaches to learning: Comparing conventional and action learning designs. *Assessment & Evaluation in Higher Education*, *30*(1), 87–101. doi:10.1080/0260293042003251770

Wood, D. (1999). Editorial: Representing, learning and understanding. *Computers & Education*, *33*, 83–90. doi:10.1016/S0360-1315(99)00026-3

Zwieg, P., Kaiser, K. M., Beath, C. M., Gallagher, K. P., Goles, T., & Howland, J. (2006). The information technology workforce: Trends and implications 2005-2008. *MIS Quarterly Executive*, *5*(2), 101–108.

ADDITIONAL READING

Albanese, M., & Mitchell, S. (1993). Problem-based learning: A review of the literature on its outcomes and implementation issues. *Academic Medicine*, *68*(1), 52–81. doi:10.1097/00001888-199301000-00012 PMID:8447896

Anderson, L., & Krathwohl, D. A. (2001). *Taxonomy for learning, teaching and assessment: A revision on Bloom's taxonomy of educational objectives*. New York: Longman.

Armarego, J. (2007). Aligning learning with industry requirements. In *Information systems and technology education: From the university to the workplace*. Hershey, PA: Idea Group. doi:10.4018/978-1-59904-114-8.ch008

Armarego, J. (2008). Constructive alignment in SE education: Aligning to what? In *Software engineering: Effective teaching and learning approaches and practices*. Hershey, PA: IGI Global. doi:10.4018/978-1-60566-102-5.ch002

Armarego, J. (2009, February). Displacing the Sage on the Stage: Student Control of Learning. In *Proceedings of the 22nd Conference on Software Engineering Education and Training* (pp. 198–201). IEEE.

Armarego, J., & Fowler, L. (2005). Orienting students to studio learning. In *Proceedings of the 4th ASEE/AaeE Global Colloquium on Engineering Education*.

Barrows, H. S. (1986). A taxonomy of PBL methods. *Journal of Medical Education*, *20*(6), 481–486. doi:10.1111/j.1365-2923.1986.tb01386.x

Benson, S. (2003). Metacognition in information systems education. In *Current issues in IT education* (pp. 213–222). Hershey, PA: IRM Press.

Biggs, J. B. (1985). The role of metalearning in study processes. *The British Journal of Educational Psychology*, *55*(3), 185–212. doi:10.1111/j.2044-8279.1985.tb02625.x

Boud, D., & Feletti, G. (Eds.). (2001). *The challenge of problem-based learning.* London: Kogan Page.

Brown, J. S., Collins, A., & Duguid, P. (1989). Situated cognition and the culture of learning. *Educational Researcher*, *18*(1), 32–42. doi:10.3102/0013189X018001032

Bull, C. N., Whittle, J., & Cruickshank, L. (2013, May). Studios in software engineering education: Towards an evaluable model. In *Proceedings of the 35ᵗʰ International Conference on Software Engineering* (pp. 1063–1072). IEEE Press.

Chafee, R. (1977). The teaching of architecture at the Ecole des Beaux-Arts. In The architecture of the Ecole des Beaux-Arts (pp. 171-202). New York: Museum of Modern Art.

Chi, M. T. H., Glaser, R., & Rees, E. (1982). Expertise in problem solving. In *Advances in the Psychology of Human Intelligence* (Vol. 1, pp. 7–75). Hillsdale, NJ: Erlbaum.

Collins, A., Brown, J. S., & Holum, A. (1991). Cognitive apprenticeship: Making thinking visible. *American Educator*, *6*(11), 38–46.

Csikszentmihalyi, M. (1988). Society, culture, and person: A systems view of creativity. In *The nature of creativity* (pp. 325–339). New York, NY: Cambridge University Press.

Davidson, J., & Sternberg, R. J. (1998). Smart problem-solving: How metacognition helps. In *Metacognition in educational theory and practice* (pp. 47–68). Mahwah, NJ: Erlbaum.

Duch, B. J., Groh, S. E., & Allen, D. E. (Eds.). (2001). *The power of problem-based learning.* Sterling, VA: Stylus Publications.

Duncan, S. L. S. (1996). Cognitive apprenticeship in classroom instruction: Implications for industrial and technical teacher training. *Journal of Industrial Teacher Education*, *33*(3), 66–86.

Elton, L. (2000). *Matching teaching methods to learning processes: Dangers of doing the wrong thing righter.* Paper presented at the 2nd Annual Conference of the Learning in Law Initiative.

Entwistle, N. (2003). University teaching-learning environments and their influences on student learning: An introduction to the ETL Project. In *Proceedings of the 10th Conference of the European Association for Research on Learning and Instruction.*

Felder, R. M., & Brent, R. (2005). Understanding student differences. *Journal of Engineering Education*, *94*(1), 57–72. doi:10.1002/j.2168-9830.2005.tb00829.x

Gott, S. P., Hall, E. P., Pokorny, R. A., Dibble, E., & Glaser, R. (1993). A naturalistic study of transfer: Adaptive expertise in technical domains. In *Transfer on trial: Intelligence, cognition and instruction* (pp. 258–288). Norwood, NJ: Ablex.

Haslam, S. A., Adarves-Yorno, I., Postmes, T., & Jans, L. (2013). The collective origins of valued originality: A social identity approach to creativity. *Personality and Social Psychology Review*, *17*(4), 384–401. doi:10.1177/1088868313498001 PMID:23940233

Hazzan, O. (2002). The reflective practitioner perspective in software engineering education. *Journal of Systems and Software, 63*(3), 161–171. doi:10.1016/S0164-1212(02)00012-2

Hitchcock, M. A. (2000). Teaching faculty to conduct problem-based learning. *Teaching and Learning in Medicine, 12*(1), 52–57. doi:10.1207/S15328015TLM1201_8 PMID:11228868

Jonassen, D. H. (2000). Towards a design theory of problem solving. *Educational Technology Research and Development, 48*(4), 63–85. doi:10.1007/BF02300500

Kuhn, S. (1998). The software design studio: An exploration. *IEEE Software, 15*(2), 65–71. doi:10.1109/52.663788

Kuhn, S. (2001). Learning from the architecture Studio: Implications for project-based pedagogy. *International Journal of Engineering Education, 17*(4/5), 349–352.

Larkin, J. H. (1989). What kind of knowledge transfers? In *Knowing, learning and instruction: Essays in honour of Robert Glaser* (pp. 283–306). Hillsdale, NJ: Lawrence Erlbaum Assoc.

Liao, C. H., Yang, M. H., & Yang, B. C. (2013). Developing a diagnosis system of work-related capabilities for students: A computer-assisted assessment. *Journal of Computer Assisted Learning, 29*, 530–546. doi:10.1111/jcal.12011

Lindblom-Ylänne, S. (2004). Raising students awareness of their approaches to study. *Innovations in Education and Teaching International, 41*(4), 405–421. doi:10.1080/1470329042000277002

Lucas, U., & Meyer, J. H. F. (2004). Supporting student awareness: Understanding student preconceptions of their subject matter within introductory courses. *Innovations in Education and Teaching International, 41*(4), 459–471. doi:10.1080/1470329042000277039

Milton, J., & Lyons, J. (2003). Evaluate to improve learning: Reflecting on the role of teaching and learning models. *Higher Education Research & Development, 22*(3), 297–312. doi:10.1080/0729436032000145158

Nelson, W. A. (2003). Problem solving through design. In Problem-based learning in the information age (pp. 39–44). San Fransisco: Jossey–Bass.

Norton, L. S., Owens, T., & Clark, L. (2004). Analysing metalearning in first-year undergraduates through their reflective discussions and writing. *Innovations in Education and Teaching International, 41*(4), 423–441. doi:10.1080/1470329042000277011

Robley, W., Whittle, S., & Murdoch-Eaton, D. (2005). Mapping generic skills curricula: Outcomes and discussion. *Journal of Further and Higher Education, 29*(4), 321–330. doi:10.1080/03098770500353342

Segers, M., & Dochy, F. (2001). New assessment forms in problem-based learning: The valueadded of the students perspective. *Studies in Higher Education, 26*(3), 327–343. doi:10.1080/03075070120076291

Smyth, R. (2003). Concepts of change: Enhancing the practice of academic staff development in higher education. *The International Journal for Academic Development, 8*(1/2), 51–60. doi:10.1080/1360144042000277937

Trigwell, K., Prosser, M., & Waterhouse, F. (1999). Relations between teachers' approaches to teaching and students' approaches to learning. *Higher Education*, *37*, 57–70. doi:10.1023/A:1003548313194

van Gigch, J. P. (2000). Metamodelling and problem solving. *Journal of Applied Systems Studies*, *1*(2), 327–336.

Waks, L. J. (2001). Donald Schon's philosophy of design and design education. *International Journal of Technology and Design Education*, *11*(1), 37–51. doi:10.1023/A:1011251801044

KEY TERMS AND DEFINITIONS

Active Learning: An umbrella term that refers to students learning by engaging in activities (as opposed to passive listening). Activities can include reading, writing, problem-solving, reflection, often in a collaborative setting.

Cognitive Flexibility: The ability to represent knowledge from different perspectives, switching to accommodate changes in the environment that knowledge applies to. This is related to the transfer of knowledge and skills beyond the initial learning environment.

Community of Practice: A group with a common interest seeks to learn from each other through and participating in shared information and experience.

Constructive Learning: The learner constructs knowledge based on an inconsistency between current knowledge and experience, usually undertaken in a social context.

Employability Skills: Within an Australian context, employability skills have been identified as those non-technical skills that enable successful workplace performance: communication, team work; problem solving; initiative and enterprise; planning and organising; self-management; learning; technology.

Higher Order Thinking: Learning taxonomies suggest some learning is foundational (to know, comprehend, apply) while other learning is a complex combination of these skills, and requires critical and abstracting abilities. These (to analyse, synthesise, evaluate) are considered higher order.

Metacognitive Skills: Involves a learner's awareness of their own learning process (rather than the content of the learning). The learner monitors self-awareness and uses it, implying higher order thinking, to control and improve the learning process.

Chapter 9
Practicing Soft Skills in Software Engineering:
A Project–Based Didactical Approach

Yvonne Sedelmaier
Coburg University of Applied Sciences and Arts, Germany

Dieter Landes
Coburg University of Applied Sciences and Arts, Germany

ABSTRACT

Software Engineering requires a specific profile of technical expertise combined with context-sensitive soft skills. Therefore, university education in software engineering should foster both technical knowledge and soft skills. Students should be enabled to cope with complex situations in real life by applying and combining their theoretical knowledge with team and communication competencies. In this chapter, the authors report findings from a software engineering project course. They argue that project work is a suitable approach to foster soft skills. To that end, the authors provide justification from a pedagogical point of view, setting project-based learning into relation to action-orientated didactics. As teaching goals, they focus on experiencing a complete development project from end to end, following a software process model that needs to be adapted to the specific situation, self-determined planning and acting, including the organization of the project, teamwork and team communication, and self-reflection on individual roles and contributions, and on the performance of the project team as a whole. In order to achieve these goals, the authors form teams of bachelor students, which are headed by one master student each. It turned out that a clear separation of roles is inevitable within the team, but also with respect to instructors. Self-reflection processes concerning the team roles and the individual competencies are explicitly stimulated and cumulate in individual self-reports and post-mortem analysis sessions. The authors share findings of how well the approaches have worked and outline some ideas to improve things.

DOI: 10.4018/978-1-4666-5800-4.ch009

1. INTRODUCTION

Software is a core ingredient of nearly any part of our everyday life. This software, however, needs to be developed by highly skilled individuals. Consequently, in order to acquire and exercise these skills education in software engineering plays an important role in university education. Traditionally, universities laid their main emphasis in software engineering education on technical skills, such as e.g. programming or testing skills. In recent years, however, it has become increasingly evident that non-technical, also known as soft, skills are equally important as software is developed in teams of individuals who need to interact with each other and various stakeholders such as, e.g., customers or users of their software.

Software Engineering requires a specific profile of soft skills that is closely related to technical expertise. Recently, the authors conducted a survey with junior and senior managers with respect to which skills they expect from graduates in the software engineering field. Respondents of the survey always emphasized the importance of soft skills (Sedelmaier, Claren, & Landes, 2013).

Undoubtedly, software development requires profound technical knowledge (Sommerville, 2011). But evidently, this is not the only thing that matters. Rather, various soft skills are also needed, e.g. the ability to work in a team of hundreds of members spread around the world or the ability to communicate with various other players in the project. All interviewees want software engineers to analyze and understand complex situations and use a creative and solution-orientated approach. Several other researchers arrived at similar results and emphasize the importance of non-technical skills (Lu, Lo, & Lin, 2011; Richardson, Reid, Seidman, Pattinson, & Delaney, 2011; Rivera-Ibarra, Rodríguez-Jacobo, & Serrano-Vargas, 2010).

Although soft skills obviously are important, our survey showed that students tend to overestimate their capabilities with respect to both technical and non-technical competencies. Soft skills are core competencies of a software engineer and for this reason soft skills should be a core part of software engineering education at universities. These issues are quite difficult to teach because there are no clear cause-effect-relationships between the didactical approach and the learning outcomes. So it is difficult for teachers in software engineering to find out which didactical approach works best. Furthermore, instructors are generally not communication experts themselves and often have neither pedagogical nor didactical education background. Many things they do are not grounded on pedagogical expertise.

Thus, preparing students for the real life in software engineering is a serious challenge in university education. One approach to bring complexity and problem awareness into university education is to use project work. Project work fosters many soft skills such as communication skills and the ability to work together in a team. Interpersonal skills cannot be trained without other people around, and project work combines these competencies with the context in which they are needed. Furthermore, project work could offer students opportunities to understand interrelationships between technical knowledge and soft skills.

Many software engineering projects fail due to at least one of the following reasons: scheduling, specifications and/or average manufacturing costs (Button & Sharrock, 1996). Button & Sharrock (1996) also state that software engineers tend to distinguish between two basic types of problems: "First, those that are due to deficiencies in the state of general engineering practice, and second, those that arise from the state of the project they were engaged in. Engineering work on any particular development thus does not involve only the resolution of the problems arising from the specific circumstances of the project itself, but also contends with problems that are recognized as generic problems of engineering work per se" (Button & Sharrock, 1996). Students hardly believe these facts. In their opinion they would do much better and lead the project to success if they were the actors. Project work in a university context gives students the chance to prove that they can really succeed while understanding the difficulties of project work and the reasons for failure.

In this chapter, we report findings from a software engineering project course that is regularly run in the final year of a bachelor program in informatics. As one of the goals of this capstone project, we particularly aim at advancing the non-technical skills of the participants in addition to the technical ones. In the following section, we first underpin our didactical approach, and then outline our teaching goals and the restrictions we have for the project course. Then we describe the approaches that we took in order to achieve our teaching goals, in particular with respect to advancing soft skills. This will be followed by a discussion of how well our approaches worked. Finally, we summarize our main points and point out further directions for our research and teaching.

2. COMPETENCIES IN SOFTWARE ENGINEERING

Core questions at the beginning of planning a curriculum are: What are the competencies a software engineer should have? Which aims do lecturers want to achieve? These questions must be answered before an appropriate didactical approach can be chosen.

An initial guideline with respect to technical skills is the Software Engineering Body of Knowledge (SWEBOK) (Abran & Moore, 2004). SWEBOK describes technical knowledge that any software engineer must possess after four years of working practice. The competencies in SWEBOK span the complete software development lifecycle from requirements engineering to software testing, implementation, and related disciplines. Furthermore, the required level of competencies is sketched on the basis of Bloom's taxonomy (Bloom, Engelhart, Furst, Hill, & Krathwohl,

1956). On this basis, lecturers get a first piece of evidence as to whether students are supposed to only remember some piece of information about software engineering, to develop a deeper understanding of the topic, or to be able to analyze a concept and disassemble it into relevant parts. A new revision of SWEBOK is expected in 2013 which will also cover soft skills.

Further recommendations for a Software Engineering curriculum can be found in "Software Engineering 2004 Curriculum Guidelines for Undergraduate Degree Programs in Software Engineering" (IEEE Computer Society & Association for Computing Machinery ACM, 2004). However, this Software Engineering Curriculum is intended to only support undergraduate software engineering education.

Both SWEBOK and the IEEE/ACM Curriculum can propose contents for a university curriculum in software engineering, but they are not sufficient as they currently only cover technical expertise on a very abstract level. Furthermore, SWEBOK addresses a target other than university students, software professionals with four years of work experience. Worst of all, however, is their lack of attention towards required soft skills for software engineering. Neither of these handbooks gives a recommendation of which soft skills a software engineer should have and what a particular soft skill exactly means.

Consequently, no appropriate recommendations for university education in software engineering exist to date. Since universities aim to prepare their students for a successful career in industry, industry's requirements do matter. In order to get a clearer and up-to-date picture of these requirements, we asked graduates and software professionals as well as managers which competencies they expect or need in their daily business, and how important several topics are.

We used SWEBOK as a guideline and conducted informal interviews about the required technical knowledge in software engineering. This provided us with information about the technical knowledge that is needed in everyday business. These results were cross-checked against the lecturers' aims and existing curricula. In this way we identified an initial set of required technical skills in software engineering.

Later, these informal interviews were analyzed a second time, now focusing on the required non-technical skills in software engineering. This process led to an initial code system of non-technical skills (Sedelmaier et al., 2013). These initial categories will be further elaborated based on grounded theory and are main ingredient of our research design (Sedelmaier & Landes, 2013a). To this end, we are currently conducting guided interviews to discover additional required soft skills in software engineering and to understand them better. For instance, it is still necessary to clarify what is meant by communication skills or the ability to work in a team.

Our research reveals which are the most important soft skills for software engineers (Sedelmaier et al., 2013; Sedelmaier & Landes, 2013b):

- Comprehension of the complexity of software engineering processes (understanding how things interact, the interrelationships between several software engineering issues; seeing the software engineering process and product in the overall context of an (unknown) company or organization)
- Awareness of problems and understanding of cause-effect relationships (understanding *why* something must be done, *why* some things in software engineering have to be done; the ability to recognize a problem before it appears)
- Team competence including communication skills (Working with others)

In a similar survey, team competence turned out to be the most important generic competence: "Employers searched for future employees with team competence including team-oriented thinking and acting, the ability to work in a team as a communicative, motivated, performance-oriented team player, or team leader. Furthermore, many employers (~66%) asked for communication competence, including the ability to handle conflicts, criticism and to find consensus, as well as the ability to use moderation and presentation techniques." (Kabicher, Motschnig-Pitrik, & Figl, 2009)

Our findings indicate that there are two categories of soft skills: context-sensitive soft skills, and generic soft skills. Generic soft skills are abilities that are largely independent of software development and are relevant for other disciplines, too. Presentation-skills are a typical example: they are equally relevant for a social worker, a businessperson or a software engineer. Presentations follow the same rules regardless of the context in which they are given. The presenter should speak to his audience and show readable slides, irrespective of whether software architecture or a new washing machine is presented. In contrast, domain-independence does not hold for context-sensitive soft skills. Skills in this category exhibit a special, unique profile in the context of software engineering. Lacking these skills leads to specific consequences in the software engineering process. If the software architecture is not presented properly, colleagues will still understand what to do. But if communication skills or the ability to solve problems or conflicts are insufficient, the whole software engineering project may fail. Therefore, it is important to understand what communication skills do mean exactly in a software engineering context. Should software engineers be able to talk in three different languages? Or should they be able to write technical documentation? Or must they recognize and solve misunderstandings arising from badly formulated requirements? And how can they formulate requirements as clearly as possible? Which communication techniques may help to cover these challenges?

The combination of various context-sensitive soft skills gives a special profile of software engineering competencies. Because context-sensitive soft skills are closely related to technical knowledge, students must be trained and advanced with respect to these skills in the context of software engineering. Only in relation to technical expertise and the software development process will students acquire an awareness of the inherent problems.

Therefore, a core question in our didactical research project EVELIN (Experimental improvement of learning software engineering) is how to foster soft skills in software engineering education at universities. Since context-sensitive non-technical competencies of software engineers have a specific characteristic in conjunction with other software engineering skills they need to be exercised in a software engineering context. Our findings will be summarized in a Software Engineering Body of Skills (Sedelmaier & Landes, 2013b).

3. PEDAGOGIGAL VIEW ON PROJECT-BASED LEARNING

Thus, one main focus in software engineering education lies in fostering soft skills in addition to technical knowledge. To that end, new didactical approaches are necessary. By only reading theoretical information, no student would be capable of more precise communication. Consequently, learner-activating didactical methods are required and should be applied in university education. According to Foppa (1975) Comenius stated that we learn

- 10% of what we read,
- 20% of what we hear,
- 30% of what we see,
- 50% of the what we hear and see,
- 70% of what we explain to others,
- 90% of what we do.

Therefore, learner-activating pedagogical theories look back on a long tradition and became even more important when psychology revealed new insights into how people learn (Lefrançois,

2006). Since the learner is no longer a black box and psychologists learned more about the determinants of students' learning processes, a change of view took place also in pedagogical paradigms.

Learning is no longer connected with punishment and trail-and-error but has developed a more individual orientation. A growing interest in individual context, experiences, and previous knowledge went along with a change of the pedagogical concept of man and the development of constructivist theories. Learners are no longer only pupils in schools but also adults, continuing to learn their whole life (Merriam, Caffarella, & Baumgartner, 2007). New ideas about lifelong learning have become prominent in pedagogical opinions. Consequently, learners are no longer seen as unmotivated children who have to be forced to learn. A new concept of man arose during the last century. Now, learners are recognized as self-responsible individuals with their own goals and reasons why they learn (Merriam, 2001). As a further consequence, traditional learning methods no longer work and new didactical approaches are necessary. These new didactical methods are learner-centered and activating. The aim of learning is no longer to just accumulate theoretical knowledge but rather to apply and transfer it to real life. Pedagogical methods should give learners a framework to experience things and to experiment in how skills can be applied. Consequently, the call for enabling students to cope with real-life-problems became louder and louder.

Project-based learning is a didactical approach which has been intensively discussed, especially in Germany and Denmark, since the 1970s. This method builds on a tradition of 300 years of history. Project-based learning can be allocated to action-orientated didactical categories with focus on real-life experiences. Going hand in hand with the changes in didactics from contents and topics towards outcome and competence-orientation, didactical approaches turned from direct instruction to activating learning styles such as project work (Terhart, 2009) or problem-based learning.

Of course the boundaries between problem- and project-based learning are unclear. Nevertheless, both approaches share several commonalities: both can be allocated to constructivism, a social learning theory (Kemp, 2011; Savery & Duffy, 1998), and both are learner-activating didactical approaches. They are characterized by the complexity of their subject. In both approaches the lecturer is a coach rather than a teacher in the common sense (Knowles, Holton, & Swanson, 2011) and in both approaches students work together in groups.

Yet, there are also several differences. Project-based learning starts with a common goal, e.g. an end product that has to be manufactured. This is a difference to problem-based learning which starts with a problem (Barrows & Tamblyn, 1980). Bransford and Stein (1993) describe problem-based learning as follows:

I = Identify problems and opportunities
D = Define goals
E = Explore possible strategies
A = Anticipate outcomes and Act
L = Look back and Learn" (Bransford & Stein, 1993)

Problem-based learning is an inquiry model. Students are given a problem that has to be solved. After receiving a problem, students organize their existing knowledge and identify areas where they need more knowledge (Savin-Baden, 2001). In the next step they plan how to attain the required information and how to solve the problem.

Conversely, in project-based learning students start work from an idea of a desired solution and then have to develop a plan of how to achieve this goal. In project-based learning the focus lies on a production process and on working together as a team. The end product drives the planning, production, and evaluation process. Therefore, goals in project-based learning play an important role. This is different in problem-based learning approaches. Problem-based learning primarily addresses problem-solving skills while project-based learning focusses on working together to come up with a joint result.

Thus, we have chosen a project-based learning approach to foster soft skills in software engineering.

Project-based learning is characterized by several features (Jung, 2010; Rummler, 2011; Siebert, 2010); it is:

- Limited in time,
- Goal-oriented,
- Result-oriented,
- Self-organized,
- Collaborative with divided responsibilities, and
- Inherently complex and interdisciplinary.

Goal- and result-oriented work requires time and resource management. Otherwise the project runs into a dead-end. Due to the inherent complexity and interdisciplinary approaches projects contain potential for conflicts.

Team members have to communicate with their superiors and peers within the team. This is one of the stumbling blocks in working together. Communication is to a very large extent determined by team members' personalities and influenced by sympathies or antipathies. Psychological research offers many studies concerning communication (Fiske & Jenkins, 2011) and the emotional component of interaction and group dynamics (Cavrak, Orlic, & Crnkovic, 2012; Tindale et al., 2002). As an example, Schulz von Thun (1996) offers a four-ear-model to distinguish levels of conveyance that can be present in one message (Schulz von Thun, 2008). His model helps to understand and analyze ambiguities within a message. Students should become aware of communication models in order to acquire the ability to reflect on team communication and to actively improve it. But communication is not only an important success factor within the team. In order to arrive at usable requirements for software, someone must communicate with stakeholders or customers. Their background normally differs from the background of the team members and they often do not understand the software engineering terms. Communication also contains a potential for conflicts in this interdisciplinary context. This may be the case when customers talk about the requirements of software and the team is not able to understand them properly. Furthermore, implicit requirements are not formulated, but nevertheless expected. So software engineers must be able to communicate with other disciplines and understand the statements of all the other parties. Project-based learning allows students to experience this in an authentic way and fosters team and communication skills. Project work emphasizes reflection on communication, interaction, atmosphere, feelings, and interpersonal relationships (Figl & Motschnig, 2008), and, consequently, fosters self-reflection.

We are interested in fostering higher-order thinking skills including critical thinking, synthesis skills, and problem-solving. This can be achieved through project work, since as a didactical strategy, this approach exposes students to situations which cannot be mastered by simply remembering factual knowledge, thus this approach fosters higher-order thinking skills (Anderson, Krathwohl, & Bloom, 2008; Bloom, Engelhart, Furst, Hill, & Krathwohl, 1956).

Project work requires transferring theoretical learned knowledge and methods to real-life problems. Rummler (2012) observed that tasks in project-oriented learning are in general more authentic and, thus, more relevant than in problem-based learning settings. Project work offers the opportunity to apply theoretical knowledge which students have already acquired. Students should develop a sense of the complete software development process from its beginning to its end. Project-based learning enables students to assemble the theoretical knowledge that they learned in somewhat isolated chunks into a co-hesive unit. Students become aware of complex interdependences and the interrelation of many distinct issues they have learned.

Project-based learning includes learning by doing and allows students to gain practical experience which is more important than theoretical knowledge: "Give the pupils something to do, not something to learn; and the doing is of such a nature as to demand thinking; [...] learning naturally results." (Dewey, 2011) Consequently, learner motivation and active engagement are stimulated by being challenged in a complex project.

Thus, project work is a good didactical approach to gain a deeper understanding of the interrelationships between various topics in software engineering expertise combined with soft skills.

4. GOALS AND RESTRICTIONS

4.1 Constraints

Students in the bachelor program in informatics at the University of Applied Sciences in Coburg can enroll in a software engineering project (SE project) in their final year. This project is offered as an elective course and is associated with 6 credits according to the European Credit Transfer System (ECTS). As such, the project is expected to put a workload of approximately 180 hours on each of the participants. This workload includes 4 contact hours per week in which the project team physically meets at the university. During these contact hours, instructors are also present so that open issues can be discussed easily. Typically, the project runs for 14 weeks from October to January. So far, we have had three iterations of the software engineering project course in the years 2011 to 2013.

Since the course is an elective, the number of participants varies from year to year, ranging from 10 to 25 students. Participants are divided in projects teams of 4 or 5 members. Normally, project topics differ between teams, although occasionally two teams work on the same problem, yet following different approaches, in particular with respect to the underlying process model.

Participants acquired solid programming skills during courses in their first and second years, and they already took a basic course in software engineering and an elective course focusing on software requirements, architecture, and testing in more detail. These two courses establish the theoretical and methodological basis for relevant software engineering topics. Yet, these topics are covered in a somewhat abstract and isolated fashion: although there are lab exercises to provide some hands-on experience with these issues, there is no common frame that ties the individual topics together. Furthermore, these two courses almost exclusively address technical competencies. Two professors are in charge of teaching these two courses; they jointly supervise the software engineering project. Starting in the most recent course, additional support from two research assistants is available.

Several general non-technical competencies have also already been practiced before students embark on the software engineering project. For instance, every student already has had to give a presentation in a seminar, and some students have taken an additional elective course on time- and self-management.

The software engineering project is intended to tie technical and non-technical software engineering skills together as a capstone project in which students traverse all steps of the development cycle, from eliciting and analyzing requirements through delivering a working and well-tested software system which is accompanied by adequate documentation, according to a firm deadline in mid-January. This implies that the project topic for each of the projects is much more complex than anything the students have had to master so far. In all iterations of the project course so far, the project topic was concerned with developing a

software system from scratch, i.e. without code or other work that is intended to be used as a basis. A maintenance project, however, which is targeted at extending and restructuring an existing system, might be an option for future iterations. Yet, this would also imply slightly different teaching goals. For instance, understanding existing legacy code on the basis of potentially thin documentation is required for maintenance, but not really necessary for developments from scratch.

In general, the instructors present several proposals for potential project topics, but students may also make proposals which need to be approved by the instructors to make sure that they comply with the teaching goals for the course. On the basis of this presentation, students decide on their favorite project topic and, thus, also choose their teammates, though only implicitly. Instructors will, however, adjust project teams in order to make sure that each team is adequately staffed and is neither too big, nor too small. As an additional restriction, each project team needs to follow a software process model. To that end, each team has to choose either a plan-driven model, e.g. Unified Process (Kroll & Kruchten, 2003) or V Model (Dröschel, 2000), or an agile process model such as SCRUM (Schwaber & Beedle, 2002) or Extreme Programming (Beck, 2000). Freedom of choice is slightly constrained by the fact that in each iteration of the project course, there should be at least one team following a plan-driven approach as well as at least one agile team. The idea behind this restriction is that students might acquire insight into a process model, even if they did not use it themselves, simply by talking to their colleagues or listening to others' final presentations.

4.2 Experiences from Previous Iterations of the Project

Adhering to a process model, regardless which one it is, implies that several roles need to be filled. Relevant roles are within the project team, such as requirements analyst, software architect, project leader, or software tester, or outside the project team, in particular the role of the customer. The instructors stipulate which internal roles should be filled within the project. The project team decides on its own which team member will take which roles. For some project topics, but unfortunately not all of them, there is a real customer, e.g. someone from other departments of the university. In the first and second iteration of the course, the instructors also adopted the role of the customer in those projects that were lacking a "real" customer.

Generally instructors are responsible for more theoretical courses on software engineering issues and for setting the general rules. They also decide on the grades that the students will earn in the software engineering project. In addition, instructors also act as coaches who guide the software engineering process, and as mentors who can be asked theoretical questions. In some past software engineering projects, the instructors also represented the customer who set requirements and offered ideas about the software that should be built. As it turned out, there was a role conflict between instructors adopting the role of a customer and their primary roles. The conflict between these roles became particularly evident in the review process. On the one hand, an instructor played the part of the customer; on the other hand, he provided guidance concerning the structure of the review process. Students never knew which role the teacher was actually playing at a specific point of time. In addition, this role conflict was also confusing for the instructor since he had to facilitate and attend the process as well as examine and grade the students' performance.

Software engineering projects demand a lot from the students. They have to keep many facts in mind and must cope with various complex issues in parallel. Their technical expertise should be applied in the context of a whole software engineering project. Fragmentary theoretical know-how must be applied and combined into something coherent in the project. Theoretical expertise must be transferred to nearly realistic situations. Students have to structure themselves in order to plan and organize the whole engineering process on their own. Furthermore, the process model cannot be applied out-of-the-box, but rather needs to be adapted to the specific situation. Within each of the project teams, one member has to fill the role of the project leader. In earlier iterations of the course, it turned out that the project leader was easily overburdened by his duties since he or she was not only responsible for management issues, but also for some of the more technical tasks. Furthermore, a lack of project management knowledge became apparent.

Students work in a team of 4 or 5 individuals for the first time. Although working in a team advances soft skills, initially this was not an explicit learning goal. Teamwork just happened "en passant" (Reischmann, 1986) and without clear instruction or explicit support, in particular with respect to team formation. As a consequence, the learning effects were neither clear to students, nor to instructors.

As one of the deliverables of the project, each project team is required to present their approach and their results in a presentation at project completion. Each team member is required to be actively involved in this presentation in order to outline the results from the perspective of her or his roles. On this occasion, it often becomes evident that a majority of the students have weak presentation skills.

4.3 Teaching Goals

Our survey revealed three main competencies a software engineer needs (see sec. 2), namely comprehension of complexity, awareness of problems, and team competence. So in a first step it was one of the goals of the software engineering project to make formal tacit teaching goals explicit, also for the students, and to reach an agreement and a shared understanding among the instructors with respect to these goals. If instructors have no explicit teaching goals it is impossible for students to achieve them and also impossible for instructors to check the achievement of objectives. "If you're not sure where you're going, you're liable to end up someplace else." (Mager, 1992) Teaching goals are necessary for instructors to design a curriculum, to decide on a didactical approach and to check the achievement of teaching goals. Especially in project-based learning the definition of goals is very important (see sec. 3) because there are no closed-question-tests or clear scales to measure the learning outcome. Only by defining goals can lecturers evaluate the success of learning. This is similar to the software development process: One cannot test software if there are no requirements to test against. Aggravating this situation, in our software engineering project two instructors act in parallel. Ideally, they have to coordinate their individual teaching goals and to pursue their goals consistently. Otherwise, it would be difficult to reach conflicting objectives.

4.3.1 Technical Teaching Goals

Originally, when we started to think about our learning concept, the explicit teaching goals concentrated on technical expertise. These teaching goals still hold after three iterations of the software engineering project course. In particular, students are to combine all the theoretical bits and pieces they have learned before in order to master a complete project, resulting in a usable executable software system plus a set of compulsory documents. They are to experience a complete software development project in a setting that is as realistic as possible, given the context of a university. In particular, bachelor students should be able to recognize how things fit together and develop an increased understand of issues that have remained abstract and somewhat intangible up to this point. This is particularly true for process models: before embarking on a project of realistic size and complexity, the need to adhere to a process model was not really clear to the majority of our

students. It is also one of the goals that students should gather hands-on experience in tailoring a "text-book" process model to their concrete needs in the project.

Another teaching goal consisted in pointing out that it is not easy to get a complete set of clear and consistent requirements, even when the customers behave perfectly cooperatively.

Although these teaching goals still hold, after two iterations we had the impression that we needed to react to our experience up to that point (see sec. 4.2). One of our findings showed that the intertwining of technical and project management issues was fairly hard to handle for bachelor students. In the third and latest iteration of the project course we had an opportunity to tackle this problem by combining bachelor and master students in the same project. This allowed bachelor students to concentrate on technical issues and experience project management in a more passive fashion, while master students were in charge of leading the project and in particular of adapting the process model to the specific situation.

In addition, we decided to address non-technical competencies to a much larger extent. Our teaching goals with respect to non-technical competencies will be detailed in the following section.

4.3.2 Non-Technical Teaching Goals

As a first change, teaching goals concerning soft skills are explicitly defined. In our new didactical concept we want to foster team skills while addressing the teaching goals mentioned above (see sec. 4.3.1). Team skills in this context include the capability to communicate with each other and, furthermore, to accept and fill a team role adequately. In addition, team skills encompass a sense for verbal and non-verbal communication as well as active listening skills and the importance for working together. Our survey supports this understanding of team skills. As mentioned above, we conducted a survey at the beginning of our research. This survey also showed a rather unrealistic and overly optimistic self-assessment of students concerning their non-theoretical skills.

Project work seems to be a good opportunity to help students come to a realistic self-assessment without megalomania. One of our teaching goals is to give students the technical expertise they are conscious of as well as the wisdom to acknowledge the experience and knowledge of others. Bachelor students should learn to integrate themselves in a team while master students should improve their leadership skills. Project work is a good opportunity to participants to integrate themselves in a team and to listen to the opinions of other team members. But project work also fosters the ability to do what a project leader decides is a good approach without questioning that approach too much. Moreover, it is not necessary that each team member knows every detail, especially when it is not relevant for his or her own work or the success of the team. Students should learn that the team leader has information and knowledge about all aspects of the project and, as a consequence, takes responsibility for all decisions. All other team members do not need to have all the information, but only the fraction that is relevant for their role. Team members have to accept their roles. Project work provides an opportunity for students to learn this at the university instead of having to experience the problems caused by not accepting their roles in the workplace. Experiencing this was helpful for bachelor students as well as for master students because at the beginning of the project they all felt that they were in some way better than all the other team members. During the project, they often learned that this feeling was a mistaken assumption.

Each student has to play an individual role in the team such as software architect, software tester or team leader. There are, however, no pre-defined role assignments among the bachelor students. Only the roles of master students are pre-defined in that they are acting as project leaders or process managers (depending on the process model) and are responsible for planning appropriate quality assurance activities. Each bachelor student has to decide which role he or she wants to fill according to his or her own individual gifts and inclinations and fill his or her role in such a way that the others can rely on his or her results.

As a teaching goal, students should clearly understand that the contribution of each student, regardless of bachelor or master level, is needed to cope with the team challenge and to achieve the team goal: produce running software. A single student, all by herself, cannot save the project if trouble develops. Students should become aware that they can only succeed if they work together as a team. The whole team is responsible for the final outcome.

One further important non-technical teaching goal is that students should organize their work independently. This means that instructors no longer act as super-project managers who overrule the "legal" project managers in some situations. Rather, students must plan their time and agree on a time schedule – there is no pre-set project plan, but rather students, and in particular master students, need to assemble the project plan with the team.

As a teaching goal, students are expected to explore various options in a specific situation and decide which options best meet their own goals. Yet, students should also have the freedom to make mistakes, learn lessons from them, eventually develop alternatives, and decide which is the best. Students are expected to try out things without being bounded by tight restrictions imposed on them by the instructors, except for one: The deadline for the termination of the project must be met under all circumstances.

An additional teaching goal is that students should acquire missing information independently if the need arises. In order to do so, they should refer to any kind of resource that they see fit, including asking the instructors for hints and advice. Instructors, however, will not provide ready-to-use-solutions but rather act as coaches who point out to the project members what they might try to find a solution.

Further teaching goals differ in detail for bachelor and master students. The focus for bachelor students lies on understanding and combining chunks of technical knowledge, which up to then have been isolated, into one big picture and on integrating in a team, which includes fostering communication skills. They should learn to accept a team leader and to concentrate on their own task.

Master students focus on leading a team. The difficulties for them are, e.g., communicating with team members, and structuring tasks and motivating team members for effective team work. Master students are responsible for the results and for meeting deadlines as well as for assuring the quality of the software. But master students must recognize that they rely on goodwill and adequate results of their team members. It is no solution to give commands, but rather team leaders must learn to convince and motivate team members to do a good job. In contrast, bachelor students should become aware that a project leader also has important issues to deal with and even a project leader accomplishes tasks for the whole team. Master students should learn to communicate in an open way and to set team goals. They must strike the balance between leading in an authoritarian manner and giving individual team members the feeling of being important for the team success.

5. SPECIFIC APPROACHES TO ACHIEVE OUR GOALS

In section 4.2, we briefly discussed some of the findings we made during earlier iterations of the SE project course. Some of the arrangements that we will outline below try to address the shortcomings that we have identified so far.

5.1 Master Students Act as SCRUM Master / Project Leader

As mentioned in the previous section, there is a pre-set role assignment for master students: they act as project leader in plan-driven process models

or process managers in agile process models. In particular, in the most recent iteration of the project course, one team was expected to work according to SCRUM, one of the most popular agile process models. Consequently, the master student in this project team acted as SCRUM master.

Regardless of the process model, master students were responsible for customizing the process model to the specific situation. In textbooks, such as Kroll and Kruchten (2003) or Schwaber and Beedle (2002), the process model is described in a fairly generic fashion, with all the options that are generally available. For instance, the Unified Process distinguishes a multitude of roles and deliverables – far too many for a relatively small project as in the software engineering project course. Therefore, the project team, led by the project manager, has to decide on which deliverables and which project roles are really important. Likewise, the SCRUM master has to organize the SCRUM process for her team, answering such questions as how long daily SCRUMs should last, how they should be held when the team cannot meet physically, or what methods should be used to express and prioritize user requirements.

5.2 Clear Separation of Roles

The role assignment for master students allows both bachelor and master students to be more focused on specific aspects of the overall project: master students emphasize management issues and quality management, while bachelor students concentrate on the technical and operative aspects, such as writing requirements, defining the system architecture, implementing the system, and testing. Master students lead the team while bachelor students integrate into the team. While bachelor students focus on technical issues concerning the operative implementation, master students constitute the interface between the team and customers or instructors.

To enable them to fulfill their roles, during the project master students were coached to reflect and improve their leading skills in a related course in which they discussed the challenges and problems they faced in their teams.

Yet, there was also a need to avoid a mixture of partly conflicting roles for the instructors. In order to do so, instructors decided to play a rather passive role as observers. They only took a more active part as coaches on methodological and technical issues when they were called in by the project team, or more specifically the project leader or process manager of one of the teams. They did, however, refrain from becoming actively involved even when they observed the project team running somewhat off track. Likewise, they tried to avoid setting restrictions and overruling proposals or decisions in the project teams. Otherwise, they would have been acting as super-project leaders and would have undermined the role of the process managers or project leaders in the teams.

In the most recent iteration of the project course, both teams selected project missions for which there were no real external customers. Therefore, customers needed to be simulated by two research assistants, who basically provided requirements for the two systems that needed to be built.

5.3 Self-Reflection on Learning Outcomes

Compared to earlier iterations of the project, instructors played a considerably less active role in the projects in the most recent iteration. As a consequence, responsibility for the learning process transitioned to the students themselves to a much larger extent. The instructors initialized and sustained this learning process.

As one element to stimulate learning, a post-mortem reflection was completed. To that end, students were asked to reflect on their own individual role in the project as well as the performance of the entire team. Reflection and metacognition are advantages of project work and are didactical methods to foster soft skills and competencies (Rummler, 2012).

Self-reflection was stimulated in a two-step process. First, each of the students had to prepare a short individual self-report that addressed issues such as

- Their roles and tasks in the project,
- Their expectations with respect to the project and the degree to which these had been met,
- Particular issues in the project that they personally would have treated differently and, from their personal point of view, more successfully,
- Which role they would have liked in the project and what they would have done differently in that role, and
- How interaction and cooperation between team members evolved during the project, including their subjective explanation for these changes.

Secondly, one week after the project was complete, the project team met with instructors and research assistants for a post-mortem analysis session lasting approximately two hours. This feedback session had an open format without strict structure and served to reiterate any possible aspect that seemed worth being discussed in the group. Instructors used information found in the students' self-reports, addressing statements mentioned in the reports, challenging some of the statements made in the reports, and asking the entire team to comment on these statements.

5.4 Grading Based on Technical and Non-Technical Issues

Grading joint work in a project team fairly which entails paying adequate tribute to individual contributions, is a complex undertaking (Farrell, Ravalli, Farrell, Kindler, & Hall, 2012; Huffman Hayes, Lethbridge, & Port, 2003). Building on and extending some of the ideas of Hayes et al. (2003)

and Farrell et al. (2012), we devised an elaborate grading scheme for the SE project.

In earlier iterations of the project, we did not pay special attention to non-technical skills, but rather focused on technical issues. This was based on the assumption that since students knew each other beforehand, they would therefore be able to work in a team to cooperate and achieve their goals, and hence these non-technical aspects were not taken into consideration in grading. This situation changed when master students were introduced to the teams: master and bachelor students did not necessarily know each other. Therefore, coming together as a team and identifying ways of accomplishing tasks were issues that needed to influence grading as well. To that end, we devised a grading scheme that covered technical and non-technical results as well as individual contributions and the results of the entire team.

In particular, we based the grading of the bachelor students on the following aspects:

- Technical quality of results (completeness, complexity of the project topics),
- Customization of and adherence to a process model,
- Individual technical contribution,
- Individual team-orientation,
- Self-reflection, and
- Final presentation.

Likewise, grading for the master students was based on

- Adaptation of the process model,
- Process quality and leadership,
- Self-reflection,
- Final presentation, and
- Resumes of team members.

Master students wrote "employer's references" for their team members. These references were not involved in instructors' grading but bachelor students knew about these certificates without knowing their contents.

6. EXPERIENCE: WHAT WORKED WELL AND WHAT DID NOT

We pursued several goals in the SE project course, in particular with respect to establishing a basis for practicing soft skills. Not all of them, however, could be achieved as we had intended. This will be discussed in more detail subsequently.

Team formation did not work as easily as we had expected. Due to different academic schedules for the bachelor and master students, the bachelor students began the project one or two weeks before the master students joined them. In particular, the bachelor students selected their technical roles and began working on the project plan without guidance of the master student as project leader. One of the bachelor students in each team had taken the role of informal project leader. This situation made it very difficult for master students to catch up; also, it took quite some time before contributions of the master students were perceived as valuable. An additional complication was caused by the fact that the bachelor students already knew each other, while the master students were perceived as outsiders for quite some time. Parts of these problems may be fixed easily in future iterations by better matching the schedules of the different programs and giving master students a chance to actually lead their team. Team formation as such, however, may deserve additional attention since it apparently does not simply happen by itself.

Furthermore, it turned out that the role of the master students in the project was not as clearly defined as we had expected. In particular, there was some debate if contacts with the customers should always be initiated through the master students in their role as project leader or SCRUM master. Also, it was unclear if bachelor students were entitled to ask for technical support from the instructors directly or whether master students should act as proxies in such a situation. Furthermore, the question of whether the project leaders and process managers were in a position to take final decisions caused considerable uneasiness; likewise, bachelor students were not at all comfortable with the mistaken idea that master students would be involved in grading them. (Actually, master students were not involved in grading the bachelor students.)

However, we achieved our teaching goal of initiating a self-reflection process of our students. During the 2-step-process of our post mortem review after the end of the project, a change in the students' self-assessment became evident. Although during the project they sometimes acted like little children defending their own toy castle, in the written exposés first indications of a development were seen. In the review discussion one week later those who were aggressive toward their team members had become remorseful and insightful. From these findings can be concluded that we achieved our teaching goal of giving students a better understanding of the importance of working together and which aspects makes working together efficient and successful. At least some of our students came to a realistic self-assessment concerning their actual competencies.

Jung (2002) explains poor monitoring of students by instructors as one stumbling block. During the running project it became evident that the coordination and consistent implementation of our teaching goals was inadequate. While one of the instructors behaved in a cooperative-coaching way the other tended to give more instructions and became more often a super-project-manager. One instructor acted more laissez-faire while the other one led in a more authoritarian manner. Furthermore, there was no coordination between the instructors as to who gave what information to which team member and why. This was disturbing for our students, especially for the master students

who had to coordinate team work and represent the team outwardly.

7. SUMMARY, CONCLUSION, FUTURE WORK

In this chapter, we reported on a software engineering course that we have run in three iterations. Although we focused on technical competencies in the beginning, it soon turned out that such a capstone project is also a very good setting for practicing soft skills. To address this type of competencies, we introduced various aspects into the project that are intended to foster the soft skills of students, in particular their abilities for teamwork and improved self-reflection on their role and their contributions to the project.

In conclusion, we are convinced that our software engineering project is a successful didactical approach to achieve our teaching goals and to foster soft skills in software engineering. Both the students and the instructors learned quite a bit. But there are still some improvements to be made in future iterations.

In order to build the teams successfully, we will bring master students into the project from the beginning. They need to receive clear instructions with respect to their mission and the opportunity to prepare themselves for organizing and leading a team professionally. Master students should actively build their teams, e.g. by applying the team roles of Belbin (2010b).

Furthermore, our master students should receive more systematic instructions and coaching in order to put them in a position to lead a team to success. The weekly meetings among master students will be supplemented with more theoretical knowledge. Possible issues are e.g.

- How to write a reference for employers,
- How to communicate in an open and constructive way and to listen actively,
- How to give and to actively elicit feedback,

- The influence of the various team roles,
- How to manage the team (Belbin, 2010a),
- The phases of group development (Tuckman, 1965),
- How to cope with and solve conflicts, and
- How to adapt a process model to a specific project.

One challenge for teachers is to coordinate their teaching goals and their behavior towards students. It is essential to act in unison.

In the next iterations of the project, a more systematic assessment of students' competencies would be helpful. To that end, a student assessment will be developed based on our framework for describing soft skills. This is necessary to evaluate our didactical approach and to collect information guiding further improvement of the project.

ACKNOWLEDGMENT

The research project EVELIN is funded by the German Ministry of Education and Research (Bundesministerium für Bildung und Forschung) under grant no. 01PL12022A.

REFERENCES

Abran, A., & Moore, J. W. (2004). *Guide to the software engineering body of knowledge*. IEEE Computer Society.

Barrows, H. S., & Tamblyn, R. M. (1980). *Problem-based learning: An approach to medical education*. New York: Springer.

Beck, K. (2000). *Extreme programming eXplained: Embrace change*. Reading, MA: Addison-Wesley.

Belbin, R. M. (2010a). *Management teams: Why they succeed or fail*. Amsterdam: Butterworth-Heinemann.

Belbin, R. M. (2010b). *Team roles at work* (2nd ed.). Oxford, UK: Butterworth-Heinemann.

Bloom, B. S., Engelhart, M., Furst, E., Hill, W., & Krathwohl, D. R. (1956). *Taxonomy of educational objectives – The classification of educational goals – Handbook 1: Cognitive domain*. London: Longmans, Green & Co. Ltd.

Bransford, J. D., & Stein, B. S. (1993). *The ideal problem solver: A guide for improving thinking, learning, and creativity*. New York: W.H. Freeman.

Button, G., & Sharrock, W. (1996). Project work: The organisation of collaborative design and development in software engineering. *Computer Supported Cooperative Work*, *5*(4), 369–386. doi:10.1007/BF00136711

Cavrak, I., Orlic, M., & Crnkovic, I. (2012). Collaboration patterns in distributed software development projects. In *Proceedings of the 34th International Conference on Software Engineering* (pp. 1235–1244). Academic Press.

Dewey, J. (2011). *Democracy and education: An introduction to the philosophy of education*. IndoEuropean.

Dröschel, W. (2000). *Das V-modell 97: Der standard für die entwicklung von IT-systemen mit anleitung für den praxiseinsatz*. München: Oldenbourg.

Farrell, V., Ravalli, G., Farrell, G., Kindler, P., & Hall, D. (2012). Capstone project: Fair, just and accountable assessment. In *Proceedings of the 17th Conference on Innovation and Technology in Computer Science Education* (pp. 168–173). Academic Press.

Figl, K., & Motschnig, R. (2008). Researching the development of team competencies in computer science courses. In *Proceedings of the 38th Annual Frontiers in Education Conference* (pp. T1A 1–6). Piscataway, NJ: IEEE.

Fiske, J., & Jenkins, H. (2011). *Introduction to communication studies*. London: Routledge.

Foppa, K. (1975). *Lernen, gedächtnis, verhalten: Ergebnisse und probleme der lernpsychologie*. Köln: Kiepenheuer & Witsch.

Hayes, J. H., Lethbridge, T. C., & Port, D. (2003). Evaluating individual contribution toward group software engineering projects. In *Proceedings of the 25th International Conference on Software Engineering* (pp. 622–627). IEEE.

IEEE. Computer Society & Association for Computing Machinery ACM. (2004). *Software engineering 2004 curriculum guidelines for undergraduate degree programs in software engineering: A volume of the computing curricula series*. Retrieved September 24, 2013, from http://sites.computer.org/ccse/SE2004Volume.pdf

Jung, E. (2002). *Projektunterricht - Projektstudium - Projektmanagement*. Retrieved July 25, 2013, from http://www.sowi-online.de/praxis/methode/projektunterricht_projektstudium_projektmanagement.html

Jung, E. (2010). *Kompetenzenerwerb: Grundlagen, didaktik, uberprüfbarkeit*. München: Oldenbourg, R/CVK.

Kabicher, S., Motschnig-Pitrik, R., & Figl, K. (2009). What competences do employers, staff and students expect of a computer science graduate? In *Proceedings of the 39th Frontiers in Education Conference* (pp. 1–6). IEEE.

Kemp, S. (2011). *Constructivism and problem-based learning*. Retrieved October 09, 2013, from http://www.tp.edu.sg/pbl_sandra_joy_kemp.pdf

Knowles, M. S., Holton, E. F., & Swanson, R. A. (2011). *The adult learner: The definitive classic in adult education and human resource development*. Amsterdam: Elsevier.

Kroll, P., & Kruchten, P. (2003). *The rational unified process made easy: A practitioner's guide to the RUP*. Boston: Addison-Wesley.

Lefrançois, G. R. (2006). *Theories of human learning: What the old woman said* (5th ed.). Belmont, CA: Thomson/Wadsworth.

Lu, H. K., Lo, C. H., & Lin, P. C. (2011). Competence analysis of IT professionals involved in business services — Using a qualitative method. In *Proceedings of the 24th Conference on Software Engineering Education and Training* (pp. 61–70). IEEE.

Mager, R. F. (1992). *Preparing instructional objectives*. London: Kogan Page.

Merriam, S. B. (2001). Andragogy and self-directed learning: Pillars of adult learning theory. *New Directions for Adult and Continuing Education*, (89): 3. doi:10.1002/ace.3

Merriam, S. B., Caffarella, R. S., & Baumgartner, L. (2007). *Learning in adulthood: A comprehensive guide* (3rd ed.). San Francisco: Jossey-Bass.

Reischmann, J. (1986). Learning en passant: The forgotten dimension. In *Proceedings of Conference of the American Association of Adult and Continuing Education*. Columbus, OH: ERIC Clearinghouse on Adult, Career, and Vocational Education.

Richardson, I., Reid, L., Seidman, S. B., Pattinson, B., & Delaney, Y. (2011). Educating software engineers of the future: Software quality research through problem-based learning. In *Proceedings of the 24th Conference on Software Engineering Education and Training* (pp. 91–100). Academic Press.

Rivera-Ibarra, J. G., Rodríguez-Jacobo, J., & Serrano-Vargas, M. A. (2010). Competency framework for software engineers. In *Proceedings of the 23rd Conference on Software Engineering Education and Training* (pp. 33–40). Academic Press.

Rummler, M. (2011). *Crashkurs hochschuldidaktik: Grundlagen und methoden guter lehre*. Weinheim: Beltz.

Rummler, M. (Ed.). (2012). *Innovative lehrformen: Projektarbeit in der hochschule: Projektbasiertes und problemorientiertes lehren und lernen*. Weinheim: Beltz.

Savery, J. R., & Duffy, T. M. (1998). Problem based learning: An instructional model and its constructivist framework. In B. G. Wilson (Ed.), *Constructivist learning environments: Case studies in instructional design* (pp. 135–148). Englewood Cliffs, NJ: Educational Technology Publications.

Savin-Baden, M. (2000). *Problem-based learning in higher education: Untold stories*. London: Society for Research into Higher Education.

Schulz von Thun, F. (1996). *Miteinander reden: Störungen und klärungen*. Reinbek: Rowohlt.

Schulz von Thun, F. (2008). *Six tools for clear communication: The hamburg approach in Englisch language*. Schulz von Thun, Institut für Kommunikation.

Schwaber, K., & Beedle, M. (2002). *Agile software development with scrum*. Upper Saddle River, NJ: Pearson.

Sedelmaier, Y., Claren, S., & Landes, D. (2013). Welche kompetenzen benötigt ein software ingenieur? In A. Spillner, & H. Lichter (Eds.), *Software engineering im unterricht der hochschulen 2013* (pp. 117–128). Academic Press.

Sedelmaier, Y., & Landes, D. (2013a). A research agenda for identifying and developing required competencies in software engineering. *International Journal of Engineering Pedagogy, 3*(2), 30–35.

Sedelmaier, Y., & Landes, D. (2013b). *Software engineering body of skills*. Unpublished manuscript.

Siebert, H. (2010). *Methoden für die bildungsarbeit: Leitfaden für aktivierendes Lehren*. Bielefeld: Bertelsmann.

Sommerville, I. (2011). *Software engineering*. Boston: Pearson.

Terhart, E. (2009). *Didaktik: Eine einführung*. Stuttgart: Reclam.

Tindale, R. S., Heath, L., Edwards, J., Posavac, E. J., & Bryant, F. B. Myers. J., Suarez-Balcazar, Y., & Henderson-King, E. (2002). Theory and research on small groups. Boston, MA: Springer US.

Tuckman, B. (1965). Developmental sequence in small groups. *Psychological Bulletin, 63*(6), 384–399. doi:10.1037/h0022100 PMID:14314073

ADDITIONAL READING

Azer, S. A. (2007). *Navigating problem-based learning*. Elsevier Australia.

Barrett, T., & Moore, S. (2011). *New approaches to problem-based learning: Revitalising your practice in higher education*. New York: Routledge.

Boud, D., & Feletti, G. (1998). *The Challenge of problem-based learning*. London: Kogan Page.

Damian, D., & Borici, A. (2012). Teamwork, coordination and customer relationship management skills: As important as technical skills in preparing our SE graduates. In *Proceedings of the First International Workshop on Software Engineering Education based on Real-World Experiences* (pp. 37–40).

Evensen, D. H., & Hmelo, C. E. (2000). *Problem-based learning: A research perspective on learning interactions*. Lawrence Erlbaum Associates Publishers.

Goodman, B., Henderson, D., & Stenzel, E. (2006). *An interdisciplinary approach to implementing competency based education in higher education*. Lewiston, NY: Edwin Mellen Press.

Grayling, I., Commons, K., & Wise, J. (2012). It's the curriculum, stupid! *Teaching in Lifelong Learning: A Journal to Inform and Improve Practice, 4*(1), 21–31.

Landes, D., Sedelmaier, Y., Pfeiffer, V., Mottok, J., & Hagel, G. (2012). Learning and teaching software process models. In *Proceedings of the Global Engineering Education Conference* (pp. 1153–1160). IEEE.

Mallow, J. V. (2001). Student group project work: A pioneering experiment in interactive engagement. *Journal of Science Education and Technology, 10*(2), 105–113. doi:10.1023/A:1009468912400

Mayer, R. E., & Alexander, P. A. (2011). *Handbook of research on learning and instruction*. New York: Routledge.

Merriam, S. B. (1988). *Case study research in education: A qualitative approach*. San Francisco, CA: Jossey-Bass.

Motschnig-Pitrik, R., Derntl, M., Figl, K., & Kabicher, S. (2008). Towards learner-centered learning goals based on the person-centered approach. In *Proceedings of the 38th Annual Frontiers in Education Conference* (pp. T1A 1–6). Piscataway, NJ: IEEE.

Savin-Baden, M. (2003). *Facilitating problem-based learning: Illuminating perspectives*. Maidenhead: Society for Research into Higher Education.

Slavin, R. E. (1995). *Cooperative learning: Theory, research, and practice*. Allyn and Bacon.

van Berkel, H. J. M. (2010). *Lessons from problem-based learning*. Oxford, New York: Oxford University Press. doi:10.1093/acprof:oso/9780199583447.001.0001

Weinberg, G. M. (1998). *The psychology of computer programming*. New York: Dorset House Pub.

Wilkerson, L., & Gijselaers, W. H. (1996). *Bringing problem-based learning to higher education: Theory and practice (No. 68)*. San Francisco, CA: Jossey-Bass.

Wilson, B. G. (1998). *Constructivist learning environments: Case studies in instructional design*. Englewood Cliffs, NJ: Educational Technology Publications.

KEY TERMS AND DEFINITIONS

Didactical Approach: A teaching method that follows a consistent process and uses predefined teaching style to educate students.

Pedagogical View: Using scientific methods to study education.

Pedagogy: The science and art of education, which includes education theories and education practices.

Project-Based Teaching: A teaching method that involves students to work on a team project.

Soft Skills: Non-technical skills, such as communication skills, management skills, and team working skills.

Chapter 10
Controlled Experiments as Means to Teach Soft Skills in Software Engineering

Marco Kuhrmann
Technische Universität München, Germany

Henning Femmer
Technische Universität München, Germany

Jonas Eckhardt
Technische Universität München, Germany

ABSTRACT

The job profile of a Software Engineer not only includes so-called "hard-skills" (e.g. specifying, program-ming, or building architectures) but also "soft skills" like awareness of team effects and similar human factors. These skills are typically hard to teach in classrooms, and current education, hence, mostly focuses on hard rather than soft skills. Yet, since software development is becoming more and more spread across different sites in a globally distributed manner, the importance of soft skills increases rapidly. However, there are only a few practical guides to teach such tacit knowledge to Software Engineering students. In this chapter, the authors describe an approach that combines theoretical lectures, practical experiments, and discussion sessions to fill this gap. They describe the processes of creating, planning, executing, and evaluating these sessions, so that soft skill topics can be taught in a university course. The authors present two example implementations of the approach. The first implementation lets students experience and reflect on group dynamics and team-internal effects in a project situation. The second implementa-tion enables students to understand the challenges of a distributed software development setting. With this knowledge, the authors critically discuss the contribution of experimentation to university teaching.

DOI: 10.4018/978-1-4666-5800-4.ch010

INTRODUCTION

Software Engineering (SE) aims at developing software-intensive systems in a systematic, methodically sound, and economic manner to master the key challenges of time constraints, budget adherence, quality and functionality. Beyond the technical questions that cover topics such as programming or testing, SE also addresses topics that deal with the organization and the management of software projects. Although both perspectives are important, today's SE education is often focused on the technical topics, as those are easier to teach (Kuhrmann et al., 2013), e.g. it is easier to teach coding and evaluating a program than to teach performing successfully in a project.

As a consequence, the current education contains a conflict between the curricula, the students' self-perception and the reality that students face when leaving the university. Especially partners from industry note that the students are usually not ready for work, even if they master a couple of programming languages. One of our partners told us: "I need to qualify a graduate for 2 or 3 extra months to make him fit." Besides company-specific knowledge (which a university is not able to teach) most partners from industry complain about missing soft skills, a missing understanding of how organizations and projects work, and missing skills regarding teamwork (Kuhrmann, 2012). In consequence, while we have a curriculum strongly addressing the technical topics, we still have gaps in appropriately educating the students on the organization and management level. We must enable students to experience that soft skills play an important role besides deep technical understanding.

Academic Curricula and Soft Skills

When teaching the required soft skills, one of the main challenges is to fit the required setting into the academic context and its constraints. For instance, since grades are given to rate the individual suc-

cess, students are often forced to be "lone fighters" and teachers set up their courses aiming at clearly separated individual performance instead of team effort. And in situations where students work in teams, we, as supervisors, often observe critical situations that are not caused by technical aspects, but by communication or behavioral issues, and we often see that students struggle in fixing such a situation. However, this is not surprising, as students often have no chance to learn about the "right" communication and interaction patterns in teams, or psychology basics that enable them to realize, analyze and solve a conflict in a team setting.

To underpin and explain the clash of interests, we give an example of typical teaching formats in an academic curriculum, in this case at the Technische Universität München (Table 1), and discuss their contribution to the above-mentioned skills afterwards.

As the collection of teaching formats in Table 1 shows, only two of the listed teachings formats address practical work. Of these, the guided research aims at scientific work and only the lab provides a context in which practical problems are addressed. However, lab courses are usually focused on teaching a specific practical topic, e.g. "Windows App Development," and are not focused on teaching soft skills. During the lab courses, students usually work in a team and might even experience the resulting phenomena; however, as lab courses are not focused on teaching soft skills, this experience is only a "by-product" and not made explicit. Hence, there are few opportunities for students to experience the importance of soft skills, and in these situations the focus is on software craftsmanship instead of understanding the team effects.

The current situation highlights this conflict: Practical work is dominated by theory, and soft skills cannot be learned theoretically. Most university curricula lack in providing space to appropriately teach soft skills, which would require courses in which students have to work together,

Table 1. Summary of typical teaching formats as implemented at the Technische Universität München

Teaching Format	Description
Lecture	A lecture is a teaching approach in which the lecturer more or less directly instructs the students. Except for some questions, lectures are not interactive; a lecture is the professor's stage on which he presents the knowledge. Lectures are usually complemented by exercises in which the students repeat and apply the theoretical contents of the previous class. Exams are either oral or written (mid and/or end term).
Seminar	A seminar is a teaching format in which students work independently on individual topics. Usually, advisors define the topics and guide the students. As a result, students give presentations and write an essay on their particular topic. The final grade is based on the evaluation of both, the presentation as well as the essay.
Lab	Practical labs/sessions/trainings (short: labs) focus on transferring theoretical knowledge into a practical environment. This teaching format is usually applied when learning programming languages. The final individual grade is based on the (set of) outcomes, e.g. software.
Guided Research	This teaching format addresses students who are interested in participating in current research. Students work as a part of the research group on research topics. The goal of a guided research project is to write a research paper and submit it to a peer-reviewed conference or workshop. Advisors, e.g. senior researchers, evaluate the student's performance in order to determine the grade. This teaching format is especially designed to support a long-term analysis, which is complemented by a Master's Thesis.
Theses	Theses are comprehensive pieces of work that students write individually to get their academic degrees, e.g. Diploma, Bachelor's or Masters' Theses.

have to deal with (critical) situations, and have to apply theoretically learned knowledge. We thus face a gap in the curricula. Although being restricted by the standard curriculum and tough schedules, courses that allow teaching and practicing soft skills are desperately needed to prepare the students for the real life outside the university campus.

Contribution

In this chapter, we describe an approach that combines theoretical lectures, practical experiments, and discussion sessions for teaching soft skills in SE courses. Over the years, we developed innovative teaching formats that allow for better balancing theory and practice. We learned that especially controlled experiments are good means to teach soft skills by enabling the students to experience the outcome of certain situations. In subsequent sections, we describe our notion of experimentation in SE and how we integrate experiments with SE courses. Furthermore, we describe two exemplary implementations of our approach (one on group dynamics and one on distributed software development) and critically discuss the contribution of experimentation to university teaching.

Although still being limited by the university's curriculum and schedule, and facing the challenge to define the right and appropriately-sized problem, our experience shows that we can confront the students with real-world problems stimulating self-learning periods, reflection, and so forth. Even though on a small scale, students can observe and, more importantly, experience the materialization of theoretical problems, e.g. team conflicts, stress and pressure, or dealing with failure.

COURSE CONCEPT

In this section, we present a concept of teaching that is based on theoretical knowledge, problem experience and reflection to enable students to understand and handle situations that require soft skills. We introduce this concept by giving a short background, a brief explanation of the basic structure and a discussion on how knowledge is assessed.

Background: Experimentation in Software Engineering

Experimentation is a well-accepted technique in SE (Wohlin et al., 2012) to, e.g. test hypotheses, back up research findings, or evaluate new methods and techniques. Therefore, experimentation can provide benefits to both students as well as lecturers, who also aim at conducting research for their professional careers (Kuhrmann, 2012). Introducing experimentation into university teaching is an instrument to involve students in classes and, therefore, is another approach to transfer knowledge. Experiments give students the opportunity to learn and experience rather than just consuming knowledge. Also, teachers have the chance to integrate their research activities with teaching. Experiments provide settings, e.g. for data collection, or hypothesis testing. Both, students and teachers, benefit: students are actively taking part in classes that deliver latest knowledge from practice and research, and teachers have active students contributing to classes and also to research. However, in SE, teaching with experiments is rare. Besides traditional classroom teaching, we can find approaches that work with practical exercises, games, simulation, and some small-scale experiments.

For instance, the Personal Software Process (PSP) (Humphrey, 2005) is the most prominent teaching approach that focuses on practical exercises. The underlying idea is to apply process principles at the level of single developers. Studies have shown that applying such processes at individual level can lead to significant performance improvements (Rombach et al., 2008). PSP exercises usually deal with relatively small and local processes, e.g. coding or testing.

Another popular area is educational games in the domain of lean and agile practices (Dukovska-Popovska et al., 2008). Here, certain practices and principles are demonstrated with the means of a game to impart knowledge. This kind of teaching typically aims at a better understanding of a specific philosophy rather than at better understanding the challenges and the consequences of acting in a particular way.

Empirical approaches focus on teaching by conducting experiments, typically as part of a regular course. The students take the role of experimental subjects to experience the effects of selected processes or techniques themselves. Typical objectives of such experiments consider comparisons of different quality assurance processes (e.g. (Kamsties & Lott, 1995)), Global Software Development (GSD, e.g. (Deiters et al., 2011), (Keenan et al., 2010), (Richardson et al., 2006)), or Software Engineering in general (e.g. (Huang et al., 2008), (Dahiya, 2010), (Bavota et al., 2012), (Pádua, 2010)). Most of the experimental treatments address the engineering level (coding and testing), but do usually not focus on soft skills.

Finally, simulation is sometimes used in teaching, e.g. in the domain of software processes (Münch et al., 2005). Here, students can make local decisions and see their global effects. Simulation is also suited for playing "what if games" and, thus, help to better understand situations. Simulations usually have a limited scope that addresses the understanding of process or project dynamics. Mandl-Strieglitz (2001) proposes a simulation-based approach to teach students project management that gives them the opportunity to make "real" experiences, similar to what we have reported in the context of GSD (Deiters et al., 2011).

These approaches focus on addressing students as individuals without taking into account the reality of today's project settings, or using students as subjects to support research without primary educational intent. In contrast to this, our approach focuses on teaching by paying attention to real-world demands of software projects, e.g. soft skills, and by also conducting research within in a structured teaching approach.

Course Structure

When it comes to experimentation in a course, we must carefully fulfill special prerequisites in order to create the desired effect. Experiments need to be prepared, conducted, and evaluated. For the preparation, we need to provide the students with the necessary background and theoretical knowledge. To this end, the course design deviates from classical designs, as one "class" may span up to three sessions. To implement experimentation, we propose a structure consisting of three parts, as shown in Figure 1:

1. A theory session that provides the students with the required background (classic lecture).
2. An experiment session in which the students experience the practical effects.
3. A reflection session in which the students discuss their experiences based on the theoretical background.

In the first block, we build up profound theoretical knowledge on a specific topic. For instance, in the case of group dynamics, this block would contain an introduction to communication in general, team structures, and different models for teamwork and communication. After this block,

the students should possess the theoretical basis of the topic and its effects.

In the second block, the main goal is that students experience problems in the situations directly. An example for this would be to create situations where teams have to cope with stressful situations. The challenge for the teacher (and the researcher) is to create an experiment that addresses a particular topic, but is concise and "pure" enough that students experience the desired effect. Furthermore, the experiments must be easy to set up in order to be regularly implemented in class. After this experimentation block, the students should have experienced the effects of the topic supporting the theoretical understanding.

Finally, the students shall combine the theoretical knowledge and the practical experiences. The reflection is done in the student group and should be moderated by the lecturer, e.g. by asking questions or by creating links between experiences and the corresponding theoretical knowledge. After this block, the students should understand the topic based on a connection between theoretical understanding and the experiences and feelings they witnessed during the experiments.

Assessing the Knowledge

University teaching involves not only teaching the knowledge, but also assessments, which must be based on the particular course goals. When teaching soft skills, course goals cannot be precisely defined and progress of students cannot be formally evaluated. Therefore, to design a test strategy regarding tacit knowledge is challenging since every student is an individual that specifically perceives experiments, and also specifically draws conclusions based on the individual background: There is no right or wrong.

Instead, our goal was that students are able to rationally analyze and understand a situation with the help of the theoretical background, derive one (possible) solution, and argue why their proposal solves the analyzed issue.

Figure 1. Schematic course structure

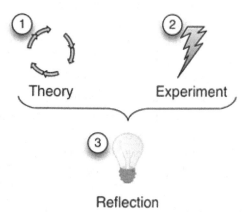

As a solution, we propose a scenario-based assessment approach in which students have to transfer their knowledge and have to develop a solution. Real-world settings in which problems of interest occur can support the creation of exams. Students are then asked to describe, analyze and solve a given situation based on their theoretical background. For instance, if an experiment on group dynamics was conducted, exams should reflect (real world) settings in which the knowledge can be applied. However, testing soft skills is not about answering questions right or wrong, it is about evaluating whether the students were able to apply the proper theoretical background (using a fitting model and understanding how the theoretical model fits to the existing situation) and understand the given situation. Furthermore, we grade students based on how well they argue that their proposed solution addresses the problem.

To give an example, one of our exams described a team conflict in detail. The students were then asked to analyze the underlying conflict (we wanted to see if students could apply the Tuckman model (Tuckman, 1965) and propose a solution. For grading, we evaluated how the students analyzed the situation, whether they could name the described problems, and if they were able to propose and explain a solution.

In the following sections, we present two examples of experiment designs. Complementing material can be depicted from the Website of the course "Agile Project Management & Software Development" (http://www4.in.tum.de/lehre/vorlesungen/vgmse/ws1213/index.shtml, password on request).

AN EXPERIMENT ON GROUP DYNAMICS

In this section, we present an exemplary experiment on group dynamics. We give some background and explain our learning goal for the students of this class. Then, we show how the course structure was implemented, how the experiment was prepared, conducted, and evaluated, and finally how students reacted to the experiment.

Context

The experiment reflects the situation that, as of today, software is usually developed in teams. However, it is well known that disturbing teams impacts the project performance, and that adding, removing, or replacing team members may cause additional stress. In today's business, teams are usually composed of experts, e.g. a brilliant coder, a gifted tester, and an architect. Therefore, expertise often relies on individuals (so-called "head monopolies") that have to collaborate to achieve the project goals. Once a team is performing, every external influence might be seen as an attack; for instance, a new architect might be seen as "intruder" and, therefore, might bother the team's harmony. However, it is a complete difference if a student knows these aspects from a theoretical lecture, or if the problem is experienced and felt in a short practical situation.

In the following, we describe an experiment, which is easy-to-organize and that allows for creating situations in which the students experience the effects of group dynamics.

Learning Goals

In order to teach students the importance of soft skills in Software Engineering, we developed an experiment in which students experience two problems that arise in teams:

1. Team size and project organization influences the performance. Strategies that work for one-person teams often do not work for teams of larger size. This situation gets even worse if the members are under stress or if a bottleneck arises.

2. Turnovers in teams affect the performance. Reducing the team size by removing team members, increasing the team size by adding team members, or replacing team members by shuffling the team impacts the performance as well as the working atmosphere in the project.

Both problems cause stress in projects that may even intensify the aforementioned problems. In detail, the students shall experience the effects of disturbed teams on the project performance by various problems; a selection is depicted in Table 2.

Experiment Preparation, Execution, And Evaluation

In the following, we provide details regarding the experiment according to the structure *theory, experiment,* and *outcome,* as proposed in the "Course Concept" section.

Theory

In the theory sessions, the students need to learn the foundations of working in teams and group dynamics. This includes psychological models, e.g. the Tuckman model (Tuckman, 1965) or the Myers Briggs model (Myers, & Myers, 1995).

Table 2. Selected problems potentially caused by group dynamics (GD)

ID	Problem Description
GD1	The larger the size of the team, the more difficult is its organization.
GD2	The larger the pressure, the more difficult is it to produce high quality work.
GD3	When a new team is created, organizational effort is needed. This effort grows with the size of the team.
GD4	When new team members join an existing team, their inclusion into the team can be expensive.
GD5	Team members who disturb the team are expensive to manage.
GD6	Bottlenecks in large teams are a subject to high risk.

Teaching material regarding this can be found on the course's Website.

Experiment

For teachers, experiments also include a preparation and a conduction phase. To prepare this experiment, we require a setting in which the effects of group dynamics can be concisely elaborated and which is easy to implement in a single teaching session. For this, we chose a sorting task. Specifically, in the experiments the students had to sort treats and document their sorted items. Prerequisites for the experiment are a large number of treats of different kind and color. We decided to use M&Ms, as they are easy to sort by color, but hard to sort by kind (and the sorting of M&Ms makes the experiment fun for the students). Furthermore, a rough estimate is needed about the expected performance and how many M&Ms are needed (an estimation can be based, e.g. on the numbers that we present in the outcomes section, but one should have in mind to add extra treats, as the students will eat them during the experiment). In addition to the treats, we need bowls as well as written instructions for the supervisors and prepared protocols. Before the experiment starts, every student must draw a number. This number will be used in the individual runs to assign students to teams.

Figure 2 shows an overview of the experiment: It consists of three sub experiments, which in turn, contain individual runs. Each of the runs is limited to only a few minutes (e.g. 2 minutes for runs 1 and 2) to create a basic time pressure. Teams remain the same within the sub experiments, but are mixed after each sub experiment.

- **Sub Experiment 1 (Warm-up):** Students get used to the task (sorting and counting) and understand the challenges and conditions. In the warm-up, group dynamics are not in the focus of the experiment; instead, students only train the task itself to mitigate the task's "complexity" as influencing

Figure 2. Experiment design

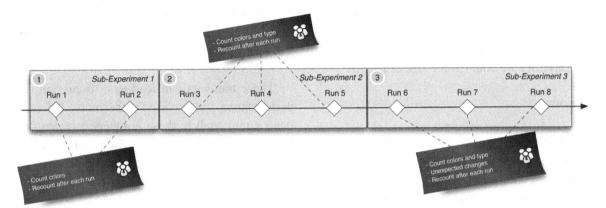

factor. This also prevents statistical outliers due to misunderstanding of the task.

- **Sub Experiment 2 (Regular team-size experiments):** Students work in teams of certain size and realize the amount of work that comes with organizing a team (and consequently how easy it is to work on your own). Also, students should realize that performance increases over time, as the team overcomes the storming phase of the Tuckman model.

- **Sub Experiment 3 (Unexpected change and bottlenecks):** Students should realize what happens if the team is unexpectedly disturbed or changed: What happens if the central team member leaves the team? What happens if team members start disturbing others and weaken the overall team performance? What happens if the team size is doubled?

All of the three sub experiments follow the same procedure:

1. The teacher randomly selects numbers to distribute the students into teams of different size (1-8 persons each team).
2. The teacher hands out bowls with mixed treats, a few empty bowls, and a form with predefined rows of the types of treats that the students are supposed to sort.
3. The teams sort the treats for a predefined time (2 – 8 minutes). The teacher informs the teams that at the moment when the time is over, the forms are collected and that they will afterwards be checked against whatever is noted on that sheet (performance, error rate, etc.).

In every run, the following steps are taken:

1. The experiment starts, the teams start working.
2. After x minutes the teacher takes away the forms.
3. The students count the sorted treats again and note the number on a second, control sheet.

The procedure of the sub experiment 3 differs slightly: Before the sub experiment starts, the teacher informs the students that there might be some instructions during the experiment and asks the students to follow any given instruction carefully. Furthermore, one team is separated from the others and is not disturbed anymore (reference team for a statistical performance baseline). During the individual runs, the teacher starts to mix up teams: For example, the teacher finds the key organizer in a team (usually the person writing down the number of treats sorted) and replaces him with a person from another team. He does this by handing a sheet of paper to the students with a text like "Please join Team A." Another example would be to change the team size in the middle of a run: For example, we analyzed the student's reactions after we doubled the size of a team.

Outcome

Right at the end of the experiment, the students are asked to fill out feedback sheets. We suggest ending the session after the feedback, as we found that students were very exhausted from the stress they experienced.

As explained above, the last block intends to foster the students understanding, by collectively bringing together the theoretical models with the practical experiences. We propose to begin with a qualitative reflection on the experiment. Questions like

1. How did the teamwork go?
2. What went well?
3. When and where did the work not go well?
4. How did you organize the work in your team?
5. Did you face any team issues during the experiment?
6. Did you experience stress?
7. Were there any differences in the organization structures of the different teams?
8. What was your overall feeling with the experiment?

can stimulate a discussion which leads to a common understanding of the individual feelings of the participants.

Afterwards, the teacher should lead the students to a quantitative reflection of the experiment by giving the students an idea of the raw data that has been collected and by asking the students how these data can be used to spot the phenomena they have mentioned in the discussion. For example, the students can analyze for each run of each sub experiment and for each team the following numbers:

1. Number of treats noted during the run on the first sheet (Step 2).
2. Number of treats noted at the end of the run on the second sheet (Step 3).

These numbers can lead to a discussion on performance and errors. Furthermore, the students could analyze these numbers between the runs, i.e. this will lead to a discussion on what happens if the teams have a learning curve or if the teams change.

Exemplary Results

Since this way of teaching is based on a concept in which we combine teaching and research, our results are twofold. First, results address the students and their knowledge about particular topics. Second, the experiments generate data for investigating research questions, or for hypothesis testing. In the following, we present exemplary results.

We conducted this experiment twice – with a group of master's students and with a group of pupils. To explain what can be expected in this experiment, we will describe results of the experiment with the group of Master's students in the following coarse-grained analysis. However, we will only discuss the overall tendencies of the statistics as this chapter focuses on the educational aspects.

We collected the following metrics throughout the experiment: (i) the number of M&Ms that were written down by each team in every single run (*Written Number of M&Ms per Person per Minute per Team per Run*) and (ii) the number of M&Ms that the teams actually sorted and counted into their bowls (*Counted Number of M&Ms per Person per Minute per Team per Run*). The second value serves as control value to assess the quality of the work. Hence, the absolute difference between the counted and written number of M&Ms gave us a third key value: (iii) the number of wrongly counted and written M&Ms (*Error per Person per Minute per Team per Run*).

The following figures show these three numbers during the three sub experiments. Figure 3, Figure 4 and Figure 5 show the development of values ((i), (ii), and (iii)) in the course of the experiment for each team.

Every curve represents the results of one team. One can see for example in the first two figures (Figure 4 and Figure 5) that the upmost line represents the number of written and counted M&M for team A, which consists of one person. To enable comparison, the measurement values displayed in both graphs have been broken down to values per person per minute.

As one can see in Figure 3, Figure 4 and Figure 5, throughout the whole experiment the results of the different teams skew at a wide range. For instance, in run 3 the number of written M&Ms ranges from 5 (Team D, 5 persons) to 23 (Team A, 1 person) written M&Ms per person per minute. Nevertheless, in Figure 3 and Figure 4 the curve progressions resemble one another and the curves seem to be shifted in direction of the y-axis, whereas the curves in Figure 5 at first glance do

Figure 3. Written number of treats per person per minute per team per run (sub experiment 1-3)

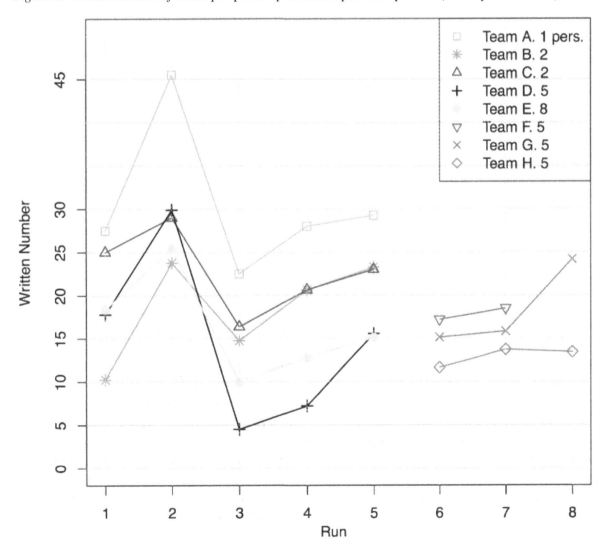

Figure 4. Counted number of treats per person per minute per team per run (sub experiment 1-3)

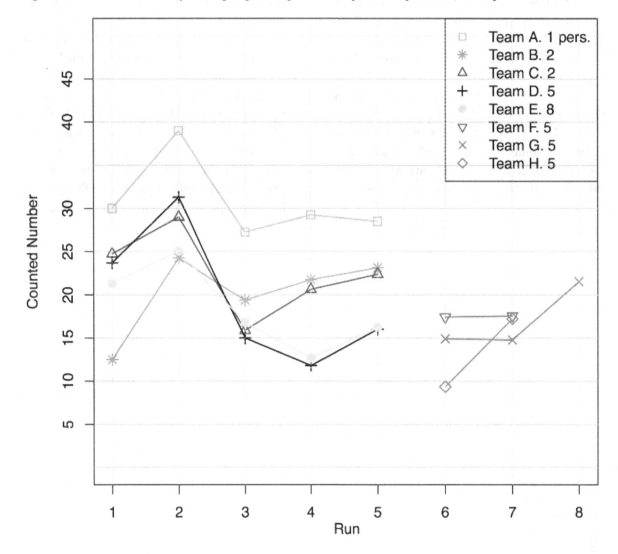

not seem to point out such a clear tendency. However, the change between the sub experiments is clearly reflected by both graphs; when changing from sub experiment 1 to 2 (between run 2 and 3) the written number of M&Ms decreases and the number of errors increases. This observation applies for all teams without any exceptions.

From these observations, we can conclude that increasing the complexity of the task and rebuilding teams seems to have an impact on the team performance as well as on the number of errors,

and that the Tuckman model also appears in such small groups performing easy tasks.

Discussion

The goal of the experiment was to let the students experience important aspects in working together as a team. From the discussions during the analysis phase, we learned that students understood the dynamics that happened in the experiment. We saw from the feedback we got after the course that through this kind of teaching, students were really

Figure 5. Error per person per minute per team per run (sub experiment 1-3)

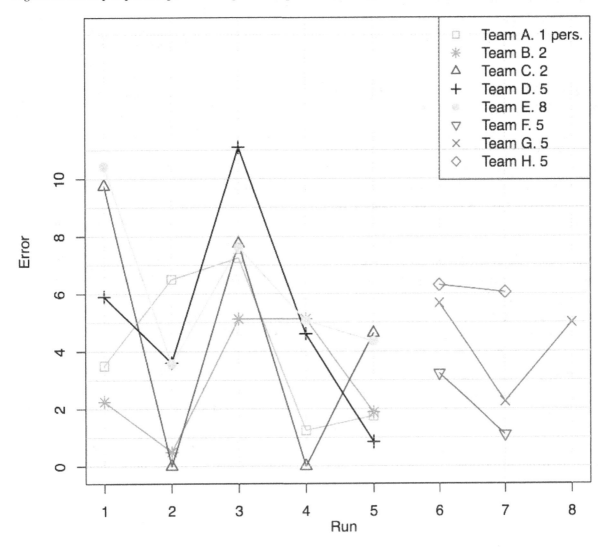

experiencing the problem, as they said "It took some time to work out a strategy" or "Confusion on too many counting possibilities. Other person didn't talk enough - communication problem," but at the same time they enjoyed the format, as one student simply wrote "M&Ms! Funny!"

Also, we think that the setup and task is highly extensible, so that the experiments focus also on other parts of teamwork. Additionally, the choice enables us to conduct the experiment easily with first-year-students and even with pupils. Lastly, the setup is really easy to reproduce and built up

(i.e. the repetition needed only approximately 2 – 3 hours of preparation including shopping), considering the intense form of learning it is able to create.

However, one could discuss whether the results are transferable to Software Engineering (external validity). Even though we do not know of a formal empirical analysis of this, from our own experience in industry software projects, we strongly argue that the same effects, which we could see happening between students, are also happening in many software development projects today.

Furthermore, similar problems are also reported from case studies in the Global Software Engineering domain (e.g. Piri et al., 2010).

AN EXPERIMENT ON DISTRIBUTED SOFTWARE DEVELOPMENT

The second experiment focuses on a different problem of today's software development. One of the reasons why software is more and more developed on a global scope is size: Today's software projects are getting increasingly large, which demands for increasing men power and team size (Sangwan et al., 2006). Often the required skills are no longer available at one place or cheaper at another, which leads to globally distributed software development.

However, as university courses often fear the organization overload, practical courses focus on working within properly organized teams. Rare examples of university projects that are globally distributed in fact turn out to be as organization-intense as teachers fear. However, we were looking for a concise setting in which teams can experience the issues of distributed teams in a quick and easy-to-create setup.

Learning Goals

In this session the students should experience the difference between developing software as an in-house team, where the whole team fits around a table, and developing software in a distributed team that has only restricted ability to communicate. Also, the team might understand the real meaning of commonly known observations, such as Conway's Law (Conway, 1968) and others (Table 3).

Experiment Preparation, Execution, and Evaluation

As described in the framework in the "Course Concept" section, the experiment is framed with a theoretic preparation and an analytic discussion.

Table 3. Selected problems in Global Software Development (GSD)

ID	Problem Description
GSD1	Indirect communication leads to incomplete information and hidden assumptions.
GSD2	Non-personal communication is very difficult.
GSD3	Conway's Law: System design follows team communication.

Theory

In a first theory session, students need to learn the foundations of teamwork in distributed software development. This includes, inter alia, team organization, communication patterns, Conway's Law, etc.

Experiment

The idea of the experiment for experiencing distributed software development is to let students develop a very simple software program in a pair of teams, and in a setting where communication is very restricted. That is, two teams work together on a task and they may only communicate via Skype or e-mail (e.g. Team 1 is only allowed to communicate via Skype and Team 2 is only allowed to communicate via e-mail).

This experiment requires careful preparation, as it relies on technical infrastructure:

- A SVNs or SVN folder for each team
- A basic project framework (e.g. interfaces or an existing Eclipse project)
- Skype or e-mail accounts for each team
- Every student must have Eclipse, SVN, Skype/e-mail ready to run
- Best: A room or an area for each team.

The basic setup is the following: The class is split up into *n* teams of a few students each. Each team must develop the same console-based application (in our case we created 18 small features in the form of user stories; (Cohn, 2004)), which the students have to implement. Each feature must be checked against a given "definition of done" in order to assess whether a feature is correctly implemented. The teams are asked to develop the features one-by-one, starting from the first, in order to be able to compare between teams.

Each team is again split up into two sub teams, which are distributed in a way that the sub teams cannot directly communicate with each other (see Figure 6). In order to force the realistic and necessary communication, we ask respective sub teams "A" to develop the backend support and sub teams "B" to create the frontend user interface (i.e. a console application).

After an explanation of each feature, the teams start to work and the teachers take notes of occurring problems and how the teams solved these problems.

Outcome

After the experiment, the teams get together in a room. Teachers should observe closely, as the students will probably immediately start discussing the issues that arose in their teams.

Again, we ask the students to reflect the issues that they experienced and look at the developed software and the implemented features together. We take notes how many features are working,

and how many features are implemented, but have issues (e.g. a feature does not fulfill requirements or code is buggy, according to the "definition of done").

Similar to the first experiment, the students afterwards reflect on how the issues are connected to the theoretical background.

Exemplary Results

We expected that the Skype team had a better solution than the e-mail team, both in terms of quality as well as quantity of features, because we considered speech and visual context to be very important. In fact, the outcome was quite the opposite: The Skype team produced far less features, which also had many issues. When looking at the teams, we understood that the individual programming skills were higher in the e-mail teams, which might have caused these results. A definitive conclusion cannot yet be drawn, as e-mail as communication instrument also bears risks regarding communication (Carmel, 1999).

However, the students could experience some of the tacit knowledge, which we explained earlier. A difficult task was in fact defining interfaces between the teams and understanding the different architectures people had in mind. As expected, the major issue for the students in this experiment was the absence of face-to-face communication. In summary, students realized that developing software in a distributed team comes with various challenges that have to be actively addressed.

Discussion

Due to the size of our group we were only able to compare e-mail and Skype communication. However, other forms could also form an interesting picture, such as direct communication (in one room), chats or through virtual worlds, such as Second Life.

Even though we were able to conduct the experiment within the estimated time slot, the

Figure 6. Schematic setup for Experiment 2

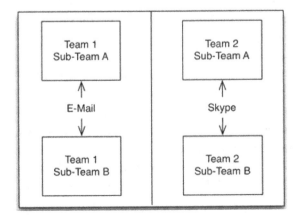

amount of preparation for this experiment is higher. Setup of tools and explanation of examples makes this experiment slightly more difficult to initiate in the first place. However, we consider that this setup time could be significantly reduced, e.g. by reusing computer labs and -software.

We created this experiment with a relatively small group of 14 students. Yet, we are very positive that the experiment can scale, given enough space to spread out. Hence, it would be interesting to see on larger scale how the different forms of collaboration impact the development.

CONCLUSION AND LESSONS LEARNED

Due to the increasing complexity of software development, and due to the fact that software is developed collaboratively across different countries with different cultures, soft skills become an indispensable part of students' qualification. Therefore, heterogeneous software development teams face the challenge to form an efficient and productive unit. Teaching in experiments is an instrument to introduce soft skill trainings into a university curriculum. Students do not only consume knowledge, but also become part of the knowledge generation. Using self-learning periods, self-reflection, and practical "doing," students mature in class. They do not only know about topics, but also have experienced the topic and associated problems, e.g. consequences of acting and not acting in certain situations. This way of teaching allows students to experience the materialization of theoretical topics in practice already at university.

In this chapter, we introduced a course structure that allows for integrating experimentation in university courses, and we discussed how experimentation contributes to the improvement of courses. The presented concept was implemented at the Technische Universität München without affecting the standard curriculum. We imple-

mented two courses using the proposed pattern: one course on software process modeling and another course on agile project management. We learned that the inclusion of experiments aids a better understanding of topics, as students consume knowledge "by the way" while working in the experiment setting. At the same time, the consumed knowledge becomes retrievable and realizable. We presented two exemplary experiments on group dynamics and global software development that were successfully implemented at a Master's course. Integrating experimentation and university courses showed to be beneficial for the students as well as for the teachers. Students learn about specific topics and, at the same time, experience, e.g. practical problems and the consequences of acting and not acting in certain situations. To this end, experimentation makes theoretical knowledge tangible and thus shows the necessity of profound theoretical background knowledge. For instance: Why should students learn about the Tuckman model? – Because effects of group dynamics impact the performance of project teams.

The experiences that we could gather until now are promising; however, a centralized pool of experiments must be created and extended in order to spread good ideas for experiments and to make experiments as easily applicable as possible. To this end, we need to define more experiments to be conductible in classes – also for replication purposes. Although having done the first transfers, the proposed concept needs to be implemented in further courses, which address Software Engineering disciplines that also rely on soft skills, e.g. quality assurance/testing or requirements engineering. From the teacher's perspective, the preparation of experiments bears some challenges (Kuhrmann, 2012), e.g. the definition of problems, or the alignment of theoretical content and experiments. However, preparing experiments carefully lays the foundation to build a knowledge base that contains easy to use teaching material, and that allows for multiple implementations of experiments, i.e. to replicate experiments. Therewith, also the "researcher in the teacher" can benefit. Experimentation can be used to, e.g. build hypotheses and initially test them, repetitively collect data in order to investigate research questions, or create and test experiment settings that can later be used in industry settings.

REFERENCES

Bavota, G., De Lucia, A., Fasano, F., Oliveto, R., & Zottoli, C. (2012). Teaching software engineering and software project management: An integrated and practical approach. In *Proceedings of the 2012 International Conference on Software Engineering* (pp. 1155–1164). IEEE Press.

Beck, K., & Andres, C. (2004). *Extreme programming explained: Embrace change*. Reading, MA: Addison-Wesley Longman.

Carmel, E. (1999). *Global software engineering teams*. Upper Saddle River, NJ: Prentice Hall.

Cohn, M. (2004). *User stories applied: For agile software development*. Reading, MA: Addison-Wesley Professional.

Conway, M. E. (1968). How do committees invent? *Datamation*, *14*(5), 28–31.

Dahiya, D. (2010). Teaching software engineering: A practical approach. *ACM SIGSOFT Software Engineering Notes, 35*(2).

Deiters, C., Herrmann, C., Hildebrandt, R., Knauss, E., Kuhrmann, M., Rausch, A., et al. (2011). GloSE-lab: Teaching global software engineering. In *Proceedings of the 6th IEEE International Conference on Global Software Engineering*. IEEE.

Dukovska-Popovska, I., Hove-Madsen, V., & Nielsen, K. B. (2008). Teaching lean thinking through game: Some challenges. *In Proceedings of the 36th European Society for Engineering Education on Quality Assessment, Employability & Innovation*. Academic Press.

Huang, L., Dai, L., Guo, B., & Lei, G. (2008). Project-driven teaching model for software project management course. In *Proceedings of 2008 International Conference on Computer Science and Software Engineering* (Vol. 5, pp. 503–506). IEEE.

Humphrey, W. S. (2005). *PSP(SM) - A self-improvement process for software engineers*. Reading, MA: Addison-Wesley.

Kamsties, E., & Lott, C. (1995). An empirical evaluation of three defect-detection techniques. In *Proceedings of the 5th European Software Engineering Conference*. Academic Press.

Keenan, E., Steele, A., & Jia, X. (2010). Simulating global software development in a course environment. In *Proceedings of the 5th IEEE International Conference on Global Software Engineering* (pp. 201–205). IEEE.

Kuhrmann, M. (2012). A practical approach to align research with master's level courses. In *Proceedings of the 15th International Conference on Computational Science and Engineering* (pp. 202-208). IEEE.

Kuhrmann, M., Fernández, D. M., & Münch, J. (2013). Teaching software process modeling. In *Proceedings of the 2013 International Conference on Software Engineering* (pp. 1138-1147). IEEE Press.

Mandl-Strieglitz, P. (2001). How to successfully use software project simulation for educating software project managers. In *Proceedings of the 31st Frontiers in Education Conference* (Vol. 1, pp. T2D–19). IEEE.

Münch, J., Pfahl, D., & Rus, I. (2005). Virtual software engineering laboratories in support of trade-off analyses. *International Software Quality Journal, 13*(4).

Myers, I. B., & Myers, P. B. (1995). *Gifts differing: Understanding personality type.* Nicholas Brealey Publishing.

Pádua, W. (2010). Measuring complexity, effectiveness and efficiency in software course projects. In *Proceedings of the 32th International Conference on Software Engineering* (vol. 1, pp. 545–554). ACM.

Piri, A., Niinimäki, T., & Lassenius, C. (2010). Fear and distrust in global software engineering projects. *Journal of Software Maintenance and Evolution: Research and Practice, 24*(2).

Richardson, I., Milewski, A. E., Mullick, N., & Keil, P. (2006). Distributed development: An education perspective on the global studio project. In *Proceedings of the 28th International Conference on Software Engineering* (pp. 679–684). ACM.

Rombach, D., Münch, J., Ocampo, A., Humphrey, W. S., & Burton, D. (2008). Teaching disciplined software development. *International Journal of Systems and Software, 81*(5).

Sangwan, R., Mullick, N., & Paulish, D. J. (2006). *Global software development handbook.* Auerbach Publishers Inc. doi:10.1201/9781420013856

Tuckman, B. W. (1965). Development sequence in small groups. *Psychological Bulletin.* doi:10.1037/h0022100 PMID:14314073

Wohlin, C., Runeson, P., Höst, M., Ohlsson, M. C., Regnell, B., & Wesslén, A. (2012). *Experimentation in software engineering.* Berlin: Springer. doi:10.1007/978-3-642-29044-2

KEY TERMS AND DEFINITIONS

Agility: Agility in software development is a philosophy aiming to performing only necessary tasks that create customer value (Beck & Andres, 2004). Agile methods are usually lightweight method bundles addressing development-centered activities, e.g. pair programming, unit testing, or continuous integration. Furthermore, agile methods follow a widely accepted set of principles and values.

Conway's Law: An adage named after Melvin Conway (introduced in 1968). It states (Conway, 1968): "organizations which design systems [...] are constrained to produce designs which are copies of the communication structures of these organizations."

Experiment: According to (Wohlin et al., 2012), an experiment (a controlled experiment) is an empirical enquiry that manipulates one factor or variable of the studied setting. Based on randomization, different treatments are applied to or by different subjects. If the assignment is not based on randomization, but on the characteristics of the subjects or the objectives, we speak of a quasi-experiment.

Global Software Development: Global software development (Sangwan et al., 2006) aims to collaboratively develop software across different locations. Development activities are carried out by distributed teams that cooperatively work on a software system.

Group Dynamics: Group dynamics (or group processes) describes behaviors and processes occurring within a group. Effects of group dynamics can occur in a group (intragroup dynamics) or between groups (intergroup dynamics). The understanding of group dynamics in Software Engineering and in software project management is an important skill, e.g. for decision-making processes, or conflict management.

Project Management: Project management as a discipline comprises different methods and techniques to organize and steer (software) projects. Project management of software projects usually comprises activities from project planning, tracking and controlling, quality assurance, change management, risk management, or team management.

Soft Skills: Soft skills as a sociological term summarize personality traits characterizing relationships with other people, e.g. social graces, communication, language, or personal habits, and that complement hard skills, e.g. (technical) requirements of a job.

Tuckman Cycle: The Tuckman cycle (Tuckman, 1965) is a model on group dynamics that occur in team building processes. The cycle comprises the phases forming, storming, norming, performing, and adjourning. Every time a team is created or changed, the team goes through these phases to establish a structure by analyzing the other peers, finding positions, and forming a productive unit.

Chapter 11
Developing Personal and Professional Skills in Software Engineering Students

Lynette Johns-Boast
Australian National University, Australia

ABSTRACT

Although industry acknowledges university graduates possess strong technical knowledge, it continues to lament the lack of commensurately strong personal and professional skills that allow graduates to apply their technical knowledge and to become effective members of the workforce quickly. This chapter outlines a research-backed course design that blends experiential learning to create an industrial simulation, the rewards of which go well beyond the usual benefits of group-project capstone design courses. The simulated industrial context facilitates the graduation of software engineers who possess the requisite personal and professional attributes. Innovations include combining two cohorts of students into one, engaging industry partners through the provision and management of projects, and implementing proven education approaches that promote the development of personal and professional skills. Adoption of the suggested practices will help institutions produce "work-ready" graduates repeatedly, year after year, even by software engineering academics who may not have received teacher training and who may not possess significant industry experience themselves.

DOI: 10.4018/978-1-4666-5800-4.ch011

INTRODUCTION

During the last century, there have been a number of changes within the domain of engineering education. Traditionally engineering education focused on the practical aspects of engineering. However, the engineering-science revolution of the last century and increasing government emphasis on research has caused a drift from this "hands-on practice to mathematical modeling and scientific analyses" (Froyd, Wankat, & Smith, 2012). During the final decade of the century, industry and academia both expressed concern about graduates' lack of professional skills, such as teamwork and communication, as well as their lack of awareness of the importance of societal influences. Industry was also concerned about graduates' readiness to enter the workforce (Benyon, 2012; Jollands, Jolly, & Molyneaux, 2012; McLennan & Keating, 2008).

This concern lead to the development of the outcomes based accreditation models that are now typical of 21st century accreditation processes. Undergraduate engineering degree program accreditation processes aim to ensure that newly graduated "Professional Engineers" are "professionally competent" (ABET, 2012-2013). In other words, that a graduate's education, training and experience has enabled the development of appropriate knowledge, skills and attitudes. However, despite this evolution of the accreditation process, industry continues to call for improvements (Litzinger, Lattuca, Hadgraft, & Newstetter, 2011).

The new accreditation processes and their associated competency-based frameworks require a "culture change among faculty" and a move away from "the science-focused preparation that has characterized engineering education since World War II" (Passow, 2007). Instead of developing curricula that are content-focused and closely aligned with the discipline, academics must now develop curricula that are not so closely identified with the specific discipline. This requires a change of approach from the more straight forward transmission of knowledge to the more challenging one of helping students to grow and acquire the critical skills they will need to succeed after university (Passow, 2007; Woods, Felder, Rugarcia, & Stice, 2000).

Complicating these required changes to engineering curricula is that "most university professors … were not taught anything about how to teach" (Felder & Brent, 2004). Instead they rely on how their "professors taught, but nobody taught them anything about teaching either. It doesn't make a lot of sense, but that's our system" (Felder & Brent, 2004). The accreditation process and universities themselves require individual programs and courses to develop students' generic graduate attributes. However, academics "charged with responsibility for developing them do not share a common understanding of either the nature of these outcomes, or the teaching and learning processes that might facilitate the development of these outcomes" (Barrie, 2004, p. 263).

Both the literature and the academics themselves consistently identify the importance of these generic attributes (Male, 2010). Nonetheless engineering academics accord the teaching of them low status. The increasing focus on research and the engineering science revolution have placed greater emphasis and importance on theory and analysis of abstract problems. Furthermore, as noted by Male (2010), the "gendered nature" of engineering and engineering education has helped bestow lower status on the "stereotypically feminine traits, such as those related to people and nurture … [while] abstract science has higher status" (p37). This has led to a marginalization of "communication, teamwork, management, definition of problems, practical engineering, and context" (Male, 2010, p. 37). Notwithstanding this, when asked to rank ABET's 11 competencies on their level of importance, academics accord highest importance to the competencies related to problem solving and communication and place only average importance on math, science and engineering knowledge (Passow, 2007).

As universities attempted to meet the demand for software engineering graduates who possess not only good technical knowledge but also appropriate personal and professional skills, they first introduced group work into one or more courses as well as capstone design courses. However, graduates of these courses have not consistently shown the expected improvement of personal and professional attributes. Such courses generally have not attempted to actually teach students any of these skills (Felder & Brent, 2004). Simply requiring students to work in small groups, without any real introduction to the skills required, does little to develop these skills. What is required is appropriate activities, support for students to ensure that they actually gain the skills associated with teamwork, and assessment of the attainment of those skills.

Potentially acknowledging the lack of interest and ability of academics to teach these skills, many institutions are now including various forms of work-integrated-learning within their degree programs. Many have introduced internships or work placements where, as part of their program, students spend time working in industry gaining work experience, usually for academic credit (Cooper, Orrell, & Bowden, 2010). When including internships and work-placements in their programs institutions have to make the decision either to extend the length of the program or to reduce the number of courses which students are required to undertake and thus reduce the content to which students are exposed. Yet again, the literature demonstrates that these courses do not consistently deliver the improved development of personal and professional attributes because institutions do not have sufficient control over learning in the workplace (Abeysekera, 2006; Orrell, 2004).

To ensure our software engineering graduates are successful in 21st century software development "teaching for professional practice should be the touchstone for future choices about both curriculum content and pedagogical strategies in

undergraduate engineering education" (Sheppard, Macatangay, Colby, & Sullivan, 2008, p. 7). To do this we need to "explicitly identify these target skills and adopt proven instructional strategies that promote those skills" (Woods et al., 2000, p. 12). Experiential learning (D. A. Kolb, 1984) is one such proven instructional strategy that we can use to help our students develop personal and professional skills. This student-centered, recursive and adaptive process facilitates the development of these skills. Students undertake authentic tasks. Then through reflection, they connect their current experience to previous experiences and learning, thus improving their ability to transfer learning from university to the workplace.

This chapter focuses on experiential learning – knowledge and understanding of which I contend will help academics design and implement courses that consistently develop students' personal and professional skills. I introduce the concepts underpinning experiential learning and describe the principal implementations within software engineering education. I then present a case study of a course which implements a blend of experiential learning approaches designed to develop strong personal and professional skills in its graduates. Since its implementation in 2004 more than 350 students have graduated from the course. Following this I discuss how regular evaluations have helped refine course design to meet learning outcomes more consistently. I demonstrate how careful curriculum design and implementation makes it possible to develop personal and professional skills in our students without employing specialist academics, increasing the length of our degree programs, or reducing the content to which our students are exposed, all the while providing a valuable and fun learning experience. I argue that unless academics consciously adopt appropriate learning approaches, any success achieved is accidental and thus unable to be repeated with any certainty.

EXPERIENTIAL LEARNING

As already mentioned, one of the proven instructional strategies that can be used to help learners develop personal and professional skills is experiential learning (D. A. Kolb, 1984). The most established model of which was developed by Kolb. Underlying Kolb's model is the assumption that we seldom learn from an experience unless we reflect upon it. This enables us to understand the experience in terms of our own goals, aims, ambitions, and expectations. Individual insights, discoveries, and understanding emerge from this process. As the pieces fall into place, the experience takes on added meaning in relation to previous experiences (Saddington, n.d.) which leads to increased satisfaction, motivation and personal development (D. A. Kolb, 1984). The process of experiential learning is portrayed as an:

idealized learning cycle or spiral where the learner "touches all the bases" – experiencing, reflecting, thinking, and acting – in a recursive process that is responsive to the learning situation and what is being learned. Immediate or concrete experiences are the basis for observations and reflections. Reflections are assimilated and distilled into abstract concepts from which new implications for action are drawn. These implications can be actively tested and serve as guides in creating new experiences. (A. Y. Kolb & Kolb, 2009, pp. 298-299)

Experiential learning is a learner-centered approach which promotes deep learning (Entwistle, 2000). It helps equip graduates with relevant discipline knowledge and skills, and builds discipline and context specific vocabulary and terminology. As well as facilitating growth of transferable skills, such as problem solving and critical thinking, it fosters work ready graduates who, in many cases, possess hands on, industry experience (Freudenberg, Brimble, & Cameron, 2010; Litchfield, Frawley, & Nettleton, 2010;

Orrell, 2004). It helps students to understand and grasp the relevance of the theoretical concepts they learn in class, to put theory into practice and to appreciate that academic success is not the only attribute required for career success. To improve transferability of knowledge and skills learned at university, students must also be supplied with the context in which that knowledge and skills may be applied. However, university curricula frequently lack focus on the context (Savery & Duffy, 2001). Experiential learning can be used to provide that context. This helps students fuse theory with practice and so improves their ability to transfer knowledge from the academic environment to the work place and vice versa.

Furthermore, experiential learning helps students develop the ability to think critically; analyze and solve complex, real-world problems; to find, evaluate and use appropriate learning resources; and when undertaken in small groups, to work co-operatively; to demonstrate effective communication skills and to use content knowledge and intellectuals skills to become life-long learners (Savery, 2006; Savery & Duffy, 2001). It builds upon existing knowledge, so helping students to construct knowledge, to 'make sense' for themselves (Savin-Baden, 2000). This process engages them in deep rather than surface learning practices (Entwistle, 2000), so that they are able to synthesize ideas and to apply that learning in new environments, rather than simply recalling facts. Moreover, experiential learning is seen as providing a mechanism by which university learning can be aligned with the professional world that students will enter upon graduation (Biggs, 2003; Biggs & Tang, 2007). To gain most benefit, experiential learning should occur within small groups of students (Dunlap, 2005) and the learning outcomes need to be explicitly identified (BIE, n.d.).

Participating in experiential learning helps develop students' work readiness. It requires students to have knowledge of the concept, ranging from simple recall or awareness through to creative thinking or evaluation, as well as the skill required to apply the knowledge in new situations. Furthermore, experiential learning helps students develop confidence. It helps develop coping skills and persistence, which leads to improved attitudes, behaviors and work readiness (Freudenberg, Brimble, & Vyvyan, 2010).

When implementing experiential learning academics need to ensure the learning experience is authentic and reflects the complexity of the environment in which students will be required to function on graduation. All learning activities should be anchored to a larger task or problem and should require linking of theory with practice. Furthermore the problem or task should be sufficiently ill-structured that it encourages experimentation and requires integration of a wide range of disciplines and subjects. The learning environment needs to be designed to support as well as to challenge students' thinking. Students need to be encouraged to test their ideas against alternative views and alternative contexts. Instead of transmitting information, academics support students as they take ownership of the overall problem and the processes they use to develop a solution. Academics need to become coaches and facilitators of learning. Furthermore assessment should focus on the processes of knowledge acquisition rather than the products of such processes. Additionally, there should be a focus on the importance of developing communication and interpersonal skills, meaning collaboration between students is essential (Savery, 2006; Savery & Duffy, 2001; Savin-Baden, 2000).

While the focus of this chapter is on the benefits to students, it should be noted that experiential learning has the potential also to benefit institutions through improved rankings. Strong disciplinary knowledge on its own is not a guarantee of employment but relevant work experience can have a positive impact on graduate employment and starting salaries, both of which lead to improved university rankings.

There are many approaches associated with experiential learning; for example, problem solving, project work, activity-based work, work and community placements, prior learning, independent learning and personal learning (Saddington, 2000). Within the domain of higher education, however, there are four principal implementations: inquiry-based learning; case-based learning; problem based learning (PBL) with its variant project based learning (PjBL); and work integrated learning (WIL).

Inquiry Based Learning

Inquiry based learning focusses on questioning, critical thinking and problem solving and is driven by the curiosity of the learner. It encourages a hands-on approach where students practice the scientific method on authentic problems. The teacher is both a facilitator of learning – encouraging and expecting higher order thinking – and a provider of information. Science courses frequently use this technique. (As the motivation of this chapter is the development of personal and professional skills for software engineering students, I have dealt only briefly with this particular approach to experiential learning: it is not widely used in software engineering education.)

Case Based Learning

Case based learning is particularly useful when attempting to develop understanding of a particular concept or issue. Students are presented with a well-constructed case (example) or scenario. The teacher is a facilitator and coach guiding and encouraging students' research and/or discussion as they attempt to understand the important elements of the problem situation. Cases are constrained and focus on the topic or issue the teacher wishes the students to understand. There are two types of cases (Queensu, n.d.):

- **Finished:** Based on facts. Cases are for analysis only, since the solution is indicated or alternate solutions are suggested. Discussion often centers on attempting to understand and explain why a particular approach was adopted.
- **Unfinished:** Open-ended and with unclear final results (either because in real life the case has not concluded, or because the academic has eliminated the final facts). As students investigate these, they must predict, make choices and offer suggestions that will affect the outcome.

Cases can be fictional (written by the academic) or original documents (news articles, reports with data and statistics, summaries, excerpts from historical writings, artefacts, literary passages, video and audio recordings, ethnographies, etc.). When creating fictional cases it is important that these are complex enough to mimic reality, yet do not have so many red herrings as to obscure the goal of the exercise. They can be difficult and time consuming to create. With the right questions original documents can become excellent problem solving opportunities. Comparison between two original documents related to the same topic or theme is a strong strategy for encouraging both analysis and synthesis. This gives opportunity for presenting more than one side of an argument, making the conflicts more complex.

Problem Based Learning (PBL)

While similar to an unfinished case, PBL allows students greater ownership of the problem and associated solution. Students are presented with an ill-formed problem that does not have a single correct answer (Helmo-Silver, 2004). Problems provide a starting point and students work towards a solution with guidance from the teacher. It is tempting to see all learning as problem-based learning, especially when that learning involves problem solving activities as do so many software

engineering, computing and information technology courses. This is not the case, however. This type of learning is predicated on students learning material that is in some way "linked to a specific curricula content which is seen as vital for students to cover, in order for them to be competent practitioners" (Savin-Baden, 2001) while PBL curricula organize content around the problem scenarios.

PBL was first introduced into academia in 1969 at McMaster University in Canada in their medical program. Despite the lack of quality evidence of its efficacy it has gained wide acceptance and is now used in many other disciplines – the multiple domains of medical education (dentistry, nursing, paramedics, radiology etc.), MBA programs, engineering, computing, economics, and architecture (Savery, 2006).

Project Based Learning (PjBL)

Sometimes also referred to, and confused with, PBL. PjBL focuses on developing a product – the project – while PBL focuses on studying and understanding a problem. Students are usually provided with a specification for a desired end-product and the learning process is tailored to follow 'correct' process and procedure. The 'problems' learners encounter as they work through their project are used as 'teachable moments'. Rather than being tutors or mentors, teachers are facilitators or coaches, providing expert guidance, feedback, and suggestions for 'better' ways to achieve the final product. The teaching – modeling, scaffolding, questioning etc. – is provided according to student need, based on the problems encountered within the context of the project. PjBL encourages students to think and act professionally. Completing a project successfully requires critical thinking, problem solving, collaboration, communication, evaluation, revision, and hard work.

Generally, when compared with non-project based courses, project based courses reverse the order of content presentation. Learning begins with an understanding of the vision and end product. This provides the context and the reason to learn and understand the concepts the course aims to teach. As the course progresses, students are introduced to the prescribed concepts and content which they are then required to apply to their project in order to complete it.

Work-Integrated Learning (WIL)

WIL is an umbrella term covering a range of approaches and strategies that involve the blending of academic learning (theory) with experience of the work-place (McLennan & Keating, 2008; Patrick, Peach, & Pocknee, 2008). Typically WIL is described as "educational programs which combine and integrate learning and its workplace application, regardless of whether this integration occurs in industry or whether it is real or simulated" (Atchison, Pollock, Reeders, & Rizetti, 2002). PjBL and WIL overlap when project work is closely related to the real world.

Frequently, WIL is implemented as an immersion experience outside the university, where students work in industry for part of their academic program, often for academic credit (Patrick et al., 2008). To be effective WIL must be implemented within the framework of a purposefully and imaginatively designed and embedded assessment. Assessment must require deliberate and intentional learning and be supported by the appropriate induction of students and supervisors (McLennan & Keating, 2008; Orrell, 2004; Washbourn, 1996). WIL exists in many forms, the principal ones being: internships, work placements, industry based learning, community based learning, clinical rotations, a sandwich year, and practical projects.

As noted previously, despite the well-documented value of such learning (Abeysekera, 2006; Orrell, 2004), quality implementation of WIL, especially in the form of internships and work placements is rare (Orrell, 2004) as universities generally do not have control over the students' work environment, their tasks and responsibilities.

Group-Project Courses: Where Do They Fit?

As mentioned earlier, software engineering programs are increasingly incorporating group-project courses, often as capstone design projects. Are software engineering academics using experiential learning? Certainly some may be. However, group work and group-project courses as implemented experience a range of problems.

While these projects do provide students with some freedom of choice in approach and development of the solution, the assessment scheme usually constrains their freedom and ensures they develop a mandated set of products regardless of project need. Group-project courses typically require students to spend an extended period responding to a complex problem, question or challenge and students are expected to learn to work with others through cooperative or collaborative learning. Usually projects are carefully planned and managed, from the belief this will best help students learn key academic content; practice collaboration, communication and critical thinking; and create quality, authentic products and presentations. Consequently, students are often presented with 'canned' projects – effectively fictional cases. Students, however, do not want work that is 'practice' or 'pretend': they want to do real work on real projects. In some instances, students may develop solutions to real-world problems posed by the institution's industry partners, but these too are usually constrained by the assessment scheme.

Furthermore, the quality of student learning in group-project courses is influenced significantly by the individual student experience of working as part of a team. For example, when students have a negative experience of teamwork they are less inclined to work collaboratively while those who have a positive experience are more likely to see the benefits of working collaboratively and of teamwork more generally (Ulloa & Adams, 2004). The teamwork experience largely colors both their feelings about the course and how well they learn.

While many group-project courses appear to be based upon the concepts of both PBL and PjBL, I suggest this is largely unconscious on the part of the course designer. As noted, many issues are encountered despite both practice and the literature suggesting that group-project courses, especially those linked to industry, are highly effective at helping students consolidate their learning, fuse practice with theory and in preparing them to face the real-world. Without knowledge and understanding of the factors of experiential learning that lead to success, academics are not in control of the process. Any success achieved is accidental, cannot be repeated, and explains why these courses have been less successful than hoped.

In the following section, I describe the design and implementation of a novel, industry-based group-project course that blends various experiential learning approaches to help deliver the overall course goals. Experiential learning concepts have been carefully and consciously built into the course and the design has been evaluated and refined over a period of years. I posit that the explicit use of these concepts underpins the success of the course. Subsequent to the case study, I show how the design implements aspects of experiential learning to achieve the goals of the course. I also provide details of the other proven educational approaches we have adopted to overcome many of the problems group-project courses face.

CASE STUDY

The Australian National University (ANU) is one of Australia's principal research-intensive universities. It has been offering degree studies in computing since the 1970s. From 2002, the ANU has offered a four-year Bachelor of Software Engineering (BSEng) degree program, accredited by Engineers Australia (EA). Degree rules require BSEng students to undertake a group-project in both their third and fourth (final) years of study. The ANU also offers a three-year Bachelor of Information Technology (BIT) degree program, accredited by the Australian Computer Society (ACS). BIT degree rules require the majority of students to participate in a group-project in their third (final) year.

Since 2004, more than 350 students have graduated from the course and 65 external/industry projects with 46 organizations and 26 university projects have been completed successfully. Depending on enrolments, we run up to 13 different projects per year. BSEng students participate as team members in one project in their third year and as team leaders in another project in their fourth year, while BIT students only participate in their third (final) year as team members. The course runs for two consecutive semesters and is one quarter of a full-time load. Each student is expected to contribute an average of about ten hours per teaching week, or around 260 hours for the whole course. Teams consist of four or five third year students led by one or two fourth year students.

At the end of each year, the teaching team holds a post course review (PCR). The PCR is used to evaluate how well the principal goals of the course have been met. Overall course goals have not changed since 2004. To assess the course we use the concepts of constructive alignment (Biggs, 2003) and Tyler's (1949) notion that the specification of behavioral change suggests learning activities which lead to increased familiarity with the associated content. As we believe that the assessment scheme is the primary driver of student behavior, the PCR focuses on refining the assessment scheme to help the course better meet its learning outcomes. Input to the PCR comes from anonymous student feedback obtained through the learning management system (sometimes also called virtual learning environment), grades, team performance based on mentors' observations, indirect feedback obtained through peer assessment and students' reflective homework, as well as feedback from industry partners. PCR evaluation has led to both major and minor changes to the assessment scheme. The most recent implementation of the course is described in this case study.

Additional detail not included here can be found in a series of papers written by Johns-Boast & Flint (2009, 2013), Johns-Boast (2010) and Johns-Boast & Patch (2010).

Course Design

A review of our group-project courses – one for third year BSEng and BIT students and another for fourth year BSEng students – confirmed that not all students were developing the required skills that would enable them to complete their projects successfully. Therefore we set about designing and implementing a capstone design course that consistently would facilitate the development of strong personal and professional skills in all our graduates. We believe strongly that learning through doing, through solving problems, is most effective at helping students build a usable body of knowledge; and that successful software engineers use problem solving skills rather than memorization and recall skills. Accordingly, after a study of the literature we decided that a suitable approach was to base our course design on the principles of experiential learning.

Each form of experiential learning comes with its own set of positives and negatives. For example, when implementing case based learning or PBL it is very difficult and time-consuming to come up with a suitably complex, yet authentic case or problem that enables the required content to be conveyed. Furthermore it becomes boring for the academics if students are working on the same problem year after year and it increases the temptation for students to see the cases, problems or projects simply as an artifice and therefore of little real relevance. Such issues are then likely to lead to a loss of motivation by both students and academics. The speed with which technology is changing is also likely to mean that such problems do not remain current for very long. Courses that involve group or team work frequently experience

problems related to assessment of individual contribution and issues such as social loafing (Albanese & van Fleet, 1985; Oakley, Felder, Brent, & Elhajj, 2004). Work-integrated learning through work placements and internships suffer especially from issues related to consistency of learning opportunities and quality across placements (Orrell, 2004).

While course design was based upon elements of experiential design, the overall goal of the course and individual course outcomes placed constraints both upon the design and the implementation of that design. The overall goal of the course was:

that we will help students develop leadership skills and become an effective member of a team which makes and implements appropriate engineering decisions related to the development of software systems that deliver measurable value to clients.

The outcomes for the individual courses were:

that fourth year students will demonstrate a high level of professional judgment and application of software engineering best practice, as demonstrated through the identification, development, use and evaluation of appropriate processes and artefacts required to provide real value to the industry partner;

while third year students will demonstrate technical competence in all aspects of the software development life cycle.

These goals and outcomes have not changed since 2004.

When designing and implementing the course we followed the advice of Woods et al (2000) and explicitly identified the skills and attributes we wished our students to develop and then set about finding proven, educational approaches and strategies that would help us develop those skills. Three key mechanisms – course organizational structure, learning environment, and assessment scheme – are used to implement the design.

Organizational Structure

To deliver the course we implemented what is effectively a virtual company – ANU Consulting Limited. The virtual consultancy company enables us to simulate employment in the software development industry. See Figure 1 – Course organizational structure.

Course academics play the role of senior management within a consulting company that responds to requests for tender. Each year course academics seek real-world projects from our industry partners which are then treated as if they have been won at tender. To carry out the development of those projects senior management (course academics) select existing staff (students) to comprise small teams, led by one or two team leaders (fourth years) and staffed with between three and five team members (third years). For two consecutive semesters, teams work together developing a solution to their industry partner's problem. Acting as senior project managers within the company are non-academic, team mentors – who have significant industry and project experience. Each team is assigned a mentor with whom they work closely, especially the team leaders.

Each team works on a different project with a different industry partner. In keeping with the industrial simulation, teams control their project and have responsibility for negotiating scope, schedule and deliverables. Project clients treat students as a team of junior consultants, frequently inducting them into company culture. Students are expected to demonstrate the same behavior and results as project clients expect from their own junior software engineers. Industry partners may also require student teams to comply with company processes and procedures related to how they run projects within their organization. This may impose a project management life cycle approach, such as Scrum or Kanban, and may require teams to use particular management and collaboration software, such as Redmine, Jira, Hudson and Yammer.

Figure 1. Course organizational structure

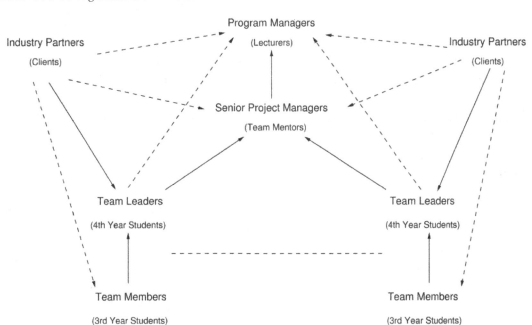

207

Although fourth year students are designated as team leaders, the separation of roles is not absolute, however, since in teams of this size there is typically a greater proportion of technical work. Nonetheless, the fourth year students are responsible for ensuring planning, estimating, scheduling and monitoring of progress actually happens, as well as being the principle contact point for project clients.

Constraints Placed on Industry Partners

The principal limitation we impose on our industry partners is that proposed projects must not be of strategic importance to the organization. The university cannot guarantee the quality of the solution, nor in fact, whether they will receive anything of value. Moreover, requiring projects to be non-critical frees both the project client and the students to take risks that might not otherwise be possible. This in turn expands the learning opportunities presented to students. Despite this limitation, since the 2009 academic year we have always had more offers of projects than we have had student teams able to complete them. Furthermore, industry partners have told us that the tasks that they provide for the students are not artificial or 'make-work' in any way.

Participation imposes certain requirements upon our industry partners. For example, the project client, or their representative, must meet with students at least once per month, preferably more frequently, and must respond to emails or other forms of communication from the students, the team mentor or course academics within an agreed time frame. Project clients are also required to attend the three project reviews and the industry showcase, where students present their year's work and their projects to their peers, industry partners and other invited industry contacts.

Learning Environment

The learning environment is composed of both formal and informal elements. The formal learning environment comprises two parts: weekly course-based lectures and workshops, and every two weeks, formal team meetings with mentors. The informal learning environment comprises regular face-to-face and virtual team meetings, held at least weekly. Given the physical space available to students at our university, teams, or sub-sets of teams, frequently meet in the same place at the same time, which encourages cross-fertilization of ideas.

Figure 1- Course organizational structure, shows communication and information flows within the virtual company. Solid lines indicate a requirement for formal communication, such as progress reports, system requirements, architecture and design etc. Dotted lines indicate informal communication and information flows. The dotted line between the teams represents the informal sharing of information about problems and their resolution that happens between teams, especially during lectures and workshops.

Lectures, workshops and mentor meetings implement aspects of case-based learning. Problems encountered by teams mimic both finished and unfinished cases. Course academics and mentors use these to provide the context for earlier learning and so help students create links with prior learning. Problems are also used to help illustrate links between theory and practice. Frequently we also use them to help students develop critical thinking skills and so to make defensible engineering decisions. Furthermore, lectures, workshops and mentor meetings provide opportunities to develop skills, such as project estimation and scheduling, and to elicit and document requirements.

Assessment Scheme

Since assessment is the key to student learning (Brown, 2004; Falchikov, 2005; Ramsden, 2003; Rust, 2002) the best way to ensure we meet the desired student outcomes for the course, was to carefully design and then continuously refine the assessment scheme. Refinements to the assessment scheme have moved it progressively from mandating and focusing on the quality of deliverables to focusing on process and teamwork. This transition was completed with the introduction of formal peer assessment in 2008, and project reviews in association with a holistic view of the quality of deliverables in 2009. The assessment scheme contains four principal assessment items:

- Project reviews at three key points during the course (45% – team assessment);
- Submission of a comprehensive 'handover' package compiled from software and artefacts completed during the course (20% – team assessment);
- An industry 'project showcase' at the end of the course, including a 20 minute presentation and a poster summarizing the project (10% – team assessment); and
- Regular reflection homework including a reflective report, submitted towards the end of the course (25% – individual assessment).

Peer assessment as implemented in this course is the assessment of students' own and their peers' performance in relation to group work. To do so we use a set of key performance indicators (KPIs). KPIs are based on the Stage 1 Competencies (EA, n.d.). At the end of each term, students are required to submit peer assessment ratings using our WebPA-x software. WebPA-x is a branched version of the WebPA software (Webpaproduct, n.d.) which collects peer assessment ratings and calculates peer assessed multipliers. These multipliers are then used to generate individual marks for all team assessment items – project reviews, final artefact submission, the project poster and presentation. Used in this way, peer assessment helps hold students' accountable to their peers for their performance.

Figure 2 – Timeframes, waypoints & reviews provides a summary description of the key focus for each of the project reviews, and the major assessment items and their timing. This should not be read as mandating or even suggesting that projects should follow a waterfall systems development lifecycle. Nothing is further from the truth. Many of the projects undertaken since 2004 have been developed using some form of iterative or agile life cycle. For example, of the 13 projects undertaken in 2012, as their project management methodology four adopted Scrum (Scrumalliance, n.d.) and another used Kanban (Peterson, 2009). The remainder of the projects sat somewhere on the spectrum from traditional through iterative to agile, with some moving to different points on that spectrum as the project progressed. Students understand that while they are expected to be able to adequately address certain topics for each project review, this is not all that they are expected to accomplish during each review period.

The project review process is reminiscent of gateway reviews (Review Process, n.d.) used in industry. Project reviews last for about one hour and broadly mirror an Agile Sprint Review meeting, and incorporate concepts of a Sprint Retrospective (Cohn, 2013). They are essentially a form of oral examination or mini-viva. Project clients, course academics, mentors and all team members actively participate in the reviews. Teams are required to present a 15-minute overview of their progress in the previous period and explain their goals and plans for the following period. The remainder of the time is spent in active discussion where course academics, mentors and project clients question the team. Teams display artefacts, demonstrate their solution and generally engage in discussion. Project reviews examine a team's

Figure 2. Timeframes, waypoints & reviews

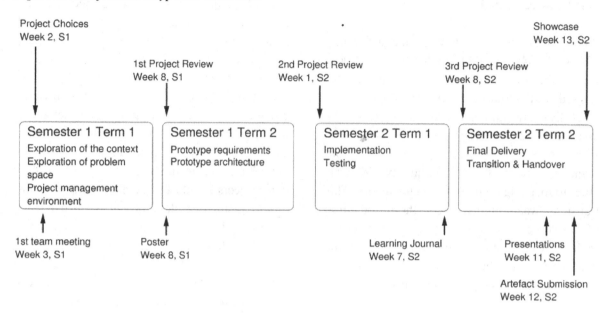

performance and project progress to provide guidance and assurance that they can progress successfully to the next stage of the project. They provide the principal mechanism for academic assessment of team conduct and project progress.

Summative assessment – both team and individual – is provided after each project review. Mentors and project clients provide formative feedback as part of the project review process but also as needed. Standards-based, holistic rubrics (Stevens & Levi, 2005) are used for all assessment and feedback on all assessment items, including project reviews. Rubrics are made available to students at the start of the course.

By design, the assessment scheme places responsibility on students to control their projects: to determine scope, schedule and deliverables. In consultation with their project's client, students are also responsible for establishing their own systems and software development lifecycle and project management environment. Experienced, industry-based mentors along with their project clients and course academics closely support student teams throughout.

IMPLEMENTING EXPERIENTIAL LEARNING TO DELIVER COURSE OBJECTIVES AND DESIRED LEARNING OUTCOMES

The course design aimed to provide an authentic learning experience for students. The most beneficial way to realize this was to simulate the industrial context. We achieved the industrial simulation by combining two cohorts of students into one course, developing an assessment scheme that gives true freedom to students to manage their projects as they determine to be appropriate, and working with real-world clients on real-world problems. The capstone software engineering project course described in the case study weaves together into a single entity, elements of four of the forms of experiential learning outlined earlier – case-based, problem-based, project-based and work-integrated learning. It implements key features of experiential learning, especially PjBL, that lead to positive outcomes. The course:

- Provides an authentic learning experience;
- Facilitates the development and practice of strong personal and professional skills;

- Provides a context for previous learning to facilitate fusing theory with practice; and
- Encourages students to become autonomous learners.

The following sections discuss how we have implemented these features of experiential learning; the outcomes we have seen; how we have met the usual challenges encountered with group-project courses; and the benefits and the issues we have encountered as a consequence of implementing our design.

Provide an Authentic Learning Experience

Our early design discussions were influenced by Dawson's "Twenty Dirty Tricks to Train Software Engineers" (2000). Many of Dawson's "dirty tricks," such as providing students with an inadequate specification, changing requirements and priorities, conflicting requirements and pressures, changing deadlines, and having naïve customers who are unclear of exactly what it is they need, are simply normal, real-world project experiences.

Simulating the industrial context has provided an authentic learning experience that has enabled the achievement of student outcomes. Simulation is an accepted and proven approach to realizing educational goals. Games and simulations have been used widely to support professional and vocational training. Real-world scenarios permit participants to practice and develop skills before taking up professional employment (De Freitas, 2006). Scenarios provide an "immersive learning" opportunity where participants have the "subjective impression that ... [they are] participating in a comprehensive, realistic experience" (Dede, 2009) which helps provide motivation. High levels of motivation need to be maintained for learning through games and simulations to be effective. Regular feedback and coaching, reflection, and active engagement help maintain motivation (Garris, Ahlers, & Driskell, 2002). In our simulation,

realism is boosted and student motivation is further enhanced through involvement of industry partners in the provision and conduct of the student projects. Industry involvement also helps ensure that student projects are relevant to the current world of software engineering. Furthermore, simulating the industrial context also ensures that we replicate the complexity of the workplace.

Combining third and fourth year students into a single cohort replicates the software development industry in Australia where frequently team leaders have only a year or two more experience than the team members they are leading. In both their third year and fourth years of study, students get to work on a 'real' project with a 'real' client and therefore experience and deal with real-world pressures, demands and issues, and experience the full software development lifecycle. Such experiences tend to mature the students quickly and thus prepare them for their ultimate careers in software engineering. Simulating the industrial context provides an environment where students have a real opportunity to learn more of the practical aspects of software engineering and therefore to be more prepared to join the workforce as qualified software engineers.

Facilitate the Development and Practice of Strong Personal and Professional Skills

To help students learn from their experience, it is important teams "create a conversational space where members can reflect on and talk about their experience together" (Kayes, Kayes, & Kolb, 2005, p. 332). Mentor meetings, class discussions, the project review process, and required reflection facilitate the creation of just such a conversation space. As part of their introduction to teamwork and the course, students learn about the need to develop an environment of trust as an integral part of the group evolving into an effective team (Costa, Roe, & Taillieu, 2001; Porter & Lilly, 1996). Lectures and workshops in week two deal specifically with team formation, personality and its impact, the importance of trust and the development of high performing teams.

Students develop communication skills and strategies that will enable them to work successfully in small teams. Successful teamwork requires students to communicate their ideas clearly to peers and to listen to others. Workshop activities, project reviews, presentations and the project showcase all require students to develop and practice public speaking skills. For many students the project and associated documentation assists them to develop further their written communication skills. Team mentors support students as they develop these and more general team working skills such as working collaboratively and providing assistance to others. Mentors frequently provide the context that enables students to understand why good interpersonal skills are important for successful project outcomes.

The need for students to communicate and work with industry partners encourages students to take a more professional approach to their activities than they may usually take when dealing with university assignments. No longer is the conduct of the project simply an item of assessment that is private between the student and the course academic. It is now something that is in the public domain which helps motivate students to meet the expectations of project clients, team members, mentors and course academics. Deadlines set by project clients and working collaboratively with their peers further encourages development of self-discipline and time management skills. Again team mentors work actively with individual students to help them develop these skills.

The simulated industrial context and the assessment scheme make students responsible for the conduct of their project. This ensures they are introduced to and have an opportunity to develop and practice the personal and professional skills required of a competent software engineer: management, communication with both peers and supervisors and teamwork in particular; as well

as ethical and other responsibilities, in addition to requiring students to develop and practice leadership. There is no pre-determined schedule of artefacts and timing for production of them. Teams negotiate these with their project client to determine a schedule and list of required artefacts that suits the individual project. This environment provides fourth year students especially with an authentic opportunity to develop and practice team leadership.

The combination of third and fourth year students into a single cohort has provided benefits over and above those that more usual group-project courses can deliver. For example, the third year students can focus on developing their technical skills and understanding how the jigsaw of their previous learning fits together, all the while observing how (or how not) to manage a team and through experience and reflection understand the importance of good project management to project success. This model has led to more consistent capability among graduating students. Students themselves see many benefits. Anonymous feedback from students includes statements such as "the course provides opportunities to learn things that aren't part of a pre-existing and highly structured course – to do something semi-original." Students also enjoy "the challenge … to get something designed and built based upon what the client wants, and it gives you a great feeling of satisfaction and pride to know that you built something from scratch."

Provide a Context for Previous Learning to Facilitate Fusing Theory with Practice

In addition to the usual and well documented benefits from undertaking group-project work (e.g. Beasley, 2003; Chamillard & Braun, 2002; Clark, 2005; Isomoettoenen & Kaerkkaeinen, 2008; Todd & Magleby, 2005), the simulated industrial context helps students fuse theory with practice and thus improves their ability to transfer knowledge from the academic environment to the work-place and vice-versa (Abeysekera, 2006; Orrell, 2004). Students also get to experience the effects of what they are being taught within their program, especially in the areas of planning, project management, requirements elicitation, design, and quality management. For example, they get to see first-hand the impact of poor project management, what happens when requirements are not clearly understood and the critical role testing plays in software release planning.

Combining third and fourth year students into a single cohort facilitates the incorporation of peer assisted learning (PAL) (Boud, Cohen, & Sampson, 1999; Topping & Ehly, 1998) which has been shown to be a powerful method for leveraging learning beyond the direct efforts of the teacher. To be effective, however, PAL requires situations that motivate the more experienced students to assist the less experienced. The assessment scheme, through shared project marks, provides an incentive to the fourth year students to improve their juniors' performance. This requires fourth year students to draw upon their experiences from when they previously completed the course. PAL, lectures and structured reflection homework provide students with an opportunity to share the experience of peers' projects and develop a community of practice which helps develop greater understanding (Brown, 2004; Falchikov, 2005).

The assessment scheme and the simulated industrial context facilitate development of critical thinking, problem solving, collaboration, and various forms of communication if students are to complete their projects successfully and gain the grades they desire. The open-ended nature of the problems our industry partners offer means that in order to deliver a high quality solution and gain high marks, teams must investigate more than one possible solution. In order to make a decision that they can defend at a project review students need to use critical thinking skills to determine an appropriate solution for their particular project. A decision to accept a solution simply because students know how to program in a certain lan-

guage, or have used a particular architecture before, becomes immediately obvious during a project review. An inability to defend their decisions logically harms the team's marks.

Encourage Students to Become Autonomous Learners

Through reflection, experiential learning in all its forms encourages students to become autonomous learners and to develop life-long learning skills. Furthermore, as noted by Savin-Baden (2000) focusing "too much on what students are able to do and on their ability to perform, could be to deny many students the vital opportunities to critique the situations and information with which they are being presented" (p. 21). Thus the assessment scheme requires regular formal and informal reflection and focuses upon the processes students adopt rather than the products produced from those processes.

The assessment scheme does not mandate required artefacts, or a schedule to be followed, nor even whether projects are to be managed using traditional, iterative or agile approaches. Instead, in consultation with project clients, students are responsible for the establishment and management of their own project management environment. Teams are required to decide upon an appropriate system and software development lifecycle according to the project itself and project client requirements, frequently adopting agile approaches. They use software tools, programing languages and project management approaches which are prevalent in industry but which are not always taught at university. Students adopt a broad range of tools for project tracking, issue management, continuous integration, code review, and configuration management and version control. Again, students do not use the majority of these tools as part of their formal degree study. Students learn about the iterative nature of software development and regularly discover that a single approach does not suit all stages of their project and thus learn the value of flexibility.

Removing most constraints from the assessment scheme and handing over complete control of their projects to students, along with focusing on individual and team reflection, self and peer assessment, and critical thinking, facilitates the development of autonomous learning. The completion of the project itself is frequently sufficient motivation for students to take ownership of the problem and engage in self-directed learning. At times, students become so engaged and motivated we have problems limiting the effort they put into their project.

Problems encountered along the way provide learning opportunities, or cases, which are used to help students develop the required knowledge and problem solving skills. As well as providing increased motivation and authenticity, involving industry partners as project clients elevates the coursework experience above that of simply completing an assignment whose impact is limited to individual students, or at most the small teams of which they are members. The simulated industrial context also teaches students about the need to work continuously and collaboratively, rather than in concerted bursts just before an assignment deadline.

Dealing with the Challenges Encountered with Group-Project Courses

As previously noted, group and team work – cooperative or collaborative learning where a small group works on a structured problem – is increasingly common in many university courses, especially in engineering, software engineering, computer science and information technology. Many accreditation standards (for example, the Washington Accord, EA, ABET, the Engineering Council of the United Kingdom, the ACS, the British Computer Society), require graduates to possess good teamwork skills as well as an ability to function effectively as part of multidisciplinary and cross-cultural teams. Nonetheless and despite the increasing use of group and teamwork, both students and academics continue to express concern about and dissatisfaction with it.

Students frequently complain that group, and teamwork is stressful, non-productive and the better students complain that it harms their marks. They complain about poor group dynamics and functioning leading to issues surrounding equity of effort and marks. Periodically they have trouble finding time to work together and may lack confidence in their own abilities. Often, many would rather do the work on their own. A significant concern of both academics and students is the existence of non-productive group behavior such as social loafing and free riding (Albanese & van Fleet, 1985; Oakley et al., 2004) where students who do not complete tasks as responsibly as their peers, nonetheless attain the same mark as their more responsible peers (Ellis & Hafner, 2007; Kaufman, Felder, & Fuller, 1999, 2000; Oakley et al., 2004).

Academics are also concerned that group work may have a negative impact upon a good student's marks and may enable a poorer student to obtain a mark higher than their level of learning and effort warrants. Moreover, accurately assessing individual student learning and contribution can be a difficult and time consuming task.

One approach that overcomes the concerns of both academics and students is peer assessment. Additionally, peer assessment provides valuable formative feedback to students that can assist with ongoing professional development, may assist students to improve future marks (Willey & Gardner, 2010) and can motivate and facilitate personal growth.

In the following section I discuss how we have implemented peer assessment. I also discuss the approach we have taken to team formation, team roles, and what factors lead to team success. I conclude this section with a discussion of how we assist students in developing teamwork skills.

Peer Assessment to Determine Individual Contribution, Motivate Participation and Facilitate Personal Growth

One of the common challenges academics face with group-project courses is "how work can be fairly and consistently assessed and how students can be encouraged to engage with the group work itself" (Gordon, 2010, p. 20). Commonly students participating in group work receive the same mark for group artefacts. This, however, does not take account of differing levels of contribution. Students generally dislike this approach. Another approach is for the academic to investigate the group work and attempt to assign marks according to their assessment of levels of contribution. This approach can be very labor intensive and does not necessarily lead to an appropriate outcome (Johns-Boast, 2010). Another common challenge facing academics and students alike is the unequal contributions of students: social loafing and free-riding.

As well as overcoming these challenges, there were two principal factors at work in our decision to use formal peer assessment as part of the course design. As already identified, research has shown that the assessment scheme drives and shapes student learning (Boud & Falchikov, 2006; Falchikov, 2005; Rust, 2002). Research also shows that cooperative learning works best when team marks are adjusted for individual contribution (Herrington & Oliver, 2000). Therefore, in order to signal to students that we consider learning to work as a productive team member to be important, we included peer assessment in the assessment scheme as the way of assessing this behavior.

The assessment scheme uses peer assessment of team colleagues to generate differential individual marks from team marks. Mentors and course academics provide feedback and mentoring to individual students and teams to help them learn from their peers' assessment and to improve overall team performance. Peer assessment provides a valuable tool to course academics, mentors, team leaders and individuals themselves to develop strong personal and team working skills. It encourages students to learn to evaluate their behavior by reflecting on how team members perceive their behavior within the team environment. It also delivers accountability to the team for individual efforts. On a number of occasions we have seen a student's behavior change dramatically through the course as they have received mentoring in response to poor peer assessments.

Anonymous student feedback collected at the end of each year as part of the annual PCR process indicates that overall students like and value the peer assessment process. They see it as a "good way to give feedback," as a way to "make us reflect on our own and our team members' performance" and that "feedback from peers can be used by each student to improve themselves." While some students see it as a method that will "motivate all students within a group to work hard" others acknowledge that it is a powerful tool which needs to be used wisely. "Having peer assessment doesn't necessarily improve the attitude of the 'lazier' team member. If that student just aims for a pass, peer assessment will not help much. This is based on my experience this year. Also, what if the team is really performing and everyone contributes equally? If 1 or 2 members get lower than the team mark, they may become disillusioned and may disrupt the harmony of the team." Students have also identified the need for the process and the outcomes of peer assessment to be transparent if it is to have full impact, because without that "students cannot be clear what they need to improve." While it is a "potentially useful indication if someone on the team is not doing as much work as they should" without transparency students are unsure "how much of an impact" it has had. Others have indicated they like that it "helps compensate for the variable level of work put in by the rest of the team" and "the feedback helps individual marks be decided fairly."

Prior to the introduction of software to collect and help generate peer assessed ratings, we used other forms of peer assessment which did not prove so successful at encouraging the development of team skills. Initially we held team meetings where we discussed how much each team member had contributed to the artefact being assessed and allocated individual marks accordingly. This led to students indicating they had all worked equally hard and then coming to see course academics afterwards to complain that individuals had not pulled their weight. When asked why they did not speak up in the meeting their answer was nearly always that they had to continue to work with the other students and they did not want to upset them. On occasion, especially when a single member of the team had contributed very little, we would have particularly contentious discussions about who had done what. Subsequently students were asked to supply this information as part of quarterly reporting, which was available to all team members. To generate individual marks academics then used this information, along with assessment of individual contributions based upon commits to the team's repository and information contained within the report. This required significant academic involvement and needed to be followed up with a team meeting to confirm marks. For further information on the implementation of assessment within the course see Johns-Boast (2010). Complaints from students about other team member's contributions all but ceased after the introduction of peer assessment based upon the concepts built into WebPA and use of the WebPA-x software .

Team Formation

Through the lifetime of the course we have moved from carefully crafting teams based on providing an even spread of academic achievement to the model we use currently where we let students express interest in undertaking certain projects and then course academics put together teams. While we have no empirical evidence to support one method over another, we have found that the students are more motivated at the start of the course if they feel that they have had some say in which project and team they have been placed.

Students submit their preference for undertaking each of the projects on offer. They also list, with reasons, any project that they would not like to work on. They may also indicate, with reasons for their choice, the students with whom they would like to work, and more importantly, those with whom they do not wish to work. Course academics form teams based on this information. Previous academic performance rarely plays a part in determining team composition. We also try to avoid having groups of more than two friends together on a single team as we have found that friendships can get in the way of good team performance. Each year we offer more projects than we have students to undertake them. Exactly which projects will run is determined by fourth year student preferences. Once we know which projects they have chosen the remainder of the team is decided, based upon third year preferences.

The process we follow when forming teams is made clear to students. We stress that it is important that we have well balanced teams and assure students that once the project is underway they will find that being on their chosen project is not as important as they imagine. Frequently fourth year students provide confirmation of this. Both the marks gained by individual students and the end result in terms of the quality of the artefacts produced show no obvious benefit from producing teams by one method or another. Nor have we seen any obvious change in grades achieved by either good or bad students. The use of peer assessment, however, has seen a greater range in individual marks being achieved.

Enrolment numbers each year determine team size. We have had greatest success, however, both in terms of learning outcomes and from the point of view of project success, when teams are between five and seven students. Our experience is that when teams are too small they do not provide the same opportunities to develop and practice leadership and project management. On the other hand, when teams are larger than seven students it becomes too easy for one or more students to 'hide' and so social loafing and free-riding are more likely to occur. In addition, productivity drops and so the value to the industry partner is not as much as might be expected.

Team Roles

Although notionally fourth year students are the team managers and leaders while third year students form the technical team, we have had teams where the team is effectively led and managed from behind, by at least one of the third year students. When this happens, course academics and mentors stress to students that this situation is also experienced in the workplace. It is very rare that when the fourth year students lack motivation, and leadership and management skills that one or more of the third year students do not take over the role unofficially. "As a team develops from a group of individuals into an effective learning system, members share the functional tasks necessary for team effectiveness." (Kayes et al., 2005, p. 333). Peer assessment recognizes this role change and accords the students appropriate peer ratings and an increased share of the team mark.

Team Success

In our experience, project success is greatest – from the students', the project clients' and the academics' points of view – when teams have a committed and engaged project client. The project client provides considerable motivation for the fourth year team leaders who in turn motivate their third year team members. Convincing students to focus on project outcomes rather than marks leads to higher overall grades and greater student satisfaction with the course and their learning. Regardless of the academic composition of the

student team, the presence of these three factors leads to success (Johns-Boast & Patch, 2010).

Teamwork Skills

Despite having participated in group work in previous courses, students are unlikely to have been taught much about team work (Gordon, 2010; Livingstone & Lynch, 2002). Recognizing that highly performing teams do not just 'appear' we use lectures and workshops to provide students with background on what it means to be part of an effective team. In the first weeks of the course, we introduce students to the concepts of personality and its potential impact upon individual and team performance. Students participate in workshops on Myers Briggs Type Indicator (Myers Briggs, n.d.) and Belbin Team Roles (Belbin, n.d.). They are also introduced to various theories relating to team formation, such as Tuckman (Tuckman, 1965; Tuckman & Jensen, 1977) and Benne & Sheats (Benne & Sheats, 1948). We also remind students of the impact that social loafing and free-riding can have on their experience and how they might deal with such behavior should they encounter it in their team (Oakley et al., 2004). Additionally we use team problems as cases for learning, explaining that the problems they encounter in their project teams are also encountered frequently within the workplace. Sometimes this provides an excellent opportunity to introduce students to employment legislation. Mentors and industry partners often discuss with teams how such problems are dealt with in the workplace.

Along with the use of peer assessment, the assessment scheme focuses on the process rather than the product which helps students develop teamwork skills. Assessing the process sends the message to students that they are expected to learn to work in teams, and that the process in which they engage to deliver their project is just as important as the product. To achieve this, the skills the students are expected to develop are clearly identified as outcomes. It is especially important that PBL and PjBL assess students' learning with an authentic form of assessment (Baxter & Shavelson, 1994; Birenbaum, 1996). The project review process provides such an authentic form of assessment. While it mirrors industry practice, it is in effect an oral examination or mini-viva that ensures all students are able to explain and justify the decisions the team has made. During a project review, teams are expected to not only answer questions from course academics, mentors and project clients, but to display and explain artefacts and demonstrate their software as it is developed. Lectures, workshops, peer assisted learning and peer assessment, help students understand and develop key skills related to teamwork and performance review.

All stakeholders are pleased with the project review process. Anonymous student feedback shows they see real value in the project reviews which provide rich formative feedback with a direct link between acting on that feedback and future progress and outcomes. For example students say the "project reviews work well because they force you to process the feedback being given and act on it (at least more than other forms of assessment I've seen). You often don't see the feedback from exams, and I often dismiss the feedback from normal assessments because it's a specific situation. But with the projects, there's a direct link between acting on the feedback and future progress and outcomes." Students also indicated that they "really like the reviews … [as] they were a great chance to show how our process was working (or not)." Additionally, students acknowledge that project review preparation and participation develops their ability to critically assess their own work as it requires them to take "a step back from the inner workings of the project, [which] gives us an objective view of our project's progress rather than our highly subjective one."

Students similarly note that they "like the idea of reflections and peer assessment. It is a good way to give feedback and reflect on the working of a team." They also like the unstructured nature of the assessment scheme saying, for example, "I liked organizing things for myself and working things out by myself. I liked not having something due every 2 weeks and how there were few well-defined deliverables." While students frequently acknowledge the nature of the work required by the course, they comment that despite sometimes encountering "a lot of stress caused by the project … it was definitely worth it."

Flow-On Benefits

While the focus of this chapter is the benefits to student learning that emerge from the simulated industrial context it is clear there are significant additional advantages from our approach.

For the Students

Many students find employment based on connections made through the course. Jobs can be with the industry partner sponsoring their project, often part-time while they complete their degree and frequently as a full-time employee on graduation. On occasion at the Project Showcase – effectively a trade show at the end of the course where students present their work to invited representatives of local industry and Government departments – students meet representatives of other industry partners who are sufficiently impressed with individual students to offer them employment.

Furthermore, over the years a number of student teams have won state and national awards for student projects from the ACS and the Australian Information Industry Association (AIIA) which is a nice addition to a student's curriculum vitae. All students are encouraged to nominate for appropriate awards but it is something which we insist that the team themselves need to drive.

For the Industry Partners

As part of the annual PCR, our industry partners are asked to provide feedback to help improve the course. Despite the occasional project that does not go to plan, our industry partners are keen to engage and we now regularly have more projects proposed than we have student teams to undertake them. Many of our industry partners are repeat clients; half a dozen of whom have undertaken three or more projects since 2009. Universally their feedback is positive. They feel that the student projects generally deliver a "good outcome that can be used" and "result in the production of 'very nice to have' capability that we would probably not otherwise have been able to introduce … [and] provide a small but worthwhile addition to the overall operational capability of the software engineering group during the course of the year." They also like that involvement with the course allows them "to stay connected with the latest trends" and "intend to continue … involvement into the foreseeable future." They like that the course "allows students to engage with industry early," that they get to "look over" the students prior to employing them, and that participation provides an opportunity for them "to give back to the ANU" as well as providing "a means … to corporately contribute to producing more highly skilled and capable engineering graduates. This in turn will be for the betterment of the profession and ultimately for the betterment of society as a whole."

Industry partners' involvement with the course is frequently "a very positive one from the point of view of the … project team and the wider company." In 2012, a first time industry partner said "we had no idea on how the setup worked but the teaching staff were fantastic in making sure we were well prepared in knowing how to earn the right to welcome a project team to the company as well as how best to engage with their students, many of whom the staff work with personally … It has been so much of a success, we want to come back for more next year. The staff have been supportive and their students superb." Industry partners consider that "the projects are an extremely valuable pedagogical tool as they provide the students with an experience of all the complexities and difficulties (and rewards!) of real-world engineering in a relatively benign and supportive environment" and view "the ANU student projects as being an innovative educational program that is extremely beneficial for all involved."

It can be argued that through the medium of these student projects the ANU annually provides more than $1 million software development assistance, mainly to start-ups and small and medium enterprises. Based on the number of hours course requirements suggest each student should spend on their project (10-12 hours for the 26 teaching weeks each year), estimated hours of industry partner involvement and including overheads for mentors and course academics, we consider that the value of each project is around $150,000. Since 2004, over 75% of industry partners sponsoring student projects have been start-up companies, individuals developing a proof of concept prior to formal start-up, non-profit organizations and small to medium enterprises.

One regular industry partner obtains a secondary benefit from participation because "the student projects also provide opportunities for junior software engineers … to participate in the projects as clients and client project managers. This is an educational and career enhancing experience for those engineers" (Johns-Boast & Patch, 2010).

For the Institution

The course also enables the institution and its academics to build relationships with industry. Individual academics themselves have also benefitted from the relationships created with industry. For example, we have had a Researcher-in-Business grant (ARC, n.d.) which grew from the organization's engagement with the academic through provision of a student project.

Challenges for Course Academics

As with all project based courses, the workload for course academics is high. The rewards, however, also are high. When faced with workload impacts it is tempting to revert to assessing the product rather than the process. Using peer assessment of quantity and quality of contribution helps minimize the effort required to generate valid, individual student marks and to keep students focused and motivated to work as a member of a team. One of the biggest hurdles that course academics need to overcome is the temptation to teach – to transmit information – rather than to facilitate students' own problem solving and investigation. Using an unstructured course schedule helps overcome this. Basing weekly lectures and workshops upon the problems that teams are encountering at that time as advised by mentors and students themselves, provides a mechanism for engaging students and encourages active learning.

Assessing team progress via the project review process is challenging for both the course academics and students the first time they participate. However, clearly identifying the goals for each project review keeps all participants focused and provides consistency despite widely varying types and difficulty of projects. Occasionally students, who are focused on marks, and therefore product, have expressed skepticism at the ability of the project review process to determine a valid assessment of team output and functioning. However, is quite clear to course academics, team mentors and project clients when a team is not functioning well and / or not working to capacity. Using a rubric to assess performance as demonstrated at project reviews and to provide formative feedback to teams enables course academics and mentors to articulate clearly the expected performance at each grade level. Rubrics also make it easier to be consistent across projects that are widely differing in complexity and type. Course academics are confident in the team marks awarded from the project review process. Active questioning

facilitates the discovery of poor quality work, lack of continuity of process and so on, in a manner that is not possible in any other form of assessment. The participation of industry clients in project reviews alleviates the need for course academics to be familiar with the intricacies of each project and ensures that teams do not simply spin a good yarn.

Experienced, industry-based mentors and project clients provide a valuable supplement for course academics who may not have much, if any, industry software development experience. Additionally, even when course academics have significant industry experience, students listen more closely to advice given by team mentors and project clients.

CONCLUSION

The need for software engineering students to develop strong personal and professional skills – problem solving, communication, teamwork, self-assessment, systems-thinking and other process skills – will not diminish in the near future. To develop these skills in our graduates, first we need to state them clearly and then to develop our courses using proven learning approaches that promote development of the targeted skills. To do so will involve many university academics changing their focus and approach to designing and teaching courses. Our courses must provide students with learning experiences that more closely reflect real-world experience (Shaw, 1992) and we must move from transmitting information to facilitating learning. Accepting that much of the technology to which we introduce our students in their first year will be obsolete by the time they practice software engineering will help academics design programs and courses that focus on ideas rather than content and allow space in the curriculum for the development of personal and professional skills.

When compared with more usual PjBL courses and WIL the simulated industrial context provides students with a remarkably broad learning experience. They get to work with two different teams on two different projects with two different project clients over a two-year period. This introduces them to a greater variety of work place cultures, project types, and a greater range of people – peers and project clients. The simulated industrial context and the project review process encourage students to develop critical thinking skills and to become autonomous learners as they must make defensible decisions based upon their investigation of possible approaches for developing a solution to the problem. Sourcing projects from industry partners and then involving them as project clients, helps students develop and improve their negotiation skills as they come to an agreement with the project client on an appropriate software development lifecycle, project management approach, required artefacts, schedule and so on. Combining two cohorts of students into one and then introducing peer assessment encourages students to develop mentoring skills, to learn from each other and to develop and practice leadership skills. Importantly, students obtain this experience without extending their time at university by the period of internship or work placement. It also means students do not miss exposure to valuable content that they might not meet elsewhere because courses have been sacrificed to make room for the internship or work placement for credit.

The innovations included in this course – two cohorts of students in combination; requiring students to take the course twice, once in third year as technical team members and then again in fourth year as team leaders; an assessment scheme that assesses the process and not the product and encourages individuals to pursue quality work in a group work environment through the use of peer assessment; and use of experienced, industry-based mentors – all work together to provide a viable alternative to more usual implementations of work integrated learning. The simulated industrial context offers students a true apprenticeship for real-life problem solving.

Clear identification of the desired attributes graduates need to demonstrate, followed by the adoption of educational approaches proven to develop those attributes, enables software engineering academics, many of whom may not have been taught how to teach and who may not be specialists in the domain of personal and professional skills, to develop those skills in their students. Development of these skills is no longer a matter of accident. It can be repeated year after year.

ACKNOWLEDGMENT

Since the course described in the case study was first taught in 2004, many people have contributed as members of the teaching team. In particular, I would like to acknowledge the ideas and contribution of my colleague, Dr Shayne Flint. I would also like to acknowledge our industry-based mentors who have unstintingly contributed to the refinement of the course and to helping students develop into competent and capable software engineers. Equally, without our industry partners this course would not have been possible, nor would it have achieved the successful outcomes it has.

In 2012, the author and Dr Flint received the Australian Council of Engineering Deans' (ACED) award for Engineering Education Excellence (Highly Commended) for this course.

REFERENCES

ABET. (2012-2013). *Criteria for accrediting engineering programs*. Retrieved November 2013, from http://www.abet.org/DisplayTemplates/DocsHandbook.aspx?id=3143

Abeysekera, I. (2006). Issues relating to designing a work-integrated learning program in an undergraduate accounting degree program and its implications for the curriculum. *Asia-Pacific Journal of Cooperative Education*, *7*(1), 7–15.

Albanese, R., & van Fleet, D. D. (1985). Rational behavior in groups: The free-riding tendency. *Academy of Management Review*, *10*(2), 244–255.

ARC. (n.d.). Retrieved from http://www.arc.gov.au/applicants/rib.htm

Atchison, M., Pollock, S., Reeders, E., & Rizetti, J. (2002). *Work integrated learning paper*. Retrieved November 2013, from http://mams.rmit.edu.au/a0o4e48729rdz.pdf

Barrie, S. C. (2004). A research-based approach to generic graduate attributes policy. *Higher Education Research & Development*, *23*(3), 261–275. doi:10.1080/0729436042000235391

Baxter, G. P., & Shavelson, R. J. (1994). Science performance assessments: Benchmarks and surrogates. *International Journal of Educational Research*, *21*(3), 279–298. doi:10.1016/S0883-0355(06)80020-0

Beasley, R. E. (2003). Conducting a successful senior capstone course in computing. *Journal of Computing Sciences in Colleges*, *19*(1), 122–131.

Belbin. (n.d.). Retrieved from http://www.belbin.com/rte.asp?id=8

Benne, K. D., & Sheats, P. (1948). Functional roles of group members. *The Journal of Social Issues*, *4*(2), 41–49. doi:10.1111/j.1540-4560.1948.tb01783.x

Benyon, J. (2012). *Response to the senate inquiry: The shortage of engineering and related employment skills, on behalf of the Australian Council of Engineering Deans Inc.* Retrieved from http://www.aph.gov.au/DocumentStore.ashx?id=14789c30-d180-4e1e-973e-592350643e39&ei=FL4WUvSuEae4iAfq6IHQCA&usg=AFQjCNE3qHqBqR8dMFlZ2HKZvDJ6ocOijQ&sig2=XEIzLBZ_xLPdcs1ukes2QQ&bvm=bv.51156542,d.aGc&cad=rja

BIE. (n.d.). *What is PBL?* Retrieved from http://www.bie.org/about/what_pbl

Biggs, J. (2003). Aligning teaching and assessing to course objectives. In *Teaching and learning in higher education: New trends and innovations*. University of Aveiro.

Biggs, J., & Tang, C. (2007). *Teaching for quality learning at university*. Berkshire, UK: Open University Press.

Birenbaum, M. (1996). Assessment 2000: Towards a pluralistic approach to assessment. In *Alternatives in assessment of achievements, learning processes and prior knowledge* (pp. 3–29). Berlin: Springer. doi:10.1007/978-94-011-0657-3_1

Boud, D., Cohen, R., & Sampson, J. (1999). Peer learning and assessment. *Assessment & Evaluation in Higher Education*, *24*(4), 413–426. doi:10.1080/0260293990240405

Boud, D., & Falchikov, N. (2006). Aligning assessment with long-term learning. *Assessment & Evaluation in Higher Education*, *31*(4), 399–413. doi:10.1080/02602930600679050

Brown, S. (2004). Assessment for learning: The changing nature of assessment. *Learning and Teaching in Higher Education*, *1*, 81–89.

Chamillard, A. T., & Braun, K. A. (2002). The software engineering capstone: Structure and tradeoffs. *ACM SIGCSE Bulletin*, *34*(1), 227–231. doi:10.1145/563517.563428

Clark, N. (2005). Evaluating student teams developing unique industry projects. *Proceedings of the 7th Australasian Conference on Computer Education, 42*, 21–30.

Cohn, M. (2013). *Agile & scrum software development*. Retrieved November 2013, from www.mountaingoatsoftware.com

Cooper, L., Orrell, J., & Bowden, M. (2010). *Work integrated learning: A guide to effective practice*. Abingdon, UK: Routledge.

Costa, A. C., Roe, R. A., & Taillieu, T. (2001). Trust within teams: The relation with performance effectiveness. *European Journal of Work and Organizational Psychology, 10*(3), 225–244. doi:10.1080/13594320143000654

Dawson, R. (2000). *Twenty dirty tricks to train software engineers*. Paper presented at the 22nd International Conference on Software Engineering. New York, NY.

De Freitas, S. (2006). *Learning in immersive worlds*. London: Joint Information Systems Committee.

Dede, C. (2009). Immersive interfaces for engagement and learning. *Science, 323*(5910), 66–69. doi:10.1126/science.1167311 PMID:19119219

EA. (n.d.). *Stage 1 competency standard for professional engineer*. Canberra, Australia: Engineers Australia (EA). Retrieved November 2013, from https://www.engineersaustralia.org.au/sites/default/files/shado/Education/Program%20Accreditation/130607_stage_1_pe_2013_approved.pdf

Ellis, T. J., & Hafner, W. (2007). Assessing collaborative, project-based learning experiences: Drawing from three data sources. In *Proceedings of the 37th Annual Frontiers In Education Conference-Global Engineering: Knowledge Without Borders, Opportunities Without Passports* (pp. T2G–13). IEEE.

Entwistle, N. (2000). *Promoting deep learning through teaching and assessment: Conceptual frameworks and educational contexts*. Paper presented at TLRP Conference. Leicester, UK.

Falchikov, N. (2005). *Improving assessment through student involvement*. London: Routledge Falmer.

Felder, R. M., & Brent, R. (2004). *The ABC's of engineering education: ABET, Bloom's taxonomy, cooperative learning, and so on*. Paper presented at the 2004 American Society for Engineering Education Annual Conference & Exposition. New York, NY.

Freudenberg, B., Brimble, M., & Cameron, C. (2010). Where there is a WIL there is a way. *Higher Education Research & Development, 29*(5), 575–588. doi:10.1080/07294360.2010.502291

Freudenberg, B., Brimble, M., & Vyvyan, V. (2010). The penny drops: Can work integrated learning improve students' learning?. *e-Journal of Business Education & Scholarship of Teaching, 4*(1), 42–61.

Froyd, J. E., Wankat, P. C., & Smith, K. A. (2012). Five major shifts in 100 years of engineering education. *Proceedings of the IEEE, 100*(13), 1344–1360. doi:10.1109/JPROC.2012.2190167

Garris, R., Ahlers, R., & Driskell, J. E. (2002). Games, motivation, and learning: A research and practice model. *Simulation & Gaming, 33*(4), 441–467. doi:10.1177/1046878102238607

Gordon, N. A. (2010). Group working and peer assessment—Using WebPA to encourage student engagement and participation. *Innovation in Teaching and Learning in Information and Computer Sciences, 9*(1), 20–31. doi:10.11120/ital.2010.09010020

Helmo-Silver, C. E. (2004). Problem-based learning: What and how do students learn? *Educational Psychology Review, 16*(3), 235–266. doi:10.1023/B:EDPR.0000034022.16470.f3

Herrington, J., & Oliver, R. (2000). An instructional design framework for authentic learning environments. *Educational Technology Research and Development, 48*(3), 23–48. doi:10.1007/BF02319856

Isomoettoenen, V., & Kaerkkaeinen, T. (2008). The value of a real customer in a capstone project. In *Proceedings of the 21st Conference on Software Engineering Education & Training*, (pp. 85–92). Academic Press.

Johns-Boast, L. F. (2010). *Group work and individual assessment.* Paper presented at the Past, Present, Future: 21st Annual Conference of the Australasian Association of Engineering Education. Sydney, Australia.

Johns-Boast, L. F., & Flint, S. (2009). *Providing students with 'real-world' experience through university group projects.* Paper presented at the 20th Annual Conference of the Australasian Association for Engineering Education. Adelaide, Australia.

Johns-Boast, L. F., & Flint, S. (2013). *Simulating Industry: An innovative software engineering capstone design course.* Paper presented at the 43rd Annual Frontiers in Education (FIE) Conference. Oklahoma City, OK.

Johns-Boast, L. F., & Patch, G. (2010). *A win-win situation: Benefits of industry-based group projects.* Paper presented at the Past, Present, Future: 21st Annual Conference of the Australasian Association of Engineering Education. Sydney, Australia.

Jollands, M., Jolly, L., & Molyneaux, T. (2012). Project-based learning as a contributing factor to graduates' work readiness. *European Journal of Engineering Education, 37*(2), 143–154. doi:10.1080/03043797.2012.665848

Kaufman, D. B., Felder, R. M., & Fuller, H. (1999). Peer ratings in cooperative learning teams. In *Proceedings of the 1999 Annual ASEE Meeting.* ASEE.

Kaufman, D. B., Felder, R. M., & Fuller, H. (2000). Accounting for individual effort in cooperative learning teams. *Journal of Engineering Education, 89*(2), 133–140. doi:10.1002/j.2168-9830.2000.tb00507.x

Kayes, A. B., Kayes, D. C., & Kolb, D. A. (2005). Experiential learning in teams. *Simulation & Gaming, 36*(3), 330–354. doi:10.1177/1046878105279012

Kolb, A. Y., & Kolb, D. A. (2009). The learning way: Meta-cognitive aspects of experiential learning. *Simulation & Gaming, 40*(3), 297–327. doi:10.1177/1046878108325713

Kolb, D. A. (1984). *Experiential learning: Experience as the source of learning and development* (Vol. 1). Englewood Cliffs, NJ: Prentice-Hall.

Litchfield, A., Frawley, J., & Nettleton, S. (2010). Contextualising and integrating into the curriculum the learning and teaching of work-ready professional graduate attributes. *Higher Education Research & Development, 29*(5), 519–534. doi:10.1080/07294360.2010.502220

Litzinger, T. A., Lattuca, L. R., Hadgraft, R. G., & Newstetter, W. C. (2011). Engineering education and the development of expertise. *Journal of Engineering Education, 100*(1), 123–150. doi:10.1002/j.2168-9830.2011.tb00006.x

Livingstone, D., & Lynch, K. (2002). Group project work and student-centred active learning: Two different experiences. *Journal of Geography in Higher Education, 26*(2), 217–237. doi:10.1080/03098260220144748

Male, S. A. (2010). Generic engineering competencies: A review and modelling approach. *Education Research and Perspectives, 37*(1), 25–51.

McLennan, B., & Keating, S. (2008). *Work-integrated learning (WIL) in Australian universities: The challenges of mainstreaming WIL.* Paper presented at the ALTC NAGCAS National Symposium. Melbourne, Australia.

Myers Briggs. (n.d.). *MBTI personality type.* Retrieved from http://www.myersbriggs.org/my-mbti-personality-type/mbti-basics/

Oakley, B., Felder, R. M., Brent, R., & Elhajj, I. (2004). Turning student groups into effective teams. *Journal of Student Centered Learning, 2*(1), 9–34.

Orrell, J. (2004). *Work-integrated learning programmes: Management and educational quality.* Paper presented at the Australian University Quality Forum 2004. Canberra, Australia.

Passow, H. J. (2007). *What competencies should engineering programs emphasize? A meta-analysis of practitioners' opinions informs curricular design.* Paper presented at the 3rd International CDIO Conference. New York, NY.

Patrick, C. J., Peach, D., Pocknee, C., Webb, F., Fletcher, M., & Pretto, G. (2008). *The WIL (work integrated learning) report: A national scoping study.* Queensland University of Technology.

Peterson, D. (2009). *What is kanban?* Retrieved November, 2013, from http://www.kanbanblog.com/explained/

Porter, T. W., & Lilly, B. S. (1996). The effects of conflict, trust, and task commitment on project team performance. *The International Journal of Conflict Management, 7*(4), 361–376. doi:10.1108/eb022787

Queensu. (n.d.). *Learning strategies.* Retrieved from http://www.queensu.ca/ctl/resources/topic-specific/casebased/learningstrategies.html

Ramsden, P. (2003). *Learning to teach in higher education.* London: Routledge Falmer.

Review Process. (n.d.). Retrieved from http://www.finance.gov.au/gateway/review-process.html

Rust, C. (2002). The impact of assessment on student learning: How can the research literature practically help to inform the development of departmental assessment strategies and learner-centred assessment practices? *Active Learning in Higher Education, 3*(2), 145–158. doi:10.1177/1469787402003002004

Saddington, T. (2000). The roots and branches of experiential learning. *NSEE Quarterly, 26*(1), 2–6.

Saddington, T. (n.d.). *What is experiential learning?* Retrieved from http://www.edb.gov.hk/attachment/sc/curriculum-development/kla/pshe/references-and-resources/ethics-and-religious-studies/experiential_learning_2.pdf

Savery, J. R. (2006). Overview of problem-based learning: Definitions and distinctions. *Interdisciplinary Journal of Problem-Based Learning and Teaching in Higher Education, 1*(1), 9–20.

Savery, J. R., & Duffy, T. M. (2001). *Problem based learning: An instructional model and its constructivist framework.* Bloomington, IN: Indian University.

Savin-Baden, M. (2000). *Problem-based learning in higher education: Untold stories.* Buckingham, UK: The Society for Research into Higher Education & Open University Press.

Savin-Baden, M. (2001). The problem-based learning landscape. *Planet,* (4), 4–6.

Scrumalliance. (n.d.). *Why scrum.* Retrieved from http://www.scrumalliance.org/why-scrum

Shaw, M. (1992). We can teach software better. *Computing Research News, 4,* 2–4, 12.

Sheppard, S. D., Macatangay, K., Colby, A., & Sullivan, W. M. (2008). *Educating engineers: Designing for the future of the field.* The Carnegie Foundation for the Advancement of Teaching.

Stevens, D. D., & Levi, A. J. (2005). *Introduction to rubrics: An assessment tool to save grading time, convey effective feedback and promote student learning.* Stylus Publishing.

Todd, R. H., & Magleby, S. P. (2005). Elements of a successful capstone course considering the needs of stakeholders. *European Journal of Engineering Education, 30*(2), 203–214. doi:10.1080/03043790500087332

Topping, K., & Ehly, S. (1998). *Peer-assisted learning.* Hoboken, NJ: Lawrence Erlbaum Associates, Inc.

Tuckman, B. W. (1965). Developmental sequence in small groups. *Psychological Bulletin, 63*(6), 384. doi:10.1037/h0022100 PMID:14314073

Tuckman, B. W., & Jensen, M. A. C. (1977). Stages of small-group development revisited. *Group & Organization Management, 2*(4), 419–427. doi:10.1177/105960117700200404

Tyler, R. W. (1949). *Basic principles of curriculum and instruction.* Chicago: University of Chicago Press.

Ulloa, B. C. R., & Adams, S. G. (2004). Attitude toward teamwork and effective teaming. *Team Performance Management, 10*(7/8), 145–151. doi:10.1108/13527590410569869

Washbourn, P. (1996). Experiential learning: Is experience the best teacher? *Liberal Education, 82*(3), 1–10.

Webpaproject. (n.d.). Retrieved from http://Webpaproject.lboro.ac.uk/

Willey, K., & Gardner, A. (2010). Investigating the capacity of self and peer assessment activities to engage students and promote learning. *European Journal of Engineering Education, 35*(4), 429–443. doi:10.1080/03043797.2010.490577

Woods, D. R., Felder, R. M., Rugarcia, A., & Stice, J. E. (2000). The future of engineering education: III: Developing critical skills. *Chemical Engineering Education, 34*(2), 1–20.

ADDITIONAL READING

Adams, R. S., & Felder, R. M. (2008). Reframing professional development: A systems approach to preparing engineering education to educate tomorrow's engineers. *Journal of Engineering Education, 97*(3), 239–240. doi:10.1002/j.2168-9830.2008.tb00975.x

Beranek. G., Zuser, W., & Grechenig, T. (2005). Functional group roles in software engineering teams. *Paper presented at the ACM SIGSOFT Software Engineering Notes.*

Boss, S. (2013). PBL for 21st century success: Teaching critical thinking, collaboration, communication and creativity. Buck Institute for Education (BIE).

Brown, J., & Dobbie, G. (1998). Software engineers aren't born in teams: Supporting team processes in software engineering project courses. In *Proceedings of International Conference on Software Engineering: Education & Practice* (pp. 42–49). IEEE.

Duch, B. J., Allen, D. E., & White, I. H. B. (1999). Problem-based learning: Preparing students to succeed in the 21st Century. *Teaching Matters, 3*(2).

Felder, R. M., Woods, D. R., Stice, J. E., & Rugarcia, A. (2000). The future of engineering education: II. Teaching methods that work. *Chemical Engineering Education, 34*(1), 26–39.

Hansen, R. S. (2006). Benefits and problems with student teams: Suggestions for improving team projects. *Journal of Education for Business, 82*(1), 11–19. doi:10.3200/JOEB.82.1.11-19

Healey, M., & Jenkins, A. (2000). Kolb' experiential learning theory and its application in geography in higher education. *The Journal of Geography, 99*, 185–195. doi:10.1080/00221340008978967

Henry, S. M., & Todd Stevens, K. (1999). Using Belbin's leadership role to improve team effectiveness: An empirical investigation. *Journal of Systems and Software, 44*(3), 241–250. doi:10.1016/S0164-1212(98)10060-2

Kerka, S. (2001). Capstone experiences in career and technical education. Retrieved November, 2013, from http://www.calpro-online.org/eric/docs/pab00025.pdf

Kiley, M., Mullins, G., Peterson, R. F., & Rogers, T. (2000). Leap into ... Problem-based learning.

Martin, A. J., & Hughes, H. (2009). *How to make the most of work integrated learning.* Palmerston North, New Zealand: Massey University.

Maudsley, G., & Strivens, J. (2000). Promoting professional knowledge, experiential learning and critical thinking for medical students. *Medical Education, 34*(7), 535–544. doi:10.1046/j.1365-2923.2000.00632.x PMID:10886636

Michaelsen, L. K., & Sweet, M. (2008). The essential elements of team-based learning. *New Directions for Teaching and Learning*, (116): 7–27. doi:10.1002/tl.330

Orrell, J. (2011). Good Practice Report: Work-integrated learning. Retrieved November, 2013, from http://www.acen.edu.au/resources/docs/WIL-Good-Practice-Report.pdf

Pfaff, E., & Huddleston, P. (2003). Does it matter if I hate teamwork? What impacts student attitudes toward teamwork. *Journal of Marketing Education, 25*(1), 37–45. doi:10.1177/0273475302250571

Rugarcia, A., Felder, R. M., Woods, D. R., & Stice, J. E. (2000). The future of engineering education: I. A Vision for a New Century. *Chemical Engineering Education, 34*(1), 16–25.

Shaw, M. (2000). *Software engineering education: A roadmap.* Paper presented at the Proceedings of the Conference on The Future of Software Engineering.

Shaw, M. (2005). *Software engineering for the 21st century: A basis for rethinking the curriculum. Institute for software research international.* Carnegie Mellon University.

Sheard, A., & Kakabadse, A. (2002). From loose groups to effective teams: The nine key factors of the team landscape. *Journal of Management Development, 21*(2), 133–151. doi:10.1108/02621710210417439

Sibley, J., & Parmelee, D. X. (2008). Knowledge is no longer enough: Enhancing professional education with team-based learning. *New Directions for Teaching and Learning*, (116): 41–53. doi:10.1002/tl.332

Strauss, P., & U, A. (2007). Group assessments: Dilemmas facing lecturers in multicultural tertiary classrooms. *Higher Education Research & Development*, 26(2), 147–161. doi:10.1080/07294360701310789

Woods, D. F. (2003). Clinical review: ABC of learning and teaching in medicine: Problem based learning. *BMJ (Clinical Research Ed.)*, 326.

KEY TERMS AND DEFINITIONS

Academic: A teacher, lecturer, professor teaching at a higher education institution; sometimes also called faculty.

Authentic Activities / Learning / Problems: Activities / learning / problems which have real-world relevance (Herrington & Oliver, 2000) where the "cognitive demands, i.e. the thinking required, are consistent with the cognitive demands in the environment for which we are preparing the learner" (Savery & Duffy, 2001, p. 4).

Competencies: Are the knowledge, skills, abilities, attitudes, and other characteristics that enable a person to perform skillfully (i.e., to make sound decisions and take effective action), in complex and uncertain situations such as professional work, civic engagement, and personal life (Passow, 2007). See also *Generic Competencies*.

Course: A single unit of study, sometimes also known as a unit, subject or module.

Gateway Review: Gateway reviews are "short, intensive reviews at critical points in the project/program's lifecycle by a team of reviewers not associated with the project/program …[to] provide an arm's length assessment of the project/program against its specified objectives, and an early identification of areas requiring corrective action" (Review Process, n.d.).

Generic Competencies: Are important for all disciplines, including engineering, software engineering and computer science.

Program: A complete, integrated program of study leading to the award of a degree. A program is composed of many courses.

Rubric: "A scoring tool that lays out the specific expectations for an assignment. Rubrics divide an assignment into its component parts and provide a detailed description of what constitutes acceptable or unacceptable level of performance for each of those parts" (Stevens & Levi, 2005, p. 3).

Student Outcomes: Describe what students are expected to know and be able to do by the time of graduation. These relate to the skills, knowledge, and behaviors that students acquire as they progress through the program. (ABET, 2012-2013, p. 2). See also *Competencies*.

Section 5
Promoting Project-Based Learning

Chapter 12
Project–Based Learning:
An Environment to Prepare IT Students for an Industry Career

Luís M. Alves
Instituto Politécnico de Bragança, Portugal

Pedro Ribeiro
Universidade do Minho, Portugal

Ricardo J. Machado
Universidade do Minho, Portugal

ABSTRACT

The lack of preparation of Software Engineering (SE) graduates for a professional career is a common complaint raised by industry practitioners. One approach to solving, or at least mitigating, this problem is the adoption of the Project-Based Learning (PBL) training methodology. Additionally, the involvement of students in real industrial projects, incorporated as a part of the formal curriculum, is a well-accepted means for preparing students for their professional careers. The authors involve students from BSc, MSc, and PhD degrees in Computing in developing a software project required by a real client. This chapter explains the educational approach to training students for industry by involving them with real clients within the development of software projects. The educational approach is mainly based on PBL principles. With the approach, the teaching staff is responsible for creating an environment that enhances communications, teamwork, management, and engineering skills in the students involved.

INTRODUCTION

During Software Engineering (SE) training, it is very difficult to provide industry-standard knowledge and skills, especially non-technical knowledge. These skills can be grouped into three main areas: management, engineering and personal. One challenge facing SE education is that the current lecture-based curriculum hardly engages students. Students often view SE principles as mere academic concepts, which are less interesting and less valuable. The reality is that Computer Science (CS) and Information Systems (IS) graduates often have to develop SE knowledge

DOI: 10.4018/978-1-4666-5800-4.ch012

and skills, especially non-technical knowledge and skills, later on, when they start their careers in industry.

The lack of preparation of SE graduates for a professional career is a common complaint raised by industry practitioners (Karunasekera & Bedse, 2007). One approach to solving, or at least mitigating this problem, is the adoption of the Project Based Learning (PBL) (Barrows & Tamblyn, 1980) training methodology. Involving students in industry projects during their undergraduate degree is a well-accepted method of preparing students for their professional careers.

In our approach we involve students from BSc, MSc and PhD degrees in Computing from our university to develop a complete software project requested by a real client. At the BSc level (Bologna 1st cycle) we involve students from Software Process and Methodologies (PMS) and Development of Software Applications (DAI) courses. At the MSc level (Bologna 2nd cycle) we involve students from Analysis and Design of Information Systems (ACSI) and Project Management of Information Systems (GPSI) courses. The curriculum integration and the pedagogical cooperation, through an integrated project between the four courses in analysis, are intended to promote students to work in a software development environment that is similar to an organization environment. Parts of the contents of these courses were also framed several times in training given by the teachers in business or industrial contexts, under protocols between the university and relevant organizations.

In this group of courses we must highlight the DAI course because of the unifying role it plays when compared to the remaining three. DAI has a learning value of 10 ECTS (European Credit Transfer and Accumulation System) and teachers of subsequent courses "expect" from students an effective ability to develop IT (Information Technology) solutions to problems with medium complexity. This main goal drives the teaching

team to adopt a set of procedures and pedagogical practices capable of dealing with the complexity inherent in managing a course of this kind.

To perform these software projects, students pursuing the same degree constitute the teams. However, they have to work in close collaboration with teams from other degrees. The teaching staff is responsible for creating an environment that enhances communications skills, team working skills, management skills and engineering skills of the students involved. It is a well-accepted fact that a competent software engineer requires a wide variety of skills in areas such as management, engineering, team working and communication (Ali, 2006; Nunan, 1999).

Another challenge is to evaluate teams and individuals who develop unique industry projects (Clark, 2005). In our case, we use "*Assessment Milestones*" distributed throughout the semester that allow us to track the students' work progress and thus avoid an end-point evaluation only.

Our approach presents an advance in SE education, in order to overcome the aforementioned challenges. The teams have the opportunity to interact with a real client. They can learn and apply SE principles through a real software project. Thus, they can evolve and improve their technical and non-technical skills. In our setting we promote a win-win approach for all stakeholders: clients, students, teachers and researchers. Clients will have state of the art projects implemented in their companies. Students can acquire technical and non-technical skills and work closely with real-world problems. Teachers will have the opportunity to teach technical knowledge authentically and realize new problems that companies are facing. Researchers can perform experiments on new and/ or existing techniques, tools and methods. This gives us the opportunity to provide guidelines for SE educators in order to improve their curricula and provide CS students with ready-to-apply SE knowledge skills.

This chapter is structured as follows: Section 1 introduces the range of problems that promoted the present work. Section 2 addresses, in detail, the current challenges in the software PBL that we are trying to respond to, as well as affording a comprehensive state-of-the-art picture that will suit the purpose of validating the strength of our approach. This section will also be dedicated to motivating the reader towards the systematic use of software PBL, which will be the subject of the chapter and that is, essentially, revealing its significance. Section 3 contains our PBL approach. In this section we will explain the environment and the integration of the different teams from all the courses involved. Section 4 shows the main results obtained from our PBL approach. We will start this section with an evaluation of the students teams method used in our approach, followed by a detailed analysis of the work projects developed by ACSI and GPSI students. It will also focus on discussing the results we are going to present in the chapter. Finally, in section 5 we will present our major concluding remarks.

BACKGROUND

SE has brought to the CS field the confluence of the process and development methodologies with the economic surroundings that are found to be indispensable to the professionals working in the industry with roles and responsibilities that go beyond the mere computer programming (Engle, 1989). In SE we find a clear concern with what is beyond purely technical issues, which have always been the ones that truly the CS devoted full attention to. In this sense, it is important for a software engineer to be educated in communication and management skills. These are skills which are vital to the software engineer's success and they cannot be left to be learned by "osmosis" (Engle, 1989).

While SE (as a discipline) is dedicated to study the software process development, IS (as a discipline) analyzes the impact of software-based systems on individuals, organization and society. However, in the scientific context there is a great divide between methods and research questions (Finkelstein, 2011; Gregg, Kulkarni, & Vinzé, 2001). The cross-fertilization between the two disciplines has allowed the cooperative and coordinated development of the engineering and requirement management, modeling and systems architecture, process development and project management approaches (Avison & Wilson, 2001; Birk et al., 2003; McBride, 2003). This is why, in the context of project technology-based solutions in problems of organizational nature, the two disciplines (and corresponding professional performances) merge into a blend of common methodologies, techniques and tools (Fung, Tam, Ip, & Lau, 2002; Hellens, 1997; Jayaratna & Sommerville, 1998) demonstrating the existence of an SE/IS convergence. The understanding of the intersection between the two disciplines allows students to develop projects of better quality, since the whole project is grounded in the knowledge of both disciplines.

When we are to deliver knowledge and skills to IT students we should be able to teach engineering and software management. The success of a software engineer is related to the ability of engineering and software management methodologies to adapt to the huge demands that professionals are subject to nowadays, which include dealing with all procedural issues of software production, along with technological competence and sensitivity to the needs and expectations of users (Platt, 2011).

The rationalization of all decisions relating to issues not explicitly technological is fundamental for a correct (effective and efficient) operation of a software development team (Dutoit, McCall, Mistrik, & Paech, 2006). Into this set of concerns and attitudes also fit the aforementioned software economics issues (Tockey, 1997). When properly reconciled with the technological dimension, *Software Engineering Management* has adopted the designation of *Software Engineering and Management* (Shere, 1988). This designation, which reinforces the existence of a management dimension within the SE (as in any other Engineering specialty), represents a break with the CS discipline and an assumption of the socio-technical nature of the SE, and its inevitable convergence with the IS discipline (Kurbel, 2008).

In our approach PMS and DAI courses intend to provide students an environment of software development projects similar to an environment in organization context. They introduce engineering and management software techniques to collect and specify the requirements of the software development projects. They also deal with the software lifecycle management issues. These two courses are mainly responsible for software development methodologies and software project planning teaching.

The ACSI and GPSI courses intend to instill in students an engineering approach to information systems development. This attitude derives from the SE/IS convergence mentioned above and is materialized in the business solutions development study (enterprise business applications). Thus, the engineering and requirements management techniques and modeling and systems architecture are complemented and the maturity and software processes improvement issues are systematized, as well as the processes management and project development.

The Department of Information Systems (DSI) of the School of Engineering (EEUM) of the University of Minho, where this research is performed, offers an educational portfolio in the Information Systems and Technologies (IST) area that covers all levels of higher education. Initial training of IST professionals is achieved through the integrated Master in Engineering and Management of Information Systems (MIEGSI). This is the main master course that has PMS, DAI, ACSI and GPSI courses in its curriculum. The challenges of keeping up with the constant evolution of IST professional training and the demands of adjusting higher education programs to the new principles and rules for higher education have been addressed through several changes in the program structure of this Master's degree. The MIEGSI results from the Information Systems and Technologies BSc degree and the Engineering and Management of Information Systems MSc degree combination. These courses were the result of the adaptation (appropriateness) of the Information Management course to the higher education model established in 2006 (the Bologna Process).

Although the current *Model Curriculum and Guidelines for Undergraduate Degree Programs in Information Systems* is IS 2010 (ACM-Curricula-IS, 2010), MIEGSI was originally structured according to the previous version of the standard, the IS 2002 (ACM-Curricula-IS, 2002).

Based on knowledge of the areas and their comparison we are able to infer that the MIEGSI is a course in the IS area, but with less depth in topics of *Information Technology* curriculum area than suggested by IS 2002. The PMS and DAI curricular units deal with some topics in great depth whose inclusion in ISBOK (*Information Systems Body of Knowledge*) has its origin in the SWE-BOK (*Software Engineering Body of Knowledge*) (Abran, Bourque, Dupuis, Moore, & Tripp, 2004). These courses belong to a curriculum plan of IS and not CS or SE. The main difference is on the software organizational nature focus (business or enterprise software applications), in conjunction with the strategy and culture of the enterprise, the existence of a great diversity of requirements sources (from traditional business stakeholders to contexts in which the architecture of information systems already appears rudimentarily conceived) and the coexistence with professionals with very different perspectives on ITs (functional consultants, information systems architect, manager of technology infrastructure, ...), all of which require a large capacity of flexibility in the implementation and management of development processes.

The publication of IS 2010 for updating the IS 2002 model resulted from the natural changes of the technological and industrial practices that the domain has been permanently subject to from its beginnings. The IS 2010 model introduced some adjustments in the courses recommended as mandatory (core courses) and significantly increased the range of elective courses. The PMS course fulfill the stipulated of *IS 2010.4 IS Project Management* course. In the case of DAI course, its syllabus covers what is stipulated in *IS 2010.6 Systems Analysis and Design* and *Application Development* courses. This last course, despite being classified as elective type, appears to be referenced as mandatory for all profiles of professional training (career track) considered by IS 2010. Notwithstanding the structural and programmatic adjustments, the author of this chapter considers that the syllabus of MIEGSI is aligned with the recommendations of IS 2010.

In the case of ACSI and the GPSI courses from MIEGSI integrated master, the study of alignment with the curriculum frameworks requires analyzing the MSIS 2006 (ACM-Curricula-MSIS, 2006). The framework belongs to the *Computing Curricula* program, which in the AIS (Association for Information Systems) leadership focuses on post-graduate education in IS. This framework presents a strict balance between curricular areas of *IT Technology* and *IS Management* treated in mandatory courses which recommends.

The ACSI course is aligned with the *MSIS 2006.2 Analysis Modeling and Design* course. Following the recommendations of MSIS 2006, the ACSI course complement the training received, particularly in DAI course, on modeling and development cycles issues of the technology-based solutions. The GPSI course is aligned with the *MSIS 2006.5 Project and Change Management* course. Following the recommendations of MSIS 2006, the GPSI course complement the training received, particularly in PMS course, on management of the development process issues, introducing the whole socio-technical dimension of the IS and considering the projects as drivers of technological and organizational change.

The curriculum frameworks under the *Computing Curricula* program were developed with the explicit aim of being adopted in university education in the USA (United States of America) and Canada. However, the use of its recommendations outside this geographical context has proven to be a useful practice, with many more advantages than disadvantages. Given that the IT area (in its various forms) is still in its infancy (and therefore remains subject to rapid evolution, which makes it extremely difficult to keep up with the various curricular offerings) and that the abovementioned two countries are at the forefront of technological development, it is a recommended practice to regularly peruse what is being said about the training of their professionals.

Despite the enormous worldwide spread of the frameworks produced under the *Computing Curricula* program, there are countries that choose to follow different paths to develop their own frameworks as a result of a concerted effort within their universities. One example is Australia, where, in some universities, an approach has been adapted to training in IS with a specific curriculum (Tatnall & Burgess, 2009). The Australian approach has evolved from a perspective called "Information Management" and for decades was designated "Business Computing" (Retzer, Fisher, & Lamp, 2003).

Another example is Germany, which has developed a body of knowledge suitable for framing how the courses in IS should be taught (K. Kurbel, Krybus, & Nowakowski, 2013). In the German language, this area has historically been designated "Wirtschaftsinformatik" which over the years has been translated into English as "Business Informatics" or "Business Computer Science." Recently the accepted term for the IS area has been "Business & Information Systems Engineering."

The variability of possible approaches in teaching, not only in the IS sub-area, but, generally, in the IT (Computing) area suggests, sometimes, the use of accreditation curriculum and professional certification as a way to introduce some order in the training of professionals and to recognize, unequivocally, the specific skills that a professional must have in a set of sub-domains required by the industry.

OUR PROJECT BASED LEARNING APPROACH

This section describes how the integration of the four courses mentioned in the previous section has been managed, as well the solutions tried to promote educational cooperation in the teaching practices context based in projects. This integration has been carried out over the last four years. More precisely, it began in the academic year 2009/2010. This integration has been managed to allow minor changes over the years and to improve some relevant aspects. It is in this controlled context that we developed our PBL approach.

The *Bologna Declaration* (European Commission, 2010) was a political commitment to achieve in the short term, a clear set of objectives recognized as fundamental to the construction of the "European Higher Education Area" and to promote "European System of Higher Education" throughout the planet.

Integrated in the changes recommended by the *Bologna Process*, the ECTS is part of a set of procedures that support the new paradigm of organizing student-centered learning (and training objectives) and move from a traditional system curriculum based on the "juxtaposition" of knowledge to a system focused on developing broad curricular areas, defined in terms of training objectives to pursue. This argument reinforces the relevance of integrating a group of courses around global goals so that we can instill in students several skills in a way not comparable with skills obtained with the same courses in isolation mode. In ECTS, the work done by students in the subject area is expressed by a numerical value that takes into account the hours of student work, in

their overall activities, including contact hours and hours spent on internships, projects, works in ground study and evaluation. The entire work done by the student in each course is expressed in ECTS credits and each ECTS credit corresponds to a total of 28 hours per semester.

The ECTS system is suited to changes in training, mainly in the development and adoption of: (1) new learning methodologies (more active and participatory), (2) horizontal capabilities and skills (learning to think, learning to learn, learning to teach), (3) specific skills of the profession, (4) general skills (communication capabilities, integration team, leadership, innovation and adaptation to change). All these dimensions were considered and are thought to be adequately incorporated in the joint project that integrates the four courses of our approach.

In many classrooms, learning is a passive activity. Students take notes during lectures and repeat the same information in the exams. When students read a chapter indicated by the teacher and they answer questions about that, the answers can be found in the chapter and are already known. Even in more experimental areas, the teachers rarely allow to students discover principles themselves; instead, the teachers present laws and techniques, and then they build exercises where students simply practice what they have been taught. This teaching is essentially based on the transmission of knowledge. The *Bologna Process* suggests switching to a school based on the students' work and the effective acquisition of skills, however, it should not put into question a proper proportion of more traditional activities also dedicated to the "simple" transmission of knowledge.

Although the *Bologna Process*, apparently "censors" the "teaching based on the transmission of knowledge," what is called contact time is no more than a series of moments in which teachers and students synchronize spatially to exchange knowledge with the perspective of developing, in students, certain skills. The *Bologna Process* should not avoid the transmission of knowledge between teachers and students, but rather aim to go further, catalyzing the emergence of new skills in students during the transfer of knowledge process. In this context, the transmission of knowledge is desirable and beneficial, so the pedagogical action in circumstances where every student has limited opportunities to interact with teachers does not foresee success. It is in the laboratory classes that transcendence of "skills transfer" can occur if each student is given the opportunity to receive knowledge. Consequently, the number of students spatio-temporally synchronized with the teachers in laboratory activities must be carefully determined, taking into account the expected level of depth of the "quasi-tangible" learning outcomes.

It was based on this change promoted by the *Bologna Process* that in the academic year 2006/07, it was decided to adopt teaching and learning practices based on the PBL (Barrows & Tamblyn, 1980) principles in the DAI course (since it is DAI course which catalyzes the main project work that integrates other courses). The teaching and learning context that PBL facilitates has been shown to be appropriate to the group of courses that integrate our environment (approach), where we want students to develop real skills in the production of technological artifacts imbued with a spirit of great rigor and methodological and procedural awareness and not simply to reproduce texts and definitions held in any book or manual timelessly accepted on any shelf or in any drawer. What drives the student to want to persist in school learning has to do mainly with the way we create and organize educational environments and all activities that we develop there.

Solving a problem according to the PBL approach requires the participation of students. The teacher helps and advises, but does not drive. Learning becomes an act of discovery as students examine the problem by investigating its base, analyze possible solutions, develop proposals and produce a final result. This active learning is not only more interesting and committed to the students but also develops a greater understanding

of the syllabus since the students themselves seek information and then use it in an active way with the skills they already hold to complete the project. This way of organizing the teaching and learning activities for skills development promoted by the four courses is considered appropriate. Svinicki & McKeachie (2011) argue that "problem-based education is based on the assumptions that human being evolved as individuals who are motivated to solve problems, and that problem solvers will seek and learn whatever knowledge is needed for successful problem solving. Thus if an appropriate realistic problem is present before study, students will identify needed information and be motivated to learn it. However, as in introducing any other method, you need to explain to students your purposes" (Svinicki & McKeachie, 2011).

PBL requires the realization by students that learning lies in the prosecution of skills, not with the teacher to tell them that they are right, but based on experimentation with the artifacts and documents that they produce. In cooperative learning promoted by PBL, students learn from each other and they work together to develop the project. This aspect is extremely important in the context of our group of courses, in which there is the involvement of students with different academic degrees and maturity. In PBL, students grow more thoughtful and are harder working than in exercises that require rote memorization. In our approach, the emulation of a real project (with a real client) forces the students to learn from a variety of different sources and to make decisions based on their own research. This process allows students to achieve more advanced levels of cognitive skills, research skills and problem solving skills.

The PBL's real academic goal is not to develop a final response to the project. The students do not find just one true answer to the problem that, instantly, they agree can be the "correct" solution. Instead, the real learning occurs through the process of solving a problem: thinking through various steps, investigating the subjects and developing one solution. With the continuing explosion of knowledge and the pace of technological change, the universities cannot continue to provide all the information to the students that they need for their lives. Increasingly, the most important skill that the university can teach students is to learn by themselves. Within the group of courses integrated in our approach, this issue arises repeatedly when we want students to understand by themselves that throughout their careers will have to learn how to use and design new development processes, notation models, standards, paradigms, frameworks, management, process improvement and project standards.

The recognition of the effectiveness of the PBL approach in engineering teaching has resulted in several initiatives (The Higher Education Academy, 2003; University of Nottingham, 2003), including scientific projects of a pedagogical nature, in order to produce guidelines for the teaching and learning activities organization under PBL principles. In some forums, the use of PBL in engineering teaching has adopted the *Project-Led Engineering Education* (PLEE) designation (Powell, Powell, & Weenk, 2003).

The project that is presented every academic year to the PMS and DAI students requires energetic learners. Nobody will give them all the necessary information, nor will the answers be found in the books. To solve this problem requires that students "discover complaints," "investigate the reasons" and develop the best way to resolve the situation. Thus, our approach, which is based on this project, presents some crucial features for creating a context of enormous catalyzing of the phenomena of teaching and learning, namely:

1. **Real client:** The existence of a real client, who interacts with, gathers and receives students in his organization, promotes in students the ability to feel, in practice, the difficulty of organizational software development with incomplete information and systematic doubts from the client, but with

explicit business support needs (Trendowicz, Heidrich, & Shintani, 2011). It also allows students to understand how the customer thinks and evaluates the effort put into the development of the solution (Burge & Troy, 2006).

2. **Project proposal versus project:** In the first semester, in the PMS course, the project proposal phase is created, in which feasibility studies are designed and carried out, with some initial incursion into requirement elicitation. This phase ends with a project proposal elaboration, including time and cost estimates and an initial definition of generic features of the solution. This separation between the project proposal phase (emulated in PMS course) and the project phase (emulated in DAI course) allows the students to understand the different requirements and planning approaches that they are required to adopt in each of those circumstances and perceive the difference between a more commercial nature (required in the project proposal phase) and a more technically oriented approach (essential in the implementation phase of the project) (Brazier, Garcia, & Vaca, 2007).

3. **Large Teams:** The high dimension teams allow the recreation of the complexity of an industrial context in terms of the multiplicity of tasks to be performed (Blake, 2005; Chaczko, Davis, & Mahadevan, 2004) and the enormous need for interaction between the different roles, promoting the development of skills of an inter-personal nature (Slaten, Droujkova, Berenson, Williams, & Layman, 2005). The development of the *soft skills* is one of the great gains acquired by students as a result of their involvement in the project.

4. **Rigor in software process development:** The adoption of a configurable software process development permits the instillation of awareness and rigor in how students decide

the management and implementation of the project (Suri & Sebern, 2004). Actually, this procedural rigor results from the RUP framework adoption with the reduced model of roles (Borges, Monteiro, & Machado, 2011), the obligation to follow the CMMI model practices and to adopt the PMBOK (*Project Management Body of Knowledge*) (PMI, 2008) and SWEBOK bodies of knowledge. With these standards, the relationship between the project management and the software process development adopted becomes explicit, allowing sensitization of the students to the dire need to own specific methodological skills in the area in order to be able to manage projects in the field of technology and information systems. Additionally, it allows the creation of a true perception of the relationship between the structures of the solution implementation and the requirements that gave rise to it (Burgstaller & Egyed, 2010), as well as between product quality and the rigor of process practices adopted by the team.

5. **Complexity of the solution:** The project can create conditions in the educational context where students are engaged in designing and building medium-size software solutions based on requirements from the "real world." This makes for the experience of conceiving an architecture which is methodologically plausible (Naveda, 1999), while developing the students' sensitivity to the effectively effort needed to conceptualize artifacts with similar complexity to those developed in industrial contexts (Wohlin, 1997).

6. **Others:** In every academic year a mechanism or new feature in project work is introduced in the second semester that cumulatively add some novelty to the operational mode of the previous year. Over the years, the following features or mechanisms were introduced (not in chronological order): alignment with SWEBOK, alignment with

PMBOK, RUP model, internal evaluation and promotion of human resources, CMMI level 2 practices, CMMI level 3 practices, real client, final presentation with a commercial focus, formalization of the product delivery, outsourcing, technological interoperability between partial solutions, consulting of outside services, use of patterns, formal documentation with RUP templates. It is under study by the teaching team, how the following mechanisms or features will be adopted over the next academic years: agile practices, reinforcement and hiring human resources, competition between enterprises and corporate bankruptcy.

Over the years it has only been possible to cumulatively manage all these features and mechanisms because the teaching staff is stable, cohesive and aligned in the way it organizes and engages with students who annually attend the four courses (about a hundred students in PMS and DAI and from about three-and four dozen students in ACSI and GPSI). The curriculum integration and the pedagogical cooperation among the four courses have also been decisive in making possible the management of all these educational processes that converge in the laboratory classes in the DAI course, since it consists of the place of convergence by the students from both levels of education that functionally associate in the project (see Figure 1): (1) the ACSI students provide external services (regarding to the functional consultant or senior analyst role) to the PMS students and GPSI students provide external services (regarding to the project facilitator, senior project manager or process engineer role) to the DAI students; (2) the PMS students emulate phenomena studied by ACSI students and DAI students emulate phenomena studied by GPSI students. Often ACSI and GPSI students choose for their dissertations some of the themes that they dealt when they worked with PMS and DAI students.

Figure 1. Integration among courses

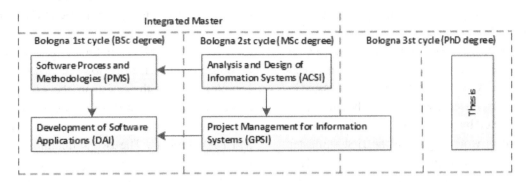

Recently, some of the PhD students in the SEMAG (*Software Engineering and Management Group*) research group (SEMAG, 2001) also began to engage with the students of the four courses, in order to carry out scientific studies supported in educational context experiments and this work has given rise to international publications (some involving ACSI and GPSI students) (Monteiro, Borges, Machado, & Ribeiro, 2012; Monteiro et al., 2013). Thus, this group of four courses is simultaneously a space of pedagogical and scientific innovation from which all stakeholders benefit (Siqueira, Barbaran, & Becerra, 2008). It is our perception that almost all stakeholders in this huge process (students, teaching staff and researchers) feel "pleasure" with the involvement in the project (Robert L. Glass, 2007; R. L. Glass, 2007), despite the great dedication and tremendous effort that is demanded from all involved.

RESULTS OBTAINED FROM OUR PBL APPROACH

The real-world project/study approach to teaching software engineering has been successful thus far. It has helped to motivate the teams and to encourage development of higher quality products by the teams. The teams took seriously the importance of the problems that they were helping to solve. The approach teaches inexperienced graduate

students many principles of software engineering and software verification and validation that they could apply to newly gained jobs and/or subsequent courses (Hayes, 2002).

Our PBL approach began to be effectively implemented in the academic year 2009/2010, although some well-controlled trials had been tested in previous years. The real client collaboration in the software project just started in that academic year. This client was located in the region of our university. The real client activity area was quite diverse. In our previous software projects we had collaborations with a factory enterprise, a non-profit institution of social solidarity, a professional handball team and a professional football team.

Over the past four years we have had, approximately, one thousand students involved in the four educational syllabuses. In each academic year, we had approximately 150 students in PMS and DAI courses and 100 students in ACSI and GPSI courses.

Evaluation of Student Teams in Our PBL Approach

It is not easy to teach the syllabus of the four courses because of their considerable size and a lack of conditions revealed by the students for realization of topics with a more abstract nature. In fact, this group of courses has a considerably high level of requirement, so the methodologies or teaching strategies to be adopted should be in accordance with the level of depth of the topics.

The students should be able to use the study objects to perform the tasks. In the case of the DAI course, this level of learning depth requires a deep involvement by the students in practical classes (in the form of exercises proposed and solved in class), complemented by the development of a software project. This work begins in the first semester under the PMS course, in groups of about five students. These groups of students emulate the project proposal context in which they perform an initial analysis of requirements and time and cost planning of the project proposal solution for a real client with whom the teaching staff establishes an annual academic collaboration partnership (Ali, 2006; Kornecki, Khajenoori, Gluch, & Kameli, 2003). In the DAI course, in the second semester, the work project is performed by groups of about 15 students. These groups of students emulate the project implementation context of a proposed solution to the same real client, which they perform a comprehensive analysis and design of the problem, construction, testing and delivery of a software solution coded in an object-oriented language (*Java* or *C#*) with relational databases (*SQL Server* or *MySQL*) and with interfaces for the Web. In both courses, student groups follow the RUP model (Bergandy, 2008) and they use the UML notation extensively. The teams use the laboratory classes to meet in the presence of teachers. In these weekly meetings, the teachers monitor the progress of the project work.

In the case of ACSI and GPSI courses, the learning outcomes with higher depth level are achieved through the involvement of students with groups of PMS and DAI courses, respectively, promoting play roles with responsibility and a high level of maturity, such as the functional consultant, senior analyst, project facilitator, senior project manager or process engineer. These two courses have different learning outcomes with deep levels corresponding to some of the emerging topics addressed and discussed in lectures, which prepare

the students with a good understanding of a set of problems that currently emerge in the area.

The learning outcomes with a more advanced level are achieved through the use of a scientific literature searches in order to study in detail some complex themes that arise in the engagement context by the ACSI and GPSI students when they participate in the project work performed by PMS and DAI students. The two semiannual project works instill a hands-on training dynamic that is essential for the two pairs of courses operation as a whole (Broman, 2010; Port & Boehm, 2001): PMS and ACSI courses in the first semester and DAI and GPSI courses in the second semester.

The university regulation concerning the evaluation of courses was changed in 2004. From that year it became possible to evaluate a student that "only" executes a project and thus dispense with the realization of an exam. Thus, the evaluation system adopted allowed the organization of all activities of the four courses around an annual educational project. This project is the integrator of all the topics listed in the syllabus and it follows a student time management approach that is able to facilitate the effective development of skills necessary for evidencing learning outcomes stipulated (Stiller & LeBlanc, 2002). Since the academic year 2006/07 this group of courses has adopted this evaluation system, which provides a much more effective management of students activities inside and outside of the university physical spaces and in a horizon that it extends over the two semesters of the academic year, according to the PBL approach.

Since DAI is the core course in our approach then we will describe in some detail the evaluation system of this curricular unit. This evaluation scheme encourages the incremental demonstration of the partial learning outcomes over the semester.

Between the two semesters in which annual project is developed, the second semester is the one that concentrates the largest effort and diversity of activities and in this sense DAI is a curricular unit of 10 ECTS. The student evaluation in DAI is carried out based on the activities undertaken by various teams within the semiannual project work in five *assessment milestones*. Each one of the five *assessment milestones* is associated with partial evaluation, four are team evaluations, and one is an individual evaluation. However, it is systematic that several students in a team obtain different classifications in the four team *assessment milestones*, since the individual score calculation results from a plurality of weighting factors from three sources of information (Clark, 2005), such as the teachers traditional evaluation, client evaluation and peer evaluation (other students of the team).

Work Projects Developed in ACSI and GPSI Courses

By academic year, the reports developed in ACSI and GPSI are edited in two separated documents which are assigned with ISSN (International Standard Serial Number). These documents contain all the student project works of both courses. These project works can be a report or a scientific paper that describes all the work performed by the students. These documents serve as consultation reference for future students. These students

can find the difficulties, limitations, the positive and negative aspects and lessons learned and experienced by colleagues of the previous years. Table 1 was created based on a detailed analysis of these documents. The table presents a mapping between the SWEBOK Knowledge Areas (KA) and the work projects developed by the ACSI and GPSI students.

As we can see in Table 1, in the 2010/2011 and 2011/2012 academic years, the 140 students enrolled in the ACSI and GPSI courses developed 130 work projects in different SWEBOK KAs. Although the majority of work projects are developed individually, some of them involve teams of 2 or 3 students.

Figure 2 shows that most of the project works are developed in the *Software Engineering Management* area, representing 42.3% of the total. The high percentage of the project works in this KA is justified by the involvement of ACSI and GPSI MSc students as a consultant/facilitator role of the DAI development teams (BSc students).

Another conclusion obtained by the Table 1 analysis is the total absence of project works in the *Software Testing* and *Software and Configuration Management* areas. Normally the tests are

Table 1. ACSI and GPSI work projects by SWEBOK knowledge area

SWEBOK Knowledge Area	Academic Year				Sum
	2010/2011		2011/2012		
	ACSI	GPSI	ACSI	GPSI	
Software Requirements	2	2	8	6	18
Software Design	3	0	7	1	11
Software Construction	0	0	1	2	3
Software Testing	0	0	0	0	0
Software Maintenance	0	0	1	1	2
Software Configuration Management	0	0	0	0	0
Software Engineering Management	11	7	11	26	55
Software Engineering Process	6	0	13	8	27
Software Engineering Tools and Methods	0	1	0	0	1
Software Quality	2	0	3	8	13
Sum	24	10	44	52	130

Figure 2. ACSI and GPSI work projects by SWEBOK KA

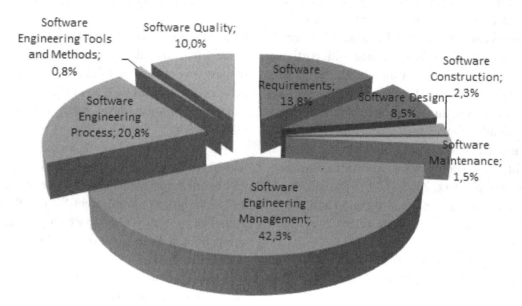

performed by the DAI students, those who develop the software applications, and until now, the MSc students have not expressed interest in this KA. Regarding *Software and Configuration Management,* the absence is justified by the fact that we are working in an educational context and some activities of that KA are not performed, such as, software configuration identification, software configuration control, software configuration status accounting, software configuration auditing, and software release management and delivery. It is also worth noting the low number of project works in *Software Construction* and *Software Engineering Tools and Methods* KAs.

Table 2 and Table 3 show the topics and subtopics of the SWEBOK KA covered by ACSI and GPSI Work Projects respectively. The numbers preceding the names of the topics and subtopics are those in SWEBOK document.

From Table 2 and Table 3, we can infer that seven and eight of the ten SWEBOK KAs are covered by the ACSI and GPSI work projects respectively. We just can find one GPSI work project in the *Software Engineering Tools and Methods* area. Specifically, this project work in-

volved the *Microsoft Project Server 2007* key features study, through two visions: one as a project manager and other as system administration. Despite the topics and subtopics diversity covered by the work projects, we cannot infer that there is a great variability of subjects between the two courses.

FUTURE RESEARCH DIRECTIONS

It is a well-accepted fact that a competent software engineer requires a wide variety of skills such as managerial, engineering, team working and communication. The problem is how to teach these skills to students within the classroom. One approach to solving, or at least to mitigating this problem, is PBL. There are many reasons for incorporating real industry projects: increased student motivation and confidence; the ability for students to explore in-depth IT areas not covered in the curriculum or only covered superficially; the development in students of increased problem-solving and critical thinking skills, communication skills and business insight.

Table 2. Topic and subtopic of the SWEBOK KA covered by ACSI work projects

SWEBOK KA	ACSI Course	
	SWEBOK Topic	SWEBOK Subtopic
Software Requirements	1. Software Requirements Fundamentals	1.1. Definition of a Software Requirement
		1.2. Product and Process Requirements
		1.3. Functional and Non-functional Requirements
	2. Requirements Process	2.3. Process Support and Management
	4. Requirements Analysis	4.1. Requirements Classification
		4.2. Conceptual Modeling
	5. Requirements Specification	5.3. Software Requirements Specification
	7. Practical Considerations	7.2. Change Management
		7.4. Requirements Tracing
Software Design	1. Software Design Fundamentals	1.3. Software Design Process
	3. Software Structure and Architecture	3.1. Architectural Structures and Viewpoints
		3.2. Design Patterns (microarchitectural patterns)
Software Construction	3. Practical considerations	3.2. Construction Languages
Software Maintenance	4. Techniques for Maintenance	4.3. Reverse engineering
Software Engineering Management	2. Software Project Planning	2.1. Process planning
		2.4. Resource allocation
		2.5. Risk management
	3. Software Project Enactment	3.2. Supplier contract management
		3.4. Monitor process
		3.5. Control process
Software Engineering Process	1. Process Implementation and Change	1.1. Process infrastructure
		1.2. Software process management cycle
	2. Process Definition	2.1. Software life cycle models
		2.2. Software life cycle processes
		2.3. Notations for Process Definitions
	4. Process and Product Measurement	4.1. Process measurement
		4.2. Software product measurement
Software Quality	1. Software Quality Fundamentals	1.3.Models and quality characteristics

As a part of our approach, the teaching team is currently studying how the following mechanisms or features will be adopted over the next academic years: agile practices, reinforcement and hiring of human resources, competition between enterprises and corporate bankruptcy. The inclusion of these features and mechanisms should be done with a great deal of reflection and sensitivity so as not to collide with ethics, morality, privacy and pedagogical issues. Another aspect to consider in this type of environmental change is the internal regulations of the university.

Another important challenge in the near future is to develop an approach that allows, in the educational context, possibly with industrial cooperation, the planning, design and implementation of software projects in geographically distributed contexts and locations in order to strive for globalization.

Table 3. Topic and subtopic of the SWEBOK KA covered by GPSI work projects

SWEBOK KA	GPSI Course	
	SWEEBOK Topic	**SWEEBOK** Subtopic
Software Requirements	1. Software Requirements Fundamentals	1.3. Functional and Non-functional Requirements
	2. Requirements Process	2.1. Process Models
		2.3. Process Support and Management
	3. Requirements Elicitation	3.1. Requirements Sources
		3.2. Elicitation Techniques
	4. Requirements Analysis	4.4. Requirements Negotiation
	5. Requirements Specification	5.3. Software Requirements Specification
Software Design	5. Software Design Notations	5.1. Structural Descriptions (static view)
		5.2. Behavioral Descriptions (dynamic view)
Software Construction	3. Practical considerations	3.1. Construction Design
		3.2. Construction Languages
		3.3. Coding
		3.4 Construction Testing
Software Maintenance	4. Techniques for Maintenance	4.3. Reverse engineering
Software Engineering Management	2. Software Project Planning	2.4. Resource allocation
		2.5. Risk management
		2.6. Quality management
		2.7. Plan management
	3. Software Project Enactment	3.1. Implementation of plans
		3.2. Supplier contract management
		3.4. Monitor process
		3.5. Control process
		3.6. Reporting (RUP Roles)
Software Engineering Process	1. Process Implementation and Change	1.1. Process infrastructure
	2. Process Definition	2.2. Software life cycle processes
	4. Process and Product Measurement	4.1. Process measurement
		4.2. Software product measurement
Software Engineering Tools and Methods	1. Software Engineering Tools	1.7. Software Engineering Management Tools
Software Quality	1. Software Quality Fundamentals	1.3.Models and quality characteristics
	2. Software Quality Management Processes	2.1.Software Quality Assurance
		2.2.Verification & Validation
	3. Practical Considerations	3.4.Software Quality Measurement

Finally it is important to emulate a software process development environment that allows for using a plurality of reference standard quality models. This work can drive us to implement an approach where the teams, in an educational context, must follow this new environment.

CONCLUSION

The approach described in this chapter to teach topics of the engineering and management process of the software process development in university courses in the IS area aims to train professionals for the industry. These students, professionals in the near future, must be able to apply the most modern methodological approaches in the analysis and design of IS based on software, as well as manage the corresponding projects and development processes, in close cooperation with the most demanding quality and maturity procedural reference models. This dual training of students in engineering approaches and methodologies and in techniques and management methods are complemented with interpersonal skills. The dynamics promoted by the teaching-learning project promoted by the integration of the four annual courses develops management, engineering and personal skills. This tripartite training promoted by our approach instills in students an attitude and not only technological learning, in the SE context. The vision of *Engineering and Management Software* applied to IS problems is what industry seeks to build. So that engineering teams are increasingly able to intervene in organizations.

Using case studies, even with awareness of the limitations of possible inferences and simplifications when compared with empirical software engineering approaches, the students will continue their studies through various courses, capitalizing on experience, awareness, insight and ability to avoid paths of less desirable decisions. By the end of their studies, trainees will be ready to integrate easily in the industrial world with developed maturity and valuable experience.

An engineer cannot be just a "mere" specialist: it is necessary to combine technical skills with the human and social dimensions. Our students should be prepared to live and work as global citizens, understand how engineers contribute to society. They should be aware that it is not enough to be technically excellent, because there are other dimensions to consider. Our approach allows delivering some important skills: (1) a basic understanding of business processes; (2) a product development with high-quality concerns; (3) know-how to conceive, design, implement and operate medium-size complexity systems and (4) communicative, initiative/leadership, teamwork, analytical and problem solving and personal abilities.

REFERENCES

Abran, A., Bourque, P., Dupuis, R., Moore, J. W., & Tripp, L. (2004). *Guide to the software engineering body of knowledge (SWEBOK)*. Los Alamitos, CA: IEEE Computer Society Press.

ACM-Curricula-IS. (2002). *IS 2002: Model curriculum and guidelines for undergraduate degree programs in information systems. Association for Computing Machinery (ACM), Association for Information Systems (AIS), Association of Information Technology Professionals*. AITP.

ACM-Curricula-IS. (2010). *IS 2010: Curriculum guidelines for undergraduate degree programs in information systems. Association for Computing Machinery (ACM), Association for Information Systems*. AIS.

ACM-Curricula-MSIS. (2006). *Model curriculum and guidelines graduate degree programs in information systems. Association for Information Systems*. AIS.

Ali, M. R. (2006). Imparting effective software engineering education. *ACM SIGSOFT Software Engineering Notes*, *31*(4), 1–3. doi:10.1145/1142958.1142960

Avison, D., & Wilson, D. (2001). A viewpoint on software engineering and information systems: What we can learn from the construction industry? *Information and Software Technology*, *43*(13), 795–799. doi:10.1016/S0950-5849(01)00186-0

Barrows, H. S., & Tamblyn, R. H. (1980). *Problem-based learning: An approach to medical education*. New York: Springer.

Bergandy, J. (2008). Software engineering capstone project with rational unified process (RUP). In *Proceedings of the 38th ASEE/IEEE Frontiers in Education Conference*. New York: ASEE/IEEE.

Birk, A., Heller, G., John, I., Schmid, K., von der Massen, T., & Muller, K. (2003). Product line engineering, the state of the practice. *IEEE Software*, *20*(6), 52–60. doi:10.1109/MS.2003.1241367

Blake, M. B. (2005). Integrating large-scale group projects and software engineering approaches for early computer science courses. *IEEE Transactions on Education*, *48*(1), 63–72. doi:10.1109/TE.2004.832875

Borges, P., Monteiro, P., & Machado, R. J. (2011). Tailoring RUP to small software development teams. In *Proceedings of the 37th EUROMICRO Conference on Software Engineering and Advanced Applications* (pp. 306–309). IEEE.

Brazier, P., Garcia, A., & Vaca, A. (2007). A software engineering senior design project inherited from a partially implemented software engineering class project. In *Proceedings of the 37th ASEE/IEEE Frontiers in Education Conference* (pp. F4D–7). IEEE.

Broman, D. (2010). Should software engineering projects be the backbone or the tail of computing curricula? In *Proceedings of the 23rd IEEE Conference on Software Engineering Education and Training*. Pittsburgh, PA: IEEE.

Burge, J., & Troy, D. (2006). Rising to the challenge: Using business-oriented case studies in software engineering education. In *Proceedings of the 19th Conference on Software Engineering Education and Training* (pp. 43–50). IEEE.

Burgstaller, B., & Egyed, A. (2010). Understanding where requirements are implemented. In *Proceedings of the 26th IEEE International Conference on Software Maintenance* (pp. 1–5). IEEE.

Chaczko, Z., Davis, D., & Mahadevan, V. (2004). New perspectives on teaching and learning software systems development in large groups. In *Proceedings of the 5th International Conference on Information Technology Based Higher Education and Training* (pp. 409–414). IEEE.

Clark, N. (2005). Evaluating student teams developing unique industry projects. In *Proceedings of the 7th Australasian Computing Education Conference* (pp. 21–30). Australian Computer Society.

Dutoit, A. H., McCall, R., Mistrik, I., & Paech, B. (2006). Rationale management in software engineering: Concepts and techniques. In A. H. Dutoit, R. McCall, I. Mistrik, & B. Paech (Eds.), *Rationale management in software engineering*. Springer. doi:10.1007/978-3-540-30998-7_1

Engle, C. B. (1989). Software engineering is not computer science. In N. Gibbs (Ed.), *Software engineering education* (Vol. 376, pp. 257–262). New York: Springer. doi:10.1007/BFb0042363

European Commission. (2010). *The Bologna process - Towards the european higher education area*. Retrieved 08-07-2013, 2013, from http://ec.europa.eu/education/higher-education/bologna_en.htm

Finkelstein, A. (2011). Ten open challenges at the boundaries of software engineering and information systems. In H. Mouratidis, & C. Rolland (Eds.), *Advanced information systems engineering* (Vol. 6741, pp. 1–1). New York: Springer Berlin Heidelberg. doi:10.1007/978-3-642-21640-4_1

Fung, R. Y. K., Tam, W. T., Ip, A. W. H., & Lau, H. C. W. (2002). Software process improvement strategy for enterprise information systems development. *International Journal of Information Technology and Management*, *1*(2-3), 225–241. doi:10.1504/IJITM.2002.001198

Glass, R. L. (2007). Is software engineering fun? *IEEE Software*, *24*(1), 96–95. doi:10.1109/MS.2007.18

Glass, R. L. (2007). Is software engineering fun? Part 2. *IEEE Software*, *24*(2), 104–103. doi:10.1109/MS.2007.46

Gregg, D., Kulkarni, U., & Vinzé, A. (2001). Understanding the philosophical underpinnings of software engineering research in information systems. *Information Systems Frontiers*, *3*(2), 169–183. doi:10.1023/A:1011491322406

Hayes, J. H. (2002). Energizing software engineering education through real-world projects as experimental studies. In *Proceedings of the 15th Conference on Software Engineering Education and Training* (pp. 192–206). IEEE.

Hellens, L. A. (1997). Information systems quality versus software quality a discussion from a managerial, an organisational and an engineering viewpoint. *Information and Software Technology*, *39*(12), 801–808. doi:10.1016/S0950-5849(97)00038-4

Higher Education Academy. (2003). *PBLE (project based learning in engineering)*. Retrieved 09-07-2013, 2013, from http://www.heacademy.ac.uk/resources/detail/resource_database/SNAS/PBLE_Project_Based_Learning_in_Engineering

Jayaratna, N., & Sommerville, I. (1998). The role of information systems methodology in software engineering. *IEE Proceedings. Software*, *145*(4), 93–94. doi:10.1049/ip-sen:19982193

Karunasekera, S., & Bedse, K. (2007). Preparing software engineering graduates for an industry career. In *Proceedings of the 20th Conference on Software Engineering Education & Training* (pp. 97–106). IEEE.

Kornecki, A. J., Khajenoori, S., Gluch, D., & Kameli, N. (2003). On a partnership between software industry and academia. In *Proceedings of the 16th Conference on Software Engineering Education and Training* (pp. 60–69). IEEE.

Kurbel, K., Krybus, I., & Nowakowski, K. (2013). *Lehrstuhl für wirtschaftsinformatik*. Retrieved 2013-07-05, 2013, from http://www.enzyklopaedie-der-wirtschaftsinformatik.de/wi-enzyklopaedie/lexikon/uebergreifendes/

Kurbel, K. E. (2008). *The making of information systems: Software engineering and management in a globalized world*. Berlin, Germany: Springer. doi:10.1007/978-3-540-79261-1

McBride, N. (2003). A viewpoint on software engineering and information systems: Integrating the disciplines. *Information and Software Technology*, *45*(5), 281–287. doi:10.1016/S0950-5849(02)00213-6

Monteiro, P., Borges, P., Machado, R. J., & Ribeiro, P. (2012). A reduced set of RUP roles to small software development teams. In *Proceedings of International Conference on Software and System Process* (pp. 190–199). IEEE.

Monteiro, P., Machado, R., Kazman, R., Lima, A., Simões, C., & Ribeiro, P. (2013). Mapping CMMI and RUP process frameworks for the context of elaborating software project proposals. In *Software quality: Increasing value in software and systems development* (Vol. 133, pp. 191–214). Berlin: Springer. doi:10.1007/978-3-642-35702-2_12

Naveda, J. F. (1999). Teaching architectural design in an undergraduate software engineering curriculum. In *Proceedings of the 29th ASEE/IEEE Frontiers in Education Conference* (Vol. 2, pp. 12B1–1). IEEE.

Nunan, T. (1999). *Graduate qualities, employment and mass higher education*. Paper presented at the HERDSA Annual International Conference. Melbourne, Australia.

Platt, J. R. (2011). Career focus: Software engineering. *IEEE-USA Today's Engineer*. Retrieved from http://www.todaysengineer.org/2011/Mar/career-focus.asp

PMI. (2008). *A guide to the project management body of knowledge* (4th ed.). Newtown Square, PA: Project Management Institute, Inc.

Port, D., & Boehm, B. (2001). Using a model framework in developing and delivering a family of software engineering project courses. In *Proceedings of the 14th Conference on Software Engineering Education and Training* (pp. 44–55). IEEE.

Powell, W., Powell, P. C., & Weenk, W. (2003). *Project-led engineering education*. Lemma Publishers.

Retzer, S., Fisher, J., & Lamp, J. (2003). Information systems and business informatics: An Australian German comparison. In *Proceedings of the 14th Australasian Conference on Information Systems* (pp. 1–9). Edith Cowan University.

SEMAG. (2001). *Software engineering and management group*. Retrieved 01-06-2013, 2013, from https://sites.google.com/a/dsi.uminho.pt/semag/

Shere, K. D. (1988). *Software engineering management*. Upper Saddle River, NJ: Prentice Hall.

Siqueira, F. L., Barbaran, G. M. C., & Becerra, J. L. R. (2008). A software factory for education in software engineering. In *Proceedings of the 21st Conference on Software Engineering Education & Training* (pp. 215–222). IEEE.

Slaten, K. M., Droujkova, M., Berenson, S. B., Williams, L., & Layman, L. (2005). Undergraduate student perceptions of pair programming and agile software methodologies: Verifying a model of social interaction. In *Proceedings of the Agile Development Conference* (pp. 323–330). IEEE.

Stiller, E., & LeBlanc, C. (2002). Effective software engineering pedagogy. *Journal of Computing Sciences in Colleges*, *17*(6), 124–134.

Suri, D., & Sebern, M. J. (2004). Incorporating software process in an undergraduate software engineering curriculum: Challenges and rewards. In *Proceedings of the 17th Conference on Software Engineering Education and Training* (pp. 18–23). IEEE.

Svinicki, M., & McKeachie, W. J. (2011). *McKeachie's teaching tips: Strategies, research, and theory for college and university teachers*. Wadsworth: Cengage Learning.

Tatnall, A., & Burgess, S. (2009). Evolution of information systems curriculum in an Australian university over the last twenty-five years. In A. Tatnall, & A. Jones (Eds.), *Education and technology for a better world* (Vol. 302, pp. 238–246). Berlin: Springer. doi:10.1007/978-3-642-03115-1_25

Tockey, S. (1997). A missing link in software engineering. *IEEE Software*, *14*(6), 31–36. doi:10.1109/52.636594

Trendowicz, A., Heidrich, J., & Shintani, K. (2011). Aligning software projects with business objectives. In *Proceedings of the 2011 Joint Conference of the 21st International Workshop on Software Measurement and the 6th International Conference on Software Process and Product Measurement* (pp. 142–150). IEEE.

University of Nottingham. (2003). *PBLE (project based learning in engineering)*. Retrieved 09-07-2013, 2013, from http://www.pble.ac.uk/

Wohlin, C. (1997). Meeting the challenge of large-scale software development in an educational environment. In *Proceedings of the 10th Conference on Software Engineering Education and Training* (pp. 40–52). IEEE.

KEY TERMS AND DEFINITIONS

Assessment Milestone: It is a partial evaluation assigned by teachers' staff to the students. This event gives to the students an orientation about their effort in the software project development.

Engineering Skills: Know how to perform requirements engineering, analysis and design, implementation, quality control and software administration.

Managerial Skills: Know how to perform resource estimation and tracking, project scheduling and tacking, software development life cycle, risk management, configuration management, release management, quality assurance and traceability management.

Personal Skills: All communication, initiative/leadership, teamwork, analytical and problem solving and personal time abilities.

Project-Based Learning: Considered an alternative to paper-based, rote memorization, teacher-led classrooms. The benefits of these strategies implementation in the classroom including a greater depth of understanding of concepts, broader knowledge base, improved communication and interpersonal/social skills, enhanced leadership skills, increased creativity, and improved writing skills.

Rational Unified Process: It is an iterative software development process framework created by the Rational Software Corporation, a division of IBM. RUP is an adaptable process framework, intended to be tailored by the development organizations and software project teams that will select the elements of the process that are appropriate for their needs.

SWEBOK: It is a *Software Engineering Body of Knowledge* guide created under the IEEE sponsorship in order to serve as a reference in issues considered relevant by the Software Engineering community.

Chapter 13

Experiences in Software Engineering Education:
Using Scrum, Agile Coaching, and Virtual Reality

Ezequiel Scott
ISISTAN (UNICEN-CONICET) Research Institute, Argentina

Guillermo Rodríguez
ISISTAN (UNICEN-CONICET) Research Institute, Argentina

Álvaro Soria
ISISTAN (UNICEN-CONICET) Research Institute, Argentina

Marcelo Campo
ISISTAN (UNICEN-CONICET) Research Institute, Argentina

ABSTRACT

Software Engineering courses aim to train students to succeed in meeting the challenges within competitive and ever-changing professional contexts. Thus, undergraduate courses require continual revision and updating so as to cater for the demands of the software industry and guarantee academic quality. In this context, Scrum results in both a suitable and a flexible framework to train students in the implementation of professional software engineering practices. However, current approaches fail to provide guidance and assistance in applying Scrum, or a platform to address limitations in time, scope, and facilities within university premises. In this chapter, the authors present a software engineering training model based on the integration of the Agile Coach role and a virtual-reality platform called Virtual Scrum. The findings highlight the benefits of integrating this innovative model in a capstone course. Not only does this approach strengthen the acquisition of current software engineering practices but also opens new possibilities in the design of training courses.

DOI: 10.4018/978-1-4666-5800-4.ch013

INTRODUCTION

The increasing complexity of the software industry, constant changes in system requirements, distributed environments and professional mobility are ordinary challenges to software developers, who are required to be competent and proactive by rapidly expanding software industries. In order to prepare undergraduate students for ongoing success in this context, Software Engineering (SE) courses must provide effective training in the application of software development practices typically implemented in large projects and organizations (Alfonso & Botia, 2005; Mahnic 2010).

In the light of the above, we structured a SE course based on Capability Maturity Model Integration (CMMI), since this model includes professional and reliable SE practices (Kulpa, 2008). In particular, we used the framework CMMI for development version 1.3 (CMMI-DEV 1.3), which aims to provide software organizations with superior project performance, client satisfaction, and quality of products and services. To support CMMI, we designed an initial version of the course based on the Rational Unified Process (RUP) (Kruchten, 2003) to develop a capstone project (Coupal & Boechler, 2005; Devedzic & Milenkovic, 2011).

Unfortunately, teaching SE to students running a software project following RUP presented several drawbacks. As it is a plan-driven development framework, RUP requires the association of project milestones with specific dates. This made students focus on meeting deadlines and delivering the agreed milestone, skipping activities in the RUP framework. Additionally, even though RUP encourages the overlapping of phases, students inevitably slid into a waterfall-like process (Osorio, 2011). Thus, it is difficult for inexperienced students to detect mistakes made in early stages until the development process reaches the final stages. Furthermore, the prescriptive nature of RUP and the definition of specific roles (i.e. project manager and business process analyst,

IBM, n.d.) to perform planning and estimating activities fail to both promote actively participation of all team members, and encourage them to make commitments regarding those activities. Since SE involves people and social interaction, these aspects are cornerstone of its reality and crucial to professional training and development. As a consequence, we moved our RUP implementation of CMMI to Scrum.

The selection of Scrum is founded on the fact that it has increased in the last few years; in 2012, 57% of respondents used Scrum or its variants, and nowadays, more than 50% of the surveyed companies are utilizing Scrum as an effective path towards agility (Versionone, n.d.). First, this agile methodology concentrates on project management practices and includes monitoring and feedback activities that focus on transparency, team effort, effective time management and personal interactions. Second, Scrum facilitates the process of satisfying requirements, for while all the requirements appear in the backlog, only a limited set is fulfilled in each Sprint. Third, the retrospective meetings and continual customer contact allow early detection of impediments and rearrangements of working plans.

In line with the popularity and effectiveness of Scrum in professional contexts, some scholars have engaged in integrating Scrum in undergraduate software engineering courses. In (Alfonso & Botia, 2005) the authors compared the results of teaching Scrum practices in an undergraduate software course to those obtained in an experience using a waterfall-like rigid process. From the experience, the authors founded that the use of Scrum resulted in lower risks and process overhead (Diaz, 2009). Chuan-Hoo Tan et al. introduced Extreme Programming (XP), Scrum and Feature-Driven Development (FDD) to initiate future software professionals in the importance of agility, flexibility and adaptability in professional contexts. The research consisted in training students in developing and delivering large-scale systems under time constraints and shifting deadlines.

The students had to carry out a project-work by following an iterative and incremental teaching approach (Tan, Tan, & Teo, 2008). Mahnic taught Scrum through a capstone project[1] and described the course details, students' perceptions and teachers' observations after the course (Mahnic, 2012a). The author concluded that the students were overwhelmingly positive about the Scrum-based course, indicating that the course fully met or even exceeded their expectations.

Although those experiences proved to be effective in the classroom, there are still challenges to be faced by professors so as to facilitate students' integration in professional contexts. On the one hand, a typical problem of current approaches is the lack of assistance to unskilled students in performing the agile practices (Dubinsky & Hazzan, 2003). Even when learning in an agile context, the students tend to write enormous documents of requirements, fall in waterfall-like issues, follow a plan instead of responding to change, and focus on delivery dates instead of product quality. On the other hand, the aforementioned approaches fail to address teaching constrains in university courses such as large classes, multiple groups working at a time as well as limited space, and tutors. Using a room for multiple teams may jeopardize the effective implementation of the Scrum process, since each team may require customized configurations of the room. What is more, accessing to the required teaching material or physical space for a personalized class may prove impracticable.

In this context, the aim of this chapter is two-fold. First, we have introduced Scrum in an undergraduate software engineering course and have complemented the Scrum roles with the Agile Coach role, which is played by the professor (Soria, Rodriguez, & Campo, 2012). This role arises as an alternative to scaffold students' progress, remove impediments, and advise on leadership and management strategies as needed to ensure success of the methodology (Hedin, Bendix, & Magnusson, 2005). Here, the Agile Coach is focused on the optimization of the teams

and is responsible for clarifying and facilitating the Scrum activities and artifacts, assessing the team's health, and providing teams with needed learning opportunities to address difficulties such as decision-making and meeting deadlines (Appelo, 2010). Second, we have presented *Virtual Scrum* (Rodríguez, Soria, & Campo, 2012a) as a solution to the challenges of implementing Scrum within university premises. *Virtual Scrum* is a prototype tool that aims to help Scrum students set up a virtual working environment in which the visual metaphors required by the methodology are effectively displayed. The tool exploits the richness of the 3D metaphor by using virtual worlds and provides a virtual working environment equipped with different visual management aids for accessing team performance and a room for Daily Meetings.

The lessons learned from our experience in an undergraduate SE course show that Scrum together with the Agile Coach role allows professors to generate learning scenarios in which students can explore and acquire non-technical skills, namely communication, negotiation, interaction with customer, adherence to a process, awareness of software quality, among others. We evidenced an increase in the coverage of the professional and reliable SE practices and enhancement in Scrum perception, resulting in high-quality delivered products. Additionally, this chapter reports on the effectiveness of *Virtual Scrum* to teach up-to-date software engineering procedures and training students in the use of valuable skills to work in professional contexts through a virtual world. Along this line, we found that *Virtual Scrum* is a viable tool to implement the different elements in a Scrum team room and to perform activities throughout the Scrum process. Finally, this virtual hands-on experience may result intrinsically motivating for students, who participate in collaborative work, and enhance their comprehension of Scrum.

The rest of the chapter is organized as follows. Section 2 describes the main concepts of Scrum and current trends in its teaching on software engineering courses. Section 3 presents our innovative training model and the support provided by *Virtual Scrum*. Section 4 describes the results of our experience and recommendations to introduce this model in different contexts. Section 5 reports the future research directions. Finally, section 6 outlines the conclusions of this chapter.

BACKGROUND

CMMI is a framework which consists of a set of best practices that address the development and maintenance of products and services. These practices cover the product life cycle from conception through delivery and maintenance (Kruchten, 2003). CMM is neither a recipe nor guarantee for success, i.e., CMMI refers to "what to do" rather than "how to do it." CMMI is organized in process areas. A process area is a group of related activities performed collectively to achieve a set of goals. Some goals and practices are specific to the process area; others are generic and apply across all process areas (Kruchten, 2003).

Introducing CMMI into the classroom allows software engineering students to deliver reliable, evolvable and quality products (Barbosa & Maldonado, 2006). The idea behind the inclusion of CMMI in the curricula is fourfold, namely to broaden the coverage of important topics in a software engineering course, to deepen the students' understanding of software engineering practices, to provide quality and process standards at organization level, and to define the key elements of an effective process and outlines how to improve suboptimal processes (Lutteroth et al., 2007). However, CMMI is often misunderstood as being required massive documentation, many layers of personnel and the use of a rigid waterfall life cycle. By following Agile Methods (AM) it is possible to obtain maturity levels with less overhead and effort (Boehm & Tuner, 2008; Alegría & Bastarrica, 2006). That is, the use of a combination between AM and CMMI results in

benefits to the business performance by exploiting the synergies of both approaches (Glazer, 2008; Glazer, 2010). The value from AM can only be obtained through disciplined use. It is worth mentioning that we utilized CMMI as a reference model for teaching software engineering practices by means of Scrum; on the other hand, the use of CMMI for a single project is attributed to time and schedule constraints within university premises. Although CMMI areas, such as the ones related to quantitative project management, causal analysis and management of product suppliers, among others, fail to be widespread explored, students explore reliable software practices such as adherence to a process and compliance with quality criteria, requirement management, and configuration management, among others, which are also trustworthy practices defined in CMMI. Particularly, we mainly emphasizes on practices of maturity level 3 associated with process definition and process improvement, quality assurance, validation and verification, requirement development, software design, configuration management, and project monitoring and control, among others, which are essential to prepare students for success in professional contexts.

The emergence of the Manifesto for Agile Software Development (Agile Manifesto, n.d.) brought changes in the software development community. The Manifesto essentially defines a new focus on the software development based on agility, flexibility, communication abilities and the capacity for offering new products and services with high value to the marketplace in short periods of time (Highsmith, 2004). Most companies are adopting Scrum to become agile smoothly and reduce overhead and bureaucracy progressively, without losing sight of the quality of the software product (Diaz, 2009; Maher, 2009).

Scrum is an iterative and incremental methodology that organizes projects to make them manageable for small, self-organized and cross-functional teams (Schwaber & Beedle, 2002), and also systematizes software projects, pursuing successful software development practices by emphasizing teamwork interaction. Scrum defines three roles: Product Owner, Scrum Team and Scrum Master. The Product Owner represents the customer and owns the Product Backlog, and works closely with the Scrum Team to provide clarification and approval on User Stories. The Scrum Team is responsible for completing the work and developing User Stories. The Scrum Master is responsible for facilitating the process, assisting the Product Owner in planning the releases and helping the Scrum Team progress by removing impediments and making resources available.

The widespread use of Scrum has evidenced a mismatch between academic instructions and software industry requirements. Beyond being focused mainly on trainings and certifications, the teaching of Scrum has become a cornerstone of undergraduate software engineering courses. However, although Scrum has been considerably used as a teaching strategy to introduce software engineering practices to students, there is insufficient educational material covering the procedures, methods and personal interactional patterns within the software industry, which would highly facilitate students' integration in professional contexts. In order to address this issue, this chapter presents guidelines and teaching strategies to successfully integrate Scrum into undergraduate and postgraduate courses as a further step to reduce the gap between software industry and academia.

Teaching Scrum

Given that the adoption of the Scrum by industry is becoming mainstream (Mathiassen & Pries-Heje, 2006; Nerur & Balijepally, 2007), it is important to consider why agile software development should be taught at university level. The iterative and incremental nature of Scrum, together with daily and retrospective meetings, allow students to receive immediate feedback on how agile practices are performed; furthermore they allow timely identification of problems in the understanding of agile practices, and to define corrective actions to be applied by the students in subsequent Sprints. In contrast to RUP, the students receive this feedback without waiting for the end of the project. On the other hand, student teams may lack the skills to manage a project, particularly when Scrum is implemented for the first time. Taking the recommendations from the Software Engineering 2004 Curriculum (CCSE, n.d.), there are skills, such as teamwork-related ones, that students should acquire, since agile software development is based on teamwork. Moreover, since agile methods support learning processes, we endorse the idea that teaching agile software development might reduce the cognitive complexity of software development processes and foster the acquisition of skills by making Scrum more comprehensible and suitable to undergraduate students.

Recently, the use of capstone projects based on Scrum has been adopted as a vehicle for teaching the basic concepts in software engineering (Mahnic, 2012a). This kind of project is developed in the classroom and supervised by professors. This strategy aims to increase student's participation in the learning process and address not only common problems found in the development of software systems but also several values proposed by the Agile Manifesto (Cano, 2011; Maher, 2009). In (Coupal & Boechler, 2005) an experience comparing a capstone project developed following an agile approach to their previous projects developed in a traditional way has been reported. In (Devedzic & Milenkovic, 2011), the authors described their eight years of experiences in teaching agile software methodologies to various groups of students at different universities. Based on the experience acquired, they recommended how to overcome potential problems in teaching agile software development. Out of the various agile approaches, Scrum has shown to be the most dominating AM in use (Kniberg, 2007; Mahnic, 2010). In (Mahnic, 2011) the achievement of teaching goals and the empirical evaluation of student's progress in estimation and planning skills using Scrum have been discussed. Also, the behavior of students using Scrum for the first time has been observed (Mahnic, 2012a). One of the benefits of Scrum is that it enables students to have a better teamwork environment and a better communication that results in high-quality products (Diaz, 2009).

All the aforementioned approaches share that, beyond the scope of the Scrum Team, there are management responsibilities such as management of financial resources, business decision-makings and management of the organization's environment (Kniberg, 2008). Following this line, we argue that augmenting the Scrum roles with the agile coach is vital in learning agile principles, particularly when students implement the agile method for the first time.

Agile Coaching

A typical problem in teaching contexts is the lack of coaching in Scrum teams. Augmenting the Scrum roles with the Agile Coach allows professors to train students in the diversity of aspects of software development, to deepen in the understanding of a Scrum-based process, and to acquire skills related to coaching and leadership (Hedin, Bendix, & Magnusson, 2005). It would be beneficial to integrate coaching in a Scrum course built on a simulation of a professional scenario, in order to provide students with an opportunity to experience the use of Scrum within a safe and controlled context. Despite, the acknowledged benefits of coaching, it has been proved challenging to implement effective agile coaching strategies even within academia.

An Agile Coach takes into account critical problems inside and outside students' teams (Dubinsky & Hazzan, 2003), solve impediments and encourage professors, team members and Scrum Masters to apply agile values. Although all the academic works mentioned above have recognized that agile coaching is a key factor, there has been insufficient analysis on the impact of coaching on the performance of software engineering students. The Agile Coach is meant to support and scaffold students' development by acting as a consultant and trainer in practices of software engineering, maintaining an appropriate balance between coaching and self-organization. It is also worth clarifying that the Agile Coach should not be involved in the project and is responsible for maintaining an appropriate balance between coaching and self-organization.

On Using Virtual Aids

In order to make Agile Coaching a reality, we agree with the idea that tools for planning, performing meetings, playing the role of customer, monitoring how students follow the Scrum process, visualizing information and managing the traceability of requirements are crucial to support Scrum practices in a teaching context. At the beginning, several 2D tools were considered for teaching Scrum; however, they fail to represent the physical development environment so as to carry out a realistic simulation of the Scrum process, which would be beneficial in a distributed context. A virtual world is an adequate solution to the problem of teaching the Scrum framework, through a capstone project, within the university premises when lack of physical space and resources prevent the creation of Scrum team rooms. Recent developments have seen Virtual Learning Environments rapidly become an integral part of teaching and learning provision in further and higher education (Callaghan, 2009). This evolution progresses strongly as educators do their best to adopt and adapt new technologies in the provision of more interactive teaching materials and learning environments which allow students the ability not only to view content but also to interact and organize it to suit their personal needs. Video games and virtual worlds are moving into the mainstream as traditional media industries struggle to keep with up digital natives and their desire for information, technology and connectivity. Researchers in the field are convinced that these technologies are potentially profitable to education (Anderson, 2008; Rodríguez, Soria, & Campo, 2012b).

In the light of the above, a 3D virtual environment allows developers to know about tasks performed by their peers, and even hold meetings regardless their physical location. Also, a 3D environment can be used to provide a physical topology of both a software project and process. This characteristic allows for a faster access to information than a 2D tool plus videoconference system to carry out meetings within a distributed team (Whitehead, 2007). This benefit may be attributed to the possibility of integrating isolated tools in a single development environment, leading to a context in which it is viable to share software artifacts (Herbsleb, 2007).

Using a virtual world has many advantages in comparison with 2D tools. For instance, students use avatars to communicate through gesture, sound, icons, text, among others (Herbsleb, 2007); and manipulate elements in the modeled room going through a 3D experience of Scrum. Additionally, the sense of immersion and the simulated team room, equipped with the Scrum artifacts and rules, are crucial to engage users in a hands-on experience of using Scrum, participate in collaborative work, and improve their comprehension of this agile method.

To sum up, the rest of this chapter explores the hypothesis that introducing the role of Agile Coach into a software engineering course and supporting the Scrum process with *Virtual Scrum* are effective teaching strategies for training students in the skills that are currently required by the software industry. In contrast to an e-learning platform, we aim to simulate a quasi-real software organization by means of our training model running on a capstone course, in which students face current challenges that occur in the software industry. On the one hand, the Agile Coach is essential to train and guide students along the course; on the other hand, *Virtual Scrum* attempts to setup a Scrum-based team room despite the limitations within university premises.

TEACHING SCRUM WITH AGILE COACHING AND VIRTUAL REALITY

The training model proposed in this chapter is the results of many years of teaching software engineering practices in the context of the Software Engineering course within the Systems Engineering BSc program at the Faculty of Exact Sciences (Universidad Nacional del Centro de la Provincia de Buenos Aires, i.e. UNICEN). The aim of the course is to engage students to develop a capstone project so as to experience a context as professional as possible. The students attending the course have been trained in software system design, object-oriented programming, operating systems and networks, and database management; additionally, they were given a previous course in which they have learnt basic concepts of software engineering. The capstone project has been designed to be completed within a course usually lasting 16 weeks and taken by the students in the ninth semester, side-by-side with other courses. The capstone project is organized and prioritized according to the Product Backlog (the master list of the desired features in the product, which are called User Stories and are grouped into short iterations called Sprints) divided into three Sprints (students are expected to work for about 2 hours a day).

Following Scrum, we assign the roles of Scrum Team and Scrum Master to students and the Product Owner role to a professor. The professor, playing the role of Product Owner, is responsible for defining and clarifying the requirements of the software product, establishing requirement priorities and validating the results obtained from each Sprint. The students, in the role of the Scrum Team, are subdivided into teams that are responsible for developing a set of user stories. The Scrum Master role is taken by different team members in each Sprint (Werner, Arcamone, & Ross, 2012). The reason for this is preserving the self-organizing principle with the aim of promoting students' experiential learning about management activities such as bridging the gap between the Product Owner and the team, cleaning the team obstacles and ensuring that the Scrum process is followed in terms of values, practices and rules. Furthermore, charging students with these responsibilities leads to optimize team dynamics and productivity, leaving students to reach their full potential (Davies, 2009).

Given that the students who take the course are in contact with Scrum for the first time, they need to be guided and oriented by professors in the use of Scrum so as to learn how to balance discipline and agility. For this reason, we complement the aforementioned Scrum roles with the Agile Coach role, played by another professor. Currently, our training model is designed to support two professors: one playing the role of Product Owner and another playing the role of Agile Coach, but our model can be easily extended to multiple professors. In case the course is given by only one professor, s/he should be able to exchange the Product Owner's hat by the Agile Coach's hat or vice versa with no conflicts. It is worth mentioning that s/he should leave no room for doubt concerning the role played in a certain moment.

As the Sprints progress, the Agile Coach, who is responsible for all the Scrum teams, should try different coaching styles. At the beginning of the course, the Agile Coach focuses on teaching the Scrum rules to be followed by the students. During the first sprint, which lasts 4 weeks, the students learn the basics of Scrum and engage with the idea of producing real value. In this context, the Agile Coach should complement the instructions with anecdotes and experiences in the software industry. During the second sprint, which lasts 3 weeks, if the Agile Coach notices that the students show signs of being trained in Scrum, s/he should suggest students to figure out possible improvements to enhance their performance; however, they need to be guided to complete the recommended practices. During the third sprint, which lasts 2 weeks, the Agile Coach assumes the students have acquired the skills expected at the end of the course, so s/he offers assistance only

when necessary and advises students to make some decisions. The second and third Sprints aim to prepare students to be extremely adaptable and foster them to think of solutions that may be away from their experience so far.

To effectively work in an agile context, each group of students in a course should be able to set up an individual Scrum-like team room, with different visual management strategies to perform the Scrum process activities (Diaz, 2009; Sutherland, Jakobsen, & Johnson, 2008): planning cards for estimating user stories, whiteboards for organizing user stories in Product and Sprint Backlog, burn-down charts for accessing team performance and room for performing Daily Meetings. In order to help students within limitations in the university premises, we propose to exploit the richness of the 3D metaphor by exploring virtual worlds. This kind of 3D environments may provide a physical topology of the organization of teamwork along a software project, which would facilitate the following of the Scrum-based process and allow accessing the artifacts, generated during the implementation of Scrum regardless the synchrony of place and time of the team. To ensure reliability, virtual worlds must be designed to be as real-life-like as possible; in this context, *Virtual Scrum* allows students to have information readily available through whiteboards that display the Product and Sprint Backlogs, and trace requirements easily through task boards that show team progress and performance metrics into the virtual world. Indeed, *Virtual Scrum* arises as an alternative to help each team of Scrum students set up a virtual working environment in which the visual metaphors required by Scrum are effectively displayed.

The communication among team members within *Virtual Scrum* is supported by chat and videoconference system in order to tackle collaborative and distributed issues. However, these mechanisms may hinder communication among team members, since they cannot experience the distinct feelings that can be sensed involving people who are together in the same place. In this context, weekly meetings arise as a face-to-face alternative in order to reinforce the relationship not only among team members, but also between team members and professors, in the role of Product Owner and Agile Coach.

Figure 1 shows our training model, in which each Sprint is organized in 4 phases, namely Organizing and Preparing the User Stories, Planning of the Product Backlog, Controlling and Monitoring the Sprint Work and Closing the Sprint. The first phase consists in building the Product Backlog, customizing the development environment and workstations to students. The second phase is focused on defining the Sprint Backlog and estimating the User Stories to be done during the sprint by using Planning Poker. During the third phase, the students develop User Stories, generate a product increment and integrate it to the working product. The fourth phase consists of feedback provided by the Product Owner, suggestions made by the Agile Coach and celebration within teams. The last Sprint also includes a phase called Delivering

the Product and Assessing Students, in which the students deliver the final product obtained from the capstone project and the professors assess the performance of the students along the course.

Organizing and Preparing User Stories

The first phase of the training model is the setup, which consists of two parts. First, the professors give formal lectures on professional software engineering practices recommended by CMMI and show how these practices can be supported with Scrum. Second, the professors provide students with *Virtual Scrum* and teach them how the tool deals with each of the phases. In addition, *Virtual Scrum* can also interact with tools for versioning code, testing classes and documenting artifacts.

After the setup, the building and prioritization of the User Stories into the Product Backlog take place. The Product Backlog is the master list of the desired product features and the professor, in the role of Product Owner, loads the User Stories

Figure 1. The training model

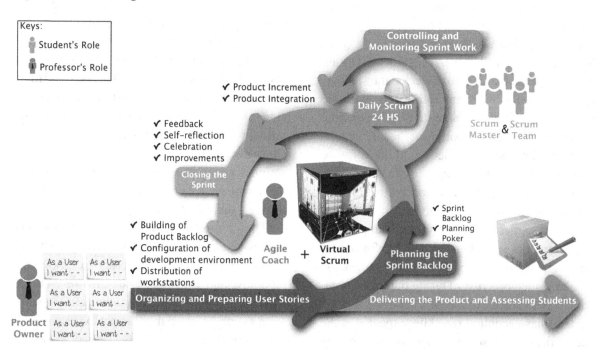

necessary to develop the capstone project. Figure 2 shows an example of the Product Backlog loaded with the user stories for *Universidad3D* (n.d.). *Universidad3D* is a virtual world under the Unity game engine designed as multi-tiered client-server architecture. The virtual world allows users to navigate the facilities of the UNICEN University, play thematic games and attend virtual courses, utilizing chat, e-mail and forum mechanisms for communication.

Universidad3D allow professors to generate different scenarios namely, building a piece of software from scratch, enhancing an already existing solution, or even reengineering an existing legacy system. In all mentioned cases, distributed Scrum teams have been given comparable user stories of similar complexity. For example, a Scrum team is in charge of incorporating features of social networks during the first Sprint, improv-

ing quality of 3D models during the second Sprint, and using Web Services to reuse and integrate heterogeneous platforms during the third Sprint.

The Product Backlog is supported by *Virtual Scrum* as a spreadsheet that contains the backlog items. The reason to use a spreadsheet is that it is a suitable tool to create a bit of history and make the work transparent. Also, this document can be easily shared and comprehended by the Scrum Team and the Product Owner because of its simplicity and accessibility (Tan, Tan, & Teo, 2008).

On the spreadsheet, the professor, wearing the Product Owner hat, clarifies and prioritizes (column "Priority" in Figure 2) the User Stories, and the Scrum Team asks the Product Owner for the necessary information by using questionnaires, interviews, and prototypes, among others. Each row of the spreadsheet represents a User Story. A User Story (US), which is identified by an ID,

Figure 2. Product backlog in Virtual Scrum

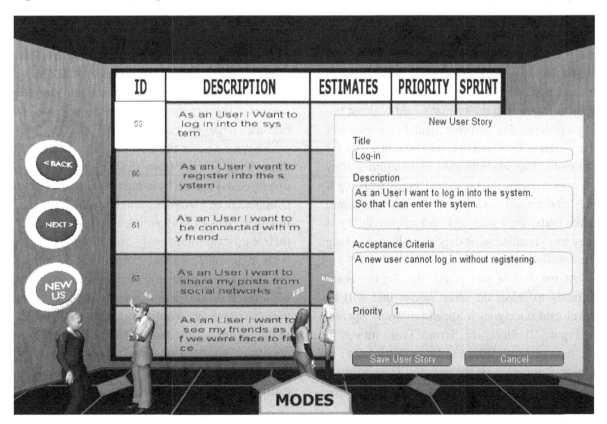

describes a desired functionality involving role ("As…"), product features ("…I want to…") and the benefit provided to the user ("…so that…"). Row 1 represents the User Story "As a User, I want to log in into the system so that I can enter the system" (ID 59).

As long as students prepare and organize the Product Backlog with the Product Owner, the Agile Coach works on fostering human relationships within the teams so each team member becomes acquainted with the strengths and weaknesses of other partners. Another important issue that the Agile Coach works on is the importance of comprehending the User Stories and the vision of the project by using the above mentioned Product Backlog. For dealing with the User Stories, the Agile Coach should help teams to refine the epics (i.e. large User Stories that rarely can be broken down into smaller ones), introduce non-functional requirements and specify acceptance criteria. For example, the Agile Coach could highlight the non-functional requirement of availability: "As a User I want to the virtual world is available 99.99% of the time I try to access it, so that I do not get frustrated." The "Acceptance Criteria" confirms when the User Story is completed; in the example, the US is tested against a user that tries to log-in into the virtual world without being registered. On the other hand, being aware of the vision is vital and derives some benefits for the team. The students gain momentum if they all work towards the same goal, enjoy working if they understand how they are contributing, and make better decisions and take responsibility if they can visualize what they are able to do along the capstone project.

At the end of this phase, the Scrum Team is ready to select the User Stories that will be developed during each Sprint (column "Sprint" in Figure 2). Here, the Scrum Team moves to the next phase to fill the column "Estimates" in Figure 2 for each of the selected User Stories, which are the intended ones to be delivered at the end of the Sprint.

Planning the Sprint Backlog

The Sprint initiates with the specification of achievable and expected outcomes of that Sprint. The Scrum Team holds a planning meeting to estimate a set of User Stories and reach a decision to start the activities of the Sprint. Considering the recommendations from the literature review (Cohn, 2004; Cohn, 2005), we selected the Planning Poker technique (Grenning, 2002; Mahnic, 2012b) since it increases individual commitment, i.e., those responsible for performing a task are the same who estimate how long this task may take (Brenner, 1996). Another motivation for our decision is that each team member must argue to support estimations (Jorgensen, 2004).

During the Planning Poker session, the team members vote the candidate estimates for the user stories using cards, which contain the story points used to estimate the US. A story point is considered to represent a working day consisting of 2 hours of uninterrupted work. As story point values, we decided to utilize the following Fibonacci sequence 0.5, 1, 2, 3, 5, 8, 13, and 20. The use of Fibonacci sequence to establish orders of magnitude is based on the idea that the ability of developers to accurately discriminate size decreases as the difference between the story points becomes larger (Miranda, 2001).

Figure 3 shows the Scrum Team estimating three User Stories from the Product Backlog displayed in Figure 2 by using the Planning Poker Component of *Virtual Scrum*. For each US, team members, using the Planning Poker View, select their cards simultaneously and those members with high and low story points have to justify their estimates by means of either chat or video-conference inside *Virtual Scrum*. Then, all team members vote again until a consensus is reached or the Scrum Master decides to finish the process after three iterations, deciding the average of story points as the estimate. For example, the resulting estimate for the US related to *log-in* was a value of

7.2 story points (Figure 3, Row 1), which meant 14.4 hours of uninterrupted work for this US.

In this context, the Agile Coach should help teams select the User Stories whose estimates are lower or equals than 13, and postponing the others for later where they can be disaggregated. Along the planning meeting, the Agile Coach assists teams to estimate the effort and communicate with the Product Owner. Regarding estimation, the Agile Coach uses *Virtual Scrum* to visualize the estimation process of each team and thus, monitor their status and skills. When it is necessary, the Agile Coach suggests teams to disaggregate complex User Stories into more simply ones to facilitate the estimation of a particular User Story. For instance, a possible suggestion could be "Be careful! You are underestimating the effort to develop that User Story. Maybe it requires to be divided into simpler User Stories."

After performing the planning session, each team has the User Stories that will be developed and load them into the Sprint Backlog. Then, each team disaggregates the User Stories into smaller tasks and the tasks are assigned among team members. Here, the Agile Coach encourages teams to elicit information from the Product Owner so as to avoid teams discarding tasks that can be vital for the development of the User Stories. At last, the team starts the development of those User Stories, which cannot be modified along the Sprint.

Figure 3. Planning Poker in Virtual Scrum

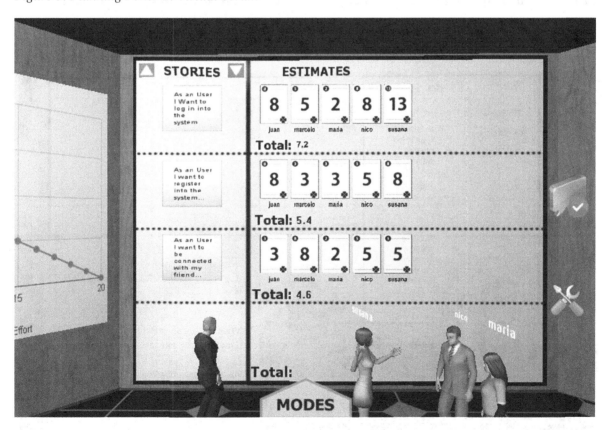

Controlling and Monitoring Sprint Work

During this phase, the Scrum Teams works on developing the tasks associated with the User Stories estimated in the previous phase. Every day of the Sprint at the same time, the team enters *Virtual Scrum* and meets in a Daily Meeting (Figure 4) to review and discuss the progress of the tasks. During a 15-minute's time-boxing period, from his/her working place at different locations each team member answers three questions through the Daily Meeting View artifact: What will you do today?, What did you do yesterday?, and Are there any impediments?. For example, a possible scenario of a Daily Meeting from a team member standpoint can be the following: *What will you do today? I will complete the HTML form for the log-in module. What did you do yesterday? I prepared the database schema for users. Are there any impediments? I need help to debug a problem with the form.* At this point, team members utilize the chat mechanism for communicating each other and can both surf the Web through a Web browser inside *Virtual Scrum* and access to a viewer of pdf documents so as to display any documents necessary to support their answers.

Remarkably, during Daily Meetings *Virtual Scrum* provides Scrum all team members together in the same place with a sense of immersion that allows them to touch a user story and talk about it, which are difficult activities to achieve by using 2D tools in combination with videoconference. Complementary to *Virtual Scrum*, there are open-source tools, such as USVN (n.d.), XWiki (n.d.), and Jira (n.d.), provided by the professors that support software engineering practices, such as configuration management,

Figure 4. Daily meeting in Virtual Scrum

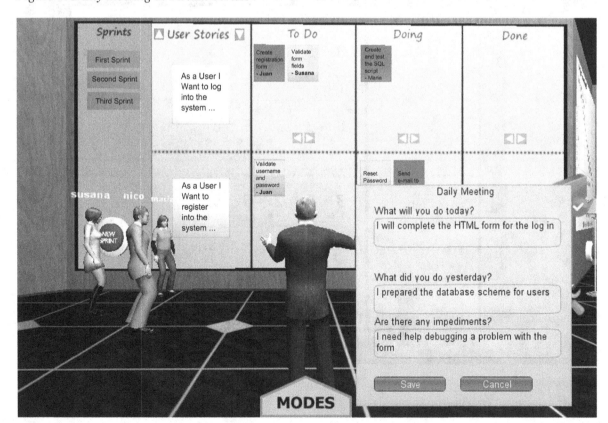

process monitoring, and project tracking, respectively. As these tools can be accessed through the Web Browser, *Virtual Scrum* acts as a project repository along with its virtual artifacts.

After answering the three questions, the Scrum Team updates the figures of the tasks in the Virtual Task Board as shown in Figure 5.

As the development of the tasks evolves, the tasks move along the DOING list on the Task Board and team members are able to load hours worked into the assigned tasks. When a task is completed by a team member and reviewed by the Scrum Master, this task moves to the DONE list. In order to promote self-organization, when a team member finishes the tasks assigned, s/he selects a new task, usually chosen when answering "What will you do today?"

Notice that the Task Board is a very powerful tool that creates communication among team members easily, since they can glance at it and review

the work in progress, completed, or pending. The Task Board both helps team members visualize tasks, and shows what they have accomplished, encouraging them to move on to new tasks. The Task Board is a useful aid to embrace the agile values such as transparency, collaboration, focus, and self-organization, among others. On the other hand, supporting a Virtual Task Board enables each team to setup a specific configuration of the work, regardless the physical location. This is a considerable issue that may jeopardize the learning process within university premises, since students are unable to fully advocate to the course due to other courses, examinations, and external links with companies.

Twice a week, the Agile Coach and each Scrum Team participate in a meeting called Weekly Meeting (up to 15 minutes long), which is face-to-face and fixed by the professor. The Agile Coach monitors students' performance in each stage of

Figure 5. The Task Board in Virtual Scrum

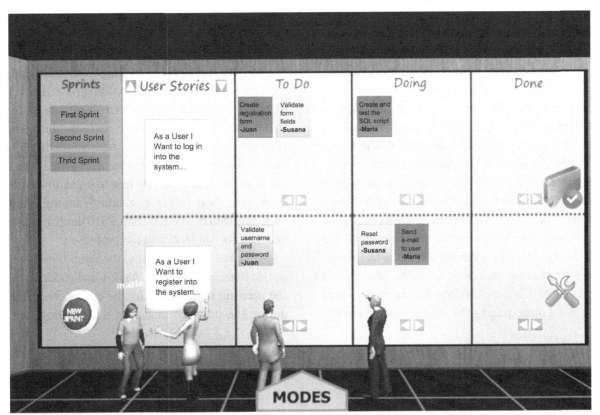

the miniature process by showing *Virtual Scrum* artifacts through a projector in order not to switch to another different Scrum environment for each team. Also, the Agile Coach guides teams in the use of prototypes to validate the user stories and obtain fruitful information to develop them. On Mondays, the students and the Agile Coach revise and analyze estimates and planning decisions; and on Fridays, the students highlight a contrast between planned tasks and executed tasks along the week. During the Weekly Meetings, the Agile Coach stimulates students to show architectural designs, user story specifications and other relevant documentation. During this assistance, *Virtual Scrum* is a crucial tool since it allows the Agile Coach to analyze how teams utilize the Task Board and how they progress on the tasks, without moving from one side to another.

Ideally, this kind of assistance should not interfere with the self-organization of the team. The Agile Coach should remain in silence waiting for the right moment to intervene. For example, the User Stories fail to be well-specified, the Burndown chart shows ups and downs, or the design document is deprecated. Either of these situations is suitable for interfering so as to discuss with the team about how to correct the course of action and address the remaining work.

It is worth clarifying that these meetings do not become Retrospective meetings. On one hand, the Agile Coach should take note of the issues observed so as to design the Retrospective meeting instead of pointing out a specific solution to the problem. On the other hand, the Agile Coach should detect competences and skills that arise spontaneously within each team. This allows the Agile Coach to be aware of strengths and weaknesses of team members; however, s/he should be cautious in emphasizing the merits in order to avoid discouraging the student effectiveness.

Closing the Sprint

The Sprint concludes with a review of the deliverables to be evaluated by the Product Owner in a Sprint Review meeting, and a Sprint Retrospective meeting to assess achievement of initial goals, review risks and carefully define the process aspects to be improved. In the Sprint Review, the Scrum Team displays the user stories completed during the Sprint by a demo. The objective of the demo is to examine the work done and get feedback from the Product Owner and other stakeholders. During the demo, the team shows how the developed features pass the acceptance tests and the Product Owner may want new features or improvements to the features, after interacting with the real software. In this context, the Agile Coach observes the demo and notes all the points to reinforce later, and identifies corrective actions to solve a particular problem in the miniature process. It is worth mentioning that in our training model, we decided to perform the demo in a face-to-face meeting. The reason for this decision stems from the necessity of augmenting *Virtual Scrum* with a desktop sharing tool to present the software product increment. The simultaneous use of both tools results in a high overhead and technological cost as well as (Oliveria & Lima, 2011) complex interaction that decreases team members' sense of presence and commitment (Paassivaara, 2009). A further reason for utilizing face-to-face meetings is that the distributed agile software development may lead to breakdowns in communication, misunderstandings among team members (Shrivastava & Date, 2010) and mistrust between the students and the professor.

After the Sprint Review, the team holds the Sprint Retrospective. The Agile Coach addresses the meeting by considering the issues observed during the previous phase. In this meeting, the

Agile Coach helps each member learns from the other in the best way; the methodology consists in pointing out a specific solution to a problem if necessary by considering the competences and skills within each team that were detected before. This meeting represents an opportunity to reflect on what occurred during the iteration, and identify the problems that may prevent the team from improving their productivity. To do so, each team member mentions the problems that need to be addressed, and the Scrum Master is in charge of monitoring the effective adoption of those solutions so as to reduce the list of impediments. The Agile Coach helps teams to reflect on few priority points to work on, and encourage team members to agree with the idea of making a list of improvements to be implemented. Following the example of the User Story "As a User, I want to log in into the system so that I can enter the system," a mistake detected by the Agile Coach was that

the students failed to consider storage of avatar' configurations, which is crucial to the load of the scene within the virtual world. The suggestion aimed at teaching the students to both improve the communication with the Product Owner and to apply elicitation requirements' techniques that they had learnt in a previous course.

To augment the monitoring process with performance indicators and help team members to adjust their estimates, *Virtual Scrum* supports a Burn-down chart, a graphic that shows the number of story points burnt during the Sprint. By using the Burn-down chart, the team knows both when the pending user stories in the Sprint Backlog should be completed and when the team deviates from the ideal effort at a specific moment. Figure 6 shows the Burn-down chart that represents the estimated story points ("Ideal Remaining Effort") and the burnt story points ("Actual Remaining Effort") during a Sprint in the *Universidad3D* project.

Figure 6. Burn-down chart in Virtual Scrum

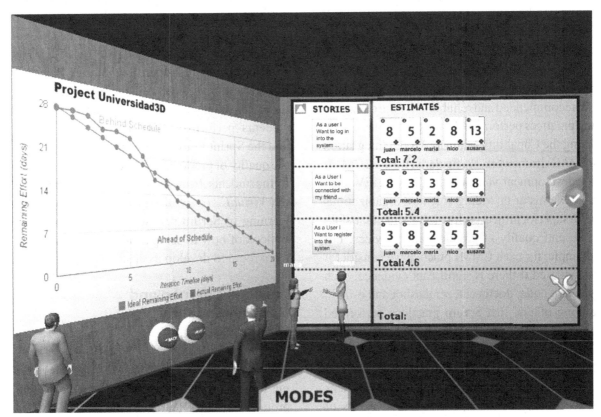

Delivering the Product and Assessing Students

At the end of the course, the teams present the final integrated product to the Product Owner, who is in charge of assessing the way students followed Scrum during the development of the user stories. In this context, there is no formal final exam, and the students' grades are calculated by the average between two criteria. Firstly, we consider that the individual mark is the same that the team mark but individual, which is determined by the number of user stories accomplished in the Product Backlog, the quality of the software and documentation developed, the fulfillment of releases and Sprint plans, and the professor's evaluations on students' cooperative work, maintenance of the Sprint Backlog, and meeting of deadlines. Secondly, we determined the individual mark by the compliance with the process, interaction with tools, artifacts fulfilled, lines of code implemented, worked hours and number of bugs solved, among others. In order to avoid subjectivity, we revise students' pieces of evidence associated to each task and assess whether each piece is satisfied or unsatisfied.

As regards the students' compliance with the Scrum process, we assess the accomplishment of the recommended and reliable software engineering practices performed during the course, by utilizing the Capability Maturity Model Integration (CMMI) framework. Particularly, we used CMMI for development version 1.3 (CMMI-DEV 1.3), which aims at providing software organizations with superior project performance, client satisfaction, and quality of products and services. For example, for each CMMI practice we consider it as satisfied if there is significant evidence, such piece of code, document, screenshot and e-mail, among others, that Scrum practices performed show that the CMMI practice is accomplished. To perform the assessment, in a previous work (Soria, Rodriguez, & Campo, 2012) we designed a mapping between CMMI and Scrum practices that stems from the proposals in the works of

(Diaz, 2009; Fritzsche & Keil, 2007; Marcal, 2008; Pikkarainen & Mantyniemi, 2006).

For instance, (Diaz, 2009) shows an empirical mapping in the context of Project Planning, Project Monitoring and Control, and Requirement Management. A general mapping between Scrum and CMMI level 2 and 3 is presented in (Fritzsche & Keil, 2007). The work presented in (Marcal, 2008) shows how Scrum allows achieving goals related to Project Planning, Project Monitoring and Control, Integrated Project Management and Risk Management. Finally, a mapping between Scrum and practices related to Requirement Management, Engineering process areas and Project Planning is presented in (Pikkarainen & Mantyniemi, 2006). Based on the aforementioned works, the coverage of the Scrum practices has been defined as the evaluation criterion for assessing students' performance. For each User Story, a practice is considered covered if there is at least a software artifact, i.e. an artifact defined in the development process designed for the capstone project, in the development repositories that evidences the practice had been completed.

To assist the professor in the students' assessment, *Virtual Scrum* provides the Task Board to evaluate the progress of the User Stories, Planning Poker View to evaluate students' estimates, Burndown chart and Daily Meeting View to evaluate students' performance and meeting of deadlines, and the Sprint Retrospective artifacts to evaluate the quality of products and documents completed by the students. In the light of the above, we believe that *Virtual Scrum* will help students improve their learning of Scrum practices and get a concrete outlook of how to setup a suitable Scrum-based team room to develop a capstone project.

Lessons Learned and Recommendations for Similar Courses

After three years of teaching software engineering to undergraduate students through capstone projects, we have learned a number of lessons about how to efficiently teach Scrum to students organized in agile teams. Our experience indicates students could effectively both acquire certain non-technical skills and perform practices related to quality assurance, monitoring and control, and estimation of user stories by means of the inclusion of both the Agile Coach and *Virtual Scrum* in the training model. In this light, it can be stated that the Agile Coach helps the students meet deadlines with high-quality processes and internalize the concept of an agile team (Soria, Rodriguez, & Campo, 2012). Furthermore, *Virtual Scrum* is suitable for teaching students how to implement different artifacts in a Scrum Team Room, and to effectively perform the practices throughout the Scrum process (Rodríguez, Soria, & Campo, 2012b). In both cases, we carried out controlled experiments with undergraduate software engineering students that were described in the works. Additionally, we found that the tool provides effective communication among developers and coordination mechanism among Scrum artifacts optimizing the synchronization of the project along the course. In the following paragraphs, different recommendations are discussed.

Avoiding Falling in Waterfall-Like Issues

In our first implementation of the course based on RUP, students usually ends up falling in a waterfall-like process, even though RUP encourages the overlapping of phases. As RUP concentrates students' efforts on eliciting all the requirements during the Inception and Elaboration phases, this resulted in inevitable delays in the early stages of the process. These delays prevented students from having enough time to design and run test cases before delivering the product, as they focus on correcting the mistakes made in early phases of the RUP process. As a consequence, the test cases are rarely run and many requirements fail to be implemented, which results in a weak coverage

of the practices related to design and implementation, verification, integration and deployment of the product.

By moving to Scrum, most of these aforementioned practices showed improvements because of the iterative life cycle that aims to work on all the aspects of software development during a Sprint. However, some students still misunderstood the concept behind the "done criteria" by assuming that a User Story is completed without fulfilling the testing stage. The concept of "done criteria" establishes that a User Story is tagged "DOING" until all the pre-established test cases have been not implemented (Williams et al., 2011).

To help students tackle this issue, the Agile Coach should take the responsibility of enforcing the concept of "done criteria," which is essential for ensuring that the User Stories developed are robust enough to handle alternative flows and/or error handling, and preventing students from misunderstanding Scrum. That is to say, the Agile Coach prompts students to complete the testing process before considering a User Story as "DONE." Rather than tracking what percentage of a new User Story is completed by a team member, the Agile Coach uses a binary "all or nothing" tag for tracking User Stories completion. In this line, we found that the Agile Coach enforces the concept of "done criteria" by making use of *Virtual Scrum* to facilitate transparency and visualization of progress within the teams, aiding students to keep the traceability of the User Stories across the Task Board.

Avoiding Following a Plan Instead of Responding to Change

As RUP relies on a rigid definition of plans, students, who usually lack the skills to describe all activities they need to undertake upfront, have to design and then follow a purely devised plan to develop user requirements. Given that the plan diverges from actual execution, a low coverage of the practices related to establishment and commitment is achieved in our initial version of the training model. In this context, obtaining commitment from all team members in this context is an issue as only some of them take part in the planning session.

Even after moving to Scrum, these problems remained; the students misunderstood how to respond to changes and thought that planning is unnecessary for agile development. The Agile Coach is central in tackling this hindrance by clarifying misconceptions, as well as guiding and monitoring achievement plan commitment and goals. One of the techniques that foster commitment is Planning Poker, in which each student must participate in the estimation of the work to be done. Most the students agreed or somewhat agreed with the idea that *Virtual Scrum* facilitates the implementation of the Planning Poker technique. This opinion may be attributed to the idea that students worked in a distributed environment and found the tool viable to estimate user stories. It is worth clarifying that students were familiar with the technique and we aimed to evaluate the impact of *Virtual Scrum* on the performance of Planning Poker.

To help students keep the plan up-to-date, the Agile Coach should encourage them to analyze the Burn-down chart at the end of each Sprint, so that they can revise and adjust the backlog estimates based on the team velocity. In this context, *Virtual Scrum* makes both individual and group work permanently visible, which allows students to trace the progress of the team, and the Scrum Master to follow each member' progress. It is possible to observe that the display of team's metrics in the 3D metaphor resulted in a useful aid for self-reflection.

Avoiding Focusing on Dates Instead of Quality

The plan-driven property of RUP emphasizes the definition of deliverables, which students misunderstand as a need to meet deadlines, sometimes disregarding the RUP process. In this context, students tend to develop User Stories as fast as possible and skip recommended software engineering practices, resulting in low quality products. Consequently, during subsequent Sprints, most the time is devoted to fixing faulty User Stories rather than developing new ones, which eventually cause delays in the releases.

When RUP was replaced by Scrum, this situation failed to improve since practices such as building plans and designing software were underestimated by students. Due to a lack of assistance in a disciplined use of Scrum, most of the students misconceived essential practices as evident in assertions such as "documentation is not necessary," "design is too hard to achieve" and "planning is a waste of time." Documenting, designing and planning are mandatory in software engineering regardless the methodology.

To balance discipline and agility along the tracking of the capstone project, the Agile Coach should encourage students to focus on developing the User Stories by performing recommended and reliable practices. Thus, the Agile Coach should prevent accepting to subsequent Sprints User Stories that fail to be implemented with the required quality. To guarantee this, we incorporate these high-quality practices into the aforementioned "done criteria" as acceptance conditions for the delivered functionality.

In this context, in order to facilitate tracking of practices, *Virtual Scrum* guarantees the availability and accessibility of software artifacts in each meeting. Moreover, *Virtual Scrum* simplifies Daily Meetings among geographically distributed students. We found *Virtual Scrum* suitable to help both students, enhancing their comprehension of Scrum, and the Agile Coach in monitoring the agile teams. From a conceptual point of view, the students stated that *Virtual Scrum* allowed them to optimize understanding software engineering concepts and problem-solving skills necessary to succeed in professional contexts.

Avoiding Writing Enormous Documents of Requirements

When students run our first version of the model based on RUP, we found that they used to write down what the professor required, instead of writing what the professor needed as a customer. The students spent a lot of time generating detailed documentation of requirements, resulting in a weak coverage of the practices related with the verification and validation of the product at the end of the project. As a consequence, the training model yielded as a result low coverage of the practices related to the identification and analysis of risks, since regular reviews of processes were rarely performed. Unsurprisingly, this situation led to weak communication among the Product Owner and team members resulted in a low coverage in the validation of the product.

In this context, an iterative and incremental life-cycle leads student to start with only the User Stories planned to be implemented within a Sprint. Focusing on a limited set of User Stories, the Agile Coach encourages students to communicate with the Product Owner and, thus, the specifications of those User Stories are complemented with the information resulting from that conversation. The Agile Coach should periodically observe the working progress and requires test cases and design documentation. Moreover, Holding Daily Meetings allowed increasing the coverage of the aforementioned practices. The Weekly Meetings held with the Agile Coach helped teams to track and

communicate noncompliance issues objectively, and ensure theirs prompt resolution. These factors resulted in more student commitment to run functional and non-functional test cases associated with a User Story, meeting successfully the "done criteria." Along this line, we found that these kind of check-point meetings facilitated internalization of the Scrum concepts, faster solutions of impediments, and guidance from the Agile Coach.

Furthermore, since *Virtual Scrum* supports the Task Board to display the User Stories, the professor, playing the role of Product Owner, can enter the virtual world, check the progress of the User Stories and answer team questions. Another advantage of using *Virtual Scrum*, over creating Scrum rooms within the university premises, is that students can also access data they need for each Sprint and exchange valuable information, vital in agile teams, without having to be in the same physical context.

To bring this section to an end, we summarize the main points. One interesting finding after our recent experience with our teaching model is a progressive increment in the coverage of all the CMMI practices by students. This increment was due to compliance with the done criteria, carefully guidance performed by the Agile Coach and improvements in project tracking. As a result, we have obtained a more homogeneous accomplishment of the software engineering practices and higher commitment among the students. Another important finding of the experience is that *Virtual Scrum* effectively supports the Scrum practices performed by both students, in the role of Scrum Team, and professors, taking the roles of Product Owner and Agile Coach.

FUTURE RESEARCH DIRECTIONS

Nowadays, there are several approaches that allow introducing agile practices within software engineering curricula. Our experience aligns with the idea of using capstone projects, which reduce the gap between professional and training contexts (Mahnic, 2010; Mahnic, 2012a; Devedzic, 2011; Schroeder, 2012). The aim of some approaches is to maximize student learning experience by using games (Lynch, 2011) or 3D interfaces (Parsons, 2010) to teach agile concepts. Other approaches propose the use of teaching models, such as agile teaching models (Hon, 2004) or frameworks that facilitate the teaching of software development processes (Hazzan & Dubinsky, 2006). Our training model takes elements of the aforementioned projects and introduces the Agile Coach role and *Virtual Scrum* support to provide tutoring and training to undergraduate students. However, students learn in many ways, and thus, professors should know the students' profile, their strengths and weaknesses in order to help them to perform agile practices. As a result, student would be able to maximize their learning experience.

From the experience of our aforementioned training model, we believe that personalization of teaching is crucial. As professors, we are planning to concentrate on teaching Scrum, complemented to XP and Kanban, catering for individual learning characteristics of students. A personalized learning approach focuses on knowing the leaner in order to offer learning scenarios that result relevant and motivating. Then, the personalization tailors these scenarios to students' needs by harnessing technology and data from the context (Ally, 2004). However, it is necessary to identify only student generated information relevant for such personalization and include irrelevant anecdotic data (Antunes, 2010). To address this issue, we will need to know the student talents and difficulties to learn of agile practices (Dick, Carey, & Carey, 2005). This knowledge will allow us to adapt their teaching strategies to student preferences (Felder & Silverman, 1988).

In this context, a line of research worth pursuing further is the study of relationships between students' performance and their learning style within heterogeneous undergraduate courses. Taking into account different learning styles within heterogeneous undergraduate courses may have an impact on learning by fostering engagement, motivation and attentiveness (Coffield, 2004; Felder, & Spurlin, 1988; Hawk & Shah, 2007). For example, Layman et al (Layman, Cornwell, & Williams, 2006) proposes to explore personality types and learning styles. Other approaches (Limongelli, 2008; Popescu, 2009; Zaina, 2011) focus on addressing various learning styles to personalize Web-based teaching contexts. A first attempt to shed some light on personalizing the teaching of Scrum on a software engineering course was carried out in (Scott, Rodriguez, Soria, & Campo, 2013) as a step towards improving the teaching of software engineering practices. In this research, we evidenced relationships between the Felder-Silverman learning styles (Felder & Silverman, 1988) of students and how they perform Scrum practices. For instance, we found a considerable influence of the processing style on the way students estimate User Stories and follow the Scrum process, and the amount of time that the students spend on the development of tasks. These findings will serve as a cornerstone for further studies to determine how the learning styles of students can be used to build a customized environment for learning Scrum.

CONCLUSION

This chapter has presented a teaching model based on a combination between Scrum, Agile Coach's role and Virtual Reality. We have discussed the design and implementation of the teaching model for introducing agile software development in a software project, focusing on both improving the learning of professional software practices and maintaining the quality of software processes. We found that teaching Scrum software development is effective, if students are involved in the development of a project rather than in traditional of-the-book classes. Moreover, facing the software engineering problems in a controlled environment gives students the required skills to work in professional contexts. We observed that this teaching strategy may facilitate the student's integration in the software industry.

It is possible to conclude that the introduction of the Agile Coach was an effective teaching strategy that allowed students to achieve higher coverage of the software practices. Furthermore, the iterative and incremental nature of the training model allowed students to experience all the aspects of software development along a Sprint. During this period, students had the opportunity to learn from their mistakes, adapt to the context, and concentrate on only the user stories for a specific Sprint. Additionally, the students revealed the benefits of using the Scrum model augmented with agile coaching, namely facilitated internalization of the Scrum concepts, faster solutions of impediments, and guidance by means of check-point meetings.

Besides, this chapter showed that *Virtual Scrum* is an alternative able to face challenges of implementing Scrum within university premises. This tool proved to be effective in enriching teaching strategies for training students in the use of valuable professional skills to work in professional contexts through a virtual world. We found that using *Virtual Scrum* was helpful in improving students' comprehension of the fundamentals of agile practices and principles of developing software with Scrum. It is worth noting that the tool outperformed students' expectations with regard to support for planning meetings, which increased students' commitment; and follow-up metrics, which allowed students to self-reflect on their performance in the Sprint Retrospective meetings. The students also provided constructive feedback on user interactions and traceability of the User Stories through *Virtual Scrum*. Based on this feedback, we intend to improve user interactions by upgrading the support of media aids, specially the avatar integration with current social networks. As for traceability of User Stories, we found that students preferred using 2D tools for

dealing with configuration management rather than a 3D representation of the artifacts.

In the light of the above, as future work, we are planning to complement *Virtual Scrum* with conventional team tools and emphasize on holding software meetings, and displaying software metrics and coverage of the Scrum practices. Another line of future work will concentrate on applying profiling techniques to provide personalized assistance to students according to their learning style, and thus, improve students' learning experience within teaching contexts.

REFERENCES

Agile Manifesto. (2014). Retrieved February 7, 2014 from http://www.agilemanifesto.org/

Alegría, J. A. H., & Bastarrica, M. C. (2006). Implementing CMMI using a combination of agile methods. *CLEI Electronic Journal, 1*(1).

Alfonso, M. I., & Botia, A. (2005). An iterative and agile process model for teaching software engineering. In *Proceedings of 18th Conference on Software Engineering Education and Training*. Academic Press.

Ally, M. (2004). Foundations of educational theory for online learning. In *Theory and practice of online learning* (pp. 3–31). Athabasca University, Canada's Open University.

Anderson, T. (Ed.). (2008). *The theory and practice of online learning*. Athabasca University Press.

Antunes, C. (2010). Anticipating student's failure as soon as possible. In C. Romero et al. (Eds.), *Handbook of educational data mining*. CRC Press. doi:10.1201/b10274-28

Apelo, J. (2010). *Management 3.0: Leading agile developers, developing agile developers*. Addison-Wesley.

Atlassian. (2014). Retrieved February 7, 2014 from https://www.atlassian.com/software/jira

Barbosa, E. F., & Maldonado, J. C. (2006). Towards the establishment of a standard process for developing educational modules. In *Proceedings of the 36th ASEE/IEEE Frontiers in Education Conference*. San Diego, CA: ASEE/IEEE.

Boehm, B., & Turner, R. (2008). *Balancing agility and discipline: A guide for the perplexed*. Reading, MA: Addison-Wesley.

Brenner, L. A., Koehler, D. J., & Tversky, A. (1996). On the evaluation of one-sided evidence. *Journal of Behavioral Decision Making*, *9*(1), 59–70. doi:10.1002/(SICI)1099-0771(199603)9:1<59::AID-BDM216>3.0.CO;2-V

Callaghan, M. J., McCusker, K., Losada, J. L., Harkin, J. G., & Wilson, S. (2009). Teaching engineering education using virtual worlds and virtual learning environment. In *Proceedings of 2009 Conference on Advances in Computing, Control and Telecommunications Technologies* (pp. 295–299). Kerela, India: Academic Press.

Cano, M. D. (2011). Students' involvement in continuous assessment methodologies: A case study for a distributed information systems course. *IEEE Transactions on Education*, *54*(3), 442–451. doi:10.1109/TE.2010.2073708

CCSE. (2014). Retrieved February 7, 2014 from http://sites.computer.org/ccse/SE2004Volume.pdf

Coffield, F., Moseley, D., Hall, E., & Ecclestone, K. (2004). *Should we be using learning styles?* Learning and Skills Research Centre.

Cohn, M. (2004). *User stories applied for agile software development*. Reading, MA: Addison-Wesley.

Cohn, M. (2005). *Agile estimating and planning*. Upper Saddle River, NJ: Prentice Hall.

Coupal, C., & Boechler, K. (2005). Introducing agile into a software development capstone project. In *Proceedings of Agile Conference* (pp. 289–297). Academic Press.

Davies, S. (2009). Appointing team leads for student software development projects. *Journal of Computing Sciences in Colleges*, *25*(2), 92–99.

Devedzic, V., & Milenkovic, S. R. (2011). Teaching agile software development: A case study. *IEEE Transactions on Education*, 54.

Diaz, J., Garbajosa, J., & Calvo-Manzano, J. A. (2009). Mapping CMMI level 2 to scrum practices: An experience report. *Software Process Improvements*, *42*, 93–104. doi:10.1007/978-3-642-04133-4_8

Dick, W. O., Carey, L., & Carey, J. O. (2005). *The systematic design of instruction*. Pearson/Allyn & Bacon.

Dubinsky, Y., & Hazzan, O. (2003). Extreme programming as a framework for student-project coaching in computer science capstone courses. In *Proceedings of IEEE International Conference on Software: Science, Technology and Engineering* (pp. 53–59). IEEE.

Felder, R. M., & Silverman, L. K. (1988). Learning and teaching styles in engineering education. *English Education*, *78*(7), 674–681.

Felder, R. M., & Spurlin, J. E. (2005). Applications, reliability, and validity of the index of learning styles. *International Journal of Engineering Education*, *21*(1), 103–112.

Fritzsche, M., & Keil, P. (2007). Agile methods and CMMI: Compatibility or conflict? *e-Informatica. Software Engineering Journal*, *1*(1), 9–26.

Glazer, H. (2008). *CMMI or agile: Why not embrace both! (Technical Note, CMU/SEI-2008-TN-003)*. Pittsburgh, PA: Software Engineering Process Management, Carnegie Mellon.

Glazer, H. (2010). Love and marriage: CMMI and agile need each other. *CrossTalk*, *23*(1), 29–34.

Grenning, J. (2002). *Planning poker or how to avoid analysis paralysis while release planning*. Retrieved from http://renaissancesoftware.net/files/articles/PlanningPoker-v1.1.pdf

Hawk, T. F., & Shah, A. J. (2007). Using learning style instruments to enhance student learning. *Decision Sciences Journal of Innovative Education*, *5*(1), 1–19. doi:10.1111/j.1540-4609.2007.00125.x

Hazzan, O., & Dubinsky, Y. (2006). Teaching framework for software development methods. In *Proceedings of the 28ᵗʰ International Conference on Software Engineering* (pp. 703–706). ACM.

Hedin, G., Bendix, L., & Magnusson, B. (2005). Teaching extreme programming to large groups of students. *Journal of Systems and Software*, *74*(2), 133–146. doi:10.1016/j.jss.2003.09.026

Herbsleb, J. D. (2007). Global software engineering: The future of socio-technical coordination. In *Proceedings of Future of Software Engineering* (pp. 188–198). IEEE. doi:10.1109/FOSE.2007.11

Highsmith, J. (2004). *Agile project management-creating innovative products*. Boston: Addison-Wesley.

Hon, A., & Chun, W. (2004). Teaching agile teaching/learning methodology and its e-learning platform. *LNCS-Advances in Web-Based Learning*, *3143*, 11–18.

IBM. (2014). *Developerworks*. Retrieved February 7, 2014 from http://www.ibm.com/developerworks/rational/library/apr05/crain/

Jorgensen, M. (2004). A review of studies on expert estimation of software development effort. *Journal of Systems and Software*, *70*, 37–60. doi:10.1016/S0164-1212(02)00156-5

Khan, F. A., Graf, S., Weippl, E. R., & Tjoa, A. M. (2010). Implementation of affective states and learning styles tactics in web-based learning management systems. In *Proceedings of the 10ᵗʰ IEEE International Conference on Advanced Learning Technologies* (pp. 734–735). IEEE.

Kniberg, H. (2008). *The manager's role in scrum*. Retrieved from http://www.scrumalliance.org/resources/293

Kruchten, P. (2003). *The rational unified process: An introduction*. Reading, MA: Addison-Wesley.

Kulpa, M., & Johnson, K. (2003). *Interpreting the CMMI*. CRC Press. doi:10.1201/9780203504611

Layman, L., Cornwell, T., & Williams, L. (2006). Personality types, learning styles, and an agile approach to software engineering education. *ACM SIGCSE Bulletin*, *38*(1), 428–432. doi:10.1145/1124706.1121474

Limongelli, C., Sciarrone, F., & Vaste, G. (2008). Ls-plan: An effective combination of dynamic courseware generation and learning styles in web-based education. In *Proceedings of the 5ᵗʰ International Conference on Adaptive Hypermedia and Adaptive Web-Based Systems* (pp. 133–142). Springer-Verlag.

Lutteroth, C., Luxton-Reilly, A., Dobbie, G., & Hamer, J. (2007). A maturity model for computing education. In *Proceedings of the Ninth Australasian Conference on Computing Education* (vol. 6, pp. 107–114). Australian Computer Society.

Lynch, T. D., Herold, M., Bolinger, J., Deshpande, S., Bihari, T., Ramanathan, J., & Ramnath, R. (2011). An agile boot camp: Using a LEGO-based active game to ground agile development principles. In *Proceedings of Frontiers in Education Conference* (pp. F1H-1). IEEE.

Maher, P. (2009). Weaving agile software development techniques into a traditional computer science curriculum. In *Proceeding on 6th Conference on Information Technology: New Generations*. Academic Press.

Mahnic, V. (2010). Teaching scrum through team-project work: Students' perceptions and teachers' observations. *The International Journal of Engineering Education*.

Mahnic, V. (2011). A case study on agile estimating and planning using scrum. *Electronics and Electrical Engineering, 5*.

Mahnic, V. (2012a). A capstone course on agile software development using scrum. *IEEE Transactions on Education, 5*(2), 99–106. doi:10.1109/TE.2011.2142311

Mahnic, V., & Hovelja, T. (2012b). On using planning poker for estimating user stories. *Journal of Systems and Software, 85*(9), 2086–2095. doi:10.1016/j.jss.2012.04.005

Marçal, A. S. C., de Freitas, B. C. C., Soares, F. S. F., Furtado, M. E. S., Maciel, T. M., & Belchior, A. D. (2008). Blending scrum practices and CMMI project management process areas. *Innovations in Systems and Software Engineering, 4*(1), 17–29. doi:10.1007/s11334-007-0040-1

Mathiassen, L., & Pries-Heje, J. (2006). Business agility and diffusion of information technology. *European Journal of Information Systems*, 116–119. doi:10.1057/palgrave.ejis.3000610

Miranda, E. (2001). Improving subjective estimates using paired comparisons. *IEEE Software, 18*(1), 87–91. doi:10.1109/52.903173

Nerur, S., & Balijepally, V. (2007). Theoretical reflections on agile development methodologies. *Communications of the ACM, 50*, 79–83. doi:10.1145/1226736.1226739

Oliveira, E., & Lima, R. (2011). State of the art on the use of scrum in distributed development software. *Revista de Sistemas e Computação, 1*(2), 106–119.

Osorio, J. A., Chaudron, M. R., & Heijstek, W. (2011). Moving from waterfall to iterative development – An empical evaluation of advantages, disadvantages and risks of RUP. In *Proceedings of 37th Conference on Software Engineering and Advanced Application* (pp. 453–460). IEEE.

Paasivaara, M., Durasiewicz, S., & Lassenius, C. (2009). Using scrum in distributed agile development: A multiple case study. In *Proceedings of the 4th IEEE International Conference on Global Software Engineering* (pp. 195–204). IEEE.

Parsons, D., & Stockdale, R. (2010). Cloud as context: Virtual world learning with open wonderland. In *Proceedings of the 9th World Conference on Mobile and Contextual Learning* (mLearn 2010). Valetta, Malta: mLearn.

Pikkarainen, M., & Mantyniemi, A. (2006). *An approach for using CMMI in agile software development assessments: Experiences from three case studies*. University of Limerick Institutional Repository.

Popescu, E. (2009). Evaluating the impact of adaptation to learning styles in a web-based educational system. In *Proceedings of the 8th International Conference on Advances in Web Based Learning* (pp. 343–352). Berlin: Springer-Verlag.

Rodríguez, G., Soria, A., & Campo, M. (2012a). Teaching scrum to software engineering students with virtual reality support. In *Advances in new technologies, interactive interfaces and communicability* (pp. 140–150). Berlin: Springer. doi:10.1007/978-3-642-34010-9_14

Rodríguez, G., Soria, A., & Campo, M. (2012b). Supporting virtual meetings in distributed scrum teams. *IEEE Latin America Transactions, 10*(6), 2316–2323. doi:10.1109/TLA.2012.6418138

Schroeder, A., & Klarl, A. (2012). Teaching agile software development through lab courses. In *Proceedings of IEEE Global Engineering Education Conference* (EDUCON) (pp. 1177–1186). IEEE.

Schwaber, K., & Beedle, M. (2002). *Agile software development wit scrum*. Upper Saddle River, NJ: Prentice Hall.

Scott, E., Rodriguez, G., Soria, A., & Campo, M. (2013). El rol del estilo de aprendizaje en la enseñanza de prácticas de scrum: Un enfoque estadístico. In *Proceedings of the 14th Argentine Symposium on Software Engineering*. Academic Press.

Shrivastava, S. V., & Date, H. (2010). Distributed agile software development: A review. *Journal of Computer Science and Engineering, 1*(1), 10–16.

Soria, A., Rodriguez, G., & Campo, M. (2012). Improving software engineering teaching by introducing agile management. In *Proceedings of the 13th Argentine Symposium on Software Engineering* (pp. 215–229). Academic Press.

Sutherland, J., Ruseng Jakobsen, C., & Johnson, K. (2008). Scrum and CMMI level 5: The magic potion for code warriors. In *Proceedings of Hawaii International Conference on System Sciences, Proceedings of the 41st Annual* (pp. 466–466). IEEE.

Tan, C. H., Tan, W. K., & Teo, H. H. (2008). Training students to be agile information systems developers: A pedagogical approach. In *Proceedings of the 2008 ACM SIGMIS CPR Conference on Computer Personnel Doctoral Consortium and Research* (pp. 88–96). ACM.

Unity 3D. (2014). Retrieved February 7th, 2014 from http://unity3d.com/

Universidad 3D. (2014). Retrieved Fenruary 7th, 2014 from http://www.isistan.unicen.edu.ar/?page_id=386

USVN. (2014). Retrieved February 7th, 2014 from http://www.usvn.info/

Versionone. (2014). *State of agile development survey results*. Retrieved February 7th, 2014 from http://www.versionone.com/pdf/2012_State_of_Agile_Development_Survey_Results.pdf

Werner, L., Arcamone, D., & Ross, B. (2012). Using scrum in a quarter-length undergraduate software engineering course. *Journal of Computing Sciences in Colleges, 27*(4), 140–150.

Whitehead, J. (2007). *Collaboration in software engineering: A roadmap*. Santa Cruz, CA: University of California.

Williams, L., Brown, G., Meltzer, A., & Nagappan, N. (2011). Scrum + engineering practices: Experiences of three Microsoft teams. In *Proceedings of International Symposium on Empirical Software Engineering and Measurement* (pp. 463–471). IEEE Society.

Xwiki. (2014). Retrieved February 7th, 2014 from http://www.xwiki.org/

Zaina, L. A. M., Bressan, G., Rodrigues, J. F., & Cardieri, M. A. C. A. (2011). Learning profile identification based on the analysis of the user's context of interaction. *IEEE Latin America Transactions, 9*(5), 845–850. doi:10.1109/TLA.2011.6030999

KEY TERMS AND DEFINITIONS

Agile Coach: A person who provides directions to the project and is responsible for removing any process impediments.

Capability Maturity Model Integration (CMMI): A framework which consists of a set of best practices that address the development and maintenance of products and services.

Capstone Project: A cooperative assignment in which undergraduates are expected to integrate the course contents in a quasi-real world experience.

Learning/Teaching Strategies: Recommended practices to improve learning and teaching experiences.

Rational Unified Process (RUP): It is a software engineering process, aimed at guiding software development organizations.

Scrum: An iterative and incremental methodology that organizes projects to make them manageable for small, self-organized and cross-functional teams.

Software Engineering Education: A set of university courses to prepare students to be professional software engineers.

Virtual Reality: A computer-simulated environment that can simulate physical presence in places within real or imagined worlds.

ENDNOTES

1.   A capstone project is a cooperative assignment in which undergraduates are expected to integrate the course contents in a quasi-real world experience (Mahnic, 2012a).

Chapter 14
A Project–Based Introduction to Agile Software Development

Marc Lainez
Agilar, Belgium

Cyrille Dejemeppe
Université Catholique de Louvain, Belgium

Yves Deville
Université Catholique de Louvain, Belgium

Jean-Baptiste Mairy
Université Catholique de Louvain, Belgium

Adrien Dessy
Université Catholique de Louvain, Belgium

Sascha Van Cauwelaert
Université Catholique de Louvain, Belgium

ABSTRACT

This chapter shows how a lightweight Agile process has been used to introduce Agile project development to young computer science students. This experience has been conducted on a project aimed at developing Android applications. The context, the process, and the results of this experiment are described in this chapter.

1. INTRODUCTION

Agile programming is an increasingly used paradigm for the development of software. Hence, in a university curriculum of computer science, it is essential that students have the opportunity to experience the Agile paradigm. At the Louvain School of Engineering (Université catholique de Louvain, Belgium), we decided to introduce Agile software development to students through a project, during the third year of their five year

curriculum. In this project, the students had to develop their own Android application.

In this chapter, we describe this experiment and report the results. The primary objective of the experiment was to assess if an Agile software development approach was suitable for young students and if the learning outcomes could be met by means of a totally project based approach. We describe a lightweight Agile process that can be used by students to experience Agile software development as well as the various tools that have been introduced to support the development of their project. At the end of the semester, all the teams came up with a functional mobile application; 13 out of the 15 apps were of high quality

DOI: 10.4018/978-1-4666-5800-4.ch014

and were distributed on the Google Play store as open-source apps.

This chapter is structured as follows. Section 2 presents the context of Louvain School of Engineering and provides a background on Agile software development. The process of this teaching experiment is described in Section 3: hypotheses (3.1), design of the process (3.2), how it has been pretested on a small set of teaching assistants (3.3), how it has been implemented for the students (3.4), and the results and evaluation of this experience (3.5). We propose directions for future research in Section 4 before the conclusion (Section 5).

2. BACKGROUND

2.1 The Context

The Louvain School of Engineering (Université catholique de Louvain, Belgium) introduced in 2012-2013 a one-semester project for its 300 third-year students of the Bachelor Degree. The students had to choose the project related to their major option. About 60 students picked the project in computer science for 2012-2013. The project was held in parallel with other courses and accounted for about 15% of the student workload of the semester.

A crucial pedagogical choice was made: the project should be a practical introduction to Agile software development. The objective was to let students develop their own Android application, in groups of four, using a lightweight Agile process. This lightweight process allowed us to focus on the key aspects of the Agile paradigm. Furthermore, we wanted to make the code available under an open-source license and to possibly release the applications to the Google Play Store by the end of the project. Besides technical skills, this project also aimed at developing cross-disciplinary skills such as modeling, teamwork, planning, management and communication.

Active learning, and more specifically problem-based learning (Boud & Feletti, 1998), is a long-standing practice at the Louvain School of Engineering. Active learning can be defined as anything course-related that all students in a class session are called upon to do other than simply watching, listening and taking notes (Felder & Brent, 2009). Essential elements in active learning are student activity and involvement in the learning process (Prince, 2004). It comes in multiple forms, among them: collaborative learning, where students work together in small groups towards a common goal, and problem-based learning, where relevant problems are introduced at the beginning of the instruction cycle and used to provide the context and motivation (Boud & Feletti, 1998; Johnson & Johnson, 1999; Prince, 2004). It may be noted that problem-based learning is often collaborative. The choice of an Agile software development approach fits very well with this commitment to active learning. Indeed, it promotes strong interactions and collaboration between students within a group and provides opportunity for rich interactions with the teaching team.

The challenge was also to introduce students to a new way of conducting projects. As a matter of fact, students tend to wait until the last days to work on result-focused projects (in contrast to process-focused projects). The development is thus done under extreme time constraints. As you may guess, this sometimes leads to disastrous results. The goal was to make the students realize that a lightweight Agile process can be used to plan and allocate the workload, giving significantly better results with very little overhead.

2.2 Agile Software Development

Commonly called Agile, this movement groups several approaches to software development and project management. It focuses on very high collaboration within the team and strong involvement of the customer in the development process by building cross-functional and self-organized teams. Each Agile framework or method uses its own set of tools as well as common recognized practices.

One of the main rules of any Agile method is always to work on the next most important feature for the customer. That way, the return on investment tends to be maximized. Good development practices and accurate use of flexible design patterns are mandatory for systems built using any Agile approach to grow and stay maintainable. In Agile projects, it is said that code quality and evolvable architecture are non-negotiable.

The currently most used Agile approaches are Scrum, Extreme Programming (XP), Kanban and Behavior Driven Development (BDD). The following sections briefly describe each of these approaches.

2.2.1 Scrum

Scrum is an iterative approach to software and product development. Scrum takes its roots in the creation of several artifacts, roles and defined rituals. Scrum suggests delivering, at regular intervals, a working chunk of the product. This product increment can potentially be deployed and used to gather feedback from the customer or the end users. This small feedback loop helps to adjust the product in case it does not meet the needs of the customer (Schwaber, 2004).

Scrum is currently one of the most popular methods, mainly because of its attractive certification system and the certified trainers working in different organizations approved by the ScrumAlliance.

2.2.2 Extreme Programming (XP)

Just like Scrum, XP suggests the frequent delivery of potentially shippable increments. XP teams tend to lower the length of iterations in order to shorten the feedback loop as much as possible. XP advocates that a customer representative and the development team sit in the same room and interact regularly. Unlike Scrum, XP advocates to follow technical practices such as continuous integration, test driven development or pair programming. Those strict practices help to improve the product maintainability and the efficiency of the team (Beck & Andres, 2004).

Despite the fact that XP is a method in itself, it is not uncommon to hear teams doing "Scrum and XP" which is a combination of XP technical practices and the Scrum way of organizing the development process, its artifacts and the team structure.

2.2.3 Kanban

Closer to the Lean manufacturing approach, Kanban is a flow management system applied to software development. Unlike the other approaches, Kanban is not based on iterations. It does not necessarily encourage the regular delivery of increments. Instead it recommends the continuous delivery of finished software components (Anderson, 2010).

One of the main goals of Kanban is to detect the bottlenecks in the value delivery chain. By showing those bottlenecks, Kanban helps to identify the issues the team (or the company) needs to address in order to progress more efficiently. The sooner those problems are addressed the better.

2.2.4 Behaviour Driven Development (BDD)

BDD, which is relatively new, tries to solve one of the most common problems teams encounter when adopting the above methods. Scrum, XP and Kanban are remarkably complete when it comes to building software the right way. However, they sometimes remain vague on how to deliver software that meets the user needs and expectations.

BDD tries to fill that gap with a specific approach to requirements using "specification by example." BDD relies on scenarios as a way to describe features and suggests developing the software "outside-in." In practical terms, it means that test cases are first designed from the user perspective. Afterwards, developers drill down to the core of the system by writing the simplest code to pass those tests. This approach can help teams to focus on end-user concerns instead of jumping to technical details.

3. THE PROCESS

3.1 Hypotheses

The 60 students that registered for this course were split into 15 groups of four students. The teaching staff was composed of four teaching assistants, one professor and a specialized Agile consultant. We felt it was necessary to have the assistance and support of an external consultant to make the experience more professional and realistic for the attendees.

This project spread over 14 weeks, but only a small part of students' time was dedicated to this course. Despite this limitation, using an Agile approach should guarantee that even if there was very little work done on the product, it is usable and potentially shippable. We expected to have a working mobile application for each group, even if it had very few features.

By using an Agile approach, we wanted to see if frequent feedback and focus on regular delivery of software increments would improve the quality of the projects, the interactions within the team and with the teaching staff.

Ideally, it would have been better if each team had a designated room where they could have met at their convenience and use physical team boards on the walls, but only one large room was available. We were afraid that this would make collaboration within a team more complicated as we had to fall back on online tools to foster collaboration.

From a technical point of view, we expected each team to deliver a working and usable Android mobile application. However, the students did not have much background in Android development, not to mention the fact that groups were mixtures of students from different orientations (engineering, computer science, professional CS education). We strongly underlined the importance of producing an application that could be released to the Google Play store and hoped, at the start of the project, that half of the groups would reach that objective.

3.2 Process Design

3.2.1 Role Definition

In this section, we define each team member's role and tasks in the unfolding of the project. The roles in this project were distributed as follows:

- **The Agile Expert:** The role of the expert was to train the teaching staff, to give crash courses on Agile software development to students, provide technical and Agile support during the whole project and to participate in the final evaluation of the teams.
- **The teacher:** The teacher played the role of client. After each iteration, each team had to give a short demo of the last running version of their application to the customer and gather feedback. The teacher also participated in the final evaluation of the projects.
- **The assistants:** An assistant was assigned to each group. His role was to follow the team through the development of their application. That means participating to meetings with the customer, facilitating the meetings at the beginning of each iteration and following the group activity during each iteration. The assistants did not participate in the evaluation of their own groups. This was explicitly stated to the students at the beginning of the project. The intention was to allow a full collaboration between the groups and their assigned assistant.

3.2.2 Tools

The students and the teaching staff relied on multiple tools throughout this project. Among them, paper posters were used in order to describe the applications using hand-made wireframes and following the template depicted in Figure 1. These posters were displayed in a large room open to all computer science students. They have helped students maintain a coherent representation of their application on the one hand and to ease discussion with other students on the other.

For managing the list of functionalities, each team used a Kanban board. A Kanban board is a visual project management tool with real-time collaboration. It enables the team to maintain constantly an overview of the projects' progress and to get work done efficiently by focusing on finishing tasks. The assistants also had access to the groups' Kanban board. In this way, it gives them the means to provide students with direct

feedback on their work planning. Here we are talking about the tool itself, It should not be confused with the Kanban Agile method presented in 2.2 which is a whole process.

Throughout the project, the tasks were sorted into the columns of the Kanban board. The first column is the backlog. It includes all the tasks that the team has thought of but which have not yet been planned in any iteration. The second column contains the tasks planned for the current iteration but which are not finished yet. The third column contains the tasks on which the students are currently working. This enables students to work even when they are not physically together and keeps track of who is working on what. The fourth column contains the tasks successfully completed during the current iteration. Finally, the last column contains all the tasks completed since the beginning of the project. Placing a task in one of the last two columns can be done only when the task has been acknowledged as completed

Figure 1. Template for paper posters describing applications

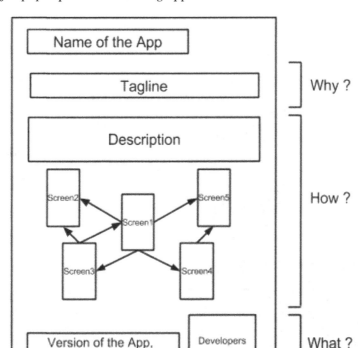

by the group members. For this project, since the context makes use of physical boards impractical, the students used Trello (n.d.), an online tool implementing Kanban principles (Figure 2).

Students developed their application with the Eclipse IDE (n.d.). We chose it for its vast and rich plugin catalog. Besides, the students used the Android SDK (n.d.) (an open-source java development kit for Android) and Mercurial (n.d.), a distributed version controlling system, to share and maintain their code. The structure of the repositories consisted of two branches. The first one, called "default," was the development branch, used throughout the entire iteration. The students used the second branch, called "master," to push functional versions of their application. Furthermore, they were encouraged to produce at least one new version on the master branch per iteration for the client demonstration. The latest functional version of their application was accessible to all the teams through a private store in order to encourage feedback between students from different groups.

Both locally using the Eclipse plugins and on a server running Jenkins (n.d.), the students continuously assessed the quality of their code with two static code analysis tools. Android Lint (n.d.) focuses on Android related conventions and provides checks for potential bugs. Checkstyle (n.d.) helps programmers write Java code adhering to community coding conventions. The remote version of the tools allowed the groups to share a common configuration.

The Jenkins server, used as a continuous integration (Fowler & Foemmel, 2006) system, was employed to automate the build of each application and configured to run the static code analysis each time a new version of the application was pushed to the repository, either on the default or master branch (Figure 3). Besides, it enables to upload the application to a private application store seamlessly.

Figure 2. Trello board example

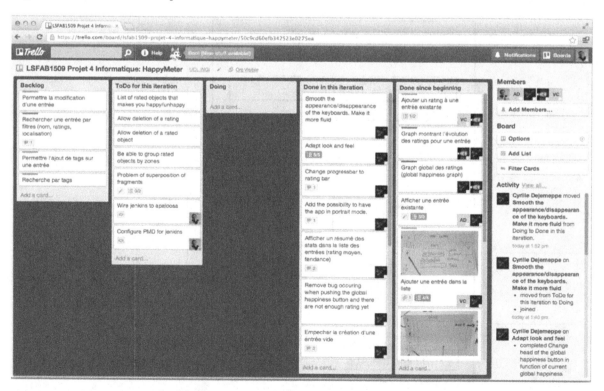

Figure 3. Jenkins showing Android Lint and Checkstyle warnings

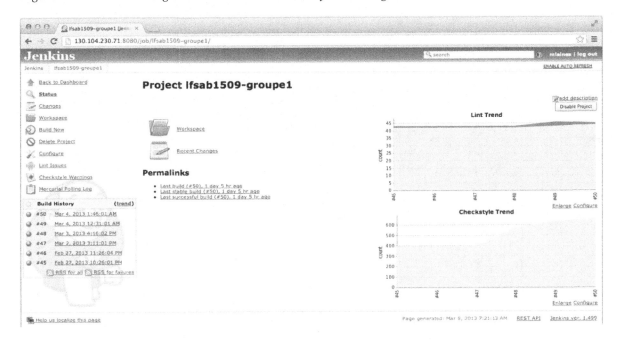

The Appaloosa (n.d.) platform enables us to set up a full-fledged private application store. This private store was used to share all applications with everyone involved in the project. Hence, each team had access to applications of other teams. The deployments of applications on Appaloosa were automatically triggered by new pushes on the master branch of repositories. Once an application was considered mature enough, it was then published to the Google Play Store (Figure 4).

Last but not least, the students have been required to license their application under an open-source license (e.g. GPL3). They were thus encouraged to use open-source libraries.

3.2.3 Practices

- **Planning poker:** At the beginning of each iteration, each group discusses the functionalities to be implemented during the iteration with the assistant. The complexity of each functionality is evaluated using planning poker. Planning poker consists in

using cards to evaluate the complexity of each functionality with respect to a reference task. Based on those evaluations, the velocity of each group is evaluated. The planning helped students have a running application exhibiting new functionalities at the end of each iteration (Cohn, 2005).

- **Commitment based planning:** One way to decide what can fit in an iteration is to take a requirement and ask the team if they think it is doable during the next iteration. If they do not agree, we try to split the functionality or we take another one and ask the same question. We then take another functionality and ask the team if they still think they can make it with this extra story. We continue until one team member thinks it is too much work. We then sum up all the complexity points of the accepted stories to estimate the team velocity for the next iteration. This practice is mostly used when there is no velocity track record for a team (Cohn, 2005).

Figure 4. Jenkins showing build states for all groups

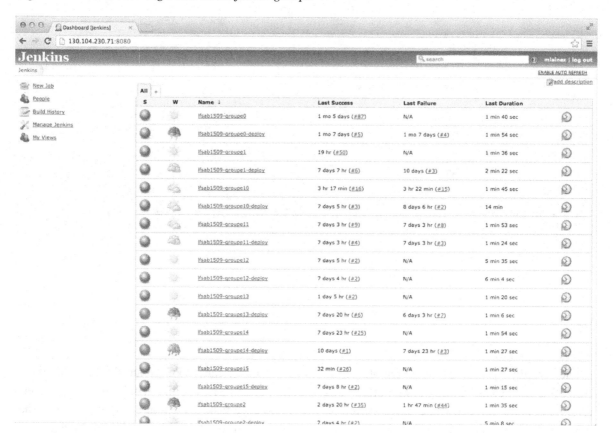

- **Demonstration to the client:** At the end of each iteration, each team gave a demonstration to the client, a role played by the teacher. It is essential for the teams to get feedback from real users and to establish a customer-team relationship. It also motivated the groups to have a running application at the end of each iteration.

- **Pair programming:** During the iterations, we recommended that students did pair programming. Pair programming consists in working by pair on the same code segment in order to improve productivity and to share knowledge on different technologies. To ease pair programming, external monitors were at the disposal of the students (Beck & Andres, 2004).

- **Retrospective:** There is always something a team can improve on. The goal of the retrospective is to identify how to improve the way of working during the iteration and to come up with actionable tasks to make things better the next iteration (Derby et al., 2006).

3.2.4 Process Overview

Scrum being the most used Agile approach in the industry, we decided it was best to stick to it as much as possible, but we also included some Extreme Programming (XP) practices and an introduction to Kanban principles. The final tailored process is divided into three phases.

Phase 1- Project Kick-Off: In order to start working on their best idea, the groups need to come up with a pool of application concepts. Each of those concepts should be presented to the teaching staff using a poster (see section 3.2.2). After the presentation, the teaching staff chooses the best and most appropriate project. From there, iterations can start in order to build the selected application.

Phase 2- Iterations: The team then gathers to brainstorm a slew of functionalities and estimates their complexity using planning poker (see section 3.2.3).

Once several functionalities have been estimated, the team needs to estimate their initial velocity. The velocity (Cohn, 2006) of a team represents the amount of complexity it can absorb during one iteration. If there is a historical record, the average of iteration velocities generally provides a good estimation, but for the first iteration, it can only be decided by the team itself. This is done using commitment-based planning (see section 3.2.3).

The functionalities then need to be prioritized to make sure that the most important tasks are tackled first. The requirements are then added in the form of cards to the team board on Trello (see section 3.2.2 Kanban).

The team is split in pairs, and each pair undertakes the task of implementing one functionality. This is simply done by assigning members to the card in Trello and moving the card representing this functionality to the "work in progress" column. It is essential that the code produced answers only to the in-progress cards. In a changing environment where requirements can evolve through time, thinking too much ahead can be harmful to the team velocity.

A functionality is considered completed when there is something to show to the customer, and the team agrees it is suitable for a demo.

At the end of the iteration, the team prepares the demo for the customer. They can only show the finished requirements. Unfinished functionalities should be dropped out of the demo. The customer is there to give feedback on the usability and accessibility, to suggest features, etc. Should the client be satisfied with the current state of the application, the demoed version is deployed on the Google play store. After the demo, the team

Figure 5. Overview of the whole development workflow

and its coach meet for the retrospective (see section 3.2.3).

Once the team has done the retrospective, they plan the next iteration, writing more requirements if necessary, based on the customers' feedback. The team usually iterates phase 2 as many times as required. In the context of this project, there were five iterations.

Phase 3- Closure: At the end of the project, a final demonstration is organized. It is followed by a full project retrospective and a last deployment on the Google play in case the customer is satisfied with this last version.

The whole process is represented in Figure 5.

3.3 Process Test

Once the process was designed, we decided to test it on a small scale in order to validate it. This test took the form of a training for the teaching assistants, whom had no experience in Agile development at that time. It was designed as a small scale version of the process itself, in which the assistants took the role of the students. Thanks to this, the assistants were able to go through what the students would experience during their project. In addition to the Agile training, this gave them some insight on the project itself, helping them to integrate complementary pedagogical skills. More generally, this test allowed them to see if the process had a good pedagogical value.

3.3.1 Training

The four assistants followed a three days training with the Agile expert. The goal of this training was to test, on the assistants, a condensed version of the real project. As for the students, the assistants had no prior knowledge of Agile development, little experience in Android and had never worked together on a project before. During this training, all the steps and tools of the project have been

used. The Agile specialist supervised the work, playing the role the assistants had to play during the actual project.

To choose an application to implement during those three days, three application proposals were made by the assistants themselves. Then a discussion with the Agile specialist resulted in the selection of one of them: "Rate your world." This application allows users to give ratings to events they attend, objects they use, meals they eat, etc. The application is also able to compute a global happiness graph for each user. As a first task, a poster of the application was made by the assistants, following the same template as the students.

The group then precisely determined the different features. Each of them was evaluated by the assistants using Planning poker, under the supervision of the Agile specialist. Then the application was developed in four iterations. The iterations were of course, shorter than the actual projects' iterations. However, the requirement of each iteration was similar to the one the students had: having a functioning prototype exhibiting new functionalities. The planning of the iterations was carried out on Trello, the Kanban application the students used. The development itself was done using Eclipse IDE using the Android SDK.

At the end of each iteration, the Agile specialist played the role of the client to provide the assistants with feedback. All along the training, the four assistants were divided into two groups, allowing them to use pair programming. The groups changed at each iteration, based on the functionalities planned for the current iteration. The code of this application was shared by the assistants using mercurial. A Jenkins server was also used. This server was used to adapt the Jenkins configuration during the training, to produce the final configuration used by the students. At the end of each iteration, the assistants published the application to show to the client on the same private application store as used during the actual project. However, the automatic deployments were not yet configured during the training.

3.3.2 Test Results

The results of this experience were positive in many respects. All the teaching staff was enthusiastic and had the opportunity to learn more about the Agile process and techniques. The training served to become better acquainted with the tools the students would use during the project. It was a chance to spot steps in the process in which the teaching staff encountered problems. Hence, the assistants were able to make the process smoother for students on these particular points.

Moreover, the test revealed some limitations of the Agile process. Frustrations or mistakes the teaching staff experienced were identified. The staff was thus able to anticipate issues and warn groups during the real project.

Finally, it gave good insight to the teaching team on what the students would be able to produce. Here are some of the teaching assistants' comments:

- All skeptical people about Agile software development should have access to this kind of training, at least to realize they are wrong regarding certain preconceived considerations.
- It's fun to see the project make progress iteration after iteration and not being forced to wait a large computer engineering period before we can begin to code.
- The cohesion of the coding group is much better when applying Agile programming.
- I prefer working in pair programming than on my own. This way, by changing pairs we all have a global understanding of the whole source code.
- The quick and continuous feedback provided directly by the client allows the application to stick with his (maybe changing) needs. This is greatly facilitated by the iteration organization of the development.

3.4 Process Implementation

Since no course in Agile development is given to the students prior to this project, all the necessary knowledge must be given to them. As described in section 3.2, the chosen development process is a reduced version of *Scrum* that integrates some concepts of *Kanban* and *XP*. As a result, we cannot simply point the students to a reference book. Therefore, we first had to teach them the concepts, as they had to put them in practice directly at the beginning of the project. In keeping with the active learning principles, some concepts were introduced before the actual start of the project and others "on the fly," after a "do-it-yourself" period.

As this was the first year the project was given, the teaching model was itself continuously tailored following Agile methodology. Throughout the course of the project, many adaptations were implemented according to feedback we got from the students (we asked for it as much as possible). This helped to improve the pedagogy for the current and future editions.

The semester was organized into five iterations. Each iteration implemented the process described in section 3.2, and lasted two weeks.

3.4.1 Semester Kick-Off

On the very first week of the semester, we quickly presented the project to the students. But mostly, we gave a presentation about the Agile Development principles and some tools the students would have to use: Trello, Jenkins, CheckStyle, Android Lint, Eclipse, Appaloosa, Mercurial, SDK Android. Indeed, in addition to concepts, the students were faced with the learning of several tools used by the industry.

The same day we organized active learning workshops. We randomly created artificial groups of students and placed them in the context of the creation of a (fake) new product: Pac-Man. This choice aimed at fostering student's acquisition of concepts and skills through the case study of a well-known application. The practices we introduced then comprised of: *dot voting, planning poker, velocity evaluation,* functional *requirements (definition, refinements and priorities, ordering and partitioning in "Must-Have" and "Nice-to-Have" groups)*. We especially wanted students to realize that it was easier to estimate the complexity of a task if it was well-defined and that discussions triggered by planning poker help improve the task definitions and their understanding by the group members.

After this first training day, groups of four students were formed. Each group had then one week to make three application proposals and create associated posters. In our point of view, this freedom in the choice of their application encouraged their involvement and strengthened their motivation. Afterwards, the different groups presented their different projects and posters to the supervising team. We then decided the best one among their three propositions. This allowed avoiding intractable projects as well as similarities between different group projects. It also provided an initial understanding of their future application, even though they were likely to be adapted throughout the semester.

3.4.2 First Iteration and Onwards

The next week, we started the first iteration. Each group met their assistant in order to get started. This first meeting was one of the most intensive as the students had to achieve the whole starting process (functional requirements, complexity evaluations, …). However, it was eased by the workshops organized the week before. Still, it took some time for some students to understand all concepts, for instance, the fact that in this framework the complexity is a relative estimation.

According to plan, the assistants met their group one week later for an intermediate meeting, in order to reinforce the comprehension of the approach by the students. For instance, some students had trouble understanding that they should not consider future task concerns during the achievement of the current task. This is a key aspect that we wanted them to learn, or at least to experience.

The demonstration of the product for the first iteration was done two weeks after its start. Some groups missed the point that each iteration product demonstration should be done as if the product was possibly the final one. This is very important in order to follow a practice like *Scrum* but also for them to learn the notion of a *finished task*.

Directly after that, the retrospectives were done. We experienced that the first retrospectives we did were the most important to help students. In fact, it is during the early period of the project that problems among groups are solved. Those problems were not only technical, but also human. We noticed that during the first retrospective some students still did not have a clear idea of the product they were about to develop. From a pedagogical perspective, this highlights the usefulness of frequent retrospectives. In this particular case, it prevents students from realizing at the end of the semester that they went in a bad direction.

From this moment, the Scrum-based iterative process begins. During the different iterations, the project was making significant progress, but also the understanding of the Agile Approach by the students was improved. Moreover, we got a better understanding of the way we should teach them this approach. In this regard, the iterative process was threefold.

After having started the second iteration, students attended another presentation about Agile Development in order to gain a better grasp of the concepts. The rest would be done in a *Do It Yourself* approach.

3.4.3 Retrospectives and Issue Resolutions

We organized a retrospective at the end of each iteration. In order to avoid repetition and to keep students motivated, we used various forms of retrospectives. Let us recall that those retrospectives served as feedback for the students and for the teaching staff.

The retrospectives played an important role in the success of the students' projects. They allowed to maintain or generate motivation, but also to resolve several problems within groups:

- **Bad teamwork**: During retrospectives, it turned out that some groups had to stop to work altogether. Some were losing efficiency because of too much verbal communication (providing possibilities for too "noisy" communication), others because they had to exchange information/ideas at a higher frequency.
- **Communication:** Some team members were not even communicating together, or were using unstructured communication systems (e.g. chatting). Retrospectives helped them to realize they did not agree on some key points, and therefore that the way they communicated had to be improved. Using Trello more frequently was one solution to that problem.
- **Organization and Planification**: Planning is a difficult task in any project management framework. Students are not spared from this problem. The final rush during the achievement of a project is something they know very well. However, thanks to regular deadlines (end of iterations) and direct feedback just after these deadlines, they got better organized and planned their work in order to do an averaged amount of work nearly every day. They were indeed able to better estimate the complexity of new tasks as well as their iteration velocity.

- **Lack of Mandatory General Consensus**: Even after an improved communication, some students did not agree on the definition of some tasks they had described together. During the retrospectives, assistants could emphasize this problem by asking someone who was not in charge of a given task to explain it to the others. The discussion was then retriggered, and the group was able to converge on a well-defined consensus. Another recurrent problem was the agreement on the notion of a *finished task* (e.g., is it implemented, tested, documented, should more be done in order to anticipate the future of the project?). Throughout the semester, the different group members refined their own definition, thanks to retrospectives.
- **Technical:** Last but not least, students faced a lot of technical problems. For instance, they were using *Mercurial* for the first time. Learning a distributed revision control tool was a bit difficult for some of them. The retrospectives allowed them to realize that; we then provided more in-depth explanations. From this, the teaching staff concluded that we should have provided a more important introduction to distributed revision control systems. This is an example of how the retrospectives helped us for the next iteration of the project. Other problems were linked to Android/Java, license compatibilities, library compatibilities or Jenkins settings. In this regard, this project allowed students to experience real problems that can occur in professional life. This is something that is sometimes neglected by some University education courses.

Those kinds of problems were detected and solved as soon as possible, so that the students were unblocked quickly, allowing them to work efficiently. Another advantage is that even if a problem was missed or only partly solved, the next retrospective could still solve it.

3.5 Results and Evaluation

As this project was implemented in an academic course context, students were asked to release precise deliverables on which they would be evaluated. Every group had to release the following deliverables:

- The three posters sketching the ideas of mobile applications among which the final one was chosen.
- The final retouched poster for the chosen application.
- An evaluation criterion chosen by the group on which the group would like to be evaluated.
- The Kanban used for the project (as students used Trello, its content was used).
- The source code for the mobile application.
- A short description and user manual of the application (similar to the description page of applications that can be found on the google Play store).
- A short video introducing/demonstrating their application.
- An auto-evaluation table filled by the group.

In addition to these deliverables, each group had to perform a short presentation in front of the teaching staff to show the results they obtained and explain the difficulties they met and how they worked during the project. Finally, students had to present an individual exam in order to pass the course.

The evaluation also took into account the involvement and motivation of the group. As such, groups whose application had been submitted on the Google Play were rewarded with a grade bonus.

In order to exchange ideas with the Agile community, a user group (User Group Agile Belgium)

around this new project and the introduction of Agile Development in the university was organized on 24th March 2013, at UCL. In addition to trigger interesting discussions with the Agile community, it provided some additional feedback from the students. As it was not part of the evaluation of the project, the bias of the feedback was reduced.

3.5.1 Results

From the 15 different groups, 13 have published their application on the Google Play. The quality of the submitted applications was of the same level as many commercial applications.

The two applications illustrated in Figures 6 and 7, namely *Bouboule* and *LLN Campus*, were of very high quality and the teaching staff asked the students to develop an iOS version (a version running on apple iPad and iPhones) of these apps.

Several groups decided to keep on developing their application after the end of the project. Such involvement of the students is quite rare and shows how much they appreciated the course.

3.5.2 Insights

This experience was enriching for the students and also for the teaching staff.

The Agile approach gave the opportunity to students to learn a new effective way to work as a group and to identify and resolve conflicts as soon as they appear. Furthermore, the ongoing follow-up of the project by a client gave students the opportunity to get an insight of what real enterprise demands can be and how they can affect a project. As the tools the students used are tools commonly used in enterprise environment they were also confronted with real world tools and could see the kind of problematic these tools involve. Finally, the contrast of this course compared to the classical academic approach of other courses brought some freshness to their training.

Several students said at the end of the project they intended to reuse several of the Agile pro-

Figure 6. Screenshots of Bouboule, a game in which the player is a sumo-ball whose goal is to push the adversary out of the playground

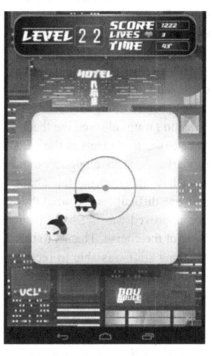

Figure 7. Screenshots of LLN Campus, an application helping students of UCL to access the university services such as course schedules, library opening hours and locations of the auditoriums

gramming mechanisms and the tools they used in the project for future assignments and personal projects. This showed us that the course was a step in the right pedagogical direction and comforted us in the idea to repeat the experiment.

From the point of view of the teaching staff, the course also brought a lot of positive new experiences. The fact that each teaching assistant was assigned as a *coach* to groups allowed the teaching assistants to be closer to students as they did not impact their marks. As such, a huge amount of feedback about how the students felt about the project, if they had difficulties and what they thought could be improved was available from the very beginning of the course. Thanks to this feedback, the teaching staff was able to rectify some matters to allow students to have a better experience of the course. The teaching staff also discovered the joys and pains of being the coach of a team on a given project. Some groups were more reactive than others and were able to manage themselves with almost no help from their coach while other groups really needed to be told explicitly what to do.

3.5.3 Improvements

As this was the first year the course was given, multiple aspects can be improved for years to come. First, the whole teaching staff now has more experience with their respective role in the project (either coach or client). Some situations when a problem occurs in a group should now be handled more rapidly to be able to resolve the problem quickly.

Several tools we used during this project were new to most of the staff and the interface provided to students to use these tools could have been clearer. In the future years, more time should be allotted for tutorials to ensure students fully understand how the tools work. Furthermore, the interface provided to the students should be easier to use and better defined.

Finally, the points on which students will be evaluated at the end of the course should be better defined and more specific to give students more insights on the specific points they should pay attention to.

4. FUTURE RESEARCH DIRECTIONS

In this chapter, we showed that software engineering education could easily be introduced early in the computer science curriculum in a pure project-based learning approach. The students were very motivated by this active learning approach. We also chose to introduce Agile software development, a real world industry practice. This project was also an opportunity to address and cover transversal skills, such as communication, teamwork, modeling and planning.

Our experience is that project-based learning is appropriate to introduce Agile software development. It is clear that such a project should be complemented with a course tackling more formal aspects of software engineering. Such a course is also provided in the curriculum. The described experiment is unfortunately too young to analyze its impact on the professional practice of our students. We, however, believe we shorten the gap between the industry expectations and what universities can provide in software engineering education.

5. CONCLUSION

The Louvain School of Engineering introduced in 2012-2013 a one-semester project for students. The form of this project, for students in Computer Science, was to develop an Android application using a lightweight Agile development process. This process has been designed to be a reduced version of *Scrum* incorporating some concepts of *Kanban* and *Extreme Programming*. The students used many industrial tools to ease the development of their application: Eclipse, Mercurial, Jenkins, Appaloosa, etc. The project goal was for the students to acquire technical skills about Android and the Agile framework as well as cross-disciplinary skills such as modeling, teamwork, planning, management and communication.

This project is perceived as a success by both the teaching staff and students. Indeed, students were able to learn, refine and master several skills and showed a huge commitment to the accomplishment of their project. Every group of students released a high quality application. These applications, for the most part, were professional enough to be published on a public store and compete with other commercial apps.

The hypothesis we stated at the beginning of the chapter that an Agile programming approach was well suited for university students has been verified. Students and supervisors were both pleased to use the Agile paradigm. From the teaching staff point of view, getting a lot of feedback and being close to student groups has helped to increase the quality of the teaching. From the students' point of view, being confronted with real world demands such as strict deadlines, team working, client feedback, tools used in the industry, etc. has allowed them to enrich their university curriculum. The fact that several students have expressed their intention to reuse the tools and Agile techniques used in the course is an achievement in itself, although a more in depth follow-up should be put in place in future experiments to make sure the course had a significant impact on their practices.

The investment of the teaching staff to set up this project course was high for this first experiment. However, the investment needed for future experiments should be reduced due to the experience gained over time by the teaching staff. The students were able to create applications that have exceeded the expectations of the teaching staff and even received positive ratings from external people.

The teaching staff recommends other teachers to run the same experiment for several reasons. Firstly, the teaching approach brings some freshness compared to traditional university courses. Secondly, early feedback in the project allows to rectify problems and to enhance the monitoring of the students. Thus, students are less likely to get lost. Finally, their involvement in such a project is rewarding for the teaching staff and gives a better overview of the students' progresses on both technical and *soft* skills.

REFERENCES

Anderson, D. J. (2010). *Kanban: Successful evolutionary change for your technology business*. Blue Hole Press.

Android. (n.d.). Retrieved from http://developer.android.com/tools/sdk/eclipse-adt.html

Android Tips. (n.d.). Retrieved from http://tools.android.com/tips/lint

Appaloosa. (n.d.). Retrieved from http://www.appaloosa-store.com/

Beck, K., & Andres, C. (2004). *Extreme programming explained: Embrace change*. Reading, MA: Addison-Wesley Professional.

Boud, D. J., & Feletti, G. (1998). *The challenge of problem-based learning*. New York: Routledge.

Cohn, M. (2005). *Agile estimating and planning*. Upper Saddle River, NJ: Pearson Education.

Derby, E., Larsen, D., & Schwaber, K. (2006). *Agile retrospectives: Making good teams great*. Pragmatic Bookshelf.

Eclipse. (n.d.). Retrieved from http://eclipse.org

Felder, R., & Brent, R. (2009). Active learning: An introduction. *ASQ Higher Education Brief, 2*(4).

Fowler, M., & Foemmel, M. (2006). *Continuous integration*. Retrieved from http://www.martin-fowler.com/articles/continuousIntegration.html

Jenkins. (n.d.). Retrieved from http://jenkins-ci.org/

Johnson, D. W., & Johnson, R. T. (1999). *Learning together and alone: Cooperative, competitive, and individualistic learning*. Upper Saddle River, NJ: Pearson.

Prince, M. (2004). Does active learning work? A review of the research. *Journal of Engineering Education*, *93*(3), 223–231. doi:10.1002/j.2168-9830.2004.tb00809.x

Schwaber, K. (2004). *Agile project management with Scrum*. Microsoft Press.

Selenic. (n.d.). Retrieved from http://mercurial.selenic.com/

Sourceforge. (n.d.). Retrieved from http://checkstyle.sourceforge.net/

Trello. (n.d.). Retrieved from http://trello.com

KEY TERMS AND DEFINITIONS

Agile Software Development: A software development process model that focuses on evolving user requirement.

Behaviour Driven Development (BDD): A software development method that is based on the test-driven approach.

Extreme Programming (XP): One of the most successful agile software development methods that promotes peer-programming.

Kanban: A flow management system that is applied to software development.

Scrum: An iterative approach to software product development.

Section 6
Engaging Classroom Games

Chapter 15
ECSE:
A Pseudo–SDLC Game for Software Engineering Class

Sakgasit Ramingwong
Chiang Mai University, Thailand

Lachana Ramingwong
Chiang Mai University, Thailand

ABSTRACT

Software development is uniquely different especially when compared to other engineering processes. The abstractness of software products has a major influence on the entire software development life cycle, which results in a number of uniquely important challenges. This chapter describes and discusses Engineering Construction for Software Engineers (ECSE), an effective workshop that helps software engineering students to understand some of these critical issues within a short period of time. In this workshop, the students are required to develop a pseudo-software product from scratch. They could learn about unique characteristics and risks of software development life cycle as well as other distinctive phenomenon through the activities. The workshop can still be easily followed by students who are not familiar with certain software development processes such as coding or testing.

INTRODUCTION

The intangibility of software makes it a highly unique product (Project Management Institute, 2008). Undeniably, the development of software has a number of different traits compared to other engineering products. In a software development project, several issues, such as potential frequent changes of requirements, low product visibility, inappropriate development models, and the need for customer involvement, are critical (McConnell, 1997; Schmidt, Lyytinen, Keil, & Cule, 2001; Tiwana & Keil, 2004). Although these challenges can be addressed in traditional lectures, it is highly unlikely that the students can actually follow and understand their practical seriousness. For example, a lecturer can describe how difficult and costly it is if a major change surfaces during the

DOI: 10.4018/978-1-4666-5800-4.ch015

last stages of development. Yet, it is not easy to imitate other associated issues such as conflicts between stakeholders, frustration and the importance of problem-solving and negotiation skills.

Indeed, the most effective method to teach software engineering is to have the students learn through an actual hands-on project. Yet, regardless of whether a traditional or agile model is chosen, the implementation usually takes days before the results can be clearly seen. Furthermore, hands-on project could become considerably less effective if the group is bigger. Additionally, since general hands-on projects mostly focus on coding, students who have less relevant skills usually feel uncomfortable and subsequently fade away from the workshop. Indeed, this could be one of the least desirable outcomes from the class.

A number of researchers have attempted to implement games in their classes in order to overcome such challenges (Caulfield, Xia, Veal, & Maj, 2011). These games can be roughly divided into two groups, i.e. traditional games and computer-based games. Traditional games involve activities in which the students can participate by using convenient physical tools such as paper, scissors, boards, cards and dice. On the other hand, computer based games uncomplicatedly refers to games which the students need to play via a computer application. Some of these games can be played in groups while others support only a single player mode. Moreover, the settings and requirements of these games generally vary. Many of these software engineering games are flexible and can be further tailored to match class objectives.

In-class competition can be an important factor to increase the workshop's effectiveness (Hainey, 2009). With this factor included, the students are more likely to put more attention to the class. They also tend to perform their actions more seriously and carefully.

This chapter introduces *Engineering Construction for Software Engineers (ECSE)*, a game that attempts to teach and simulate a complete con-

cept of software development life cycle in only two and a half hours. Instead of developing real software, the students are instructed to build a model house from corrugated plastic board. During the activity, the participants can learn basic knowledge of software engineering and the Software Development Life Cycle (SDLC). Although software development skills are not required, it can greatly benefit the team. ECSE also implements a currency and resource management system in order to increase fun and competitive factors. The students are required to plan and appropriately allocate their budget. There is no limitation on implementation strategies and approaches. The winner is, undoubtedly, the group which makes the most profit from the entire process.

BACKGROUND

The software development life cycle (SDLC) varies based on the nature of software developers and software organizations. The classic SDLC consists of five major phases i.e. requirements, design, construction, testing, and maintenance. This entire cycle can be further tailored based on business needs. Common modifications of the SDLC include the expanding, grouping, and revolving of existing phases as well as adding specific activities such as initiation, prototyping, and retrospection.

On the other hand, agile software development models have their own manifesto. They place emphasis on frequent working product delivery and do not follow the classic SDLC sequence. Agile practitioners value changes, interactions, and collaboration above plans, tools, and contracts (Beck et al., 2001). Yet, agile and traditional developments inevitably share similar activities. Indeed, in the same way as the implementation of the traditional SDLC, agile processes can also be tailored based on business needs and development environments.

In education, games and many other physical-based activities are commonly used to enhance the student's experiences (Caulfield, Xia, Veal, & Maj, 2011). These alternative approaches can be even more effective than traditional lectures. Games are widely used in various fields of study. For example, *Beer game*, a game which focuses on supply chain management of a beer production, is arguably one of the most popular simulations in both education and the industry sectors (MIT, n.d.). *Innov8* is an intriguingly visualized Web-based game (IBM, n.d.). Several case studies showed that it can provide insightful business process management experience. In addition, *Platform War*, a single player online game that imitates a business scenario of harsh competition in console gaming industry, offers an exciting practice on various marketing strategies (MIT Sloan, n.d.). Other games involving change management, production management, system security, and environmental management are also available online (Forio, n.d.).

Similar to other disciplines, in software engineering, many games are invented to simulate the software development process. For example, *Robocode, the Incredible Manager, Problems and Programmers*, *SimSE* and *MIS Project Manager* are a few of the many notable games. They are described in more detail below.

Robocode is a game which aims to improve the coding skills of software engineers (Sourceforge, n.d.). The players are required to write code for a program that orders a tank to act according to specific battle scenarios. The ultimate goal of the player is to destroy other enemy tanks in the battle arena. This game mainly focuses on the coding and testing activities. Yet, the players need to intuitively repeat most of the SDLC phases a number of times in order to continuously improve their code.

The Incredible Manager is a simulation-based game developed for project management training based on adventure and puzzle (Dantas et al. 2004). Similar to other project management games, students play a role of project manager who plans and controls software projects within schedule and limited budgets. In the beginning, players are presented with the project description documentation. Then, they create a project plan and send to stakeholders for acceptance. When the project is executed, players must be aware of its behavior and take actions where necessary. The participants learn the cycle of project development throughout the activity.

Problems and Programmers is a traditional card-based game aiming to simulate a software development process based on the Waterfall Model (Baker, Navarro, & Hoek, 2005). Students are divided into groups. Each group needs to define the most appropriate financial, management, and strategic policies in order to maximize their development results. Randomness is one of the key competitive elements in this game. The player might encounter different development situations based on their preceding decisions. One playthrough of this game is equal to one cycle of the SDLC experience. High development skills are not required since the players do not need to perform the actual development activities.

SimSE is a stand-alone computer game which emphasizes on simulations of different SDLC models (Navarro & Hoek, 2007). The players can choose either the Waterfall Model, Incremental Model, Code Inspection Model, Rapid Prototyping model, Rational Unified Process, or Extreme Programming. Simulation environment, such as facility setup, human resources, and development tools varies towards different development models. The students need to understand concepts of the selected models and appropriately allocate their resources in order to maximize their score. Similar to the previous game, the players do not need to possess high software engineering skills, because no actual development activities are performed.

SimjavaSP is a software project management and software process simulation game (Shaw & Dermoudy, 2005). In SimjavaSp, projects are modeled by given activities, properties and events. Students are given a role as project manager to develop a software project with allocated resources.

Each game ends when the project is completed or the manager run out of time or money. The students found the game useful, particularly at the introductory level. Though authors claim that the game is appropriate for teaching software process concepts, the game is more into project management than software process as little details on software process are shown.

Second Life, unlike others, is a 3-D online software engineering game that supports multiple players. The game requires students to have prior knowledge of all phases in traditional software engineering (Ye, Liu, & Polack-Wahl, 2007). Players have to play in one of the six engineer roles, and interact with the others through game. The team is evaluated at the end of the game, not only by the product, but also the collaboration. The ability to communicate is a prominent aspect of this game. However, players sometimes suffer from the delay due to network connection, and the software and interface need a considerable amount of improvement. There is also a limitation when competing as the scoring criteria are not balanced for different sized teams. The students respond that the game only slightly helps them improve their knowledge on software engineering process. They also stated that even with improved graphics, the performance of the game might not increase much. This can be implied that a virtual environment, especially close to reality one, does not play an important role in understanding of software developing process.

MIS Project Manager focuses on the project level of software development (Hear-See-Do, n.d.). Therefore, the players need to possess certain knowledge on not only SDLC but also on project management. This game is a single-player Web-based game. The players are required to define their strategies and choose appropriate options based on unexpected situations that might surface in each of the development phases. Each option results in predefined outcomes which contribute to the final game scores. No actual software developments skills are required to play the game.

Most of these aforementioned games interestingly attempt to replicate a cycle of software development. The players are required to follow the predefined objectives, requirements, and development steps. The amounts of effort required for each game are varied, i.e. 1.5 hours for Problems and Programmers, 2.5 hours for SimSE and 3-4 hours for Incredible Manager. Although most computer-based simulation games do not take long to play, the results of using these games heavily rest on the ability of games to simulate and provide appropriate feedback to provoke knowhow, understanding, learning, and eventually awareness. Another limitation of predefined games is that options and their outcomes are largely predefined. This is arguably different from actual development scenarios when many unexpected issues can surface at any time, by any stakeholder, and the development team needs to perform unforeseen actions in order to solve such challenges. Tide-turner action such as customer negotiation, change of development model and outsourcing are not applicable in these games. This restricts the outcomes of learning, especially problem-solving skills that are better stimulated via case studies (Hainey et al., 2011).

As an alternative to those activities, this book chapter proposes a physical game which focuses on software development cycle. It aims to provide an activity which offers maximum flexibility and adaptation while delivers essential knowledge and awareness to the participants within a short period of time.

ECSE: A PSEUDO-SDLC GAME FOR SOFTWARE ENGINEERING CLASS

Overview of the ECSE

Engineering Construction for Software Engineers (ECSE) is designed to simulate a cycle of software development. Important elements, e.g. requirement analysis, development model selection, and customer negotiation, are the main focus of this game. In ECSE, the students are required to physically build a "house" from lightweight disposable materials, such as corrugated plastic, cardboard, and sticky tape. They do not need to possess knowledge on SDLC to participate in this game. Yet, appropriate implementation of SDLC could significantly improve the outcome of the activity. Throughout the game, the students will learn the entire cycle of software development. Although ECSE is originally designed for software engineering students, it can be efficiently implemented to educate other non-software engineering attendances on SDLC.

The main pedagogical objective of ECSE is to develop awareness and understanding of SDLC under a real life environment. The participants are expected to understand the importance of each SDLC processes and major problems which usually surface during software development. The students will learn to foster their team working, communication, and resource management skills as well as attitudes towards changes, risks, and conflicts. The difficulty of the game and the level of learning outcome can be totally controlled. The moderator can choose either introductory, intermediate, or advance level of simulation is needed based on current knowledge of the attendances.

One major advantage which ECSE offers is the sophisticate nature of requirements in software engineering. Generally, software engineering students might not have much chance to encounter changes since typical assignments come with frozen requirements. However, in this game, requirements not only change, but they can also be difficult to elicit, prioritize and manage. Important risks such as lack of customer involvement, lack of staff experiences, and gold-plating can be flexibly included in any appropriate parts of the activity.

The building of a house from corrugated plastic board imitates a software development process interestingly well. Both the developers and customers are instructed to perceive the house as a software product. Unlike actual real estate, the plastic house can be built from either top-down, bottom-up, or separately as modules.

Each part of the house can be individually shown to the customer when it is ready. Each house part can be dissembled, reused, or refactored if needed. This flexibility allows either traditional or agile development to be implemented. Also, the students can position themselves as a classic functional-based or a cross-functional team. This provides an arguably limitless set of development strategies and approaches for the players.

"True requirements" are the heart of this game. It is critical to let the participants focus on learning of development experience and subsequently solving a number of practical development issues which are not easily described and inspired by words. During the game, the students are divided into groups. Each group is given an envelope of true requirements defined by the customer from the beginning. These requirements contain the only actual functions which will be tested during the acceptance test. The main and the most challenging mission of ECSE is to elicit and accordingly develop these true requirements.

ESCS is essentially not a resource management activity. However, in order to provide fun and competitiveness, an equal but limited amount of budget is allocated to each group of students. An additional loan can be requested if needed. Each group gains profits from meeting the true requirements. Ultimately, the group which completes the game with the highest remaining capital wins the game. Attendance from any field of study can participate in this activity. Yet, students with adequate software engineering skills, development techniques, and SDLC knowledge are likely to be more competitive.

Stakeholders

Four groups of stakeholders play important roles in ECSE. Firstly, the participants are positioned as software engineers in a developer team. Each team consists of 5-6 players. The players are free to determine their development model and their team structure. Their responsibilities are assigned based on their own choices. For example, if an agile model is implemented, each team member might be positioned as a cross-functional staff. As aforementioned, although the developers do not need to possess actual technical skills, they should have basic knowledge on software development.

The customer is the second party who defines requirements and settles the acceptance. In this game, there can be more than one customer. For example, the customers who are expecting the house might be the father and the mother of two children. Despite the clearly defined true requirements, each customer may perceive the detailed features of their intended product and prioritize them differently. The customers are expected to answer every question asked by the developers. Certainly, sometimes they can make errors and uncertainties while other times they can be very helpful and dependable. Each meeting with customers often results in additional requirements or changes. The relationship between the development team and the customers evolves throughout the game. Indeed, the customers are the key stakeholder in ECSE. They are responsible for continually leading the game direction, introducing unexpected situations at the most impactful time, as well as adjusting the overall difficulty level of the game. Therefore, this role should be ideally played by a person with customer relationship experience in actual software development.

The third and the forth stakeholders in ECSE are the store owner and the bank, respectively. These stakeholders play a minor but important role in this game. The store owner is responsible for supplying each development team with equipment and tools. The bank is an optional stakeholder who is appointed when the loan is enabled. Both of these roles can be semi-automated if an honor code is applied.

Equipment and Tools

ECSE requires basic office materials and tools such as corrugate plastic boards, sticky tape, cutters and scissors. During the activity, all of these materials and tools can be bought from the suppliers.

Corrugate plastic board is the best option as the main material in ECSE, because it is light, strong, and can be easily cut into shapes. Additionally, this board is colorful and less expensive. Other feasible replacements include colored cardboard and similar materials. The variation of colors tends to be one of the key learning experiences from this workshop. Corrugated plastic board is used to make both the structure and decoration of the house. In ECSE, a 30 x 40 cm sheet of corrugate plastic board is just adequate to build the house.

Obviously, the main purpose of sticky tape is to stick the plastic board together. Nevertheless, sticky tape can also be used as a replacement for the plastic board in some cases, i.e. windows or transparent components. Other optional tools include staplers and pins for joining the materials. Wooden chopsticks can also be used in order to further fortify the structure.

Cutters and scissors are the main tools for cutting the plastic board. Although their purposes are the same, the implementations are largely different. Indeed, some participants may be more familiar with one tool than the other. This situation is similar to tool selections in software development.

Since ECSE is a competitive game, all purchasable tools and materials are available in a limited quantity. Each of them is sold in the first-come first-serve fashion. As a result, the teams need to approach the store quickly, although this rarely happens in real world software development. Yet, the purchase can be done any time during the game if there are remaining stocks. Trading between teams is not prohibited. Sample costs of tool and material are proposed in the next section.

Game Modes and Settings

Time, budget and functionality are the three main success factors for software development. They are also the key setting for this workshop. Table 1 illustrates two possible ECSE game modes.

Table 1. Basic ECSE modes

Game mode	Normal	Hard
Initial setting		
Group member	5-6 person	5-6 person
Starting budget	$200	$100
True requirements	4 items	6 items
Payment for each true requirements	$500	$400
Total possible payment	$2,000	$2,400
Size of the house	20 cm x 20 cm	20 cm x 20 cm
Time	1.5 hours (3 virtual months)	1.5 hours (3 virtual months)
Maximum bank loan	$1,000	$1,500
Interest rate	10%	10%
Late delivery fee	$100 per minute (1 virtual day)	$100 per minute (1 virtual day)
Tools and materials		
Corrugate plastic board (40 cm x 30 cm)	$100 each	$100 each
Cutter	$100 each	$100 each
Scissors	$50 each	$50 each
Chopstick	$50 a pair	$50 a pair
Sticky tape	$50 a roll	$50 a roll

As listed in Table 1. The main differences between these two models are the starting capital and number of the true requirements. With more requirements and less budget, the hard mode is considerably more difficult than the normal mode. It can be clearly seen that in the hard mode, the players are mostly forced on making a bank loan unless they share the material with another team. Half an hour in real-time presumes one month in ECSE.

ECSE is very flexible. All of the settings can be adjusted to suit the length of the workshop and the number of participants. However, the relationship between the starting budget and the number of group members needs to be balanced. Each team should have just enough starting capital to afford only one sheet of corrugate plastic board and a few pieces of tools. In this way, the team needs to plan their procurement, cash flow, and subsequent usage of equipment carefully.

The one and a half hour of development is optimum. If a team is unable to deliver the product before this deadline, an extension can be requested. Unless the negotiation is success, a fine will be charged based on the delay of delivery. A suggested rate of fine is $100 per minute or one virtual day. This huge amount of fine forces the team to cautiously consider this choice.

Implementation

The implementation of ECSE can be divided into three phases, covering approximately two and a half hours as in Figure 1. The first phase of thirty minutes involves introduction and planning. The instructor spends the first fifteen minutes giving instructions on how to play, describing objectives of the workshop, acceptance policy, game setting, and costs of equipment. Another fifteen minutes are given to the participants to form their group, assign responsibilities, choose tentative development model, plan resources and, purchase their equipment. An envelope containing all of the true requirements is then presented to each team. Yet, they are not allowed to open it at this stage.

After the introduction phase, the development phase commences. Each team has ninety minutes of real time or three virtual months to complete their implementation. Their main activities in this phase include requirements elicitation, design, development, and testing. This can be done sequentially in a traditional approach, cross-func-

302

Figure 1. Three phases of ECSE

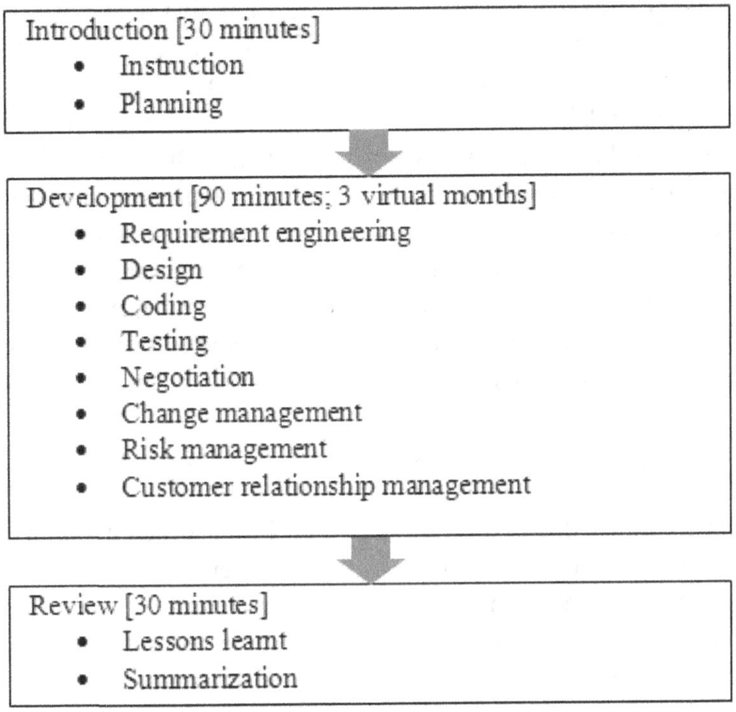

```
┌─────────────────────────────────────────────────┐
│ Introduction [30 minutes]                         │
│     • Instruction                                 │
│     • Planning                                    │
└─────────────────────────────────────────────────┘
                        ▼
┌─────────────────────────────────────────────────┐
│ Development [90 minutes; 3 virtual months]        │
│     • Requirement engineering                     │
│     • Design                                      │
│     • Coding                                      │
│     • Testing                                     │
│     • Negotiation                                 │
│     • Change management                           │
│     • Risk management                             │
│     • Customer relationship management            │
└─────────────────────────────────────────────────┘
                        ▼
┌─────────────────────────────────────────────────┐
│ Review [30 minutes]                               │
│     • Lessons learnt                              │
│     • Summarization                               │
└─────────────────────────────────────────────────┘
```

tionally in an agile fashion, or even randomly in an ad hoc style. The customer should be available and accessible all the time during the workshop. The product can be delivered prior to the deadline if needed. Additionally, in order to simplify the business aspect of this game, the payment is rewarded when the house is completed. Moreover, as previously stated, if the team is unable to meet the deadline, they are allowed to negotiate for extension and pay a heavy late delivery fee. Other relevant activities such as change management, risk management and customer relationship management also contribute to the team's success.

The last phase of the workshop includes thirty minutes of process reviewing. The true requirements are finally revealed. The instructor summarizes his/her observation and the key achievements of the workshop. A representative from each team gives a quick presentation on their lessons learnt and key success or failure issues. One of the main questions they need to answer is what they would do if they were to repeat the game.

True Requirements and Key Learning Points

True requirements depict the key learning points and objectives of the workshop. This is to be determined by the instructors based on their perception of the participants. True requirements should be related to situations and experiences which are not effectively enlightened in traditional lectures and can be frequently encountered in practices. The level of difficulty of the true requirements can be varied. A balance between easy and difficult true requirements is essential. In fact, at least one of the true requirements should be easily achievable, which serves as reward to encourage the students. Furthermore, the true requirements should be determined based on relevant local cases and customer behavior. Examples of the true requirements and their learning points for Thai software engineering students are described as follows:

- **Two floors building (easy):** It is natural that the developers will ask how many levels of the house are to be built. It is also easily achievable if the customers clearly define this from the beginning. Yet, in the context of software development, gold-plating from the customers is expectable. Many customers would state that they want as many functions as possible within the budget without realizing that they only need a few cores. In this game, the customer may request a three story house before changing to the needed two floors later. The students will ultimately realize the importance of requirement engineering, change management, and negotiation.

- **Red walls and yellow roof (easy):** Another common inquiry about the house is the color of the walls and the roof. Some students might go into details such as different paint in each room. Yet, the color of the wall is usually only for decorative purposes. In this game, color is very costly. The team which overly focuses on color is likely to unnecessarily spend much more on materials. This lesson is to teach the participants to logically balance their costs on cosmetics and core functions.

- **A back door (easy):** Back door is unquestionably an essential part of a house. However, it is often overlooked. In most customers' point of view, a back door must be mandatory and they do not need to formally inform the engineer. This scenario is similar to software development when certain functions are not identified in the requirement document, because the customer thinks they are automatically included. For example, a customer describes that record A can be added and updated by the administrator. In fact, he or she might also think that the keyword "updating" already includes deleting. As a result, the function "delete" is not included in the document. This simple miscommunication can cause major disruptions later during software development. This true requirement will remind the students to not underestimate basic requirements.

- **Three sleeping rooms which are connected to bathrooms (medium):** Usability is one of the key elements to customers' satisfaction. A system which provides adequate functions but is not user-friendly can result in a disruption. In ECSE, the floor layout which maximizes customers' usage is very important. An example setting is a scenario of a four-person family where three bedrooms are needed. One large bedroom is required for the parent and another two bedrooms for each child. For usability reasons, all of them must be directly connected to at least one bathroom. Any corridor which appears to separate the bedroom from the bathroom would result in a failure of the acceptance.

- **A bedroom spacious enough to place a mouse (medium):** In many cases, the customers might need some of their rooms to be extra-large. The word "large" is ambiguous and might be perceived differently by the customers and the developers. In this case, the true requirement clearly identifies that the main bedroom must be large enough to fit a computer mouse or notebook battery. Indeed, this requirement could not be easily tested until the floor is completed. Fail to achieve this goal can result in a major renovation of the entire floor.

- **A hip roof (difficult):** Hip roof is a roof style which all four sides lean downwards to the walls (see Figure 2). Although this

Figure 2. A hip roof

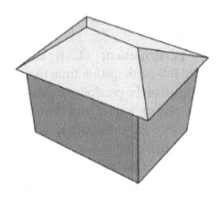

type of roof is not uncommon, it is unlikely for software engineering students to recognize. Indeed, most students would be more familiar with a flat, single-sloped or gable (two sides leaning downwards) roofs. This is similar to a scenario where an uncommon technology is involved. Furthermore, although the customers know what they want, they are unable to clearly explain their requirement due to their limited technology knowledge. All they can assist is confirming that a design or finished roof is acceptable or not.

- **A convenience store on the ground floor (difficult):** It is natural for engineers to perceive that a house they are constructing will be used for living. However, this can be only partly true. In ECSE, the customers have two major goals for their house, i.e. accommodation and trading. In fact, a house with a convenient store of the ground floor is not uncommon in Thailand. As a result, most utilizable space on the ground floor is restricted as a store front and a storage room. In order to achieve this objective, the software engineers need to start their SDLC by truly understanding the customers' core business objectives as well as the ultimate goals of their product.

Evaluation, Observation, Feedbacks and Lessons Learned

ECSE was successfully implemented for six times during 2011 to 2013 in the Department of Computer Engineering, Faculty of Engineering, Chiang Mai University, Thailand. It was organized for five times in the undergraduate class of software project management. ECSE was also participated once by graduate students in logistics and supply chain management in the class of information systems for logistics. Approximately twenty students participated in each batch. As a team, the students successfully learn the nature of software development, its associate risks, as well as essential soft skills. Most of the students were able to achieve easy requirements. In contrast, only a few groups of participants managed to clear the difficult true requirements. Some of the teams went bankrupt since they were unable to satisfy the customers. Yet, the feedbacks from the classes were exceedingly positive. Several students highlight this workshop in the course evaluation as "ECSE efficiently gave me a nice touch of software development in real life" and "The last workshop summarized the software engineering path really well." Major lessons learned from observation of these implementations are listed as follows:

- **Agile vs traditional:** Due to the nature of frequent delivery, the participants who adopt agile development appeared to be more successful in discovering the true requirement than others. On the other hand, most students who either intentionally or unintentionally implemented traditional development tended to be less effective in handling changes of requirement. Most traditional teams started their processes rather slow thus they needed to rush their development during the final virtual month.

Final products from such teams were largely flawed and were unable to fulfill the requirements due to the lack of customer involvement. Many students stated that they would definitely include more of agile elements in their future implementation.

- **True requirements:** The students were all excited and enthralled when the true requirements were revealed despite being unable to fulfill all of them. Easy true requirements such as a backdoor were interestingly overlooked by many students. Many of the participants admitted that this workshop effectively made them understand the importance of customers' business goals. They additionally realized that common sense and perceptions of stakeholders can be largely different.

- **Communication and relationship management:** In real world software development, communications between stakeholders is crucial. Likewise, communications plays a very important role in ECSE. Although communications regularly led to changes requirements, they are essential. Students also noticed that frequent communications greatly improved their relationship with the customers. Indeed, communications and relationship management are key elements which subtly contributed to the success of practical software development. These elements are especially important in patronage communities such as in most Asian countries.

- **Adaptable difficulty:** Since almost all directional activities are controlled by the customers, the difficulty level of the game can be adjusted to suit each group's proficiency. For a group with high skills and using a proper approach, the customers can portray themselves to be inexperienced, misleading, erroneous or demanding. In a worse case, a customer can even request for changes on their approved requirements.

This adaptable difficulty plays an important role for the competitiveness of each session.

- **Full participation:** ECSE interestingly gained full participation from the students. During all of the previous workshops, none of the players were found distracted. This is arguably very difficult to achieve in a traditional lecture session. Based on observations, many other game-based workshops, especially the computer based, have struggled to gain full attention from the participants.

- **Efficient simulation of the entire SDLC:** The participated students designated that ECSE was a highly efficient workshop for the replication of the entire SDLC processes. The development of a house can be practically perceived as a software development process with no trouble. Many essential elements of software development, especially involvement of customers, are also efficiently simulated.

- **Software engineering students vs non-software engineering students:** Basic performances of software engineering and logistics students do not significantly differ. Since none of the logistics students have previous knowledge on SDLC, they naturally followed a waterfall process. Participation between developer and customer was kept minimized. No prototype was offered. Although their resource management skills were excellent, almost all focus of the development was unnecessarily spent on cosmetics. Likewise, a number of software engineering students who implemented a waterfall development also suffered from similar problems. In contrast, software engineering students who adopt concept of frequent delivery significantly performed better on these issues and ultimately achieved more true requirements at the end.

FUTURE RESEARCH DIRECTIONS

The main limitation of ECSE is it largely depends on the experience and capability of the customers a.k.a. the moderator. As a result, every decision in the game can be very subjective. In order to reduce this weakness, a semi-structured guideline based on actual case studies could be developed. A variation of ECSE might be developed in order to target different audiences. This could be even more effective if a community of stakeholders is formed in order to exchange lessons learnt, settings and true requirements based on local customizations.

More alternative physical-based pedagogical activities needs to be further developed in order to provide other insightful perspectives of the real world to students. There have been little attempts to develop tools which highlight certain important software development elements such as risk management and quality management. Such research would benefit a large number of software engineering related classes.

CONCLUSION

ECSE is a game which simulates a software development life cycle. Instead of building a real world software product, the participants are required to construct a house. In this way, the students can contribute and benefit from the workshop without any need of high technical skill. The workshop lasts for two and a half hours. Its main objective is to educate students on real world nature of software development which involve uncertainties, communication and especially changes. The game is approachable by various levels of attendances.

Throughout this activity, the students learn many aspects of software development. They also learn how important the communication is and how to deal with variety of customers. Although the main goal of the participants seems to be

making profits, the actual lesson of this game is the finding of the "true requirements." Finding of the true requirements can subtly teach the students on many issues which are not easily described in traditional software engineering lectures. For example, the game convinces the participants on the importance of the business goal over minor requirements. Moreover, the students are convinced that sometimes common sense and general rules of thumb might not work in real life. The students are allowed to open the "true requirements" envelope at the end of the workshop, so they can evaluate their understanding of the requirements.

ACKNOWLEDGMENT

The concept of ECSE is inspired by Suradet Jitprapaikulsarn Ph.D., Department of Electrical and Computer Engineering, Naresuan University, Phitsanulok, Thailand. Customization, improvement and optimization of this game are accomplished with kind assistance from undergraduate computer engineering students, Department of Computer Engineering, Chiang Mai University, Chiang Mai, Thailand. This book chapter is kindly proofread by Dr. Kenneth Cosh, Innovation Warriors Research Group, Department of Computer Engineering, Chiang Mai University, Chiang Mai, Thailand.

REFERENCES

Baker, A., Oh Navarro, E., & Van Der Hoek, A. (2005). An experimental card game for teaching software engineering processes. *Journal of Systems and Software*, 75(1), 3–16. doi:10.1016/j.jss.2004.02.033

Beck, K., Beedle, M., Van Bennekum, A., Cockburn, A., Cunningham, W., Fowler, M., & Thomas, D. (2001). *Manifesto for agile software development*. Academic Press.

Caulfield, C., Xia, J., Veal, D., & Maj, S. P. (2011). A systematic survey of games used for software engineering education. *Modern Applied Science*, *5*(6), 28. doi:10.5539/mas.v5n6p28

Dantas, A. R., de Oliveira Barros, M., & Werner, C. M. L. (2004). A simulation-based game for project management experiential learning. In *Proceedings of the 16th International Conference on Software Engineering and Knowledge Engineering* (pp. 19–24). Academic Press.

Forio. (n.d.). Retrieved from http://forio.com

Hainey, T. (2009). Games-based learning in computer science, software engineering and information systems education. *Computing and Information Systems Journal*, *13*(3), 1–13.

Hainey, T., Connolly, T. M., Stansfield, M., & Boyle, E. A. (2011). Evaluation of a game to teach requirements collection and analysis in software engineering at tertiary education level. *Computers & Education*, *56*(1), 21–35. doi:10.1016/j.compedu.2010.09.008

Hear-See-Do. (n.d.). Retrieved from http://hear-see-do.com/

IBM. (n.d.). *Solutions.* Retrieved from http://www-01.ibm.com/software/solutions/soa/innov8/index.html

McConnell, S. (1997). *Software project survival guide*. Microsoft Press.

MIT. (n.d.). *Beer game.* Retrieved from http://supplychain.mit.edu/games/beer-game

Navarro, E. O., & Van Der Hoek, A. (2007). Comprehensive evaluation of an educational software engineering simulation environment. In *Proceedings of the 20th Software Engineering Education & Training Conference* (pp. 195–202). IEEE.

Project Management Institute. (2008). *A guide to the project management body of knowledge*. Project Management Institute.

Schmidt, R., Lyytinen, K., Keil, M., & Cule, P. (2001). Identifying software project risks: An international Delphi study. *Journal of Management Information Systems*, *17*(4), 5–36.

Shaw, K., & Dermoudy, J. (2005). Engendering an empathy for software engineering. In *Proceedings of the 7th Australasian conference on Computing Education* (vol. 42, pp. 135–144). Australian Computer Society.

Sloan, M. I. T. (n.d.). *Simulations.* Retrieved from https://mitsloan.mit.edu/LearningEdge/simulations/platform-wars/Pages/default.aspx

Sourceforge. (n.d.). Retrieved from http://robocode.sourceforge.net

Tiwana, A., & Keil, M. (2004). The one-minute risk assessment tool. *Communications of the ACM*, *47*(11), 73–77. doi:10.1145/1029496.1029497

Ye, E., Liu, C., & Polack-Wahl, J. A. (2007). Enhancing software engineering education using teaching aids in 3-D online virtual worlds. In *Proceedings of the 37th Frontiers in Education Conference* (pp. T1E–8). IEEE.

ADDITIONAL READING

Boehm, B. (1986). A spiral model of software development and enhancement. *ACM SIGSOFT Software Engineering Notes*, *11*(4), 14–24. doi:10.1145/12944.12948

Braude, E. J., & Bernstein, M. E. (2010). *Software engineering: Modern approaches*. Wiley.

Ceschi, M., Sillitti, A., Succi, G., & Panfilis, S. D. (2005). Project management in plan-based and agile companies. *IEEE Software*, *22*(3), 21–27. doi:10.1109/MS.2005.75

Dyché, J. (2001). *The CRM handbook: A business guide to customer relationship management*. Addison-Wesley Professional.

Humphrey, W. S. (2005). *PSP(SM), A self-improvement process for software engineers*. New Jersey: Addison-Wesley Professional.

Jaramillo, C. M. Z. (2009). Teaching software development by means of a classroom game: The software development game. *Developments in Business Simulation and Experiential Learning*, *36*, 156–164.

Norman, D. A. (2002). *The design of everyday things*. Basic Books.

Pressman, R. S. (2010). *Software engineering: A practitioner's approach*. McGraw-Hill.

Razmov, V. (2007). Effective pedagogical principles and practices in teaching software engineering through projects. In Proceedings of the *37th Annual Frontiers In Education Conference-Global Engineering: Knowledge Without Borders, Opportunities Without Passports* (pp. S4E–21). IEEE.

Schwaber, K. (2004). *Agile Project Management with Scrum*. Microsoft Press.

Schwalbe, K. (2007). *An introduction to project management*. Course Technology Cengage Learning.

Sommerville, I. (2010). *Software engineering*. Addison Wesley.

KEY TERMS AND DEFINITIONS

Game-Based Teaching: A teaching approach that involves students to play well-designed games, through which students can learn knowledge and experiences on a specific subject.

Project Management Skills: Some non-technical skills that are needed in a software project, including project planning, scheduling, budgeting, and resource allocating.

Software Development Life Cycle (SDLC): A model that describes the process that software is created, maintained, and evolved.

Software Development Life Cycle (SDLC) Game: A game that simulates the software development life cycle.

Stakeholder: Key players in a software project, including customers, users, developers, managers, etc.

Chapter 16
Teaching Software Engineering through a Collaborative Game

Elizabeth Suescún Monsalve
Pontifical Catholic University of Rio de Janeiro, Brazil

Allan Ximenes Pereira
Rio de Janeiro State University, Brazil

Vera Maria B. Werneck
Rio de Janeiro State University, Brazil

ABSTRACT

This chapter addresses the application of computer games and simulations in order to explore reality in many educational areas. The Games-Based Learning (GBL) can improve the teaching and learning experience by training future professionals in real life scenarios and activities that enable them to apply problem-solving strategies by putting into use the correct technique stemming from their own skills. For that reason, GBL has been used in software engineering teaching. At Pontifical Catholic University of Rio de Janeiro, the authors have developed SimulES-W (Simulation in Software Engineering), a tool for teaching software engineering. SimulES-W is a collaborative software board game that simulates a software engineering process in which the player performs different roles such as software engineer, technical coordinator, project manager, and quality controller. The players can deal with budget, software engineer employment and dismissal, and construction of different software artifacts. The objective of this chapter is to describe the approach to teaching software engineering using SimulES-W and demonstrate how pedagogical methodology is applied in this teaching approach to improve software engineering education. The teaching experience and future improvements are also discussed.

DOI: 10.4018/978-1-4666-5800-4.ch016

INTRODUCTION

The goal of software engineering education is to provide students the necessary methods and techniques (and later software professionals) to develop quality software (Sommerville, 2007). Teaching software engineering requires not only considering the theoretical aspects but also the principles and methodologies for software development and maintenance, including the aspects concerning the practice and application of that knowledge.

Traditional classes, even if they use real-world projects, cannot always simulate the decisions software engineers have to deal with in their daily activities. Thus game-based learning has been successfully used as a support tool in several areas, including software engineering in the traditional classroom (Bollin, Hochmuller, & Mittermeir, 2011, Hainey et al., 2011, Drappa & Ludewig, 2001, Jain & Boehm, 2006, Birkhoelzer, Navarro, & van der Hoek, 2005). Within this context, this chapter presents our experience of teaching software engineering using SimulES-W, a collaborative game, in which players perform different roles, such as software engineer, technical coordinator, project manager and quality controller. This game has been implemented over two years in both undergraduate and graduate courses, in which the learning objective is accomplished through teaching strategies.

With SimulES-W we can introduce both general and specific software engineering knowledge and apply an educational component that allows the practice of simulating. Students using SimulES-W can identify the systematic approach for developing disciplined and qualifying software, hence understand aspects related to software quality and maintenance. Thus students perform different roles in which each student has to deal with the project budget and the hiring of software engineers. The game also has a system of concepts and problem cards that are used to improve the game itself or used to block the other players' moves. These cards contain the theoretical software engineering knowledge that must be analyzed and applied by the students. Moreover, the game also has an activity for building the software product. The students would need to construct software artifacts required by the project, make inspections, fix errors if they appear in some artifacts, and pack and deliver the product. While at the same time, they need to pay attention to the budget assigned to the project. The first student who is able to construct the software without having any problems, will win the game.

The main advantage of using SimulES-W regards customization. It can be used to teach a specific software engineering aspect or a comprehensive software engineering project as the instructor assigns it. The game application in the classroom results in the most important discussion. After the game is concluded an analysis activity is performed to help identify elements that can contribute to students' knowledge improvement. Furthermore, the feedback received from the students helps improve both the game and the strategy activity.

We are aware of the difficulty regarding the leading changes in teaching methodology, both for teachers and students. The teachers are the motivators in this process, and the students have to understand they can learn using a game. Naturally, it is very motivating knowing the preference for using games as a teaching and learning mechanism, which is our greatest driving force. To cover these aspects, this chapter is divided into six sections. Section 1 contains the introduction. Section 2 describes the background of software engineering education with games, Section 3 presents the SimulES-W, Section 4 introduces our methodology to teach software engineering with SimulES-W, Section 5 presents our teaching experience with SimulES-W in the Software Engineering discipline in the Computer Science Program at the State University of Rio de Janeiro. Conclusions are in Section 6.

SOFTWARE ENGINEERING EDUCATION WITH GAMES

Games have been regarded as an alternative teaching method in the area of software engineering. Research shows they are highly engaging and can help students' effective learning (Hainey et al., 2011). The literature examined in this work also suggests positive impacts related to learning value, skill enhancement, motivational impacts, acquisition and content understanding. For example, Shaw & Dermoudy (2005) point out that students have little empathy for, or affinity with the fundamentals of software engineering practice when it is first introduced. In fact, GBL can provide a constructivism learning environment where learners can practice formulating requirements specification through requirements elicitation and learn by doing it (Aldrich, 2005). In a similar manner, Bollin, Hochmuller, & Mittermeir (2011) advocate that teaching software engineering should include social skills, economic planning and responsibility, as well as ethical concerns and evaluation. These skills can be better taught by means of simulations and/or games, which could capture partial situations of real-world projects.

Games for teaching are being implemented in classrooms. To name only a few, Sweedyk & Keller (2005), Barros & Araújo (2008) and Qin & Mooney (2009) used games in small projects when teaching software engineering. However, they did not simulate situations of big and complex systems. There are also other more elaborate games proposed, such as those presented by Ford and Minsker (2003), Jain and Boehm (2006), Oh Navarro and Van der Hoek (2005), Shaw & Dermoudy (2005), Waraich (2004), Zhu, Wang, & Tan (2007), which are used to teach software engineering, computer science and information systems. Table 1 shows a summary of the main features of the aforementioned approaches.

To describe some of these games we begin with AMEISE (Bollin, Hochmuller, & Mittermeir, 2011). This game allows students to exercise their software project management skills and reflect on their decisions. These decisions are based on incomplete and uncertain information and feedback, which are provided by the system. Also, they suggest that GBL could be used efficiently to supply and simulate real processes.

Problems and Programmers (PnP) is a card game and the players' goal is to be the first to complete a common software project (Baker, 2003). Players use strategies, according to their own preferences and available cards (software engineer cards, concept cards and problems cards). The concept and problem cards are the heart of the game, providing problems or solutions to the players. The PnP allows the players to analyze their own strategies as well as the other players', hence enabling a discussion about the use of the concept and the problems of the cards. Several other games based on PnP are also the source of inspiration for the different versions of SimulES up to SimulES-W, which are described below.

SESAM (Drappa & Ludewig, 2001), according to its authors, works as a simulation system and is able to execute models. It is focused on teaching software management. The basic idea of the game is to create a software process model with particular data, which is then simulated by the system. Quantitative data is generated based on the user selections for the specific project. As a result of the simulation the process and the user choices can be analyzed.

SimVBSE (Jain & Boehm, 2006) is a game for students to better understand value-based software engineering. It begins with what the stakeholders have defined as a critical factor for its success and the preference values. The users (students) have to identify what stakeholders regard as a critical success factor and determine a strategy to balance the critical factors with the other preference values.

SIMSE (Birkhoelzer, Navarro, & van der Hoek, 2005) is an interactive single-user educational software that simulates a software engineering process. Therefore, it guides students through different software processes, in which students

Table 1. Summary of games to teach software engineering

Purpose	Year	Tools Available	Applicability	Scope	Supported SE Phases	Authors' Proposal
TREEZ	In use since 2003	Source code available	Algorithms and recursion	Reinforce students´ knowledge of tree traversal techniques	It is not supported	The object of the game is to move through the tree, and visit and output the nodes in their correct order
SimVBSE	In use since 2003	It is not specified	Teach software engineering value.	Represent a real software project	Development based on prototypes.	Identify the stakeholders in the system with what they perceive as critical success factors and values, all of that in a simulated setting.
AMEISE	In use since 2002	Client/Server system.	Teach Project Management.	Simulation environment is used in the context of general software engineering or management courses in order to experiment with project management skills.	This is an extended version of SESAM.	Educational model is based on experimental learning. This implies that trainees should be able to learn from their own mistakes.
PnP	In use since 2003	It does not have.	Teach software engineering.	The players compete to finish their projects while avoiding the potential pitfalls of software engineering.	It is supported.	It is intended to simulate the software development process from conception to completion.
SESAM	In use since 2001	It is not specified	Teach Project Management.		Documentation related to specification, architecture, design and definition and implementation of work environment.	Create a model of Software development process and run it using a simulation system.
SIMSE	In use since 2006	This game has different versions and all are Java supported.	Teach Software Engineering Process.	Software engineering project.	Modeling as it has different game version.	Complete a software engineering project.
SimulES	In use since 2007	It does not have.	Teach software engineering.	Represents a real software project to create a software product.	It was modeled using scenarios.	Complete a software engineering project.
SimulES-W	In use since 2010	Web version in Java	Teach software engineering.	Represents a real software project to create a software product.	It was modeled using i* (intentional modeling)	Complete a software engineering project.

have to deal with budget, project time and other difficulties that can arise when the simulation is running. At the same time students must make decisions that could affect the project positively or negatively. The idea in this game is to finish the project within the stipulated time and budget.

SIMULES-W

SimulES-W is the digital version of SimulES (Figueiredo et al., 2007), an educational board and cards game, which is an evolution of the ideas of the PnP game. Different from PnP, SimulES-W does not have any specific development process and the development process can be explored pedagogically during the game; for instance one player can use an agile approach whereas the other can follow a waterfall process. Exploring different development processes can be organized by the instructor, such as an after-game discussion about development approaches.

The main precondition to play SimulES-W is the player should be either a software engineering student or a person with basic knowledge and involved in software engineering. SimulES-W is a multiplayer game and the player who first completes the software product with the quality and budget as defined in the project card, wins the game (Figure 1).

The resources used during the game are the main board (Figure 2), the individual boards (Figure 3), the project cards (Figure 1), the software engineer cards (Figure 4), the problem cards (Figure 5), the concept cards (Figure 6), the white and the gray artifact cards (Figure 3, part b) and the dice. The main board (Figure 2) presents the rounds and moves, displays the project card (Figure 1) and allows the online players to interchange messages. The individual board (Figure 3) represents the software development team of each player, where they can place their software engineers in the columns and the artifacts (white and gray cards) in the cells (Figure 3, part a). These artifacts cards are the software artifacts built by the software engineer, which may contain defects (bugs) (Figure 3, part b). The white artifact cards cost twice the price of gray cards, but they are less likely to have bugs. The rows of the individual board represent different artifact types: requirements, design, code, trace and help.

Figure 1. Project card

Project Chose	Its Modules						
Project							
Id	Description	Budget	Complexity	Name	Quality	Size	Status
7	Expert Committee is an open multi-agent system. It supports the management of submissions and review of articles submitted to conferences or workshops. The system has different activities such as sending papers, assignment of a paper to a reviewer, selection of reviewers, notification of acceptance and rejection of papers.	180	2	Expert Committee	1	1	1

Project Chose	Its Modules				
Modules					
Module	Requirement	Design	Code	Trace	Help
1	2	1	2	1	1

Figure 2. Main page of SimulES-W

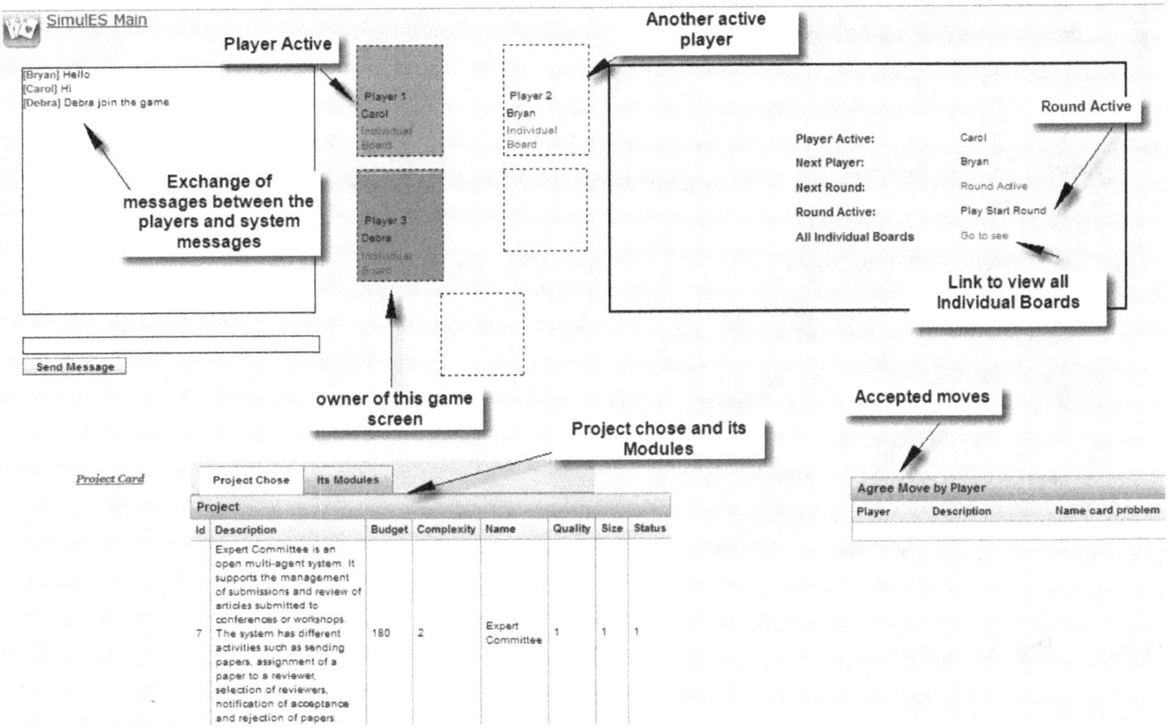

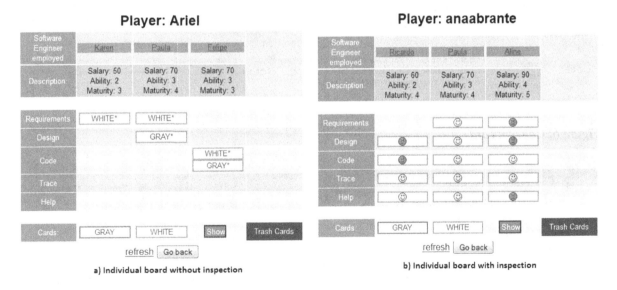

Figure 3. Individual board

The project card (Figure 1) sets the game objective and is composed of name, description, complexity, size, quality, budget and modules.

Each module has a minimal composition of artifacts of a given type. A project must have at least one module; the size defines the number of mod-

Figure 4. Software engineering card

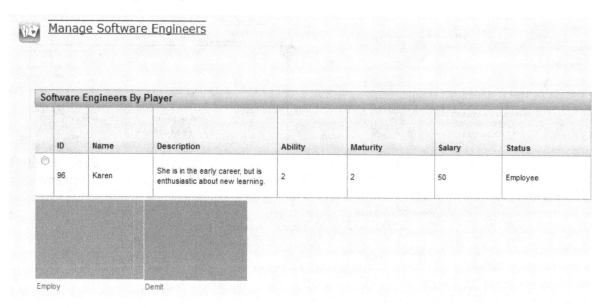

Manage Software Engineers

Software Engineers By Player

	ID	Name	Description	Ability	Maturity	Salary	Status
○	96	Karen	She is in the early career, but is enthusiastic about new learning.	2	2	50	Employee

Employ   Demit

Figure 5. Problem card

Choose Problem to Submit

	ID	Name	Description	Type	Reference	Rule
◉	85	No Test Automation	In the next round of actions, adversary's software engineers must be penalized at a point time for each artifact built or corrected.	Problem	[Korel, 1990] [Binder, 1999]	Design Artifact <= 2 and Code Artifact >= 3
○	87	Inability to Reuse	Persistent problem: All Software Engineers of Adversary must be penalized in one point of time for each artifact constructed, inspected or corrected.	Problem	[Jacobson et al., 1997] [Krueger, 1992] [Sommerville, 2000]	Requirements Artifact <=1 and Code Artifact >=4

Figure 6. Concept card

My Own Cards

	ID	Name	Description	Reference	Rule	Type	Category
○	84	Language	Use this card to add two help Artifacts and one trace Artifact to one software engineers.	Use of natural language and respect the private languages of the context where the software will be used		Concept	CCM3

ules. The project complexity is related to the value of white and gray artifact cards, i.e., the complexity times the points a software engineer needs to complete an artifact. The gray artifact is half value of the white card. For example, if the project complexity is 2 then the software engineer spends 2 time points to build a white artifact card; for the gray card, 1 time point is enough, because the white card is more expensive but has fewer bugs than the gray one. In this example the project is size 2, so the final product has 2 modules. For module 1, the players have to build 2 requirements artifacts (artifact cards), 1 design artifact and 1 code artifact. Module 2 has 2 design artifacts, 2 traceability artifacts, 1 help artifact and 1 code artifact. Each artifact is built either with a white or gray card. The quality attribute shows the maximum number of defects allowed in a module. The budget attribute determines the amount of money available to be spent in hiring software engineers.

Figure 4 represents a typical software engineer cards that has a name, a description of his/her personality, a salary, which is related to the budget, an ability, which is related to the project complexity, and a maturity, which is used in concept and problem cards. The ability is the number of time points (productivity) the software engineer has to spend in each round, so it defines the number of white and gray artifacts cards that can be produced by this software engineer. In Figure 3, white artifact cards cost 2 and gray cards cost 1, so Silas (Figure 4) with the ability of 4 time points and project complexity 2 can build 2 white artifact cards or 4 gray artifact cards. Therefore, if the player has higher-skilled engineers, the player will then have higher productivity, which he/she will build more artifacts, and could therefore finish earlier.

The problem card (Figure 5) addresses typical problems in software engineering and they should be understood by players. The idea is that the players should think about these problems and decide how to deal with the conditions stated in the card. A concept card (Figure 6) is a way to neutralize the problem card or to get advantage over the other players. Concept cards have a reference to the literature, a description and usage instructions. Thus, the players can learn more about an issue they are interested in. The player can then use the concept card to either block a problem card or to improve the software engineering performance. Categorization (upper right corner of concept and problem cards) indicates that cards can be targeted to a specific topic. For example, if the instructor identifies the topic which should be used in the training is requirements engineering management then he/she could produce cards to address it, and

as a result the discussion generated when the players interact with the game will be on this topic.

SimulES-W has different rounds where players execute their moves, such as start and manage problems, perform actions, and submit product. Figure 7 uses an SDSituation Diagram (Padua & Cysneiros, 2006) to illustrate these rounds and their sequence. When the game starts, one project must be chosen from those that are available ([T1] in Figure 7). All players roll the dice and the one with the highest dice number chooses the project and starts the game. Furthermore, the information about the project is displayed in the middle of the main board and visible to all players. Next, each player assembles an individual board and picks one software engineer from the stack of software engineering cards.

In the "play round to actions" (T2), each player with the information about his/her software engineers (skills and salary) and the information in the project card (size, complexity and budget) uses a software engineer to build artifact, inspect artifact, correct artifact and integrate artifacts in a module (see [T2] in Figure 7). In the action *build artifact*, if the player builds with white artifact cards he/she will spend the time points according to the complexity in the project card, but if he/she builds with gray cards he/she will then spend half of the time points. However, white artifact cards (5 cards to 1 defect) have a lesser defect rate than gray artifacts cards (3 cards to 2 defects).

The *inspect artifact* action regards turning up an artifact card under the responsibility of a software engineer, disclosing its quality status (with or without a bug – see Figure 2, part b). The inspection cost is fixed by 1 time point per card if it is performed by the same software engineer that built the artifact or 2 time point per card if it is performed by another software engineer. The correct defect action has to be performed when the software engineer inspects an artifact card and finds a defect ("bug"). When a defect is corrected he/she spends 1 time point if it is performed by the same software engineer that built the artifact and 2 time point if it is performed by another software

Figure 7. SimuES-W SDSituations

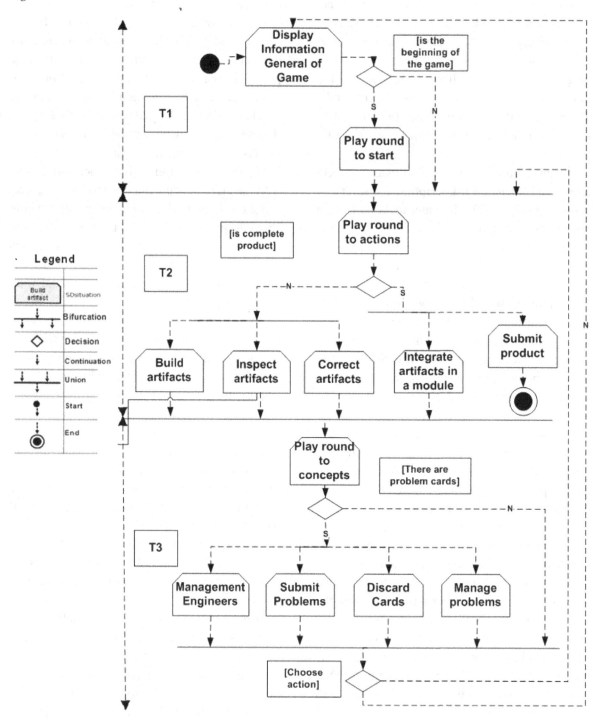

engineer. *Integrate artifacts in a module* action has to be performed before the player submits the product. This situation takes place when the player has built all types of artifacts required in a module (Figure 3). The player can choose the artifacts

available in his/her individual board considering the artifact types described in the project card to compose a module. The artifacts can be originated from different software engineers (columns in the individual board).

In the "play round to concepts" [T3] in Figure 7, each player rolls the dice once. The resulting dice number allows the player to buy concept/problem cards. These cards (concepts and problems) are shuffled and piled upside down in the main board. If the dice shows a number greater than or equal to 3, then software engineering cards can also be bought. The number of software engineer cards will be the difference between three and the dice result. As an illustration, if Mary rolls the dice and the result is 2, then Mary could buy 2 problem/concept cards. On the other hand, if Mary rolls the dice and the result is 4, then Mary could buy 3 cards (problems and concepts) and 1 software engineer card. Thus, the higher the number from rolling the dice, the more resources the player will have. Here is where luck comes into play. At this point the player has to think about team composition: the number of software engineers is limited to the overall budget (see Figure 3), that is the sum of software engineers salaries displayed in the Individual Board. This sum has to be lower than or equal to the project budget. This implies the possibility of hiring and firing software engineers (project management skills). Note that there is an educational purpose in making students deal with real world issues (hiring/firing) by means of simulations.

In [T3] the player uses concept and problem cards. So during the game, the player can receive problem cards from the other players. These cards, when received, are to be used in the next round. The objective is to sabotage the game of other players. However, if the player has one card which invalidates some problem cards (a concept card), he/she will be able to use it and the action described in the problem card will not affect his/her game, then he/she must discard both cards. On the other hand, if the player does not have any card that invalidates the problem cards, this problem will be applied to his/her game. At this point in the game, the educational goal for the players is to discuss whether concept card actually invalidates card problem. A player using a concept card has to build an argument as to why that card neutralizes the problem card. This argument can be discussed, but it will only go into effect if all players agree. As mentioned before, this discussion can be mediated by an instructor.

The "submit round" [T4] can be performed when the player's turn begins. When the player integrates one module he/she can submit the product. Then the other players have to inspect those artifacts that are not inspected (faced up). The module will be accepted if the number of remaining bugs is lower than or equal to the quality attribute number in the project card (Figure 3). The player who first finishes all the modules within the quality required by the project card wins the game.

SimulES-W was conceived for teaching software engineering in general. Alternatively, it can also be configured to focus on a particular knowledge subject. Since cards can be configured, we can use problem and concept cards tuned to the topic of interest. We can also configure project cards to only deal with certain artifacts. For example, we use problem and concept cards (Figure 4 and 5) to code artifacts, requirements, requirements management, and design.

The original version of SimulES-W, SimulES, differently from the Problems and Programmers (PnP) game (Baker, 2003), does not impose an order to types of artifacts, therefore the player can choose, for instance, to start with design or code artifacts and produce requirements artifacts later. Thus the game does not embrace a particular production process. Requirements, code, help, trace, and design artifacts must exist. But the player decides how to approach them, as required by the project profile (Figure 3), which lists the types of artifacts needed to complete the game. SimulES-W has the necessary elements to emphasize or generalize the knowledge to be transmitted and can also enact particulars of the software process as per concept and problem cards, making it a powerful and useful tool (Monsalve, Werneck, & Leite, 2011).

Software Development

SimulES-W is a Web-based implementation with a MVC (Model-View-Controller) design pattern; it is used to separate the business logic, interface, and control. Development details are found in (Monsalve, 2010). We have had courses in which students had to participate in the development of new features for SimulES-W. It was proposed that elements of the application needed to be improved.

The current version of SimulES-W is in a stable state, thus, we can run and play all levels of the game. Some details of this process have been published (SimulES-W PES, 2011). The interesting idea about this option regards providing opportunities for students to perform software developments and apply concepts as well as suggest innovative improvements for the project as a whole. This option offered to students is also coupled to the idea of motivating them in activities that simulate a work environment in a software project. That means, in their professional life they could work improving or developing software or work in a specific functionality. Professionals in the area of software projects often reach advanced stages and do not necessarily start development from the initial stage. These types of activities generate discussions about the best way to reach a solution.

There are cases in which students have more skills in software development than others. But those with less experience are also interested in participating to gain experience, acquire knowledge and to investigate how a problem or implementation of specific functionality could be sold. For these activities students generally create working groups. Finally, in the classroom they explain how the solution was given and show how it works. Furthermore, we also have used SimulES-W in classrooms to create content based on subjects being taught and which are in the course curriculum. Students create groups to discuss the contents that will be produced, which are basically concept and problem cards related to software engineering.

Using observation as a requirements elicitation technique, we have found areas for improvement in the game, student playability and behavior, performance and interaction when they play. In (Monsalve, Werneck, & Leite, 2010) an experience is presented where this technique was used so the development team understood the game dynamics.

Feedbacks have been very important for the evolution of the game. These improvements have allowed us to show new students a more refined and consistent game according to the current educational needs. Feedbacks have not only come from student participation but also from those who have been directly involved in developing the game. Several of these feedbacks are reported in (Monsalve, Werneck, & Leite, 2010; Monsalve, 2010; Monsalve, Werneck, & Leite, 2011). They describe not only feedbacks but also their ensuing improvements. The main idea is to give students complete freedom so they can judge the teaching models applied to them, reflect on what they have learned and thus help future students who will use the game.

Test

We have conducted tests to evaluate usability, understanding, content and learning. Below we briefly describe each of these items.

Usability

In the early stages when SimulES-W was being developed, we applied usability tests. This resulted in applying the user interface design, navigation, task execution, general use of the tool and subsequent implementations. The students were first asked to execute specific tasks on initial prototypes. These tasks were activities of each game round. The activity was reported verbally indicating students understood what they were doing during the entire game. Basically, usability test were based on the approach presented in (de Souza, 2005; de Souza et al. 2010) called Semiotic Inspection Method (SIM) used to verify the interactive qualities of specific systems and design strategies. It basically examines a large diversity of signs that the users are exposed to when they interact with computing artifacts. In other words, it is used to evaluate interactions between user and interface and the sings are images, words, colors, dialog structures, and etc., which are presented to the user when he/she uses the systems.

Finally, the students fill out questionnaires with additional information in order to assess how the designer-to-user meta-communication message gets across to users through the system's interactive interface. In short, usability testing has been important to the SimulES-W project as they help define user interface, and additionally, it has provided us a perspective about students' behavior with tools, navigation and how they perform certain tasks.

Testing Game Acceptance

As mentioned above, we performed some tests prior to the Web implementation of the game. Some of these assessed motivation and acceptance of the game. The SimulES board game was used, and it was implemented after acquiring positive results. On the other hand, with several activities involved SimulES-W, we sought to correlate the questions to identify students' acceptance and motivation (Monsalve, Werneck, & Leite, 2010; Monsalve, 2010; Monsalve, Werneck, & Leite, 2011). Generally, students are asked about their opinions about the activity and contents of a game, and tools and teaching methodologies.

Knowledge Tests

One of the most important activities of SimulES-W is to create tests and then determine if they help students in acquiring knowledge – given that this is the main goal of the game. Formally, we have conducted an experiment with SimulES-W. The study was aimed at identifying whether students acquired knowledge with the game. To this end, the activity was divided into several parts. A group of students participated in the activity using SimulES-W and another group of students participated in a traditional class. And at the end of the course all students had to take an exam.

This comparative study showed how SimulES-W can be used in a software engineering course as a tool to improve learning, and the positive results evidenced that game-based learning could effectively improve students' course performance; this experience also confirmed what others have reported about how games establish a motivating and enjoyable environment for students. With regard to learning, it was found that the use of SimulES-W did not compromise the students' ability to acquire knowledge and it can be an additional form of learning in software development. In addition, games have the added value of offering a more motivating and enriching learning process while also entertaining. Furthermore, in education some tools can confuse students rather than help them improve their knowledge. However this experiment showed evidences that SimulES-W improves the students' general software engineering knowledge.

For that reason, new evaluations using SimulES-W based on educational levels were performed with specific objectives to producing further empirical evidence of learning efficiency with the game. With regards to this reason the next section shows in detail the results of a new experiment.

METHODOLOGY TO LEARN SOFTWARE ENGINEERING WITH SIMULES-W

Our approach proposes a collaborative work environment around SimulES-W. In the classroom the students are engaged and with the teachers looking for a common goal for their joint contributions. That is, when students participate in the SimulES-W project, they can do so from different working fronts which are related to their interests and skills. Therefore, we provide students with different choices and they can choose the ones they want to participate in, such as software product development, creation and execution of tests, content creation and play-game activities. The latter requires that students engage as the product's end users, give their opinions, feedback and be evaluated about the knowledge acquired and the effectiveness of the game. Therefore we offer the students additional teaching, collaborative, active and motivating practices.

Interestingly, we agree with constructivist learning theories. Some authors argue that knowledge is built by the learner, not supplied by the teacher (Bruckman, 1998). Kafai and Resnick (1996) suggest that new knowledge could be acquired more effectively if the learners were engaged in constructing products that were personally meaningful to them. We understand constructivism as a way to allow students to interact with elements that will be present in their professional life, since the student participates in the construction of his own knowledge base. In this context, the student not only acquires the ability to develop new skills but also apply that knowledge to new situations, such as in his professional future. In fact, this approach allows students to interact in specific situations, stimulates the "know" and "know-how." On the other hand, the role of the teacher can change; he/she could be a moderator, a coordinator, or a facilitator or in some cases

simply another participant. Empirical evidence shows that students learn more effectively when they do this cooperatively. For this reason we are proposing a strategy with SimulES-W which can be sufficiently broad to motivate students and encourage them to interact as they play, finding out how to better develops their skills. Therefore, the question is, how can students participate in the SimulES-W project?

Figure 8 illustrates our approach to teach with SimulES-W: (i) Plan the classes using SimulES-W, (ii) Create the concepts to be included in the SimulES-W data (Cards) and the define the settings of the game, (iii) Develop all details in a new SimulEs-W instance, (iv) Play the game, and (iv) Analyze the results of the game.

Planning the Lectures

The first activity is planning. It includes determining how to apply SimulES-W during the course and identify student skills. The plan includes determining the software engineering subject that will be available in SimulES-W, the references and how many lectures will be dedicated to the game.

Creating Content

One of the most important features of SimulES-W is that it can be configured. As presented in (Monsalve, Werneck, & Leite, 2011), SimulES-W can focus on teaching a particular subject according to the students' profile, or professor's motivation. In addition, we have also used SimulES-W in classrooms to create content based on subjects being taught and in the course curriculum. So in the creating activity, the professor can create concept and problem cards to address the subject he/she defined to teach using SimulES-W, define the projects, especially their size so they can be constructed during the lessons.

Figure 8. SADT diagram, which shows the teaching methodology using SImulES-W

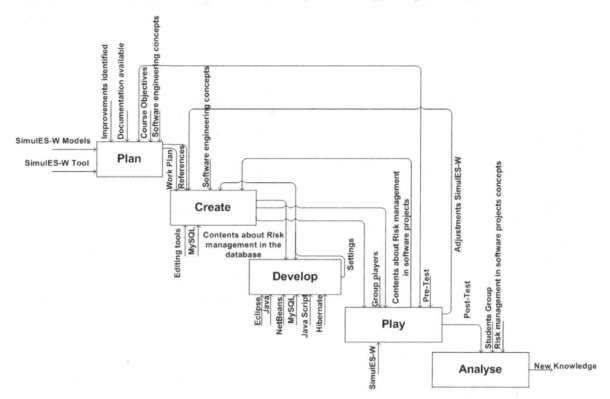

Developing

This activity regards building all the components needed for the game to be executed, including new data in the SimulES-W database for the settings of the games and for the system environment to be created.

Playing

This particular activity has been done in different ways. For instance each student can play individually, or in groups with his own board, depending on the number of students in the course and the learning objectives specified in the lectures plan. The students in a group can collaborate with each other to build a software product. The goal of this activity is for the students to learn the content of the lectures and then hold discussions with the professor during the game.

Analyzing

This activity analysis is fundamental to conclude this process and include all the test and knowledge about the game, especially the events and discussions during the game. The students' performance during the game and tests used to guide the application of SimulES are statistically analyzed.

TEACHING EXPERIENCE WITH SIMULES-W

During the *Planning stage*, in the first half of 2012, an experiment was performed with the class (63 students) in the software engineering course of Computer Science Department at the State University of Rio de Janeiro. Next, the themes of the course were chosen, i.e., the specific topic to teach in that class. In this case, the theme is risk management in software projects. This experiment was designed to study how SimulES-W could influence students learning regarding these concepts. After that, during the *Creating stage* we chose the source and collected and prepared the information related by creating specific concept and problems cards. The project and software engineer cards were also edited for the specific features of the class. Afterwards, during the *Developing stage* the contents were added to the SimulES-W database. Finally, some tests were conducted to check the contents.

The *Planning* of the student Groups defined that the experiment has a two-level treatment. The first treatment group corresponds to attending classes and receiving game instructions. The second one corresponds to the other group, the one that played the game without having participated in the class about the subject. We designed a test with ten multiple-choice questions to measure the students' knowledge on the concepts of risk management in software projects (Figure 9).

The 63 students in the class were divided into two groups, according to the fundamental principle of randomization to ensure the comparability of the groups. Each group with its activities was divided by days, participating in specific activities, with the exception of 1 day when both

groups responded to the test before playing. The detailed plan for implementing the experiment is described below.

In *Playing*, students received information related to SimulES-W, specifically on risk management in software projects. They also received instructions and details about the game: origin of the game, historical review, basic rules, dynamic rules, goals and main screens. They were then motivated to use the tool. Instructions about the tool including navigation, interface features and execution of actions were explained during this session.

In *Playing* SimulES-W, the activity took place in a classroom with a teacher and two instructors who guided the students and answered questions related to the activity. During the activity we emphasized the concepts and the problems related to subject.

In *Analyze*, on the 1st day, only 37 of the 63 students participated in the test application. Because of this it was necessary to control the participation of students in specific days to avoid contamination of the experiment. Thus, only the students of Group 1, who attended the 1st day activity, participated on the other days. The same control was applied to the students in Group 2. The students who participated in the activities playing

Figure 9. Days of activities applied to the students groups

Group 1:
• 1st day of activity – Application Test.
• 2nd day of activity – Exposure class.
• 3rd day of activity – Start the activity playing with SimulES-W.
• 4th day activity – Conclusion of the activity playing with SimulES-W and at the end, application of the same test given on the first day.

Group 2:
• 1st day of activity – Application Test.
• 2nd day of activity – Start the activity playing with SimulES-W.
• 3rd day of activity – Conclusion of the activity playing with SimulES-W and at the end, application of the same test given on the first day.

Table 2. *Distribution of the number of correct answers per question for each student before and after the activity with the SimulES-W*

	Test BEFORE		Test AFTER	
	Group 1	Group 2	Group 1	Group 2
	12	10	11	11
	14	13	8	14
	17	9	13	9
	16	9	14	12
	12	12	15	17
	11	12	15	16
	13	12	12	14
	13	10	12	10
	8	9	6	7
	15	14	16	8
	7	15	9	15
	9	12	7	8
	13	13	12	8
		14		11
n	13	14	13	14
Mean	$\overline{Y_1} = 12{,}308$	11,714	11,538	11,429
Standard Deviation	2,983	2,016	3,205	3,298

with SimulES-W (both in group 1 and in group 2) were reorganized to create subgroups in the game. Thus, group 1 and group 2 were respectively 13 and 14 students – totaling 27 students (sample size). At the end of the experiment, 54 tests were completed. The number of correct answers per question BEFORE and AFTER the activity with the SimulES-W was recorded, respectively, in Table 2. This test was applied, before and after that they had played with SimulES-W.

The statistical analysis took into account the assumption of normality. With this, we applied the following tests presented in Table 3. Table 2 shows that group one at the beginning was somewhat better than group 2. After the activities their means differences were almost the same and the comparison of results between the groups were confirmed.

The R platform was used to perform the statistical analysis. The results obtained by applying

Table 3. *Tests applied to the data obtained in the experiment*

	Parametric tests	Non-parametric tests
Comparison of results same group	T-test	Wilcoxon test for paired data
Comparison of results between groups	T-test for independent samples	Wilcoxon-Mann-Whitney test for independent samples

the platform can be seen in Table 4. With these results we found that the differences were not statistically significant (probability of significance: p-value ≥ significance level: α) in all tests. For example, students who attended class had very close scores and in some cases even lower than of the students who did not attend class. Figure 10 illustrates the score distribution in the PRIOR test group and Figure 11 the AFTER test scores

It was then observed that for the treatment effect (attend class / receive instructions) there was no statistical significance for the study in question. In other words, no significant difference in the scores of the groups, indicating that both were consistent and showed values very close to the average score. The results showed no rejection evidence for the game. The statistical analysis of these tests shows that there is no significant difference in the scores of the groups. This type of results was also presented in Ebner and Holzinger (2007), which suggested that there is evidence showing that the learning result from using

games is at least equivalent to the results from learning by the traditional method.

However, to be used effectively in a course or module would require a different qualitative study, with other enforcement strategies and with other applied levels. Reflecting on the experiment and evolution to improve the relevance of the results, an estimate for the optimal size of the sample groups was calculated. Thus, to conduct a new experiment following the design proposed in the current experiment should include at least between 11 and 15 students per group. This is part of our future study plan.

CONCLUSION

Games have emerged as alternative education methods intentionally, as they promote students participation and collaboration. In this chapter, we have shown how SimulES-W, an educational board and cards game, is used in software engineering education under different learning environments,

Table 4. The results of t-test

Parametric tests	Non-parametric tests
T-test:	Wilcoxon test for paired data:
[group 1 before X group1 after]	[group 1 before X group 1 after]
t = 1.011, df = 12, p-value = 0.332	V = 59, p-value = 0.359
95 percent confidence interval:	
-0.8885229 2.4269844	
mean of the differences =	[group 2 before X group 2 after]
0.7692308	V = 37, p-value = 0.7551
[group 2 before X group group 2 after]	
t = 0.3203, df = 13, p-value = 0.7539	
95 percent confidence interval:	
-1.641643 2.213071	
mean of the differences =	
0.2857143	
T-test for independent samples:	Wilcoxon-Mann-Whitney test for independent samples:
[group 1 after X group group 2 after]	[group 1 after X group 2 after]
t = 0.0878, df = 24.942, p-value = 0.9307	W = 94.5, p-value = 0.8836
95 percent confidence interval:	
-2.468559 2.688339	
sample estimates:	
mean of x = 11.53846	
mean of y = 11.42857	

Figure 10. Diagram of extremes and quartiles of student scores in group 1 (n=13) and students from group 2 (n=14) in the before test

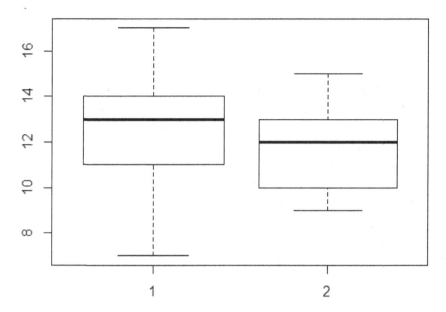

Figure 11. Diagram of extremes and quartiles of student scores in group 1 (n=13) and students from group 2 (n=14) in the after test

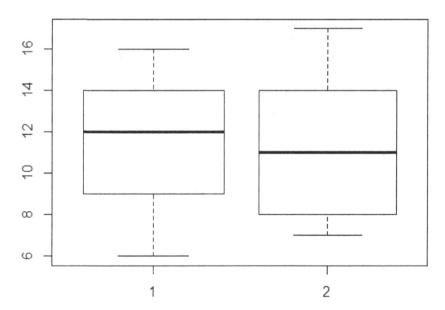

where students not only can play games, but also can perform software development and improvement activities. An experiment was designed to verify the effectiveness of SimulES-W on student learning regarding the concepts of risk management in software projects.

We are aware that there is still much more ground to cover and that more tests must be carried out. However, this work with SimulES-W demonstrated the usefulness of games in education. They stimulate students to explore ideas by forcing them to create theories and thinking about tests. Moreover, they enable students to take on social roles and assume collaborative tasks. This last feature is reflected not only by students but is also observed in the group of participants around the game, such as teachers, designers and developers.

Notwithstanding, games have an important potential in formal or informal education and therefore we should better understand their capacity and diversity as a tool. Additionally, it is important to consider games as a modern and useful learning method as they do not disturb or cause any disadvantage to the learners. Based on our experience they are the most preferred learning modes from the students' point of view.

In the works cited we have shown the limitation of traditional teaching methods. Games are very promising – however it is important to focus on the main teaching challenge, which is the connection between theory and practice. As highlighted in Park (2012), students are currently more familiar with technology, and technologies are changing generational values, student profiles and educational goals. Our experience of using SimulES-W shows computer game-based learning are fun, participative and effective in software engineering education.

Pedagogy is responsible for creating a more efficient and transparent education. In this context, the progress of teaching has increased the role of students, who are no longer passive agents. Therefore our future research focus is to look into pedagogy in order to use games in view of transparency, where game-based-learning might improve this process and provide a balance between teachers and students.

ACKNOWLEDGMENT

We want to thank Professor Julio Cesar Sampaio do Prado Leite for his collaboration in this work.

REFERENCES

Aldrich, C. (2005). *Learning by doing: A comprehensive guide to simulations computer games, and pedagogy in e–learning and other educational experiences*. Hoboken, NJ: Wiley.

Baker, A. (2003). *Problems and programmers*. (Honors Thesis). Department of Informatics, School of Information and Computer Science, University of California, Irvine, CA.

Barros, M., & Araújo, R. (2008). Ensinando construção de software aplicada a sistemas de informação do mundo real. In *Proceedings of the 1st Forum on Education in Software Engineering*. Campinas, Brazil: Academic Press.

Birkhoelzer, T., Navarro, E., & van der Hoek, A. (2005). Teaching by modeling instead of by models. In *Proceedings of the 6th International Workshop on Software Process Simulation and Modeling*. St. Louis, MO: Academic Press.

Bollin, A., Hochmuller, E., & Mittermeir, R. T. (2011). Teaching software project management using simulations. In *Proceedings of the 24th Conference on Software Engineering Education and Training* (pp. 81–90). IEEE.

Bruckman, A. (1998). Community support for constructionist learning. *Computer Supported Cooperative Work: The Journal of Collaborative Computing, 7*, 47–86. doi:10.1023/A:1008684120893 de Souza, C. S. (2005). *The semiotic engineering of human–computer interaction*. Cambridge, MA: MIT Press.

de Souza, C. S., Leitão, C. F., Prates, R. O., Bim, S. A., & da Silva, E. J. (2010). Can inspection methods generate valid new knowledge in HCI? The case of semiotic inspection. *International Journal of Human-Computer Studies*, 68(1/2), 22–40. doi:10.1016/j.ijhcs.2009.08.006

Drappa, A., & Ludewig, J. (2001). Simulation in software engineering training. In *Proceedings of the 23rd International Conference on Software Engineering* (pp. 199 – 208). ACM.

Ebner, M., & Holzinger, A. (2007). Successful implementation of user–centered game based learning in higher education: An example from civil engineering. *Computers & Education*, 49(3), 873–890. doi:10.1016/j.compedu.2005.11.026

Figueiredo, E., Lobato, C., Dias, K., Leite, J., & Lucena, C. (2007). Um Jogo para o ensino de engenharia de software centrado na perspectiva de evolução. In *Proceedings of Workshop on Education in Computer (WEI – 2007)* (pp. 37–46). Rio de Janeiro, Brazil: WEI.

Ford, C. W., & Minsker, S. (2003). TREEZ – An educational data structures game. *Journal of Computing Sciences in Colleges*, 18(6), 180–185.

Hainey, T., Connolly, T., Stansfield, M., & Boyle, E. (2011). Evaluation of a game to teach requirements collection and analysis in software engineering at tertiary education level. *Computers & Education*, 56(1), 21–35. doi:10.1016/j.compedu.2010.09.008

Jain, A., & Boehm, B. (2006). SimVBSE: Developing a game for value–based software engineering. In *Proceedings of 19th Conference on Software Engineering Education and Training* (pp. 103–114). IEEE.

Kafai, Y., & Resnick, M. (1996). *Constructionism in practice: Designing, thinking, and learning in a digital world*. Mahwah, NJ: Lawrence Erlbaum.

Monsalve, E., Werneck, V., & Leite, J. C. S. P. (2010). Evolución de un juego educacional de ingeniería de software a través de técnicas de elicitación de requisitos. In *Proceedings of the 13th Workshop on Requirements Engineering* (pp. 12–23). Cuenca, Ecuador: Academic Press.

Monsalve, E., Werneck, V., & Leite, J. C. S. P. (2011). Teaching software engineering with simulESW. In *Proceedings of the 24th Conference on Software Engineering Education and Training* (pp. 31–40). IEEE.

Monsalve, E. S. (2010). *Construindo um jogo educacional com modelagem intencional apoiado em princípios de transparência*. (Master Thesis). PUC–Rio.

Oh Navarro, E., & Van der Hoek, A. (2005). Design and evaluation of an educational software process simulation environment and associated model. In *Proceedings of the 18th Conference on Software Engineering Education and Training* (pp. 25–32). IEEE.

Padua, A. O., & Cysneiros, L. (2006). Defining strategic dependency situations in requirements elicitation. In *Proceedings of the 9th Workshop on Requirements Engineering* (pp. 12–23). Academic Press.

Park, H. (2012). Relationship between motivation and student's activity on educational game. *International Journal of Grid and Distributed Computing*, 5(1).

Qin, S., & Mooney, C. H. (2009). Using game–oriented projects for teaching and learning software engineering. In *Proceedings of 20th Annual Conference for the Australasian Association for Engineering Education* (pp. 49–54). Engineers Australia.

Shaw, K., & Dermoudy, J. (2005). Engendering an empathy for software engineering. In *Proceedings of the 7th Australasian Computing Education Conference* (Vol. 42, pp. 135–144). Newcastle, Australia: Academic Press.

Simul, E. S.-W. PES (2011). *Simuleswpes's blog*. Retrieved from http://simuleswpes.wordpress.com/

Sommerville, I. (2007). *Software engineering*. Upper Saddle River, NJ: Pearson.

Sweedyk, E., & Keller, R. M. (2005). Fun and games: A new software engineering course. *ACM SIGCSE Bulletin, 37*(3), 138–142. doi:10.1145/1151954.1067485

Waraich, A. (2004). Using narrative as a motivating device to teach binary arithmetic and logic gates. In *Proceedings of the 9th Annual SIGCSE Conference on Innovation and Technology in Computer Science Education* (pp. 97–101). Leeds, UK: ACM.

Zhu, Q., Wang, T., & Tan, S. (2007). Adapting game technology to support software engineering process teaching: From SimSE to Mo–SEProcess. In *Proceedings* of the 3rd *International Conference on Natural Computation* (Vol. 5, pp. 777–780). IEEE.

ADDITIONAL READING

Cagiltay, N. (2007). Teaching software engineering by means of computer–game development: Challenges and opportunities. *British Journal of Educational Technology, 38*(3), 405–415. doi:10.1111/j.1467-8535.2007.00705.x

Claypool, K., & Claypool, M. (2005). Teaching software engineering through game design. *ACM SIGCSE Bulletin, 37*(3), 123–127. doi:10.1145/1151954.1067482

De Freitas, S., & Oliver, M. (2006). How can exploratory learning with games and simulations within the curriculum be most effectively evaluated? *Computers & Education, 46*(3), 249–264. doi:10.1016/j.compedu.2005.11.007

De Jong, T., & Van Joolingen, W. R. (1998). Scientific discovery learning with computer simulations of conceptual domains). *Review of Educational Research, 68*(2), 179–20. doi:10.3102/00346543068002179

Garris, R., Ahlers, R., & Driskell, J. E. (2002). Games, motivation, and learning: A research and practice model. *Simulation & Gaming, 33*(4), 441–467. doi:10.1177/1046878102238607

Hung, P. H., Hwang, G. J., Lee, Y. H., & Su, I. (2012). A cognitive component analysis approach for developing game–based spatial learning tools. *Computers & Education, 59*(2), 62–773. doi:10.1016/j.compedu.2012.03.018

Keys, B., & Wolfe, J. (1990). The role of management games and simulations in education and research. *Journal of Management, 16*(2), 307–336. doi:10.1177/014920639001600205

Kiili, K. (2005). Digital game–based learning: Towards an experiential gaming model. *The Internet and Higher Education, 8*(1), 13–24. doi:10.1016/j.iheduc.2004.12.001

Kirriemuir, J., & McFarlane, A. (2004). *Literature review in games and learning. Futurelab report*. Bristol: Futurelab.

Prensky, M. (2003). Digital game–based learning. *Computers in Entertainment, 1*(1), 21–21. doi:10.1145/950566.950596

Ruiz, J. G., Mintzer, M. J., & Leipzig, R. M. (2006). The impact of e–learning in medical education. *Academic Medicine, 81*(3), 207–212. doi:10.1097/00001888-200603000-00002 PMID:16501260

Squire, K. (2005). Changing the game: What happens when video games enter the classroom. *Innovate: Journal of online education, 1*(6).

Squire, K. (2006). From content to context: Videogames as designed experience. *Educational Researcher, 35*(8), 19–29. doi:10.3102/0013189X035008019

Squire, K., & Jenkins, H. (2003). Harnessing the power of games in education. *Insight (American Society of Ophthalmic Registered Nurses), 3*(1), 5–33.

Squire, K. D. (2003). Video games in education. *International Journal of Intelligent Games & Simulation, 2*(1), 49–62.

Yee, N. (2006). Motivations for play in online games. *Cyberpsychology & Behavior, 9*(6), 772–775. doi:10.1089/cpb.2006.9.772 PMID:17201605

KEY TERMS AND DEFINITIONS

Collaborative Software Game: A computer game that involves two or more players.

Game-Based Learning: A learning approach derived from the use of computer games that possess educational value to engage students or teach complicated concepts, theories, or principles.

Pedagogical Methodology: Using appropriate scientific methods in education to promote learning of new concepts considered important for the development of thought.

Software Engineering Education: The discipline to educate students with software engineering principles.

Software Engineering Process: A model to describe how software is developed.

Section 7
Experiencing Case–Based Teaching and Problem–Based Learning

Chapter 17
Digital Home:
A Case Study Approach to Teaching Software Engineering Concepts

Salamah Salamah
The University of Texas – El Paso, USA

Massood Towhidnejad
Embry Riddle Aeronautical University, USA

Thomas Hilburn
Embry Riddle Aeronautical University, USA

ABSTRACT

While many Software Engineering (SE) and Computer Science (CS) textbooks make use of case studies to introduce difference concepts and methods, the case studies introduced by these texts focus on a specific life-development phase or a particular topic within software engineering object-oriented design and implementation or requirements analysis and specification. Moreover, these case studies usually do not come with instructor guidelines on how to adopt the introduced material to the instructor's teaching style or to the particular level of the class or students in the class. The DigitalHome Case Study aims at addressing these shortcomings by providing a comprehensive set of artifacts associated with the full software development life-cycle. The project provides an extensive set of case study modules with exercises for teaching different topics in software engineering and computer science, as well as guidance for instructors on how to use these case modules. In this chapter, the authors motivate the use of the case study approach in teaching SE and CS concepts. They provide a description of the DigitalHome case study and the associated artifacts and case modules. The authors also report on the use of the developed material.

DOI: 10.4018/978-1-4666-5800-4.ch017

INTRODUCTION

Case studies were first used in the Harvard Law School in 1871 (Tomey, 2003) and since then much study and research have been done on the effectiveness of the use of case studies in teaching and learning (Davis & Wilcock, 2013; Grupe & Jay, 2000; Herreid, 1994). Case studies have been particularly effective in teaching professional practice and have been widely used in such fields as business, law, and medicine. The main advantage of using case studies in the classroom is that they allow the instructor to create a realistic environment and context for the study of real problems in an academic setting. A case study is based on actual events and documents, or at least pseudo-real events and documents. The case study typically describes a problem, using a scenario format providing the context and summarizing key issues and events related to the problem. The scenario is usually supplemented with background material (setting, personalities, sequence of events, and problems and conflicts), artifacts, and data, which is relevant to the situation depicted to help immerse the audience in the real-world situation to be discussed, studied, or analyzed.

While case studies can be effective when used in a prescriptive, teacher-centered teaching environment, they are most effective when used in an active, student-centered approach, where the session instructor acts as a facilitator or coach. Case studies are of special value in problem-based learning, concentrating on the development of problem-solving skills, self-directed learning, and teaming skills. The State University of New York in Buffalo Website provides examples of excellent case studies in science and engineering.

The use of case studies in education has shown great success in medicine, law, and business. However, this teaching style has seen little use in computing education. For example, currently, the SUNY-Buffalo Web site contains over 350 case studies in science and engineering, but only five of these are concerned with some aspect of computing. We suspect that one of the principal reasons for this lack of usage of case study-based teaching is the shortage of sufficient material for this purpose. Many software engineering and computer science textbooks use case studies to illustrate concepts and techniques including an Airline Reservation System and a Household Alarm System (Lethbridge & Laganière, 2001), Fireworks Factory (Metsker, 2006), Picadilly Television and Ariane-5 (Pfleeger, 2005), and SafeHome (Pressman, 2007) these case studies often lack the following:

- Realistic artifacts (often space does not allow providing a complete requirements or design document),
- Completeness (covers only a portion of the life-cycle, and not an end-to-end), with a focus on design and implementation,
- Ability to decouple from the text and apply in ways not intended by the author,
- Techniques for integration into course activities or into the curriculum as a whole,
- A scenario format that would motivate students to get engaged in problem identification and solution, and
- Guidance to the instructor on how to use the case study to teach a course topic or concept.

In previous work (Hiburn & Towhidnejad, 2007; Hilburn, Towhidnejad, & Salamah, 2008) we introduced the DigitalHome case study as a way to address these shortcomings by providing a complete set of artifacts associated with software development as well as providing case modules (mini-case studies addressing different aspects of the DigitalHome project) that can be used by faculty in teaching different subjects in a computing curriculum. In this chapter, we provide a detailed description of the DigitalHome case study material developed so far. In addition, we highlight our experiences in using the case study material in different software engineering courses at our institutions.

BACKGROUND

Engineering Case Studies

The Accreditation Board for Engineering and Technology (ABET) requires that all engineering programs involve their students in "a major design experience based on the knowledge and skills acquired in earlier course work and incorporating appropriate engineering standards and multiple realistic constraints" (ABET 2013). Usually, students are grouped into teams to work on a semester or year-long senior engineering development project. This experience is too often isolated from the rest of the curriculum and does not form a real-world basis for the entire curriculum. Although, students may be exposed to elements of engineering practice in their foundation engineering courses, they often arrive in their senior project course with disjointed view of engineering practice and have an insufficient appreciation and understanding of the complexity and multifaceted nature of real-world development environments. In addition, students often lack sufficient team, analysis, and communication skills. Therefore, it is imperative that engineering curricula introduce professional and real-world education throughout a curriculum. Many programs now introduce small team-based engineering projects in the early part of the curriculum. Case studies can be used to support and extend such experiences.

An excellent way to address the challenges of introducing real-world exposure through a curriculum is the use of a robust lifecycle engineering case study – that is, a case that engages the student in the engineering of a complex system throughout its "life": system definition, preliminary and detailed design, implementation, verification and validation, and operation and maintenance. The literature describes a variety of ways that case studies can be used to support learning and research, sometimes referred to as the "case method" (Friedman & Sage, 2004; Herreid, 1994; Davis & Wilcock, 2013). Although as

mentioned above, case studies should be viewed as active learning tools; the case method can be mixed and matched with other pedagogies such as lectures, guided discussions, and project work. Case studies can be used to supply the background needed for specific problems and design projects; serve as subjects for class discussions; or they can be used to motivate further study, and to identify and formulate research problems.

DigitalHome, a Software Engineering Case Study

For several years the authors have been involved in a case study project that focuses on the development of a DigitalHome (DH) system (Salamah, Towhidnejad, & Hilburn, 2011). The DigitalHome Project, when completed, will cover the complete life-cycle development of a software product (project management, requirements analysis and specification, design, implementation, testing and maintenance).

The DH Project is based on a scenario about a real-world, but fictitious, company, HomeOwner, which is the largest national retail chain serving the needs of home owners. HomeOwner, based on market and technology research, decides to invest in the development of a "smart" house, the DigitalHome system. The development team of the DH is a fictitious group of developers with diverse cultural as well as technical backgrounds. The DH has following features:

- The DH system shall allow any Web-ready computer, cell phone or similar device to control a home's temperature, humidity, lights, security devices, and the state of small appliances.
- The communication center of the DH system shall be a personal home owner Web page, through which a user can monitor and control home devices and systems.
- The DigitalHome shall contain a master control device that connects to the home's

broadband Internet connection, and uses wireless communication to send and receive communication between the DH system and the home devices and systems.

- The DigitalHome shall be equipped with various environment sensors (temperature sensor, humidity sensor, power sensor, contact sensor, etc.). Using wireless communication, sensor values can be read and saved in the home database.

- The DH system shall include programmable devices (thermostats, humidistats, contact sensors, and small appliance and lighting power switches), which allows a user to easily monitor and control a home's environmental characteristics from any location, using a Web ready device.

- The DH system shall include a DH Planner, which provides a user with the capability to direct the system to set various home parameters (temperature, humidity, and on/off appliance and lighting status) for certain scheduled time periods.

- The interaction between the DH and the surrounding environment shall be simulated. Figures 1 and 2 below show the DH Planner GUI and the DH simulator

The Case Study Project focuses on developing a complete set of artifacts associated with software development (e.g., Requirements Document, System Test Plan, ...) as well as case modules (a "mini-case studies"). These artifacts and case modules are related by being part of and derived from a single case, the development of a single software product. Each case module relates to an artifact or activity involved in the development of the product. In addition, each case module is

Figure 1. DH planner GUI

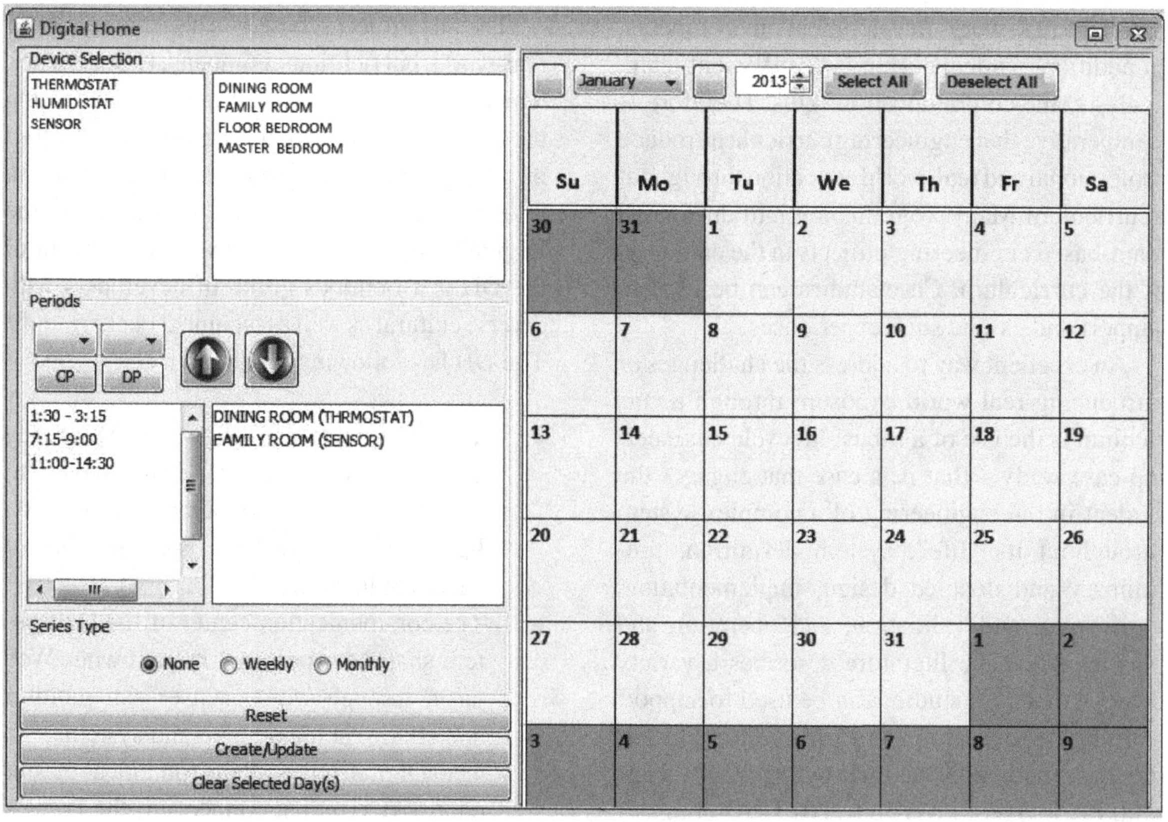

Figure 2. DH simulator

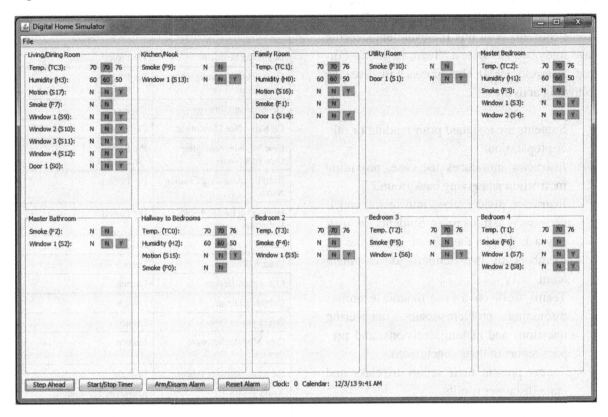

framed as part of a product development narrative, using a scenario format, which involves characters and incidents that would be part of an actual software development project (e.g., formation of a software project team, interaction with upper management, customer and user interviews, writing a use case description, formal inspection of a software artifact, designing a class interface, a design walk-through, system testing, etc.).

The DigitalHome Case Study is intended to cover the complete life-cycle development of a software product (project management, requirement analysis and specification, design, implementation, testing and maintenance). The initial phase of the case study project, concentrated on building a foundation for full development: research into case study teaching; identification of a case study problem; creation of a scenario framework; description of a launch of the software development team; development of a software development plan to be used as part of the case study; and development of several related case modules.

The initial phase of the case study project concentrated on building a foundation for full development: research into case study teaching; identifying a case study problem; creating a scenario framework; describing the launch of the software development team; fashioning a software development plan to guide development of the DigitalHome System; establishing a development process; creation of a DigitalHome need statement; analysis, modeling and specification of the DigitalHome requirements; development of a system test plan; development of a software architecture; and specification of the system components.

We have developed a set of DH mini-case studies/modules which are designed to engage students in active learning software engineering activities related to development of the Digital-Home System. With this life-cycle engineering case study, we advocate a "Team Learning Format" (Herreid, 1994). Our approach involves the following activities:

- Students are assigned prior reading or other preparation.
- Instructor introduces the case, providing motivation and giving background.
- Instructor divides class into teams and if necessary, assigns roles. Students are usually asked to be the role of one of the six members of the fictitious development team
- Teams work on a case module exercise: discussing problems/issues, answering questions and making decisions, and prepare report of their conclusions.
- Teams present their report to class and class discusses results.

THE DIGITALHOME MATERIAL

The development of the DH cases study has consisted of writing scenarios, developing DH artifacts, case modules and exercises. Thus far, Project Inception, Project Launch and Planning, and Requirements Analysis and Specification have been completed. In this section we provide a sample case module and exercise to better describe the content and directions for use of such modules. Table 1 shows a listing of the currently available Case Study material. The case study material is available for download and use at: http://www. softwarecasestudy.org/. Each case module consists of the following parts:

Table 1. Current DigitalHome case study material

DigitalHome Item	Software Development Phase
Beginning Scenario	Pre Project
DigitalHome Development Team Bios	Pre Project
Development Strategy	Pre Project
Customer Need Statement	Pre Project
Case Module: Assessing Customer Needs	Pre Project
Exercise: Assessing Customer Needs	Pre Project
High Level Requirements Definition	Pre Project
Launch Script	Launch
Launch Scenario	Launch
Conceptual Design	Launch
Context Diagram	Launch
Development Process	Launch
Case Module: Software Process	Launch
Exercise: Software Process	Launch
Artifact: SRS 1.2	Requirements Analysis
Artifact: SRS 1.3	Requirements Analysis
Use Case Model	Requirements Analysis
Case Module: SRS Inspection	Requirements Analysis
Inspection Process	Requirements Analysis
Inspection Defect Log	Requirements Analysis
SRS Checklist	Requirements Analysis
Inspection Summary Report	Requirements Analysis
Exercise: Requirements Inspection	Requirements Analysis
Case Module: Unit Testing	Testing
Exercise: Unit Testing	Testing
Case Module: Operational Profile	Operation and Maintenance
Exercise: Creating an Operational Profile	Operation and Maintenance
Case Module: Software Team Problems	Others
Exercise: Software Team Problems	Others

- **Case module title:** A descriptive title of the module and its intended use.
- **Students' prerequisite knowledge**
- **Learning objectives:** A list of students' abilities based on the completion of the case module
- **Keywords:**
  - ○ **Case study artifacts:** A list of DH material to be used in the case module and the accompanying exercise
  - ○ **Case Study Participants:** Members of the fictitious team taking part in the case module and exercise
  - ○ **Scenario:** A detailed description of the context of the real-world settings to be used in the case module and exercise.
  - ○ **Exercise:** Description of the exercise and deliverables of the students participating in the case module. The students are also provided an exercise booklet separate from the case module, which is intended for the use of the instructor only.
  - ○ **Resource information:** Any other reading or helpful material outside the scope of the DH material
  - ○ **Teaching notes:** Suggestions and alternatives on how to use the case module for teaching the specific topic.

Table 2 shows the Software Requirements Specifications (SRS) case module. We also refer to this module in later sections of this chapter.

The case module presented in Table 2 is intended for the use by the course instructor only. Each case module is accompanied by an exercise that is handed to the students for completion by the end of the session. The *Software Requirements Inspection* exercise is shown in Table 3.

For the exercise above, the students are handed an exercise booklet package to complete by the end of the session. The complete forms can

be found at www.Softwarecasestudy.org. The complete list of available material is displayed in Table 1.

USE OF THE CASE STUDY MATERIAL

Over the past four years, the case study material has been used in multiple courses in computer science and software engineering at multiple institutions such as Embry Riddle Aeronautical University (ERAU), the University of Texas at El Paso (UTEP), and Milwaukee School of Engineering (MSE). This included the use of developed artifacts as well as case modules. Table 4 shows the complete listing of the case study material that has been used in computer science and/or software engineering courses at Embry Riddle Aeronautical University. In the table, courses in bold letters are computer science courses while the rest are software engineering ones. The rest of the section sheds highlight on and provides results of the use of the Inspection Case Module and Exercise in multiple courses across ERAU and UTEP.

The Software Requirements Inspection Case Module

It is now well understood and accepted that software quality is a major influence on software cost and schedule, and on the ultimate success of a software product. Experience through the years has shown that quality cannot be an afterthought, but must be "engineered" into the product throughout the software lifecycle. Although the use of proven techniques and tools for project planning, requirements engineering, design, and construction can significantly reduce the injection of defects, human error and lack of capability play a significant role in the defect density of development artifacts. The most influential factor in addressing quality,

Table 2. Software Requirements Specifications (SRS) case module

Case Module: SRS Inspection
Prerequisite Knowledge: Understanding of Basic Elements of a Fagan Software Inspection Process.
Learning Objectives
Upon completion of this module students will have increased ability to: 1. Appreciate and understand the roles and responsibilities of an inspection team 2. Work as a member of an Inspection Team. 3. Work more effectively as part of a team. 4. Assess the quality of a Software Requirements Specification (SRS). 5. Describe problems in specifying the requirements for a software product. 6. Explain the inspection process. 7. Describe the value of the Fagan inspection process. 8. Gain experience working with and generating inspection artifacts.
Keywords: Customer Needs, Software Requirements Specification, Fagan Software Inspection
Case Study Artifacts
1. DH Customer Need Statement 2. DH High Level Requirements Definition (HLRD) 3. DH Background Scenario 4. DH Team Biographical Sketches 5. DH SRS, Version 1.3
Inspection Package
• Inspection Process Description • SRS Checklist • Defect Log • Inspection Summary Report Form
Case Study Participants
• The DH Team • Jose Ortiz, Director, DigitalHomeOwner Division of HomeOwner, Inc.
Scenario
In late August of 2010, HomeOwner Inc. (the largest national retail chain serving the needs of home owners) established a new DigitalHomeOwner division that was set up to explore the opportunities for equipping and serving "smart houses" (dwellings that integrate smart technology into every aspect of home living). In August and September of 2010, the Marketing Division of HomeOwner conducted a needs assessment for a DigitalHome product that would provide the computer and communication infrastructure for managing and controlling the "smart" devices into a home to best meet the needs and desires of home owners. The Marketing Division produced two documents: the *DH Customer Need Statement* and the *DH High Level Requirements Definition* (HLRD). In early September 2010, a five person team was assembled for the project and started a "project launch." After project planning was completed the team began work on requirements analysis and specification. The first version, 1.0, was completed in early October 2010 and versions 1.1 and 1.2 were completed by late October. In consultation with Jose Ortiz, the team has decided to carry out a formal Fagan inspection of the SRS, version 1.2. Jose has agreed to act as a customer on the inspection team, Michel Jackson is the author, Disha Chandra will be the moderator and other roles will be assigned in the overview meeting.
Exercise
1. As preparation for the case method, ask each student to read the Case Study Artifacts listed above. 2. Divide the class into a set of small teams (4-5 people). Each team takes on a role from the DH Team or as Jose Ortiz and then follows the inspection process: a. Planning: The inspection is planned by the moderator. b. Overview meeting: The author describes the background of the work product, describes the inspection process, and reviews the inspection forms. c. Preparation: Each inspector examines the work product to identify possible defects. d. Inspection meeting: During this meeting the reader reads through the work product, part by part and the inspectors point out the defects for every part. e. Rework: The author makes changes to the work product according to the action plans from the inspection meeting. f. Follow-up: The changes by the author are checked to make sure everything is correct.

continued on following page

Table 2. Continued

Resource Information
• [Fagan,1976] Fagan, M.E., Design and Code inspections to reduce errors in program development, 1976, *IBM Systems Journal*, Vol. 15, No 3, Page 182-211. (http://www.mfagan.com/ibmfagan.pdf) • [Fagan, 1986] Fagan, M.E., Advances in Software Inspections, July 1986, *IEEE Transactions on Software Engineering*, Vol. SE-12, No. 7, Page 744-751. (http://www.mfagan.com/aisi1986.pdf) • [Humphrey 2000] Humphrey, Watts S., *Introduction to the Team Software Process*, Addison-Wesley, 2000.
Teaching Notes
• This case module could be used in different level courses (from a software level introductory course in software engineering to an upper level or graduate course in requirements engineering or quality assurance.). • Since an appropriate inspection rate is 2-3 text pages per hour, it would best not to assign the entire SRS to each inspector, but rather some subset – e.g., 6-9 pages per inspector. • Assuming an adequate student preparation for the exercise, allowing students about four hours each for the exercise should be sufficient: assuming three hours for preparation and inspection, and one hour for the inspection meeting. • It would be beneficial to follow the exercise with a twenty to thirty minute discussion concerning the student team results. Some key points to include in the discussion are the following: • Discuss how closely the inspection process was followed: How well did the team conduct each phase? How well did students carry out their assigned inspection role (i.e., moderator, author, inspector)? • What was the quality of the inspection outline (NOTE: it is very helpful if the instructor completes the product inspection, and compare the students outcome to the instructor outcome) • How well was time managed? • Address any personal conflict or egoistic attitude displayed. • Student team members should be cautioned about a few things: • Leave their ego outside of the meeting room • Their job is to identify defects, not fix them. • The SRS is a Requirement document and not a Design specification; so inspect appropriately and check that design elements have not been introduced by the author or as a result of inspection. This does not include design constraints specified by the customer. • Come prepared to the inspection meeting – students need to spend appropriate effort during their individual inspections. • It is critical that students understand the data they need to collect (time and defect data), how to use the forms provided, and the importance being careful and accurate in collecting and recording data. If students do not properly collect and record inspection data it will be difficult to assess the effectiveness of the inspection. • If the course involves actual student development teams, this exercise could provide a good team building experience: it could be carried out near beginning of a course; it does not require deep technical knowledge; and everyone can participate. • If the time and the size of the class permit, the course instructor could take on the role of Jose Ortiz. • If the class is divided to multiple teams, if possible, it is best to have each team conduct their inspection in a separate room, thus eliminating distracting noise generated by other teams. • It would be best to hold the inspection meeting in class, since it would allow the instructor to observe this part of the inspection process. For a class of 20 or more, it would be good to have two instructors observe the inspection meeting. • A typical team size to accommodate all roles would be 4-5 members, with a maximum of 8 in order to accommodate a productive inspection meeting. • It would be best if all team members are inspectors. The author and moderator could inspect but not include their inspection data in the meeting discussion.

cost, and schedule issues is the early removal of defects; removing defects in the early phases of development can reduce the time and cost of defect removal by 10 to 100 fold, as compared to removal in test or in the field (Fagan 1976, Fagan 1986).

One of the most effective, proven techniques for defect removal is software inspection. In (Fagan 1976), Fagan states that software inspections at IBM have "enabled higher predictability than other means, and the use of inspections has improved productivity and product quality." (Parnas & Lawford 2003) assert that "In addition to finding errors in code and related software documents, inspection can also help to determine if coding style guidelines are followed, comments in the code are relevant and of appropriate length, naming conventions are clear and consistent, the code can be easily maintained, etc.." Although Fagan inspection techniques are used widely in industry, there are still many in the commercial and academic computing communities who do not understand their value, misuse the technique, or are unaware of the technique.

Table 3. Software Requirements Inspection exercise

Requirements Inspection Exercise
Scenario
In early September of 2010, HomeOwner Inc. (the largest national retail chain serving the needs of home owners) established a new DigitalHomeOwner division that was set up to explore the opportunities for equipping and serving "smart houses" (dwellings that integrate smart technology into every aspect of home living). In September and October of 2010, the Marketing Division of HomeOwner conducted a needs assessment for a DigitalHome product that would provide the computer and communication infrastructure for managing and controlling the "smart" devices into a home to best meet the needs and desires of home owners. The Marketing Division produced two documents: the DH Customer Need Statement and the DH High Level Requirements Definition (HLRD).   In early September 2010, a five person team was assembled for the project and started a "project launch." After project planning was completed the team began work on requirements analysis and specification. The first version, 1.0, was completed in early October 2010 and versions 1.1 and 1.2 were completed by late October.   In consultation with Jose Ortiz, the team has decided to carry out a formal Fagan inspection of the SRS, version 1.2. Jose has agreed to act as a customer on the inspection team, Michel Jackson is the author, Disha Chandra will be the moderator and other roles will be assigned in the overview meeting.
Learning Objectives
Upon completion of this exercise students will have increased ability to: • Appreciate and understand the roles and responsibilities of an inspection team   • Work as a member of an Inspection Team.   • Work more effectively as part of a team.   • Assess the quality of a Software Requirements Specification (SRS).   • Describe problems in specifying the requirements for a software product.   • Explain the inspection process.   • Describe the value of the Fagan inspection process.   • Gain experience working with and generating inspection artifacts.
Reading Assignment
Read the below Case Study Artifacts: • DH Customer Need Statement   • DH High Level Requirements Definition (HLRD)   • DH Background Scenario   • DH Team Bios   • DH Launch Scenario
Inspection Package
• Inspection Process Description   • SRS Checklist   • Defect Log   • Inspection Summary Report Form
Exercise Description
1. As preparation for this exercise, read the Case Study Artifacts listed above, except for the DH SRS. The DH SRS will be reviewed as part of the inspection process in part 3 below.   2. You will be assigned to a small development team (like the DH team).   3. Your team is to take on the role of the DH Team and prepare for the inspection of a Software Requirements Specification.   a. First, review the inspection package materials (list above).   b. Then follow the Inspection Process and any special instructions from your teacher.   4. Notes: a. Inspectors should take care to closely review their assigned portion of the SRS. A review rate of 2-3 pages per hour is reasonable.   b. Each team member must be conscientious about recording all the time spent during the inspection process.   c. The Rework and Follow-Up phases of the Inspection Process are not part of this exercise. Also, the part of the Exit Criteria "Software work product has been revised, removing identified defects" does not apply in this exercise.

Table 4. Use of the DigitalHome case study material

DigitalHome Case Study Item	Course Description
DH Beginning Scenario	Introduction to Software Engineering Software Quality Assurance Software Reliability
DH Team Bios	Introduction to Software Engineering Software Reliability
Dh Customer Need Statement	Introduction to Software Engineering Software Quality Assurance Software Reliability
DH High Level Requirements Definition	Introduction to Software Engineering Software Quality Assurance Computer Modeling and Simulation Software Reliability
DH SRS Version 1.3	Introduction to Software Engineering Software Analysis and Design Software Quality Assurance Files and Database systems Computer Graphics Software Reliability
Case Module: Software Inspection	Introduction to Software Engineering Software Quality Assurance Software Reliability
Case Module: Operational Profile	Software Reliability
Case Module: Software Team Problems	Introduction to Software Engineering
DH Simulator Document	Computer Modeling and Simulation
Case Module: Unit Testing	Software V&V
DH Simulator Design Document	Computer Modeling and Simulation
DH Simulator Detailed Design and Code	Computer Modeling and Simulation

An early effort of the DigitalHome case study was the analysis and specification of the software requirements for the system, which resulted in a Digital Home Software Requirements Specification (SRS). In addition to the specification of the functional requirements, the SRS includes a description of user characteristics, development constraints, the performance environment, and nonfunctional requirements specifying performance, reliability, and safety and security requirements. The SRS also includes a use case model. As part of the development of the requirements specification, the team developed a case module for inspection of the SRS (shown in previous section of this chapter). The learning objectives for the inspection case module go beyond assessing the quality of the SRS, but are intended to address critical software engineering education goals: appreciating and understanding the problems in specifying requirements; learning to work as part of a team; and using and following an inspection process. Although the SRS artifact is the chief focus of the inspection, the others (such as the background scenario and need statement) provide the setting and context to create a realistic environment for professional and effective software inspection exercise. The inspection package provides the sort of forms and tools used in a mature inspection process and help guide the students not only in performing an effective inspection, but in understanding how a best practice works.

Use of the Inspection Case Module in Class

The SRS Inspection Case Module (and variations of it) has been used in a variety of courses and workshops: a sophomore-level introductory course in software engineering; an undergraduate course in software quality assurance; an undergraduate course in software requirements; a graduate course in software architecture; a graduate course on Software V&V; a faculty workshop in software process; and a short course in software reliability.

In fall 2009 and spring 2010 the SRS Inspection Case Module was used in three software engineering classes (two sophomore level classes and one senior level class), in which nine inspection teams were formed to carry out the module exercise. The inspections concentrated on the functional requirement statements and used the Inspection Package described in Table 1: Inspection Process Description, SRS Checklist, Defect Log, and Inspection Summary Report Form. The Report Form includes summary data about the inspection. The inspection process and materials are based on the work of Fagan (Fagan 1976) and Humphrey (Humphrey 2000).

Although all teams followed the inspection process, and collected and recorded inspection data, some elements of the process were performed weakly and not all data was accurate or complete. Table 5 shows the data for one team, which could be considered typical. In the comments section of Table 5 we compare the team's results with benchmark goals for inspection data (based on work by Humphrey 2000). Although the defect removal rate and the overall inspection rate seem reasonable, the team identified only 10 major defects, while the DH project team had previously reviewed the SRS and identified over 20 major defects (many purposely seeded in the SRS). Two types of defects commonly missed were compound statements containing multiple requirements, and the testability problems with some requirements statements. One conclusion might be the inspection team was ineffective; however, we viewed this more as an education exercise, not strictly a quality assurance activity. By engaging students in a "close" reading of the SRS, we helped them to understand its meaning, to determine the degree to which it addressed the customer need statement, to evaluate the correctness, clarity and precision of the requirements statements, and to identify missing features. We should also note that the collection and analysis of team inspection data provides the teacher with excellent information on how the inspection was conducted and the degree to which the case module objectives were reached.

We believe this sort of "reading" is a critical, first step in helping students (or professionals) prepare for requirements analysis and specification tasks. Anyone that has been an observer in a

Table 5. SRS inspection team data

Inspection Feature	Team Data	Comment
Requirements Size	7 Pages	The team only reviewed a functional requirements section of the SRS.
Major Defects Found	10	Correction of a major defect either changes the program source code or would ultimately cause change in the program source code.
Major Defects Missed	2	Total defects of 12 were estimated using the "capture-recapture" method (Humphrey 2000)
Total Inspection Time	14.2 hours	9.2 hrs for preparation time and inspections time; one hr inspection meeting (times 5 people).
Defect Removal Rate	0.7 def/ hr	A benchmark goal is 0.5 def/hr (Humphrey 2000)
Inspection Rate	0.49 pg/hr	A benchmark goal is < 2 pg/hr (Humphrey 2000)

student inspection meeting has seen the degree to which a structured inspection activity engages students in disciplined reading and understanding of the inspection artifact: students discuss and debate the correctness, understandability and completeness of artifact elements; they reach an understanding of what the artifact should and should not contain; and they get a real sense of the meaning of "quality." In (McMeekin, Konsky, Change, & Cooper 2009) research was conducted on the utterances of students while engaged in code inspection teams: the research demonstrated various levels of learning beyond simple knowledge of the code content. This is a result similar to what we observed in the SRS inspections and that was illustrated in the defect logs and the inspection summary reports.

While inspecting an SRS does not make one a requirements analyst, it is an important first step; and it promotes the type of understanding of what requirements are, how difficult they are to get right, and their importance. This sort of learning is essential to any student who aspires to be a professional software developer. We believe that using case studies such as the inspection case module provides the perfect setting and engaging exercise for students to gain better grasp of software engineering concepts.

The experiences of inspection teams also provided valuable input to the DH project team of how well the SRS Inspection Case Module worked. In analysis of the overall inspection experience we did see ways in which the case module could be improved. We reviewed all inspection forms to insure they were understandable and convenient to use and we verified that we had included checks for compound statements and requirement testability in the inspection checklist. This motivated several changes to the module: revision of the inspection process and the SRS Checklist to improve clarity and completeness, and addition of a comment in the Teaching Notes and Exercise Booklet about the need to emphasize the importance of properly collecting and recording inspection data. We also think a requirements tutorial offered before the use of the case module would be helpful, especially for beginning students. The tutorial could include the following: the purpose and importance of an SRS; the typical structure and content of an SRS; a description of the inspection process, data, and forms; and examples of well-written and poorly written requirements.

SUMMARY AND FUTURE RESEARCH DIRECTIONS

The use of case study-base teaching has shown success in disciplines such as medicine and law. The DigitalHome case study project's goal is to enhance the use of case studies throughout a computing curriculum. This is done through the development of case study material that can be tailored to multiple courses and teaching techniques. The case study materials include case modules for teaching software development topics such as requirements analysis, inspections, OO design and construction, and testing among others. Case modules include teaching notes to guide instruc-

tors in teaching the target topic. Case modules are also accompanied by class exercises that employ scenarios to simulate real world environment and increase students' interest in the topics.

Although there is still much work to be done to complete the DH case study, we hope the material developed and our experiences with it provide insight into the value of such an approach. We believe the content, organization, and spirit of the DigitalHome Case Study provides a model for the development of other engineering life-cycle case studies. As a minimum, the DigitalHome can easily serve as a baseline for case studies/ modules for Computer Engineering, Electrical Engineering, and System Engineering. Hopefully, this discussion will broaden the reach of this work into other fields of engineering with a goal of building a community of collaborators that will contribute more significantly to the development of case studies, artifacts, and exercises.

REFERENCES

Accreditation Board for Engineering and Technology (ABET). (2013). *Criteria for accrediting engineering programs*. ABET Inc.

Davis, C., & Wilcock, E. (2013). Teaching materials using case studies. *UK Centre for Materials Education*. Retrieved from http://www.materials. ac.uk/guides/casestudies.asp

DigitalHome Case Study Project. (2013). Retrieved from www.softwarecasestudy.org

Fagan, M. E. (1976). Design and code inspections to reduce errors in program development. *IBM Systems Journal*, 15(3), 258–287. doi:10.1147/ sj.153.0182

Fagan, M. E. (1986). Advances in software inspections. *IEEE Transactions on Software Engineering*, 12(7), 744–751. doi:10.1109/TSE.1986.6312976

Friedman, G., & Sage, A. (2004). Case studies of systems engineering and management in systems acquisition. *Systems Engineering*, 7(1), 84–96. doi:10.1002/sys.10057

Grupe, F. H., & Jay, J. K. (2000). Incremental cases: Real-life, real-time problem solving. *College Teaching*, 48(4), 123–128. doi:10.1080/87567550009595828

Herreid, C. F. (1994). Case studies in science: A novel method of science education. *Journal of College Science Teaching*, 23(4), 221–229.

Hilburn, T., & Towhidnejad, M. (2007). A case for software engineering. In *Proceedings of the 20th Conference on Software Engineering Education and Training* (pp. 107–114). IEEE.

Hilburn, T., Towhidnejad, M., & Salamah, S. (2008). The DigitalHome case study material. In *Proceedings of the 21st Conference on Software Engineering Education and Training* (pp. 279–280). IEEE.

Humphrey, W. S. (2000). *Introduction to the team software process*. Reading, MA: Addison-Wesley.

Lethbridge, T. C., & Laganière, R. (2001). *Object-oriented software engineering: Practical software development using UML and Java*. New York: McGraw Hill.

McMeekin, D. A., von Konsky, B. R., Chang, E., & Cooper, D. J. (2009). Evaluating software inspection cognition levels using bloom's taxonomy. In *Proceedings of the 22nd Conference on Software Engineering Education and Training* (pp. 232–239). IEEE.

Metseker, S. (2006). *Design patterns in java*. Reading, MA: Addison-Wesley Professional.

Parnas, D. L., & Lawford, M. (2003). The role of inspection in software quality assurance. *IEEE Transactions on Software Engineering*, 29(8), 674–676. doi:10.1109/TSE.2003.1223642

Pfleeger, S. L. (2005). *Software engineering*. Upper Saddle River, NJ: Pearson.

Pressman, R. S. (2007). *Software engineering: A practitioner's approach*. New York: McGraw-Hill.

Salamah, S., Towhidnejad, M., & Hilburn, T. (2011). Developing case modules for teaching software engineering and computer science concepts. In *Proceedings of Frontiers in Education Conference (FIE), 2011* (pp. T1H-1). IEEE.

Tomey, A. M. (2003). Learning with cases. *Journal of Continuing Education in Nursing*, 34(1). PMID:12546132

ADDITIONAL READING

Bolinger, J., Herold, M., Ramnath, R., & Ramanathan, J. (2011). Connecting reality with theory - An approach for creating integrative industry case studies in the software engineering curriculum. In *Proceedings of the Frontiers in Education* (pp. T4G-1). IEEE.

Bolinger, J., Yackovich, K., Ramnath, R., Ramanathan, J., & Soundarajan, N. (2010). From student to teacher: Transforming industry sponsored student projects into relevant, engaging, and practical curricular materials. In *Proceedings of the Transforming Engineering Education: Creating Interdisciplinary Skills for Complex Global Environments* (pp. 1–21). IEEE.

Clarke, S., Thomas, R., & Adams, M. (2005, January). Developing case studies to enhance student learning. In *Proceedings of the 7th Australasian conference on Computing education: Vol. 42* (pp. 101–108). Australian Computer Society.

Ellet, W. (2007). *The case study handbook: How to read, discuss, and write persuasively about cases*. Harvard, Business School Press.

Fuller, A., Croll, P., & Limei, D. (2002). A new approach to teaching software risk management with case studies. In *Proceedings of the 15th Conference on Software Engineering Education and Training* (pp. 215–222). IEEE.

Ludi, S., & Collofello, J. (2001). An analysis of the gap between the knowledge and skills learned in academic software engineering course projects and those required in real projects. In *Proceedings of the 31st Annual Conference of Frontiers in Education,* (Vol. 1, pp. T2D-8). IEEE.

Lutz, M. J., Hilburn, T. B., Hislop, G. W., McCracken, W. M., & Sebern, M. J. (2003, November). The SWENET project: Bridging the gap from bodies of knowledge to curriculum development. In Frontiers in Education (Vol. 3, pp. S3C–S37).

Parrish, A., Hale, D., Disxon, B., & Hale, J. (2000). A case study approach to teaching component based software engineering. In *Proceedings of the Thirteenth Conference on Software Engineering Education and Training* (pp. 140–147). IEEE.

Sivan, A., Leung, R. W., Woon, C. C., & Kember, D. (2000). An implementation of active learning and its effect on the quality of student learning. *Innovations in Education & Training International, 37*(4), 381–389. doi:10.1080/135580000750052991

Varma, V., & Garg, K. (2005). Case studies: The potential teaching instruments for software engineering education. In *Proceedings of the 5th International Conference on Quality Software* (pp. 279–284). IEEE.

Weinberg, G. M. (1971). *The psychology of computer programming.* New York: Van Nostrand Reinhold.

Wohlin, C., & Regnell, B. (1999). Achieving industrial relevance in software engineering education. In *Proceedings of the 12th Conference on Software Engineering Education and Training* (pp. 16–25). IEEE.

KEY TERMS AND DEFINITIONS

Case Study: Realistic example used to illustrate a concept or technique. It involves the application of knowledge and skills, by an individual or group, to the identification and solution of a problem associated with a real-life situation.

Case Module: A mini case study which focuses on a specific aspect of a larger case study. It includes a realistic scenario, specific individual roles, team exercise, and instruction notes.

Software Inspection: A structured formal review of a software artifact by a team of software developers (or students).

Chapter 18

Incorporating a Self–Directed Learning Pedagogy in the Computing Classroom:
Problem–Based Learning as a Means to Improving Software Engineering Learning Outcomes

Oisín Cawley
The National College of Ireland, Ireland

Stephan Weibelzahl
The National College of Ireland, Ireland

Ita Richardson
University of Limerick, Ireland

Yvonne Delaney
University of Limerick, Ireland

ABSTRACT

With a focus on addressing the perceived skills gap in Software Engineering (SE) graduates, some educators have looked to employing alternative teaching and learning strategies in the classroom. One such pedagogy is Problem-Based Learning (PBL), an approach the authors have incorporated into the SE curriculum in two separate third-level institutions in Ireland, namely the University of Limerick (UL) and the National College of Ireland (NCI). PBL is an approach to teaching and learning which is quite different to the more typical "lecture" style found in most 3rd level institutions. PBL allows lecturers to meet educational and industry-specific objectives; however, while it has been used widely in Medical and Business schools, its use has not been so widespread with computing educators. PBL is not without its difficulties given that it requires significant changes in the role of the lecturer and the active participation of the students. Here, the authors present the approach taken to implement PBL into their respective

DOI: 10.4018/978-1-4666-5800-4.ch018

programs. They present the pitfalls and obstacles that needed to be addressed, the levels of success that have been achieved so far, and briefly discuss some of the important aspects that Software Engineering lecturers should consider.

INTRODUCTION

Where is the engineering in software engineering (SE)? While there are many technical skills required in the analysis, design, development and implementation of software systems, ask an IT professional to characterize their profession, and you might just as likely solicit the response that they see themselves as an artist, as opposed to a scientist. Given that there is undoubtedly an important design (some might even say creative) element within the practice of SE, it would be reasonable to expect that our SE graduates are also supported in developing non-technical skills.

In addition, if we look at what the academic world has defined under the banner of SE, we clearly see the necessity to arm our graduates with many non-technical skills. Wasserman's eight notions (Wasserman, 1996), for example, include a software process element, which is fundamental for an effective discipline of SE. This software process element focuses on quality through the organization and discipline within the various SE activities. The Software Engineering Body Of Knowledge (SWEBOK)[1] is currently adding an additional knowledge area, titled "Software Engineering Professional Practice," which includes "… subareas of professionalism, group dynamics and psychology, and communication skills." Clearly there is a growing understanding within academia that such "softer" skills play an increasingly important role in the successful outcome of SE projects.

The experiences of two of the authors bears witness to a lot of what has been identified above. OC and IR spent 25 years between them working on SE projects, large and small, in both small and multi-national companies. Their experiences have shown that while technical knowledge is a requirement for much of the SE life cycle, other non-technical skills had been seen to be increasingly important as systems grew in complexity and the business functions became less tolerant with overdue and over-budget projects. Systems complexity, in this sense, is not only a technical concern but also relates to the change in team dynamics as the number of stakeholders and project participants increase. This type of complexity requires oral, written, interpersonal and team working skills that some authors argue our graduates are not being adequately equipped in when compared to their technical abilities (Davies, 2000; Cotton, 1993; Connor and Shaw, 2008). We have recognized that, when using Problem-based learning in our classes, we can provide students with these technical and non-technical skills.

WHAT IS PROBLEM-BASED LEARNING?

"Problem-based learning (PBL) is apprenticeship for real-life problem solving, helping students acquire the knowledge and skills required in the workplace" (Dunlap, 2005). PBL has a long "intellectual history" with its origins in the "philosophies of rationalism and American functionalism" (Dewey, 1929; Schmidt, 1993). Current day PBL emerged in the 1950's and 1960's in Case Western Reserve University and McMaster University respectively (Prince & Felder, 2006). In the late sixties, Howard Barrows joined the faculty at McMaster University in Canada. During that time he collaborated with others and developed the approach to learning now called Problem-based Learning (Schmidt & De Volder, 1984). By the early seventies, Problem-based Learning was installed as a total approach to learning and instruction in the Faculty of Health Science at McMaster, with Barrows as its main proponent (Schmidt & De Volder, 1984; Schmidt, 1993b; Barrows, 1986; Barrows & Tamblyn, 1977). Inspired by the success of McMaster, universities around the world introduced Problem-based Learning into their curriculums. These include Maastricht University in the Netherlands, Newcastle University in Australia, the University of New Mexico, Harvard and Sherbrooke University in Canada. This resulted in widespread "cross fertilisation" and networking between the major universities (Barrows, 2000). Problem-based Learning has now spread well beyond the realm of medical education and is now being practiced in other disciplines such as business and engineering (Tan, et al., 2000; Tan, 2003). A number of leading universities now have dedicated PBL Websites. Coupled with this, leading journals on engineering education have dedicated entire issues to PBL (Prince & Felder, 2006).

By the 1980's and 1990's, the global economy was changing, increasing the focus on organizational performance, organizational structures and organizational change in general (Hales, 2007; Hallinger, Philip, & Bridges, 2007). Third level institutions started to come under pressure to respond to this level of industrial change and to produce graduates that were capable of operating in this changing environment (Hallinger & Bridges, 2007). A number of universities and practitioners responded to the challenge by implementing Problem-based Learning as a basis of their learning and instruction (Hallinger & Bridges, 2007). Maastricht University established its school of Economics and Business Administration adopting Problem-based Learning as its primary educational philosophy. This was revolutionary, as no examples of Problem-based Learning existed for Economics and Business Administration prior to this (Gijselaers, 1995). Similarly, Milter and Stinson from Ohio University established an MBA programme in the early 1980's also using Problem-based Learning (Milter & Stinson, 1995). Early in 1987, Stanford University, School of Educa-

tion implemented their Masters programme for administrators also using Problem-based Learning (Bridges, 1992). In the early 2000's the University of Colorado introduced a capstone course on software engineering selecting Problem-based Learning as their method of instruction (Dunlap, 2005). Nelson (2003) explores how he successfully taught software development to graduates also using PBL (Nelson, 2003; Prince & Felder, 2006). In the mid 2000's the University of Limerick (Ireland), and the National College of Ireland implemented Problem-Based Learning at varying levels within their institutions. These implementations ran from full curriculum implementation to single modules across a range of disciplines, including, medical, business, civil engineering and software engineering.

Given the widespread use of Problem-based Learning, it is not surprising that a number of variants have emerged over the years. By the mid 1980's, the term Problem-based Learning was being used extensively in a wide range of educational methods (Barrows, 1986). Consequently many attempts have been made to explain the concepts of Problem-based Learning (De Graaff, 2003). Barrows (2000) focused on the concepts of student-centered learning, small groups, the teacher as facilitator and the importance of the problem. Barrows alluded to his version of Problem-based Learning as authentic Problem-based Learning (aPBL) (Barrows, 2000; Barrows, & Wee, 2010).

While the Barrows Model (2000) has its origins in the medical profession, it has expanded into many different educational disciplines and has evolved into a distinct educational method (Barrows, 2002; Hmelo-Silver & Barrows, 2006). Barrows (2000) consistently reiterates his core model but was aware of the many variants of Problem-based Learning that had evolved since its introduction into medical education in the mid 1960's (Barrows, 1996). However he continued to remain faithful to his core model which contained the following characteristics (Barrows, 1998; Barrows & Tamblyn, 1980; Barrows, 2000):

- Learning is a student centered approach
- Learning happens within the small collaborative group using a structured process
- The teacher operates as a facilitator
- The problem is encountered first and is the main stimulus for learning
- Clinical problem solving skills are developed through interaction with the trigger or problem
- It is through self-directed learning that new information is accumulated

Student-centeredness has its foundation in the theory of social constructivism (Hmelo-Silver & Barrows, 2006). Problem-based Learning facilitates the social construction of knowledge as the learners work through ill-structured real world problems (Schmidt, 1993). Students assume responsibility for their own learning, and work collaboratively in small groups that are not teacher-centered (Barrows, 1998).

Barrows (2000) is very specific regarding the authenticity of the problem. This is also consistent with Dewey (1929) thinking that the problem should reflect real life events and should be the "starting point for learning" (Dewey, 1929; Schmidt, 1993). Barrows stressed the importance of "real patient problems" that the student will face in a work related environment. Barrows (2000) argues that without authentic problems that challenge the students, it will be impossible to develop proper "problem-solving skills." Hmelo-Silver (2004) agrees with Barrows and alludes to the fact that it has been through cognitive research and experience that Problem-based Learning practitioners have been able to identify the characteristics of good problems (Hmelo-Silver, 2004). However one life-long learning skill that is also developed in a Problem-based Learning environment is Self-Directed Learning (Barrows, 1986; Barrows, 2000).

SELF-DIRECTED LEARNING AND SMALL GROUPS

"Self-Directed Learning is an important approach for Professionalism" (Lahteenmaki & Uhlin 2011). How could self-directed learning contribute to the development of the Software engineers in terms of their level of professionalism? Barrows (2000) argues that teachers should trust the student to do their own Self-Directed Learning and dig out the material required to problem-solve. Gijselaers and Schmidt (1990) argue that the quality of the problem is of significant importance to the self-directed learning process. They argue that it impacts the amount of time that the student spends on self-study (Gijselaers & Schmidt, 1990). Perrenet et al. (2000) argue that engineering unlike medicine, has a hierarchical structure and care needs to be taken in the case of the self-directed learning process. Students should not be allowed to by-pass any critical topics as incorrect learning of fundamental concepts may impact their understanding of future concepts (Prince & Felder, 2006). Lahteenmaki and Uhlin (2011) argue that reflection plays a large part in the self-directed learning process. They explore the principles of cognitive psychology argued by Gijselaers (1996) to explain that learning is a construction from prior knowledge and that reflection plays a large part in the learning process (Lahteenmaki & Uhlin, 2011). While students spend time on self-study, they also work collaboratively in small groups. Barrows model (2000) suggested a group size of five to eight - or even nine - students (Barrows, 1996). However, Gijselaers (1996) uncovered situations where the group size was increased to twelve (Gijselaers 1996). Barrows (1996) accepted a large group size, but only under particular circumstances and in a very controlled environment (Barrows et al., 1986). A new phenomenon has arisen in Problem-based Learning which may be called "Small Group Creep": adding one more group member because it will not make a difference (Gijselaers, 2011). The concern here is that the benefits attributed to Problem-based Learning and small group learning will be lost in the interest of institutional economics and cost saving. This could affect other Problem-based Learning resources such as those of the facilitator.

PROBLEM-BASED LEARNING IN SOFTWARE ENGINEERING

Software Engineering can be seen as a technical subject in which students are expected to develop skills such as programming, software and systems design, architecture design and networks. However, these skills are no longer sufficient for a world of work which requires software engineers to collaborate with others, to understand problems and to work in cross-functional domains with which they would not be familiar. Richardson & Hynes (2008) argue that curriculum developers need to provide both content and processes that develop specific sector skills. In so doing, institutions would go a long way to preparing students for the commercial environment that they are facing into (Richardson & Hynes, 2008).

Therefore, the education of Software Engineers for the 21st Century requires more innovative approaches then the traditional didactic method of teaching (Vat, 2006). Traditional Software Engineering courses are often accused of stifling students' independence and imagination (Vat, 2006). In some cases, tutors have selected projects and team leaders but by in large have ignored the application of real world problems (Shim et al., 2009). Today's Software Engineering graduates require a wide range of characteristics including, teamwork, ability to work under pressure, customer focus and the desire for continuous learning and self-oriented learning (Shim et al., 2009; Vaughn,

2001). Therefore, it is easy to understand why software engineering students feel that software engineering is complex, requiring as it does social skills as well as technical competencies (Shim et, al 2009). One pedagogical approach which can address the challenges facing software engineers is Problem-based Learning (Dunlap, 2005; Vat, 2006; Shim et al., 2009). While Vat (2006) argues that software engineering education has always used well-defined problems, a change in mindset is required. He suggests a need for collaboration, skills development and lifelong learning as opposed to the fixed stop-start nature of current educational practices.

At Colorado University the developers of the capstone course selected PBL as a method of instruction as they considered that there was a strong line between the Software Development Life Cycle (SDLC) and PBL (Dunlap, 2005). Their aim was to expose the students to the real world of software engineering. This involved interaction with a real client, the formation of a software engineering project team and the preparation of the request for a proposal (RFP). In their course design Dunlap and her team matched the stages of the Barrows model to the SDLC model as they considered both models reflected the type of activities Software Engineers would be exposed to in a real life project (Dunlap, 2005). Richardson and Delaney have also reported on their use of PBL for educating MSc students in software process quality (Richardson & Delaney, 2009, Richardson & Delaney, 2010).

Introducing problem-based learning into the software engineering classroom takes time and commitment not only from the tutors' and students' point of view, but also from the institutions as a whole. Preparing software engineers for the 21st century may not be easy but the positives will outweigh the negatives. This could be achieved by using innovative and inductive teaching methods such as PBL.

THE PBL IMPLEMENTATIONS

Introduction

In this section we describe in detail two case study PBL implementations. Although both implementations advocate the same learning and teaching pedagogy and have comparable class sizes, it is important to point out that they are performed in different organizations, with differing student profiles and assessment strategies. The NCI case study was focused at an introductory class (2nd and 3rd year computing) while the UL case was more advanced (MSc and 4th year computing). PBL assessment within NCI case was confined to 40% of the module marks, while in the UL case it was 100%. Interestingly both cases had a good mix of international students and also students with some work experience, with the MSc course in UL being the most culturally diverse. These differences are worth bearing in mind, since they affect the way in which both lecturer and students interpret and engage in the learning process. Dahlgran and Dahlgran (2002) argue that the learning outcomes have a significant influence on the students study strategies. Through their empirical research on three academic programmes at Linkoping's University in Sweden, they have shown that not all academic programmes use learning outcomes in the same way. The variation of how the learning outcomes were used by the students and their intended use by faculty unearthed a potential difference in "educational culture" and student's interpretation of problem-based learning. This may be an aspect that could be explored in our future research.

Department of Computer Science and Information Systems, University of Limerick, Ireland

OverviewPBL has been used by one of the authors (IR) as the method of teaching Software Quality and Software Process Improvement to MSc in Software Engineering and 4th year BSc in Computer Systems classes since academic year 2009/2010. Having initially introduced PBL to a second-year undergraduate class (Richardson & Delaney, 2009), she recognized its potential as a teaching method for more senior classes within the department.

The Software Quality and Software Process Improvement modules were initially taught to MSc and 4th year students for 2 hours over 12 weeks with supplementary 1 hour tutorials as required. Lectures were generally presented on PowerPoint slides. Although discussion was encouraged, the lecturer did most of the talking. Inter-student interaction was minimal. Journal and conference research papers were assigned as reading material, but were rarely read by students. Up to two lectures were presented by guest lecturers, generally from a software engineering project manager. Within this environment, classes were seen as theoretical, uninteresting for lecturer and students, and students found it difficult to understand software quality and process concepts. The lecturer did not observe that students understood the topic nor its importance, and was concerned that they completed the module without an in-depth understanding of what is really meant by 'software quality'! Assessment for the module was divided between a team project (40%) and final exam (60%). Project teams were self-selected, worked outside of class time, and presented a final paper at the end of semester. During the semester, the project was never discussed in class, and any learning was not shared within the class. There was no record of individual involvement in the project, nor was individual's participation identified. The project was normally a case study requiring domain knowledge which the students were unlikely to have, such as manufacturing or finance. The final exam dealt with concepts presented in class. While those students who did the assigned reading performed well in the exams, there was no incentive for students to actively research for the module. No advantage was taken of student background and experience.

Software Quality and Software Process Improvement PBL Modules

In the Department of Computer Science and Information Systems at the University of Limerick, PBL for Software Quality and Software Process Improvement (SQ/SPI) modules has been implemented five times since academic year 2009/2010, twice with 4th year classes and three times with MSc classes. Class sizes have ranged from 14 to 28 students who come from a variety of backgrounds - full-time/part-time students, many years/little or no industry experience, Irish/international students, prior/no prior PBL experience and native/non-native English speakers.

As previously stated, the success of the PBL curriculum is dependent on the development of a good problem. Potential problems were considered, focusing on the requirements for an engaging and interesting problem which would motivate the students to look for a clear and deep understanding of SQ/SPI concepts. It should also relate to a familiar situation allowing students to focus on solving the problem rather than on understanding the domain. For these reasons, e-Health software quality research was identified. As IR (lecturer) was researching e-Health, use of this topic would be beneficial to her facilitation of the module, also having the advantage of bringing her research to the students. The problem trigger was presented to the students during the second week of the module. It involved the students viewing an online video titled "Emergency Department – A Day in the Life"[2]. The students were then asked to develop and write the software quality plan for a hospital.

As the video commences, a patient is taken in from ambulance on a trolley into a hospital. This is the last time we see any patient. The focus is on hospital computer hardware systems, such as bedside monitors, and on staff discussions around computer screens. Just watching this video a few times in class allowed for discussion around the use of computing equipment and medical devices in hospitals, the realization that where there is

hardware software is also present, and further discussion on software quality required by safety-critical healthcare systems.

PBL's introduction led to changes in class organization. A two-hour lecture was used. Students were split into groups of four, with three or five students in some groups depending on numbers. International students were considered. Depending on class make-up, groups in some classes consisted of students from one country, while in others, there was a requirement that groups would be global, with a mix of nationalities and language in the group. During each scheduled session, students joined their groups immediately to work on the problem. The lecturer's role changed to that of a facilitator. She circulated between the groups, discussing issues that arose, ensuring that all groups worked towards a relevant software quality plan by directing them towards relevant research. On occasion, she gave 10-15 minute lectures on specific topics. For example, one lecture ensured that students understood the characterization of processes as Organization, Management, Engineering, Customer-Supplier and Maintenance processes, thus removing the exclusive focus on Engineering processes. Additionally, at the end of class, group discussions were summarized during a short 5-10 minute discussion with all students.

During group discussions, as in PBL theory, students filled specific roles: discussion leader, recorder, observer and team member. They kept minutes of meetings and reviewed these each week. Actions from the previous week were discussed and students circulated papers that they read since the previous session. They had Internet access and freely viewed papers or other information they needed during the meetings. Some groups stayed in the classroom to conduct meetings; others moved to the adjacent café to hold their meeting and discussion. At the end of each meeting, actions for the following week were distributed. Discussions within the groups were varied and interesting. Students discussed personal situations where they had seen software and hardware system use in hospitals. These included observations at the lack of concern for privacy of patient data, the lack of integration of patient data, and the copying of data from medical devices to paper charts. In the literature, while they found that regulation is integral to the production of medical device software, they noted that regulations are not observed within many healthcare situations. In addition to these discussions, to ensure an understanding of quality requirements, and to give students an insight into the hospital quality system, a clinical quality auditor from the Health Service Executive visited the class after they had researched the problem for 4 weeks. She gave a short presentation followed by a 90 minute question and answer session with the students. She was able to give them further examples regarding how software is used within hospitals and what development practices are used there.

The problem-based learning modules have been continually assessed with no final exam. For assessment, a group paper (25%) and two presentations to the full class (12.5% content, 12.5% individual presentation skills) are required to demonstrate the students' knowledge of the concept of software quality. This knowledge includes the ability to discuss regulations and software processes. Presentations are reviewed by IR and another lecturer who is familiar with the topic. For class participation (10%), IR observes whether students are bringing knowledge from external sources and how well they engage with other group members. Students are also orally examined individually four times, each worth 2.5%. An example would be to have individual students present the group's progress to date. The final part of the assessment is a presentation of an individual portfolio (30%). This includes summaries of papers read, a personal reflective journal, meeting minutes, and an outline of individual project participation.

A summary of the student and lecturer experience is described in the next two sections. This summary was collected from discussion within the classes, formal interviews with some students, reflective journals kept by students and lecturer, and informal feedback from individual students.

Improvements through PBL

SQ/SPI classes are now very interactive with input from students and lecturer alike. Previous industrial experience, medical experiences and international experiences are brought into the discussion and learning by the students, one of whom has described PBL as *a very interesting and innovative way to learn*[3]. Additionally, students regularly receive feedback from their peers, from the lecturer as facilitator and from their assessments. It has been very important to students to have the subject matter expert (clinical quality auditor) available for the question and answer session. This gives them an opportunity to meet someone who is working at the coalface, who is very knowledgeable regarding the importance of good software quality in healthcare and presents an understanding of the difficulties that arise when quality is sub-standard. This session, held at a pivotal point in the module, has been recognized by students and lecturer alike as being invaluable to the groups' progress.

Both students and lecturer are enjoying the classes, and they have given the *students ...an opportunity to get to know the rest of my classmates better...* They have *a real sense of solving a problem*, and they are learning *from each other in a "Student" way*, while also putting *in more work...* Through reading and reviewing academic papers, discussion, peer learning, facilitation and the short lectures given, student knowledge has increased. This is obvious through assessments and reflective journals, and was not observed when this module had been taught previously. Students themselves recognize this: *Personally, I believe that I have learnt more through PBL in the first 8 weeks than I would have in a standard classroom environment.* They also notice that *...the things you learn through ...stay with you longer.*

Student attendance has improved, and students are very conscious of disrupting their group if they are unable to attend for a particular reason. Students work consistently, and each week it is noticeable that the groups are progressing with their projects. Students have been reviewing academic papers, which is a requirement for this level, but something which has not obviously been undertaken in the past. In addition to learning about SQ/SPI, students have the opportunity to acquire soft skills, which have become very important for software engineering students. We presented evidence previously (Richardson et al., 2011) that students' skills in communication, team working, problem-solving, decision-making, leadership, management and time management have improved through participation in the PBL SQ/SPI module. In addition they brought a work ethic and motivation to the module that was not seen previously.

Difficulties Implementing PBL

Although this case describes where PBL was introduced with groups of senior students (MSc and 4th year), they had not normally attended any PBL module previously. Therefore, it can be difficult to get students into the process at first. This was particularly true when the class was mainly international and not native English speakers. In some cases, their prior education seemed to be very much at odds with what was required here, for example, self-directed learning, and students found this concept difficult to grasp. This required intensive work as facilitator to get the project started within the class. All students who participate have to understand their role within the group, the roles were rotated from week to week. However, this caused problems and maybe if they retained the role for a longer time period there could be some continuity and people could get immersed in the given role. There is also recognition that their active participation in the problem was the key to their learning and

when people did not become involved sharing of the knowledge is reduced. This is also true of group projects, and in the PBL situation due to the interaction in class and regular feedback can be more controlled than in the traditional classroom. However, once students realized that lack of participation caused significant problems and was being actively monitored by the lecturer, their work rate improved and consequently their progress in the module improved.

Additionally, there was a requirement to carry out assessments throughout the semester. This consisted mainly of oral examination and observation of the students in their work. As this was not the normal way of assessment, this proved quite difficult for the lecturer.

Another concern was whether this concept suited all those involved in the class. We recognize that the same learning technique many not be universally successful, and this was also noted by the students: *I don't think it suits some people in my group.*

PBL within SQ/SPI

Using PBL within the SQ/SPI module should allow for the:

- Provision of an understanding of software quality and software process improvement concepts;
- Provision of soft skills such as teamwork, communication and problem solving;
- Introduction to up-to-date research, demonstrating how this could potentially be useful to students' in the future.

Taking each of these points into account, the implementation of PBL into the module has been successful. It is not without its difficulties, and within the context of class profile, the mode of implementation sometimes has to be modified as the module progresses. However, when one considers this compared to the traditional lecturing mode, lecturers can see that PBL shows up the difficulties experienced much earlier in the module, and changes can be made before the final examination, which is often where lecturers realize that students have not attained the knowledge they strive to impart.

School of Computing, the National College of Ireland, Dublin, Ireland

Overview

In the School of Computing at National College of Ireland we were faced with the same problems that many Higher Education institutions seem to struggle with. While students did well in exams and continuous assessment, employers of graduates felt that some students lacked communication and problem solving skills that are essential for the job roles they were offering. We were looking for a structural change in our teaching that would help to develop these skills further in our students. Facilitated by a visiting researcher who is an expert in PBL, we conducted some preliminary trials in 2009. Starting from the academic year 2009/2010, we converted several modules to PBL including subjects like Programming, Software Engineering, Artificial Intelligence and Personal & Professional Development. We were hoping to enhance students' skills development, but also to increase their motivation by applying new concepts to real life problems.

Today, we are delivering a range of modules through PBL to about 300 students each semester. In this section we summarize our experience with the implementation of PBL and reflect on the issues that may arise. We begin with an experience report which describes a typical implementation of PBL for Software Engineering. We provide data on how students experience the PBL process and how their assessment results are affected. To conclude we discuss the perceived strengths and weaknesses of using PBL for software engineering education and illustrate the barriers encountered so far.

Exemplary PBL Implementation Experience

This section summarizes one author's (OC) experiences with the implementation of PBL in a Software Engineering module. The module 'Introduction to Software Engineering' was taught to a combination class of second year students on the BA in Management of Technology in Business (BAMTB), and Higher Certificate in Computing (HCC) courses. In total there were 48 students (BAMTB 21, HCC 27). The module had three contact hours per week, which would typically be allocated as a two hour lecture and a one hour tutorial. To incorporate a PBL approach, the assessment strategy included a project component. The project was worth 40% of the final grade and, using a self-directed learning approach, required the class to work in groups and submit group projects.

Most students would already have had some introduction to the concepts of PBL from their first year, however, during the introductory session it was clear that some students did not know what the learning approach entailed, and other students seemed interested in (re)hearing the historical and theoretical background to the pedagogy.

Following this introductory session, the imperative was to form the class into groups. This was found to be a somewhat difficult task. A very important consideration is the size of the groups, with literature suggesting group sizes of 4 or 5 being effective (Delaney & Mitchell G., 2002). The lecturer allowed slightly larger groups to form not fully realizing the possible consequences this can have. The average group size came to 5.8 members. He also allowed the class to form their own groups, and as is to be expected some students were not able to find a group to join. In the end OC had to form a new group composed of just 3 people.

Central to a PBL approach, a trigger was introduced for the project which was the YouTube video of the cinematic trailer for the computer game 'Assassins Creed: Revelations'[4]. The students were then told that they had to design and build (to a prototype stage) a computer game for their project.

The formative assessment strategy consisted of specific software engineering artifact delivery every two weeks. The schedule was as follows:

- **Week 3:** Group Submission of a Requirements Specification (20%)
- **Week 5:** Group Submission of the Analysis Diagrams (20%)
- **Week 7:** Group Submission of the Architecture and Design (20%)
- **Week 9:** Group Submission of Prototype implementation (15%)
- **Week 11:** Group Submission of Test Plan, Unit Testing (20%)
- **Week 13:** Group Submission of a presentation and demonstration of a working prototype (5%)

Each two-week period began with a specific trigger indicating the deliverable for that section of the project. For example the first deliverable was for requirements specification, and the trigger was a 'Dilbert' type cartoon depicting a manager telling a developer they do not have time to gather product requirements so they should just start developing the system. Similar triggers were used for the other phases, but care was always taken to ensure the content of the trigger was both relevant and instructive for the students.

The trigger session was followed by a one week period allocated for students to work within their groups towards an understanding, and development, of the particular deliverable. OC, as facilitator, monitored the groups' progress, discussed any specific questions the groups had, and held short impromptu clarification/instructive sessions if a particular issue identified was relevant to the whole class.

During the second week, the lecturer delivered what was referred to as a 'landscape lecture' on the relevant topic. This would include any theoretical and practical components of the subject. Tutorial sessions were also scheduled where the students were required to generate the necessary artifact for a sample project, thereby assisting them in their PBL projects.

As part of a formative assessment strategy, each of the deliverables was reviewed by the lecturer and feedback given to each group in the following week. It is important to ensure feedback is given as scheduled and is constructive in nature. With the time demands of a typical lecturer at NCI being quite high (due to teaching multiple courses), it can at times be difficult to adhere to that process, but if the lecturer misses or delays feedback it can disrupt the learning process since the students start to move onto the next deliverable.

One other important aspect to the PBL project was that each team was asked to keep a journal of the group's activities. They were asked to record the important group activities such as team meetings and who attended, questions/topics that they felt they needed to research or ask the lecturer about, the tasks assigned to each team member, and any team issues that might have arisen. The journal was to be updated weekly and would be used at the end of the course to assist in the marking process. To facilitate this we made use of Moodle's online group folder functionality which allowed the lecturer to set up individual group access to a private area on Moodle. Groups were only able to see their own journal entries and each member was able to add their own comments to their group's journal.

Student Experience

Although most students engaged fully in the process initially, there were some negative attitudes which quickly began to surface. The realization that they really would not be getting direct answers to their questions was something that they were unaccustomed to and dissatisfied with. All student questions were listened to and guidance was given by the lecturer, however, feedback from the students clearly indicated that they felt they needed more direction. The skill required of the facilitator is to be able to balance this student desire to be told the answer, with the PBL methodology which calls for guidance, discussion, and explorative study by the students.

The students worked in groups and would assign tasks to each other for research or development, and bring the results of that back to the next group meeting and/or update the group's journal on Moodle. Learning was evident through this but as expected, some groups worked better than others and at times group members felt they needed to consult with the lecturer about the lack of engagement from other team members. An aspect of Software Engineering is being able to work within groups and deal with these types of issues, so as facilitators we need to encourage the groups to resolve these types of issues internally within the group. Interestingly, some students reported that they did not necessarily want to get any team member in trouble, but the fact that a lack of engagement from other members could affect all their project marks, was something they were not prepared to tolerate. This aspect of the project, individual versus group marking, is something we will return to in the next section.

The final deliverable for the group was the full set of updated SE artifacts and a group presentation of their project and demonstration of their prototype. This was both challenging and enjoyable for the students. The challenge came in pulling all the individual contributions together into a cohesive package. As with many industry SE projects, multiple team members will have been assigned individual tasks which will need to be integrated into the final product, so this gives the students some practice in this area. The enjoyable part came in the form of the group working together on developing and presenting the prototype they had designed. A sense of pride was clearly

evident as the team members became inventive and resourceful in developing and choreographing the presentation. Again, this was an excellent exercise in a typical SE prototype demonstration to stakeholders. This, however, was one occasion where the group size became an issue. With larger groups it was difficult to ensure that everyone contributed to the presentation, and it therefore becomes more difficult for the lecturer to assign individual marks.

Assessment

Assessing a PBL module, in this case the project, can be a difficult task. The nature of the PBL learning process is inherently about learning and working within a group context. The Irish Third Level examination process, however, is about individual marks and we therefore have to find some way of allocating individual marks to each student. The lecturer's approach was to use a combination, thereby rewarding a good group effort but also rewarding the individual contributions which show evidence of achieving the learning outcomes of the module. This was achieved by utilizing a detailed grading rubric which broke the deliverables down into specific components with allotted marks. For example Figure 1 depicts the requirements engineering section of the rubric showing the breakdown of the marking scheme for that deliverable.

Having learned from assessment difficulties in previous years, at the start of the project the class was instructed to break up their proposed project into functional elements and assign one to each team member. They were each to deliver all the requisite artifacts for their own part of the project but present them all together into a cohesive final deliverable. This way the lecturer was able to allocate individual members marks based on what the group submitted. The students had access to the marking rubric from the start, and were therefore in no doubt about how the project

Figure 1. Sample from grading rubric

		Group Total	Team Members			
			Member 1	Member 2	Member 3	Member 4
Requirements Engineering (20%)	Detailed requirement descriptions - 4	0	3	3	3	3
	Based on IEEE template (Format, Clarity) - 4	0	2	2	2	2
	End users identified - 4	0	4	4	4	4
	Functional Requirements identified - 4	0	2	0	3	3
	Non-Functional Requirements identified - 4	0	2	0	3	3
	TOTAL	0	13	9	15	15

would be assessed. The group presented their project in a stand-up presentation and the lecturer posed questions to individual group members to ascertain their involvement and depth of knowledge. If it was evident that the group worked cohesively together and each member demonstrated competence in the various aspects, then a single group mark was awarded and each member received the same mark. However, if this was not the case then the project deliverables were examined in more detail to ascertain what mark each member should be awarded.

Lecturer Experience

PBL requires a mindset change in the lecturer. The first thing to understand is that the lecturer's role shifts towards that of a facilitator of a student-centered learning process. In fact, the process should not only be student-centered but student-driven. Accordingly, the lecturer needs to encourage students to seek the knowledge they require by getting them to pose questions, discuss different aspects of the topic within their groups, and assign roles and tasks to each group member for individual research. This is a very different role to the common didactic (lecture) style of teaching, and it requires perseverance. There is a great temptation to give the answers to student questions as this brings immediate satisfaction to the student and lecturer. However, this bypasses the learning process inherent in a PBL process.

An additional activity which the lecturer performed, similar to the groups' journals, was to keep his own journal of events or observations which he as the facilitator experienced. Since this also requires some personal reflection, some effort is needed to remember to keep adding to the journal. It proved to be beneficial in enhancing the academic review process at the end of the module.

Progression

In the following Semester, students encounter a follow-up module to "Introduction to Software Engineering" called "Object-Oriented Software Engineering." As students are already familiar with the PBL process at that stage, the induction session can be reduced. The module is focused on the Unified Modeling Language (UML). Accordingly, groups are required to produce sets of UML diagrams for a given system. They receive a series of landscape lectures as input, providing an overview of a particular method, but students then have to explore the details on their own initiative and find out how each method applies to their specific project. In line with the requirements of the previous module, students have to document the learning process and reflect on their learning in an on-line forum. A total of 40% of the total mark was assigned to the PBL project, the remaining 60% were assessed through an exam.

Student Feedback and Performance

We collected feedback from students on their learning experience as well as assessment results for the "Object-Oriented Software Engineering" module. Of the 51 students in the 2011 cohort, 25 responded to an on-line questionnaire sent out at the end of the Semester. The questionnaire comprised a set of rating questions on their experience and progress on a five-point scale (1: "not at all" to 5: "very much"). We also asked them open-endedly to list what they liked about the module and how it could be improved.

Overall, the feedback was not as positive as we would have expected initially. On the one hand, they indicated that they felt the approach promoted teamwork and that they participated actively in discussions (x=3.91). On the other hand, they rated the improvement of their critical thinking, problem solving and communication skills as neutral. They frequently referred to the change in lecturer behavior. When asked how to improve the module, students suggested that "guidance from the tutor could be improved" and "lecturers to assist more in solving problems." Moreover, when comparing their experience to a traditional course, students felt that they had learnt the subject less thoroughly (see Table 1).

This is however in contrast to the actual results. As can be seen from Figure 2, the actual assessment results based on similar exams and similar continuous assessments remained stable in comparison to the baseline of 2009. A one-way ANOVA reveals that the observed differences in the overall results and in exam results are within random variation. The only statistically significant change was in fact an increase of the continuous assessment results between 2009 against 2011 ($p<0.012$) and 2012 ($p<0.023$). This means, that students feel that they learn less, but their actual results remain largely the same.

Table 1. Results of feedback questionnaire for the module Object-Oriented Software Engineering on a five-point scale (1: "not at all," 5: "very much")

	N		Mean	Median	Std. Deviation
	Valid	**Missing**			
Have you found the topic interesting?	25	0	3.04	3.00	1.34
Have you enjoyed the topic?	25	0	2.60	3.00	1.41
Did focusing on real problems make the topic seem more relevant to your interests?	25	0	2.88	2.00	1.24
To what extent was teamwork promoted?	23	2	3.91	4.00	1.20
To what extent did you learn from one another?	24	1	3.29	3.50	1.30
Did you participate in the group discussions?	25	0	4.60	5.00	0.58
To what extent did your critical thinking skills improve?	25	0	3.12	3.00	1.20
To what extent did your problem-solving skills improve?	24	1	3.04	3.00	1.20
To what extent did your communication skills improve?	25	0	3.04	3.00	1.34
Would you like to participate in more PBL modules?	25	0	2.04	1.00	1.31
How well did you learn the technical material associated with this topic?	25	0	2.56	3.00	1.23
Considering the material you learned, do you think you learned it more or less thoroughly than you would in a conventional course?	24	1	1.79	2.00	0.93

Figure 2. Assessment results for the module Object-Oriented Software Engineering between 2009 (baseline) and 2012, for the average overall result, the average exam result and the average continuous assessment (CA) result. Error bars indicate 95% confidence intervall.

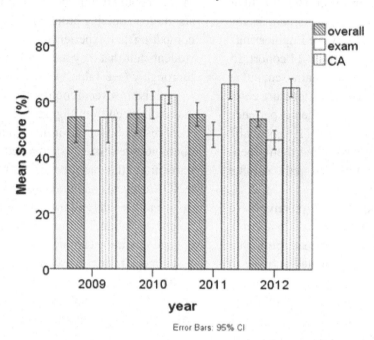

DISCUSSION

The Right Attitude

Barrows (1998) argues that student-centered learning can be "destroyed if not weakened" (pp 630-633) if it is bolted onto an otherwise traditional based curriculum. Within UL and NCI this is what was done with other computing modules, and indeed some of the NCI case study module, being delivered through a typical lecture style approach. As described, some negativity concerning the PBL process was experienced, however, this is not usual with the introduction of PBL, "…in practice, the self-directed learning of students is sometimes confined by the teacher's limited understanding of the learning styles, past learning experiences and aspirations of the students." (Chung & Chow, 2004). We feel it is important to consider the students' background when examining this. When we view it from within the

context of the Irish Primary and Secondary education system, where the predominant pedagogy tends towards authoritarian didacticism, then we may understand where this student frustration originates. Within Ireland, students at third level typically will not have encountered PBL before, whereas in some other countries they have made strides to incorporate it into the curriculum for secondary and even primary schools (Belland, 2010), (Kolodner et al., 2003). Holland, for example, is home to Maastricht University (MU), one of the first Universities to teach solely via the PBL method. Students who apply to MU are in no doubt about the teaching approach they will encounter. We suggest, therefore, that it would be interesting to compare the attitude of third level students to the PBL process between Ireland and Holland for example. This would be informative in terms of understanding how prior knowledge and exposure to the PBL process can help or hinder its use within the third level setting.

The 'Problem' with PBL

Both case studies clearly reported that students were looking for more direction. We feel this is partly a consequence of the previous point. While it is difficult for a lecturer (now facilitator) to do this, the provision of information other than that which is critical to get students starting work on the problem should be avoided. With a well-structured problem they should be able to reach their learning outcomes independently. The importance, therefore, of the problem trigger is something which needs to be highlighted. "Well designed and authentic problems are crucial to the success of PBL" and should be "…authentic, engaging, deliberately loosely-structured, linked to learning outcomes and key concepts, multidimensional, and graduate attributes and professional practice focused" (O'Grady et al., 2013). This requires some careful consideration within a SE context, where practitioners are not equipped with an established list of worked examples as they are in the Medical profession. Nonetheless, the literature for PBL in the SE domain is expanding and a growing body of knowledge on problem development and validation techniques is developing.

Breaking the Rules

Although PBL advocates self-directed learning, it is interesting to note that both case studies here incorporated short and specific lectures as part of the process. At certain points it was thought necessary to delve into a particular point to either clarify something, share the knowledge with the whole class, and/or direct the class in some manner. However as tutors that have embraced an inductive teaching pedagogy we would argue that students have different learning styles (Kolb & Fry, 1975; Prince & Felder, 2006). We used different interventions within our PBL cycles to scaffold and support the students in their learning process. Another proponent of student scaffolding is Lev Vygotsky. Within his social constructivist

view of development he argues that through collaboration and dealing with real problems true learning takes place (Harland, 2003). Vygotsky also argues in favor of the zone of proximal development, explaining that if you expose students to learning without the proper scaffolding i.e. outside their development zone then you will lose them altogether (Harland, 2003; Prince & Felder, 2006). Therefore one form of scaffolding in our PBL cycle was the use of short lectures.

Importance of Assessment

What is the best way to assess within a PBL environment? This is an interesting question given that PBL is learning within a group environment, but we as educators must provide individual assessments. Do we assess at the group level where each team member receives an equal grade? Such an approach evokes calls for fairness from students who feel they get penalized for the poor work of others. Indeed in an ever more competitive workplace for graduates, hard working students query how they can outperform their classmates if their individual effort is not being fully rewarded. Or should we assess solely at the individual level? In this case there is an argument for students to concentrate on individual learning, counter to the team working skills we would also like them to develop.

In defining the process of assessment Huba and Freed (2000, p. 8) explain that it is "the process of gathering and discussing information from multiple and diverse sources in order to develop a deep understanding of what students know, understand, and can do with their knowledge as a result of their educational experiences; the process culminates when assessment results are used to improve subsequent learning" (Huba & Freed, 2000; Levia & Quiring, 2008). In both the traditional teacher centered and Problem-based Learning approach, assessments fall into two main categories namely formative and summative (Levia & Quiring, 2008). However assessment within a student centered pedagogy such as PBL needs to be carried out in a different manner then those of a traditional teacher centered environment (Ramsden, 1992; Knight, 1995; Levia & Quiring 2008). Within a PBL cycle students establish their own learning outcomes, therefore, regular assessment within this process is required to ensure that the students achieve their course objectives (Mitchell & Delaney, 2004; Levia & Quiring, 2008). In addition, because students are working on authentic problems representing real world issues, an authentic assessment strategy concentrating on the development of critical thinking and higher order skills development is required (Tai & Yuen, 2007). There are many variations of an authentic assessment strategy: these include, assessing performance within the tutorials process, the generation of a portfolio analysis and the preparation of a reflective learning journal and finally peer and self-assessment (Hart, 1994; Phillips, 2005; Tai & Yuen, 2007). Assessments of this nature require students to engage in collaborative practices with strong team and communications skills in order to reach a resolution to a complex problem (Tai & Yuen, 2007).

The cases presented in this chapter have each included both group and individual assessments. In addition, both modules were taught as part of overall courses in which there are marks given for both individual and group work. This helps to ensure a balance within the modules and the courses.

Extending the Programme

Rolling out PBL on a larger scale, for example across a School, Department or Faculty, is a different prospect than a single module pilot implementation by enthusiastic lecturers. When proposed in NCI it was noticed that some faculty members were hesitant to adopt the new approach. Two support workshops were organized to introduce faculty to PBL and to help them convert their modules. Nevertheless, some lecturers felt

that either their subject area was not suitable for delivery through PBL or that negative reaction of students in other modules had put them off. It is important that all possible support is given to faculty new to the PBL approach.

Within NCI, to assist faculty members in getting started with PBL, two new support mechanisms were developed. First, a PBL induction session was designed that would familiarize students with the PBL process, establish ground rules in the groups and assign the roles. It consisted of a set of problem solving and communication exercises where students could practice their skills and become aware of the difficulties that can arise in group work (Weibelzahl & Lahart, 2011a). Secondly, a "PBL toolbox" was developed: Each group receives a deck of 30 cards. Each card refers to a key concept or group activity that has been introduced in the induction session. Group members and facilitators can "play" these cards during discussion to bring the group back on track or to facilitate better learning (Weibelzahl & Lahart, 2011b). Lastly, a Web-based resource was created that makes all the exercises available online and searchable (see Figure 3). Lecturers can select the skills they want to address in their induction session and then choose from the available exercises. Lecturers can also rate and comment on resources. New resources can be added through an on-line interface. Currently, there are about 100 exercises available.

Similarly, the continued implementation and development of Problem-based Learning at UL was formalized and strengthened with the development of a Community of Practice (CoP) in 2011. Through the CoP a series of staged workshops were run to train faculty and tutorial staff in the concepts of PBL. In order for these to be fully effective, the CoP identifies and invites workshop facilitators with first-class national and international expertise in the area of PBL in general, and problem (trigger) design in particular. The PBLCoP is in the final stage of developing a CoP Website. This Web site will allow the dis-

Figure 3. Screenshot of the web-based PBL Induction resource available at http://pbl.ncirl.ie/

PBL Induction Resources

The Problem-Based Learning Induction Resource is a forum for sharing resources to support learners in developing the skills necessary for successful participation in PBL activities.

These skills comprise communication, team work, problem solving and critical thinking.

Feel free to use them for your PBL induction sessions as you see fit and tell us about your experiences.

Browse resources   Add a resource

PBL Induction Resources is a project by
Orla Lahart & Stephan Weibelzahl
at National College of Ireland
last update 24/06/2011

semination of PBL news through the University of Limerick and the wider community. PBL-related journal articles have been gathered together and will be presented on the CoP Website in a single repository; this is in addition to plans for future workshops. Funding for the creation of the PBLCoP and Web site were made available through a quality in teaching initiative (QIFAC) at the University.

ACKNOWLEDGMENT

This research is supported by the SFI Principal Investigator Programme, grant number 08/IN.1/ I2030 (the funding of this project was awarded by Science Foundation Ireland under a co-funding initiative by the Irish Government and European Regional Development Fund), and supported in part by Lero - the Irish Software Engineering Research Centre (http://www.lero.ie) grant 10/ CE/I1855.

REFERENCES

Barrows, H. S. (1984). A specific, problem-based, self-directed learning method designed to teach medical problem-solving skills, self-learning skills and enhance knowledge retention and recall. In H. G. Schmidt, & M. L. DeVolder (Eds.), *Tutorials in problem-based learning* (pp. 16–32). Van Gorcum.

Barrows, H. S. (1986). A taxonomy of problem-based learning methods. *Medical Education*, *20*, 481–486. doi:10.1111/j.1365-2923.1986. tb01386.x PMID:3796328

Barrows, H. S. (1996). Problem-based learning in medicine and beyond: A brief overview. *New Directions for Teaching and Learning*, *68*, 3–12. doi:10.1002/tl.37219966804

Barrows, H. S. (1998). The essentials of problem-based learning. *Journal of Dental Education*, *62*(9), 630–633. PMID:9789484

Barrows, H. S. (2000). *Problem-based learning applied to medical education*. Southern Illinois University School of Medicine.

Barrows, H. S. (2002). Is it truly possible to have such a thing as DPBL? *Distance Education, 23*(1). doi:10.1080/01587910220124026

Barrows, H. S., Myers, A. N. N., & Williams, R. G. (1986). Large group problem-based learning: A possible solution for the 2 sigma problem. *Learning, 8*(4), 325–331.

Barrows, H. S., & Tamblyn, R. M. (1977). The portable patient problem pack: A problem-based learning unit. *Journal of Medical Education, 52*(12), 1002–1004. PMID:926146

Barrows, H. S., & Tamblyn, R. M. (1980). *Problem-based learning: An approach to medical education*. Berlin: Springer Publishing Company.

Barrows, H. S., & Wee, K. N. L. (2010). *Principles and practice of a PBL*. Springfield, IL: Southern Illinois University School of Medicine.

Belland, B. R. (2010). Portraits of middle school students constructing evidence-based arguments during problem-based learning: The impact of computer-based scaffolds. *Educational Technology Research and Development, 58*, 285–309. doi:10.1007/s11423-009-9139-4

Boud, D. J., & Feletti, G. (1998). *The challenge of problem-based learning*. Kogan Page.

Bridges, M. E. (1992). *Problem-based leaning for administrators. ERIC Clearinghouse on Educational Management*. University of Oregon.

Chung, J., & Chow, S. (2004). Promoting student learning through a student-centred problem-based learning subject curriculum. *Innovations in Education and Teaching International, 41*, 157–168. doi:10.1080/1470329042000208684

Connor, H., & Shaw, S. (2008). Graduate training and development: Current trends and issues. *Education + Training, 50*, 357–365.

Cotton, K. (1993). *Developing employability skills*. School Improvement Research Series. Retrieved from http://educationnorthwest.org/Webfm_send/524

Dahlgren, M. A., & Dahlgren, L. O. (2002). Portraits of PBL : Students' experiences of the characteristics of problem-based learning in physiotherapy, computer engineering. *Instructional Science, 30*, 111–127. doi:10.1023/A:1014819418051

Davies, L. (2000). Why kick the L out of learning? The development of students' employability skills through part-time working. *Education + Training, 42*, 436–445.

De Graaff, E. (2003). Characteristics of problem-based learning. *International Journal of Engineering Education, 19*(5), 657–662.

Delaney, D., & Mitchell, G. G. (2002). PBL applied to software engineering group projects. In *Proceedings International Conference on Information and Communication in Education* (pp. 1093-1098). Government of Extremadura.

Dewey, J. (1929). *The quest for certainty*. New York: Minton.

Dos Santos, S. C., Da Conceicao Moraes Batista, M., Cavalcanti, A. P. C., Albuquerue, J. O., & Meira, S. R. L. (2009). Applying PBL in software engineering education. In *Proceedings of 22nd Conference on Software Engineering Education and Training* (pp. 182-189). IEEE.

Dunlap, J. C. (2005). Problem-based learning and self-efficacy: How a capstone course prepares students for a profession. *Educational Technology Research and Development, 53*(1), 65–83. doi:10.1007/BF02504858

Gijselaers, W., & Schmidt, H. (1990). Towards a causal model of student learning within the context of problem-based curriculum. In Z. Nooman, H. Schmidt, & E. Ezzat (Eds.), *Innovations in medical education: An evaluation of its present status* (pp. 95–114). Springer.

Gijselaers, W. H. (1995). Perspectives on problem-based Learning. In W. H. Gijselaers, D. T. Tempelaar, P. K. Keizer, J. M. Blommaert, E. M. Bernard, & H. Kasper (Eds.), *Educational innovation in economics and business administration: The case of problem-based learning*. Kluwer. doi:10.1007/978-94-015-8545-3_5

Gijselaers, W. H. (1996). Connecting problem-based practices with educational theory. *New Directions for Teaching and Learning, 199*(68), 13–21. doi:10.1002/tl.37219966805

Gijselaers, W. H. (2011). Problem-based learning (PBL) today and tomorrow. In *Proceedings of FACILITATE Conference: PBL 3.0 in Transformative Times*. FACILITATE.

Hales, C. (2007). Moving down the line? The shifting boundary between middle and first-line management. *Journal of General Management, 32*(2), 31–56.

Hallinger, P., & Bridges, M. E. (2007). *A problem-based approach for management education: Preparing managers for action*. Berlin: Springer. doi:10.1007/978-1-4020-5756-4_2

Harland, T. (2003). Vygotsky's zone of proximal development and problem-based learning: Linking a theoretical concept with practice through action research. *Higher Education, 8*(2), 263–272.

Hart, D. (1994). *Authentic assessment: A handbook for educators*. Reading, MA: Addison-Wesley.

Hmelo-Silver, C. E. (2004). Problem-based learning: What and how do students learn? *Educational Psychology Review, 16*(3), 235–266. doi:10.1023/B:EDPR.0000034022.16470.f3

Hmelo-Silver, C. E., & Barrows, H. S. (2006). Goals and strategies of a problem-based learning facilitator. *Learning, 1*(1), 21–39.

Huba, M. E., & Freed, J. E. (2000). Learner centered assessment on college campuses: Shifting the focus from teaching to learning. *Community College Journal of Research and Practice, 24*(9), 759–766.

Knight, P. (1995). *Assessment for learning in higher education*. Psychology Press.

Kolb, D. A. (1984). *Experiential learning: Experience as the source of learning and development*. Upper Saddle River, NJ: Prentice Hall.

Kolodner, J. L., Camp, P. J., Crismond, D., Fasse, B., Gray, J., & Holbrook, J. et al. (2003). Problem-based learning meets case-based reasoning in the middle-school science classroom: Putting learning by design(tm) into practice. *Journal of the Learning Sciences, 12*, 495–547. doi:10.1207/S15327809JLS1204_2

Lähteenmäki, M. L., & Uhlin, L. (2011). Developing reflective practitioners through PBL in academic and practice environments. In *New approaches to problem-based learning: Revitalising your practice in higher education* (pp. 144–157). Academic Press.

Levia, D. F., & Quiring, S. M. (2008). Assessment of student learning in a hybrid PBL capstone seminar. *The Journal of Geography, 32*(2), 217–231.

Milter, R. G., & Stinson, J. E. (1995). Educating leaders for the new competitive environment. In *Educational innovation in economics and business administration: The case of problem-based learning administration*. Boston: Kluwer.

Mitchell, G. G., & Delaney, J. D. (2004). An assessment strategy to determine learning outcomes in a software engineering problem-based learning course. *International Journal of Engineering Education, 20*(3), 494–502.

Nelson, W. A. (2003). Problem solving through design. *New Directions for Teaching and Learning*, (95): 39–44. doi:10.1002/tl.111

O'Grady, M., Barrett, G., Barrett, T., Delaney, Y., Hunt, N., Kador, T., & O'Brien, V. (2013). Reflecting on the need for problem triggers in multidisciplinary PBL. *AISHE-J: The All Ireland Journal of Teaching & Learning in Higher Education, 5*(2).

Perrenet, J. C., Bouhuijs, P. A. J., & Smits, J. G. M. M. (2000). The suitability of problem-based learning for engineering education: Theory and practice. *Teaching in Higher Education, 5*, 345–358. doi:10.1080/713699144

Phillips, L. (2005). *Authentic assessment: A briefing*. Retrieved from http://www.clubWebcanada.ca/l-pphillips/edarticles/assessment.htm

Prince, M. J., & Felder, R. M. (2006). Inductive teaching and learning methods: Definitions, comparisons, and research bases. *Journal of Engineering Education, 95*(2), 123–138. doi:10.1002/j.2168-9830.2006.tb00884.x

Ramsden, P. (1992). *Learning to teach in higher education*. London: Routledge. doi:10.4324/9780203413937

Richardson, I., & Delaney, Y. (2009). Problem based learning in the software engineering classroom. In *Proceedings of the 22nd Conference on Software Engineering Education and Training* (pp. 174–181). IEEE.

Richardson, I., & Delaney, Y. (2010). Software quality: From theory to practice. In *Proceedings of the 7th International Conference on the Quality of Information and Communications Technology* (pp. 150–155). Academic Press.

Richardson, I., & Hynes, B. (2008). Entrepreneurship education: Towards an industry sector approach. *Education + Training, 50*(3), 188–198.

Richardson, I., Reid, L., Seidman, S. B., Pattinson, B., & Delaney, Y. (2011). Educating software engineers of the future: Software quality research through problem-based learning. In *Proceedings of the 24th Conference on Software Engineering Education and Training* (pp. 91–100). IEEE.

Savin-Baden, M. (2000). *Problem-based learning in higher education: Untold stories*. Society for Research into Higher Education.

Schmidt, H. G. (1993). Foundations of problem-based learning some explanatory notes. *Medical Education, 27*, 422–432. doi:10.1111/j.1365-2923.1993.tb00296.x PMID:8208146

Schmidt, H. G., & De Volder, M. L. (1984). *Tutorial in problem-based learning*. Van Gorcum.

Shim, C. Y., Choi, M., & Kim, J. Y. (2009). Promoting collaborative learning in software engineering by adapting the PBL strategy. *World Academy of Science. Engineering and Technology, 53*, 1157–1160.

Tai, G. X. L., & Yuen, M. C. (2007). Authentic assessment strategies in problem based learning. In *Proceedings ascilite Singapore* (pp. 983–999). Ascilite.

Tan, O. S. (2003). *Problem-based learning innovation*. Singapore.

Tan, O. S., & Little, S. Y. Hee, & J. Conway, E. (2000). Problem-based learning: Educational innovation across disciplines. Singapore.

Vat, K. H. (2006). Integrating industrial practices in software development through scenario-based design of PBL activities: A pedagogical re-organization perspective. *Issues in Informing Science and Information Technology, 3*.

Vaughn, R. B. (2001). Teaching industrial practices in an undergraduate software engineering course. *Computer Science Education, 11*(1), 21–32. doi:10.1076/csed.11.1.21.3844

Wasserman, A. I. (1996). Toward a discipline of software engineering. *IEEE Software, 13*, 23–31. doi:10.1109/52.542291

Weibelzahl, S., & Lahart, O. (2011a). Developing student induction sessions for problem-based learning. In *Proceedings of International Conference on Engaging Pedagogy*. Academic Press.

Weibelzahl, S., & Lahart, O. (2011b). PBL induction made easy: A web-based resource and a toolbox. In *Proceedings of FACILITATE Conference: Problem-Based Learning (PBL) Today and Tomorrow*. Dublin, Ireland: FACILITATE.

ADDITIONAL READING

Albanese, M. (2000). Problem-based learning: Why curricula are likely to show little effect on knowledge and clinical skills. *Medical Education, 34*(9), 729–738. doi:10.1046/j.1365-2923.2000.00753.x PMID:10972751

Albanese, M. A., & Mitchell, S. (1993). Problem-based learning: A review of literature on its outcomes and implementation issues. *Academic Medicine, 68*(1), 52–81. doi:10.1097/00001888-199301000-00012 PMID:8447896

Allen, D. E., Duch, B. J., & Groh, S. E. (1996). The power of problem-based learning in teaching introductory science courses. *New Directions for Teaching and Learning,* (68): 43–52. doi:10.1002/tl.37219966808

Barrett, T., & Moore, S. (2011). *New approaches to problem-based learning: Revitalising your practice in higher education*. New York, London: Routledge.

Bigelow, J. D. (2004). Using problem-based learning to develop skills in solving unstructured problems. *Journal of Management Education, 28*(5), 591–609. doi:10.1177/1052562903257310

Buckley, F., & Monks, K. (2007). Responding to managers' learning needs in an edge-of-chaos environment: Insights from Ireland. *Journal of Management Education, 32*(2), 146–163. doi:10.1177/1052562907299625

Butun, E., Erkin, H. C., & Altintas, L. (2008). A new teamwork-based PBL problem design for electrical and electronic engineering education: A systems approach. *International Journal of Electrical Engineering Education, 45*(2), 110–120. doi:10.7227/IJEEE.45.2.3

Colliver, J. A. (2000). Effectiveness of problem-based learning curricula: Research and theory. *Academic Medicine, 75*(3), 259–266. doi:10.1097/00001888-200003000-00017 PMID:10724315

Coombs, G., & Elden, M. (2004). Introduction to the special issue: Problem-based learning as social 1nquiry--PBL and management education. *Journal of Management Education, 28*(5), 523–535. doi:10.1177/1052562904267540

Dangerfield, P. (2006). Problem-based learning principles & application to practice : An example from a practitioner-orientated course. *Learning, 3*(2), 47–57.

De Graaff, E. (2003). Characteristics of problem-based learning. *International Journal of Engineering Education, 19*(5), 657–662.

Deignan, T. (2009). Enquiry-based learning: Perspectives on practice. *Teaching in Higher Education, 14*(1), 13–28. doi:10.1080/13562510802602467

Dolmans, D. H. J. M., De Grave, W., Wolfhagen, I. H., & Van der Vleuten, C. P. M. (2005). Problem-based learning: Future challenges for educational practice and research. *Medical Education, 39*(7), 732–741. doi:10.1111/j.1365-2929.2005.02205.x PMID:15960794

Duch, B. J., Allen, D. E., & White, H. B. (1996). Problem based learning : Preparing students to succeed in the 21st Century. *Learning*, 1–5.

Engel, C. (2009). An Internet guide to key variables for a coherent educational system based on principles of Problem-Based Learning. *Teaching and Learning in Medicine*, *21*(1), 59–63. doi:10.1080/10401330802384888 PMID:19130388

Gijbels, D. (2005). Effects of problem-based learning : A meta-analysis from the angle of assessment. *Review of Educational Research*, *75*(1), 27–61. doi:10.3102/00346543075001027

Hmelo-Silver, C. E., Duncan, R. G., & Chinn, C. A. (2007). Scaffolding and achievement in problem-based and inquiry learning: A response to Kirschner, Sweller, and Clark (2006). *Educational Psychologist*, *42*(2), 99–107. doi:10.1080/00461520701263368

Kirschner, P. A., & Clark, R. E. (2006). Work: An analysis of the failure of constructivist, discovery, problem-based, experiential, and inquiry-based teaching. *Learning*, *41*(2), 75–86.

Leary, H., Walker, A., & Shelton, B. E. (2013). Exploring the relationships between tutor background, tutor training, and student learning: A problem-based learning meta-analysis. *Interdisciplinary Journal of Problem-based Learning*, *7*(1), 3–15. doi:10.7771/1541-5015.1331

Linehan, M. (2008). *Work-based learning graduating through the workplace*. Cork: CIT Press.

Schmidt, H. G., Vermeulen, L., & Van der Molen, H. T. (2006). Longterm effects of problem-based learning: A comparison of competencies acquired by graduates of a problem-based and a conventional medical school. *Medical Education*, *40*(6), 562–567. doi:10.1111/j.1365-2929.2006.02483.x PMID:16700772

Schmidt-Wilk, J. (2010). Signature Pedagogy: A framework for thinking about management education. *Journal of Management Education*, *34*(4), 491–495. doi:10.1177/1052562910376508

Smith, G. F. (2003). Beyond critical thinking and decision making: Teaching business students how to think. *Journal of Management Education*, *27*(1), 24–51. doi:10.1177/1052562902239247

Strobel, J. (2009). When is PBL More Effective ? A meta-synthesis of meta-analyses comparing PBL to conventional classrooms. *Interdisciplinary Journal of Problem-based Learning*, *3*(1).

Van Berkel, H., & Schmidt, H. (2000). Motivation to commit oneself as a determinant of achievement in problem-based learning. *Higher Education*, *40*, 231–242. doi:10.1023/A:1004022116365

KEY TERMS AND DEFINITIONS

Formative Assessment: The monitoring of student learning to provide ongoing feedback that can be used by instructors to improve their teaching and by students to improve their learning.

Learning Outcomes: Defining the knowledge, skills and abilities that students should possess following a particular educational experience.

Problem-Based Learning: A student-centered pedagogy in which students learn about a subject through the experience of problem solving.

Self-Directed Learning: A process by which individuals take the initiative, with our without the assistance of others, in diagnosing their learning needs, goals, resources and strategies.

Software Engineering: The application of a systematic, disciplined, quantifiable approach to the development, operation, and maintenance of software.

ENDNOTES

1   http://www.computer.org/portal/Web/sWe-bok

2   http://www.youtube.com/watch?v=-xrrk-XhgVc

3   Direct quotes from student and lecturer feedback are presented in italics within the text.

4   http://www.youtube.com/watch?v=4K39UWxdm0U

Section 8
Meeting Industry Expectations

Chapter 19
Bridging the Academia–Industry Gap in Software Engineering:
A Client–Oriented Open Source Software Projects Course

Bonnie K. MacKellar
St. John's University, USA

Mihaela Sabin
University of New Hampshire, USA

Allen B. Tucker
Bowdoin College, USA

ABSTRACT

Too often, computer science programs offer a software engineering course that emphasizes concepts, principles, and practical techniques, but fails to engage students in real-world software experiences. The authors have developed an approach to teaching undergraduate software engineering courses that integrates client-oriented project development and open source development practice. They call this approach the Client-Oriented Open Source Software (CO-FOSS) model. The advantages of this approach are that students are involved directly with a client, nonprofits gain a useful software application, and the project is available as open source for other students or organizations to extend and adapt. This chapter describes the motivation, elaborates the approach, and presents the results in substantial detail. The process is agile and the development framework is transferrable to other one-semester software engineering courses in a wide range of institutions.

DOI: 10.4018/978-1-4666-5800-4.ch019

MOTIVATION

Most computer science programs offer a software engineering course and view it as a critical link in ensuring the career-readiness of computer science graduates. However, too often this course is taught in terms of abstract principles, failing to engage students in real-world software experiences. Many of the skills required in industry are best learned by hands-on practice, such as the need for effective communication among developers, or the need to interact with a non-technical client. Thus, students who have never engaged in a hands-on project in software engineering enter the workforce with gaps in their skills.

It is, however, difficult to bring a significant software development experience into the confines of a one-semester course in academia. The most common approach has been to introduce a "toy project," which is a small project designed by the instructor, and have students work in teams to complete the project by the end of the semester. The advantage of this approach is that students will ideally learn to work in teams and share responsibility for developing a codebase. The disadvantages are that the project may be oversimplified, and students gain no experience interacting with clients or with code written by others.

Another approach is to work with local companies in the private sector who sponsor *proprietary client-oriented software projects*. This has been used successfully by a number of schools, especially larger programs that already have established linkages with companies (Judith, Bair, & Börstler, 2003; Tadayon, 2004; Tan & Jones, 2008). Another setting that favors this approach is an internship course with the projects being developed onsite at local companies. The advantage is that students gain experience with real clients with high stakes in real projects. However, these projects are often standalone, one-off projects since companies may be reluctant to have students work on their internal codebase, or to develop mission critical software. This means that it may be difficult to get enough time and attention from personnel at the company while the students work on the project. Also, the project will normally become the property of the company, meaning that it cannot be freely shared with other schools trying to adopt a similar approach.

A third approach is to engage students in Free and Open Source Software (FOSS) development by having them contribute to *a large and active open source project*, such as Linux or Mozilla (Marmorstein, 2011; Ellis, Morelli, DeLanerolle, & Hislop, 2007). The advantage of this approach is that instructors and students can gain from the mentoring achieved through communication with the project's professional developers, and in some cases they contribute marginally to the "live" code base or the user documentation. The disadvantages of this approach are that most ongoing projects are large and complex, their developers may not be accessible, and given the time it takes to come up to speed in the project, students may gain little practical experience in a one-semester course.

A fourth approach, which occupies a middle ground between the proprietary client-oriented project model and the full-scale FOSS project, is to engage students in FOSS development via a relatively small project that fits in a one semester course, with a local nonprofit organization as the client. Local nonprofits are often happy to collaborate on these projects since they may have needs for mission-critical software systems that are not well met by the commercial software industry, yet they have limited technology budgets. Thus, it is relatively easy for an instructor to locate and collaborate with a local nonprofit. However, many instructors may still be unsure of how to get started or how to organize such a course.

This chapter describes our collective experience with the fourth approach, which we call *client-oriented free and open source software development* (CO-FOSS). The big advantage of treating client-oriented open source projects is the very openness of the project. An open source project developed in the context of one course for one client can be reused, extended, and adapted for new clients by subsequent iterations of the same course, or even by courses at different institutions. By providing not just the codebase but the course organization itself as an open source project, a collection of such projects can be built up to be used as models at different institutions. In addition, the tools and practices of open source projects provide a readymade infrastructure for software project courses.

Service learning projects for nonprofits have been used in a number of other software engineering courses and have been reported in these papers (Olsen, 2008; Poger & Bailie, 2006; Venkatagiri, 2006); however, these projects have not taken advantage of the reusable and extendable capabilities of the CO-FOSS approach. An example of a service learning approach that has been turned into a model for other universities is EPICS (Coyle, Jamieson & Oakes, 2005). This was developed for engineering schools and thus is not specific to computer science. In the EPICS model, interdisciplinary teams of eight to twenty students are assembled to work with community organizations on engineering problems. The teams are vertically integrated, consisting of freshmen through seniors. This model has been expanded to a number of universities and an implementation of it in a software engineering context was reported in (Linos, Herman & Lally, 2003). EPICS is similar to our model in that it provides a structure for community-oriented service learning projects, providing advice on team formation, identifying clients, structuring communication, and so on. However, it differs from our approach because it does not leverage the power of open source development, meaning that it is more similar to the proprietary client-oriented model in many

ways. In addition, the vertically integrated teams are a challenge to integrate into many computer science programs.

However, the CO-FOSS approach has some challenges. It may be difficult for an instructor who has never done this type of project before to develop the course, to manage the project, and to attend to the post-course activities that accompany placing the new software into production. Therefore, we will present a framework for developing this type of course. In the CO-FOSS approach, one of the most important goals is that the students be able to actually complete a working prototype within the boundaries of a semester. To this end, the approach is highly structured, with weekly goals for completion. The instructor (rather than the students) develops the requirements. Since CO-FOSS projects are open source, architectures and code can be reused and leveraged into new projects. And because open source projects are visible to the world, students are motivated to achieve a higher level of quality. The openness of the course artifacts also means that instructors in other institutions can easily adapt existing projects to meet their clients' needs. Allen Tucker first developed this framework as part of the Humanitarian FOSS project (Morelli, de Lanerolle, & Tucker, 2012) and has taught numerous iterations of a software engineering course this way. Bonnie MacKellar and Mihaela Sabin adapted this framework to courses at their universities by taking into account students' characteristics that are specific to their institutions (MacKellar, Sabin, & Tucker, 2013). The result is a model for teaching software engineering that brings real world experience to a wide variety of institutions. In this chapter, we will present this framework, our experiences with adaptation, and a set of guidelines and best practices that instructors can use to integrate CO-FOSS development into their courses.

INTRODUCTION TO THE CO-FOSS MODEL

The CO-FOSS model has two major elements: 1) a *process* and 2) a *product*. The process is agile, participatory, and open. The product is real, functional, and useful to the mission-critical operations of the nonprofit client. This section elaborates each of these elements in turn.

The *process* of CO-FOSS development that we advocate is carried out by a team of developers, which necessarily includes a client representative who understands the manual activities that the software will replace. Because the process is agile, the client has frequent (weekly) opportunities to interact with each element of the software as it comes on-line and to provide feedback to the developers on which features work well and which do not. In turn, the developers can take that feedback into account as they refine and extend the software during the following week.

The team leader is typically the course instructor, who evaluates the work of the student developers each week and takes client feedback into account while preparing the following week's assignment. All of this activity is facilitated by a weekly meeting, either in-person or through an interactive video conference, where the developers demonstrate and explain their work to the client, the client provides feedback, and the instructor takes notes that inform the project's next steps.

The *product* of a CO-FOSS project begins with a complete requirements document and an initial codebase. The instructor must complete the requirements step before the beginning of the semester. This is a departure from the usual software engineering course, where students often work on requirements as well as software development. An alternative is a two-semester software development course, in which requirements elicitation and engineering concepts and practices are taught in the first semester, prior to designing and implementing the software in the second. The initial codebase may be the result of a prior similar open source development project, or it may consist only of an initial set of domain classes springing out of the requirements document. The final product is a real and viable software artifact, which fully implements the requirements that had been laid out at the beginning of the project, and which may be extended and adapted by other students for different clients in the future.

Our experience with nonprofits as clients for open source projects like this suggests that many nonprofits exhibit similar software needs that are not particularly well met by commercially available software at a price that the nonprofits can afford. For example, many nonprofits use numerous volunteers to help realize their mission. These nonprofits need software that assists them with scheduling volunteers into calendar shifts using the idea of a master schedule. Volunteers typically like to use simple repeating patterns (like "every other Thursday in the 12-3 shift" or "the first Monday of each month from 8-12") but commercially-available calendaring tools do not support this sort of master scheduling in an easily customizable way. Thus, *Homebase* was initially developed in 2008 to support the volunteer scheduling needs of the Ronald McDonald House in Portland, ME. Since then, the *Homebase* design and code has been reused and adapted for other clients that have similar volunteer scheduling needs: the Ronald McDonald House in Wilmington, DE, the Second Helping food rescue organization in South Carolina, and the Mid-Coast Hunger Prevention Program in Brunswick, ME. Because it is open source, *Homebase* can evolve over time and can be easily adapted and reused for new projects with similar needs. We have found that reusing the software architecture and underlying code for a successful project can greatly simplify an instructor's task of developing a new project for a different client.

Moving to a more detailed level of granularity, we find that the process of developing and delivering a one-semester CO-FOSS course can be divided into an eight-step framework:

1. Pre-course activity
2. Curriculum design: syllabus and milestones
3. Structuring client communication
4. Team formation and task assignment
5. Developer communication and code sharing
6. Writing user and developer documentation
7. Evaluation of team members' contributions
8. Post-course activity

The following section explains each step in this framework.

STEPS TO DEVELOPING AND DELIVERING A CO-FOSS COURSE

Pre-Course Activity

Client Sponsorship

Preparing a software development course so that students can have a real-life collaborative experience with a client and develop a useful software product during the course of a single semester requires significant effort on the part of the instructor. First, the instructor must find a willing nonprofit client and identify a specific software project. Professional or personal associations with people who are familiar with the day-to-day operation of the nonprofit (such as the executive director, operations manager, or a particularly active board member) can be very helpful in this effort. If the college has a service learning office, personnel there will often have contacts with local nonprofits. The local reputation of the nonprofit in the community can also provide leads. If the instructor has already worked with a CO-FOSS project for another client or wants to work with an existing CO-FOSS project from another school, it is best to locate nonprofits with similar needs or in a similar sector. The existence of a similar successful project for a previous client often heightens interest within a targeted nonprofit client. Once a client is located, the instructor must identify a specific

employee who is responsible for the operations which the software can enhance, excited about helping with the development of such software, and able to dedicate time (a few hours before the semester begins and an hour a week throughout the semester itself) to work with the instructor and the student development team as they design and develop the software itself.

It is very important that the project be designed to ensure success. That is, the most important goal of the course should be that the students actually complete a working prototype for the software within the 13-week boundaries of a semester. This goal usually means that the instructor and the client representative make some hard choices about what will and will not be included in the final product at the end of the semester before the project begins. These decisions should inform the next step of the pre-course activity, that is, requirements elicitation and documentation.

Requirements Document

The instructor must work with the client to elicit requirements and tailor a project that both serves the needs of the nonprofit and can be completed within the context of a normal semester course by a team of students. The result of this activity will be a requirements document that spells out in substantial detail the functional, technical, and user interface requirements of the software, as well as the eventual use to which it will be put when it is completed. The requirements document must provide enough detailed domain-specific information so that students can identify initial development tasks almost from the get-go. This information should include, for example, initial domain classes for which representative attributes have been identified; a selection of programming and database languages appropriate to the project; and an overall architecture for the software so that students will clearly understand where each of their modules will fit within the larger product. Links to examples of our requirements documents can be found in the Resources section at the end of the chapter. Typically, a layered architecture that separates the user interface from persistent data representation and manipulation provides a good starting point for many Web-based applications.

While the idea of providing the requirements document to the students may seem surprising since many traditional software engineering courses involve the students in writing requirements as well as developing the software, we find that undergraduate students typically do not have the experience necessary for writing adequate requirements for a client. They either become bogged down in the task, leaving too little time for development, or they produce a requirements document that ends up having little to do with the final system. This has also been noted by Cheng & Lin (2010), who have developed a guided approach to the software engineering class project that is similar to ours, but without the client and open source focus. In their approach, the instructor also develops specifications and an overall design.

On the other hand, the presentation of a requirements document at the beginning of the course does not limit student-client interaction or student participation in requirements development. In the first case, students in this course interact weekly with the client to gain clarity on the details of requirements so that they can develop appropriate code for the weekly assignment. In the second case, the fact that the whole process is agile ensures that requirements evolve alongside the coding. So students are in fact first-class participants in the refinement of requirements, even though they do not develop the initial version of the requirements document.

Curriculum Design: Syllabus and Milestones

The syllabus can be structured in a way similar to that of other computer science courses. That is, each week of the semester, students are expected to complete a specific set of learning objectives and demonstrate their learning by completing an assignment that reflects those objectives. But there are two main differences: 1) the learning objectives and assignment are drawn directly from the requirements document, and 2) students normally must collaborate and share a common code base to complete the assignment.

The first week or two in the semester are dedicated to team formation and setting up for collaboration. Each team must initialize a shared code repository for the project using a designated version control system, such as Subversion, GIT or Mercurial. Each student must also set up his/her own computer with a development environment that supports the programming tools chosen for the project. That environment must also be interfaced with the shared code base, so that each student can integrate his/her own work with teammates' work. This start-up step is not trivial for the student team, since students are not generally familiar with code sharing from earlier courses. Depending on the particular version control system and development environment chosen, sufficient tutorial materials should be made available to students so that they can independently set up their own computers to work effectively in the course. Links to supporting materials for Mercurial and Eclipse, for instance, can be found in the Resources section.

During each subsequent week in the course, each student team member develops, unit tests, and commits a new piece of the software to the repository. These weekly assignments are determined by the instructor, who takes into account client feedback from the previous week's assignment. The layered architecture of the software design facilitates task sequencing. That is, when the domain model, user interface, and database controller layers are distinguished, they can be helpful in organizing the weekly assignments in a course. In this case, it is often preferable to design and unit test the domain classes in weeks 3-5, the database modules in weeks 6-8, and the user interface modules in weeks 9-11 of the semester. That leaves weeks 12-13 free for students to perform more detailed integration testing, develop user documentation, and address other special circumstances that inevitably arise during the semester. The final exam date for the course provides a good opportunity for the student team to deliver the completed software prototype to the client in the setting of an oral presentation. Often, other members of the client organization are invited to this presentation, such as the executive director, other staff members, or even some board members.

Structuring Client Communication

As should be evident from the above discussion, client feedback is essential as each weekly assignment is completed and demonstrated. Students need feedback to determine the extent to which their coding efforts successfully addressed the assigned task. The instructor needs feedback to help assess the project's overall progress and to determine in detail the shape of the following week's assignment.

It is not practical for the client to meet face-to-face with the student team members at each weekly class session. Moreover, it may not even be practical for all the student team members or the instructor to be physically present at each team meeting. On the other hand, "virtual attendance" at each team meeting should be required of all students and the instructor and the client representative. Virtual weekly attendance among all team members can be ensured if everyone uses a visual teleconferencing medium, such as Skype or Google Hangout, to facilitate it.

The second ingredient to ensure client feedback at weekly team meetings is access to a shared "sandbox" server that can be used to demonstrate the current state of student progress with the project. This server can be provided by either the university or an independent Internet service provider. The important point is that at each team meeting, the client can actively view and work with the partially-developed software under the verbal guidance of the students who are developing it.

The dynamics of the weekly team meeting among the students, the instructor, and the client representative are as follows. Each student on the team presents to the client a brief description of what he/she accomplished during the prior week, and shows the client how that works. The client tries it out using the "sandbox" server and provides feedback on the spot. The instructor takes notes and uses the presentation and the feedback as factors in designing the following week's assignment. Typically, a week's assignment has two parts – one part that cleans up issues identified by the client in the team meeting and the other part that makes progress developing some additional aspect of the software. At the end of the development period, a more or less complete prototype will thus emerge. Critically, the software prototype that results from this process is not the product of the student developers working in a vacuum with the original requirements document. Rather, it is the product of weekly client feedback and immediate adjustment to that feedback. This is the essence of the term "agile development," brought to life within the setting of a real software project.

Team Formation and Task Assignment

Team formation and task assignment constitute one of the key tasks when designing this type of course. The task of team formation involves some key decisions:

1. How large will each team be?
2. What tasks will be assigned to each team?
3. Who will be assigned to each team? Who makes that decision?

Team Size

There are many factors beyond the instructor's control that impact these decisions. For example, student preparedness and the size of the client project will impact the size of the teams. The number of enrolled students, as well as the number and scope of available projects will also have an impact. However, the framework for CO-FOSS projects is flexible enough to incorporate varying team structures. For example, at Bowdoin College, there were several projects, and well-prepared students, so teams were structured by project. At St John's University, on the other hand, the instructor had already been using a way of structuring the course so that all students work on one project. This works well in that environment because the students are less well-prepared and can specialize in areas where they feel most comfortable, such as writing the help system, developing the database, or working on the user interface. At UNH Manchester, components of the project are partially developed in other courses. For example, students learn database model, design, and implementation techniques or how to integrate a Web-based user interface with database services in the Database Design and Development and Advanced Web Authoring courses.

Task Assignment

The architecture of the system plays a big role in the structure of the teams and the tasks they are assigned. This is another decision point that is eased by following the practice of building upon an existing open source project from past semesters. The *Homebase* architecture consists of a database, domain objects, a help system, and user interface. Thus, it is clear that students will be allocated to each of these tasks. If smaller teams are working at multiple smaller projects, and are following the Bowdoin model of focusing on separate layers at different points in the semester, then each student will end up working on each layer. This means that

all students will need to gain expertise in every aspect of the system. This works best when all students are very well prepared for the course, and have seen topics such as database design and SQL before taking this course.

An alternative structure, used at St John's and UNH Manchester is to use the entire class as a team if the project scope is large enough. In this organization, the large team is split into subgroups along system architecture lines. If students are not familiar in advance with many of the project technologies, this organization has the advantage that students do not have to learn many new technologies at once. Also, students can be grouped according to their strengths. For example, the group working on the database layer can be composed of students who have already taken the database course. The advantages of this type of team organization are detailed in (MacKellar 2011). The disadvantage is that some layers depend on other layers. This means that interfaces between layers must be carefully specified and adhered to, and stubs for testing purposes may need to be developed.

Allocating Students to Teams

The CO-FOSS framework does not impose many constraints on the task of assigning students to teams, so the instructor has several choices to make in forming the teams. There has been quite a bit of research, both within computer science and in other disciplines, on the best way to form teams for group projects. Richards (2009) detailed many of the considerations in a survey. For example, students may collaborate more effectively if they are allowed to choose their own teams (Grundy, 1996) but when this is permitted, the more capable students are likely to end up together in one group. If it is important to distribute students of different abilities across the groups, then the instructor should assign students to groups. Criteria such as GPA in past courses or surveys of student can be used to determine placement (McConnell,

2006). Other criteria, such as gender, ethnicity, or simple time availability may come into play as well. Even issue tickets have been used as a way to form teams (Coppit & Haddox-Schatz, 2005). If the teams will consist of students working on all aspects of their project, as detailed above, then either student-based or instructor-based assignment can be used, and any of the various considerations can be used.

If the students will be working in larger teams, and assigned to specific components in the system architecture, then the instructor should do the assignment, and take their skills and interests into consideration. At St John's, the students are surveyed at the beginning of the semester, and also submit resumes. The instructor then assigns students to task groups in a way that minimizes project risks, much as a project manager in industry does. Thus, a student who indicates experience with SQL will be assigned to work on the database layer, whereas a student who has used PHP in the past will be assigned to the user interface layer, and a student who is familiar with QA, perhaps from an internship, will be assigned to the testing group. At UNH Manchester, the instructor solicits input from students in the first class about their academic and professional experiences, self-reported computing strengths, and areas of interests in their future careers. Two other sources of information are taken into consideration: student transcripts and evaluation from faculty members who know the students from their classes. Gathering this information is possible in a small department with less than 100 majors and a climate with frequent and meaningful interactions among faculty. This method of assigning development roles to students by the instructor only emphasizes learning to work in teams on large software systems rather than learning an array of specific technologies, such as PHP and SQL. The method is also inclusive of various talents and interests and avoids having students distracted by inexperience with a specific technology.

Developer Communication and Code Sharing

One of the advantages of approaching the software engineering course as an open source project is that open source development comes with a set of standard practices and tools which tend to work very well for students. The tools themselves are open source, so they are affordable in an academic setting. Open source work practices, which evolved from the need to support a highly asynchronous, distributed group of developers, also work well for students because the practices do not require as much face-to-face development effort.

What are these tools and practices? In general, most open source projects use the following tools (Fogel, 2005):

- A project repository
- Mechanisms for team communications, usually mailing lists and real time chat channels
- A version control system
- An issue tracking system

Since there are so many open source projects, a number of standard sites and tools have appeared to meet these needs. These sites constitute a ready-made toolbox for software engineering educators, greatly easing their task since they are already set up and integrated.

The Project Repository

Open source projects, like any other software project, need to reside somewhere. Not just the code, but design documents, installation instructions, build sequences, and records of defects must be maintained. Open source projects, however, are not usually tied to any one organization and must be accessible to developers and other contributors on a worldwide basis. Thus, quite a bit of effort has gone into building sites that allow sharing of project artifacts. These sites are referred to as project repositories, and often contain a number of related features, such as version control, bug tracking, and wikis. Examples of currently popular project repositories include SourceForge, Google Code, and GitHub. These repositories are free for open source projects. Obviously, the fact that they are free is very appealing for cash strapped universities. But they also are convenient for a number of other reasons. Since version control and bug tracking are typically integrated, the instructor is freed from needing to install and maintain complex software. Since they are designed to be shareable among distributed contributors, repositories allow students who may not be able to attend face-to-face meetings to still collaborate. And since these repositories host many projects, including some very famous ones, students can browse the repository site, see lots of interesting projects, and see that people much like themselves are contributing. This can add to their sense of connectedness with the discipline of software engineering and the open source development community.

Version control systems are critical to projects that require more than one developer to work on code in a controlled fashion. There are a number of version control systems in common use. Interestingly, most of them are open source systems. Examples include CVS, Subversion, Mercurial, and Git. A project repository will offer one or more of these systems, and the instructor, when setting up the project, must choose one. Using a project repository forces the students to use the version control system since they must interact with it to place their code into the repository.

Another typical feature of open source projects is the use of an online issue tracking (or bug tracking) system. These are used extensively in industry as well. In the open source world, these systems are usually integrated into the open source project repositories. For example, the issue tracker on Google Code integrates with Google Groups, which is an online discussion system, so that when an issue is entered, a message will automatically go out to the discussion group for the project, and when the issue is resolved, another message

will be generated. This is easy for an instructor to configure, and helps ensure that everyone in the class is aware of the current status of all bugs in the software.

Online Messaging Systems

A characteristic of open source projects is that developers are distributed geographically, and are not likely to work on the project during set business hours. This means that tools to support asynchronous, online communication are critical to the success of the project. Very commonly, mailing lists are used. As mentioned above, support for such mailing lists may be integrated into the project repository. Mailing lists allow developers who are both geographically and temporally separated to maintain a conversation. Mailing lists are superior to regular email because they are topic-specific and they allow all of the developers to stay in the loop. Another tool that is used to support synchronous conversations is Internet Relay Chat (IRC), although many projects also use free videoconferencing tools such as Skype and Google Hangouts.

There are some significant advantages to using online messaging systems in a software engineering course. First, using these tools means that students do not have to hold as many face to face meetings, which can be a huge problem for non-traditional students with family commitments, commuter students, and even residential students who are taking a heavy course load. Students also appreciate the ability to search the conversations on the mailing list, making it less necessary to take careful meeting notes. And finally, mailing lists are very useful from the instructor's point of view, because he or she can monitor the conversations, getting a better idea of where the students are having trouble, which groups are not communicating very well, and what the various project statuses are. It is even possible to mine the conversations to create more careful analyses of student conversations (MacKellar, 2013).

Writing Developer and User Documentation

There are two types of writing that software developers usually engage in: writing documents aimed at other developers, and writing documents aimed at users. Software engineering courses usually try to have students engage in both types of writing. A very common set of documents produced during a traditional course would include a requirements document, a system design, a user manual or help system, and code level documentation. This follows the needs of the standard waterfall method, but in a one-semester software engineering course, this also can be very rushed. It is not clear how well students learn to write a requirements document in the two or so weeks that are typically devoted to the process in a traditional one-semester course.

In the CO-FOSS model, the requirements are already developed before the course starts, and if the course project is an iteration of earlier work, the architecture is also already defined. Thus, the bulk of the writing in the course is concentrated on two sets of artifacts: technical documentation aimed at other developers, and user documentation. The technical documentation consists of system design documents in the form of use cases and class designs, as well as comments in the code itself. The user documentation consists of an online help system.

Generally, open source projects have had a reputation of being poorly documented both in terms of developer-oriented documentation and user-oriented help systems and manuals (Madsen & Nürnberg, 2005; Meneely, Williams & Gehringer, 2008). This is not a desirable outcome for a software engineering course or for a client-oriented project, particularly when the clients are small nonprofits where there may not be a lot of technical expertise. This is an area where the instructor's guidance is vital to establishing effective standards for documentation. The client's feedback is also very important to help students develop a good help system, since most students have never written about or even thought about their software from a client perspective before.

Because of the open source nature of the CO-FOSS model, all artifacts are publicly available. This means that other students as well as the client can access, comment on, and even improve the documentation as needed. It also encourages the students writing the documents to concentrate on quality, since they know that other people will see their work. The code repository consists of an introductory page explaining the point of the project. The repository can also have a project wiki, which contains information of use to developers and users, such as how to install the software, and possibly instructions on writing new code modules. The use cases and class design are posted on the code repository, either in the wiki or as a downloadable document. Most open source repositories also contain an issue tracker and a discussion forum, and even those can be seen as forms of documentation.

Since students have usually not done a lot of technical writing before this course, they require a lot of guidance, structure, and examples when completing this part of the project. Students can be directed to other CO-FOSS student projects to find examples of good writing. The instructor can discuss these projects with students and point out ways in which the writing is effective or not effective. Even more powerfully, as a given client oriented open source project evolves over time, students will be able to work with existing help systems and design documents, and to use these as templates for their current project. Once an instructor has established a consistent framework for the open source project, it can be reused again and again, and students will be able to see a wide variety of examples, discovering what works and what does not work. For an example of an on-line help system developed by students, see rmh.myopensoftware.org and login as Admin1112345678 (same password) and hit the **help** tab.

Evaluation of Team Members' Contributions

Evaluating student contributions in team projects is one of the thorniest issues encountered by instructors of project-based courses. There is a large body of literature, and many competing ideas, on how to do this task effectively. The most common approaches are 1) individual grades based on peer evaluations 2) individual grades based on instructor evaluations 3) a group grade that is assigned to all group members based on project success 4) mixed approaches incorporating the previous three approaches in various ways.

One of the problems that make it difficult to evaluate group projects is that individual contributions need to be measured to avoid the "free-loader" effect: that is, students who fail to make any meaningful contribution towards the project but who end up passing because they share a group grade. Research has found that students prefer individual grades in order to prevent the free-loader effect (Farrell, Ravalli, and Farrell, 2012). Thus, it is common to assess students with a mixture of approaches; assigning a group grade as well as an individual grade, weighting each in some fashion. Farrell finds that the group grade is often determined objectively (how many project objectives were achieved while the individual grade is often determined subjectively (how much of a contribution did a particular student make?).

At Bowdoin College, the project grade measured the degree to which the team completed all the project milestones. Individual grades were assigned in relation to this project grade in order to recognize differences among different students' contributions. At St John's University, students received individualized grades that took into account the overall group success, the students contribution to the group effort as measured by proportion of code written and activity at meetings and on the message board, and the quality of the student's code. At UNH Manchester students also received individualized grades that measured the degree to which assigned tasks were successfully completed. These tasks included: work on the software system artifacts (whether code or documentation), documenting the development process (team meeting agendas and minutes, participation on the team's reporting on status of the team's artifacts, and writing self-evaluations to reflect on progress and challenges with each student's individual contribution.

Post-Course Activity

A significant challenge to CO-FOSS projects is to ensure that the resulting software artifact is delivered and supported in a timely way to the nonprofit organization that helped develop it. This is a challenge because, just as in the case of pre-course CO-FOSS activities, this activity falls outside the normal expectations of a university faculty member.

Our experience in meeting this challenge is varied. At Bowdoin, for example, the instructor has dedicated his own time to ensuring that the software is delivered and properly supported through the first several months of its use. The quality of the software artifacts developed so far has been so high (in the case of Homebase and Homeplate, for example – see discussion in the next section) that the need for ongoing software support, once it is put into use at the nonprofit, has been minimal. At St John's and UNH Manchester, instructors have also devoted significant time to bringing the system up to production standards.

In the long run, the need for ongoing support for a successful CO-FOSS project can be met by the establishment of a partnership between the non-profit and a local software firm that has the capability and interest to provide that support on a cost-effective basis. For each of the Homebase and Homeplate projects, the non-profit has partnered with a local software firm that provides support at a reasonable cost.

We recognize that sustainability is a major challenge for any community service project involving student developers, simply because at the end of the semester both the students and the instructor typically move on to other courses and priorities. On the other hand, it is mandatory to the integrity of the CO-FOSS model that a support structure be put into place that facilitates the creation of client-software firm partnerships once a semester project is completed and the software is ready for deployment.

To this end, the authors are participating in the establishment of a new organization called the Non-Profit FOSS Institute (NPFI). The purpose of the NPFI is to facilitate the planning, creation, execution, and ongoing support for new CO-FOSS projects like the ones discussed in this chapter. Its 13-person Advisory Board represents all three types of participants in such projects – non-profits, instructors, and software firms. The Board members all have significant experience with, and enthusiasm for, developing CO-FOSS at a national level.

Key to the success of the NPFI is to facilitate and support the formation of "triads," each triad having an instructor, a non-profit client, and a local software firm. Facilitation includes supplying support materials – code bases, requirements document examples and templates, and other teaching materials that will help an instructor get started with a new CO-FOSS course. Because NPFI is in an early stage of development, we cannot say much more about it at this writing. We can say that in the spring of 2014, the Board plans to launch the http://npfi.org Web site, which will provide many more details about the mission and organization of the NPFI. Readers who are interested in becoming associated with NPFI in the future are encouraged to visit this Web site and become a member. NPFI membership will be open to all instructors, non-profits, software firms and others who embrace the CO-FOSS model.

CASE STUDIES: THE EXPERIENCE AT THREE SCHOOLS

The three authors have significant experience in adapting the CO-FOSS framework to different institutional and nonprofit environments. Overall, we have found this framework to be robust in the sense that it is adaptable to a variety of academic and nonprofit settings. However, an instructor wishing to use this framework in his/her own course must take into account local differences in student population, institutional support, and nonprofit availability and willingness to participate in such a project. The case studies in this section detail how the three authors customized the model to meet their local needs.

Case Study 1: *Homebase*

Developing the Framework with a Client-Server Application at a Small Private College

Bowdoin College is a selective residential liberal arts college with a fully-developed Computer Science Department, averaging about 10 majors per class. Students are enrolled full-time at the College, and computer science majors typically take four courses per semester. Many majors also choose to double major with mathematics, economics, or one of the sciences. At the time students enroll in the software development course, they are typically juniors and have taken a significant number of computer science courses, including data structures and algorithms, and have done a good deal of programming in different languages.

In Spring, 2008 Allen Tucker developed and taught the software course using the CO-FOSS approach for the first time. Inspired by the H-FOSS model (Morelli, DeLanerolle, & Tucker, 2012), this course aimed to develop on-line volunteer scheduling software for the Ronald McDonald

House in Portland, Maine. Four brave students enrolled in and completed the course, all seniors (three CS majors and one economics/math major who had a lot of programming experience). The software that they completed was dubbed "Homebase" by the RMH staff at some point during the semester.

Prior to the beginning of the semester, the instructor met with RMH staff to gather information that would contribute to a requirements document for the *Homebase* project. The requirements document included a description of the then-current manual volunteer calendar scheduling process, a description of the methodology and tools that would be used to develop the software, a few screen-shot sketches of the desired user interface, and a time-line of milestones that had to be met to complete the project by the end of the semester.

The goal was to develop a complete working prototype for *Homebase*, including on-line user help and a week of training. The development team included the four students who would do the major programming, the instructor who would manage the project by giving weekly assignments and overseeing the sandbox server, and a client representative who would test the partially-completed software and provide weekly feedback to the students. Throughout the semester, each milestone was adjusted as a result of the weekly team meeting and feedback from the client. By the end of the semester, the client knew exactly what she was receiving as a software tool, making this a truly agile process.

The software architecture for *Homebase* may be its most important characteristic as a model for other CO-FOSS projects that use a similar teaching strategy. The client-server architecture can be viewed as having layers – the domain classes, the database modules, the user interface modules, and the user help modules. Confined by the limits of a single semester, the course can naturally flow by developing these four layers in order as a series of 3-week chunks, beginning with students developing and unit-testing the domain

classes identified in the design document. This initial chunk breeds a vocabulary of terms that can be shared unambiguously between the students and the client – in the case of *Homebase*, terms like "Shift," "Week," and "Volunteer Availability" take on specific meanings that facilitate communication throughout the remainder of the development process.

The outcome of this project was a fully functional online volunteer scheduling module that integrated with the RMH Website and replaced the manual scheduling system during the summer of 2008. The instructor spent significant effort during that summer making the software "bullet-proof" so that it ran reliably and correctly on a 24/7 basis. Volunteers and House staff uniformly praised the software for its ease of use, security, and accessibility from anywhere there was a Web browser. The students completed this course knowing that they had not only experienced a real-world software development task but also made a significant service learning contribution. Since its completion in 2008, *Homebase* has been used as a starting point for similar software development activities, both at Bowdoin and at other universities and has been updated and expanded by two more teams of Bowdoin students in Spring 2012 and Fall 2013. A link to the current version of *Homebase* and its requirements document is listed in the Resources section at the end of this chapter.

Case Study 2: *Homeplate*

Adapting the Framework for a Mobile Computing Application

The CO-FOSS framework discussed above was used again at Bowdoin College in 2011 to develop a room scheduling module called *Homeroom*, and again in 2012 to develop a module called *Homeplate* for volunteers to record pickups and drop-offs for a food rescue and distribution organization called Second Helpings, in South Carolina. The *Homeplate* project was significant because it used the CO-FOSS framework for developing the domain, database, and user interface layers for an easy-to-use Web-based client-server application. But it was also significant because it provided a server-side platform with which an independent Android application could later be developed and deployed.

The server side of *Homeplate* was developed by a team of three Bowdoin students during the Spring 2012 semester, and deployed soon after the end of the semester. The *Homeplate* software architecture was essentially the same as that of *Homebase*, and we reused several key modules of *Homebase* in this new project. During the summer, one student worked with the instructor and the client to develop the Android tablet app that could be carried on the Second Helpings trucks themselves. This app facilitated volunteers' recording of food weights at the time the food was being picked up and dropped off from the trucks. To attempt this Android app as part of the semester project alongside the *Homeplate* software would have created both conceptual and practical overload. In general, it is always important for an instructor launching a CO-FOSS project to assess what can be accomplished in a semester, and trim the project appropriately in a way that ensures student success with a high probability.

The Android app runs on a tablet and sends and receives scheduling and weights data from the *Homeplate* server via FTP when the tablet is in a free wi-fi zone. This strategy avoids requiring Second Helpings to purchase expensive data plans to accompany their tablets, which would be a deal-breaker for most charitable nonprofits. The Android app was developed on a Java-like platform, which is provided freely through developer.android.com. Lots of tutorials about Android development are freely available on the Web, so that the student, the instructor, and the client representative were able to fully develop and deploy this enhancement during an 8-week period in the summer of 2012.

The Android app was deployed by Second Helpings on all 5 of their trucks in September of 2012, and has been running successfully ever since then. Volunteers (most of whom are retirees) remark that this app is extremely intuitive and easy to use. The success of *Homeplate* and its accompanying Android app has recently been reported in local a local newspaper article (Bredeson 2013) which highlights the essential role that students played in their success. Instructors interested in more details about *Homeplate* design, development, and source code downloads can access the *Homeplate* link in the Resources section.

Case Study 3:
RMHRoomReservationMaker

Adapting the Framework at an Urban University with Students of Diverse Backgrounds and Abilities

St John's University (SJU) is a large, urban university whose students are ethnically diverse; a large percentage are Pell-eligible and first in their family to attend college. Many students work off campus or have significant family responsibilities. Software engineering is a required course in the computer science major. Students enroll in the software engineering course with various backgrounds – some have done little programming and others know quite a bit. In order to meet the needs of this diverse group of students, we usually center the software engineering course around one larger project, on which the entire class collaborates. This organization, which is described in more detail in (MacKellar, 2011), is based on the idea that students work in smaller groups organized around a project role, such as testing, development of the database, or help system development. This is a common organization in real world software projects.

In previous years, the project for the course was always a "toy project," designed strictly for the class and without a real world client. We wanted to improve the course by working with a real world client and by bringing more open source process into the course. We had a client in mind, the Ronald McDonald House of Manhattan, where there were several potential projects available. We were aware of the CO-FOSS through the Humanitarian Free and Open Source project. One of the CO-FOSS projects at Bowdoin College, *Homeroom*, was very similar to a project request from RMH Manhattan. The framework seemed to be a good approach; the challenge for us was adapting it to work with our students and with our whole-class project approach.

The first phase, the pre-course activities, was heavily facilitated by working within the CO-FOSS framework and making use of the architecture and various components of *Homeroom*, a pre-existing open source project that had been developed at Bowdoin College. There were some major differences in requirements between the two projects, so our project became one of extending the *Homeroom* code rather than simply adapting it. Since we had never worked with a real client for our software engineering course before, we used the *Homeroom* project as a model in many ways. One important way in which *Homeroom* served as a model was simply as a guide for sizing our project. Choosing a project of an appropriate size and scope is critical to success with client-facing projects, so being able to compare our potential project to an already successful one was critical.

We also used the organization and resources from the Bowdoin College projects quite extensively. We used the extended project description in (Morelli, Tucker, & DeSilva, 2011), materials on the project Website, and discussions with Dr. Tucker to become familiar with the structure and the decision points. We followed the organization as much as possible since it provided a successful structure.

However, there were significant areas where adaptations had to be made. This became most apparent during the curriculum design phase. Our student body is quite different from the stu-

dents at Bowdoin College. The class was larger, with 25 students. Few of them had ever written a program longer than a hundred lines, and none had ever collaborated with another student on a software project. The *Homeroom* software was written in PHP, a programming language with which most of the students had little experience. A central concern in this course was to fill in their missing skills at the same time they were developing a system design for the project. To teach students the skills they would need for the project, the first half of the semester provided in-class labs on the basics of PHP, version control, and databases. Thus, students were not able to begin project implementation until about halfway into the course.

The project involved the room scheduling process at the Ronald McDonald House of Manhattan. As mentioned before, significant differences in requirements meant that we could not simply modify *Homeroom*. In particular, RMH-Manhattan needed two interfaces, a more complex workflow for processing room reservations that involved an approval process and a lot of automated email, and an audit system that tracked all changes made to reservations. Therefore, we reused the overall architecture and the lowest level layers of *Homeroom* – the database and domain object layers – but had to do an entirely new business logic layer. The project ended up spanning two semesters. In the first iteration of the course, the students completed a skeletal prototype. During the second iteration, a new group of students worked to close out issue tickets, complete functionality, and bring the system to a point where it might be deployed.

The differing approach, both in terms of the whole-class project approach and the necessity to spend time teaching students the needed skills, meant that the syllabus structure and task assignment phases had to be modified. As mentioned earlier, students spent the first half of the semester learning the technology and tools. This was done via a series of labs, including code reading exercises working with the *Homeroom* codebase from Bowdoin. Students began development at midsemester. The project work was organized into 2 multiweek chunks, similar to the organization of the *Homebase* project. Since all students in the course were working on the same project, they were organized into teams based on functionality. Students were chosen for teams by the instructor, who used the criterion of "least risk' in making the assignments, based largely on information on resumes submitted by the students in the first week.

Since the students only had half the semester to devote to project work, the project required two iterations of the course to be finished. In the second iteration of the course, the same overall schedule was used, with students learning the tools and the codebase in the first half of the semester. In the second half, the students worked on the project in feature oriented teams again, with tickets from the issue tracker assigned to the various groups.

The Google Code open source software repository was used to host the project. This repository has version control built into it, as well as an issue tracker which we used extensively. Because the students were mainly commuter students who do not spend much time on campus, a project specific discussion board was set up, which was used extensively by all of the students. Most communication relevant to the project happened either in class or on the discussion board. All tickets from the issue tracker were automatically forwarded to the discussion board so that students would all be aware of any bugs or problems. All code commits were also forwarded to a second discussion board which all students subscribed to.

Although finishing the project required two course iterations rather than one, this was not unexpected due to the fact that the students had weaker skills and less time to spend on the project than was the case at Bowdoin College. The open source codebase for *Homebase* and *Homeroom* served as both a model and as a set of classes and modules that could be adapted to the differing requirements of RMH-Manhattan. Even more importantly, the CO-FOSS course organization and materials were critical to our success; it is not likely we would even have attempted a project of this scale with a real client without a "recipe" which could be adapted to our needs.

Case Study *4: DONATE*

Adapting the Framework for a Course at an Urban, Commuter College with Transfer Students Constrained Financially and by Work and Family Commitments

Affordability has been a compelling reason for adopting FOSS in the computing curricula and for equipping the computing labs with support infrastructure in the Computing Technology program at University of New Hampshire at Manchester (UNH Manchester). Using FOSS systems and services, however, is just the first step in taking advantage of how FOSS development principles and practices can impact students learning. In this case study we describe the experience with adapting the CO-FOSS framework for an upper-level elective course in the B.S. Computer Information Systems (BS CIS) major at UNH Manchester. The major requirements are structured into a two-layer core (eight courses), integrative and professional experience (four courses, including internship and capstone project courses), a self-designed concentration in an application domain (four courses), and three computing electives.

An overarching challenge for teaching CO-FOSS development in the BS CIS program is a collection of hard constraints placed on students by their work, family obligations, and other commitments. When time on campus is reduced to class meetings only, students are forced to do project work once a week, typically the night before the class meeting, with almost no time to coordinate their work with other team members. These constraints are compounded by the students' uneven academic preparedness. A large majority of students (70%) have more than 50 credits transferred from local two-year colleges, where (1) projects were individual endeavors; 2) student exposure to FOSS principles and practices was limited or non-existent, and 3) prior programming experience did not include algorithm design and using abstraction to tackle more complex problems.

In Spring, 2012, the *Homebase* approach was used in the Web Application Development course, an upper-level elective course. Eleven students enrolled in the course: four undergraduate CIS majors, two graduate IT majors, and five continuing education students. The continuing education students were engineers from a local company in Manchester, who were interested in open source technologies and gaining experience developing Web-based services for their company's in-house applications. The client was YWCA of Manchester, with whom the UNH-Manchester program had collaborated since Fall 2008 on various projects and student service learning activities. The project addressed the client's need for an information system that tracks donations from individuals and organizations. Like the other nonprofits mentioned in this chapter, YWCA cannot afford to buy software and pay developers or consultants. Open source is the only feasible approach that they have for developing a donation tracking system.

At the beginning of the semester, the project's prototype had a code base with the same layered architecture as the *Homebase* software. The project was a combined result from two other courses with student projects of smaller and dedicated scope – one database course and one Web authoring course, both at sophomore level. The prototype's back-end had a functional MySQL database, although incomplete, with a well-designed schema, sample data, and scripts to install, populate, and query the database. The front-end had a single use case implemented, i.e., viewing donors and searching donors by name. The prototype was staged on a virtual machine that runs on UNH-Manchester's server. The code base is hosted by Google Code.

In the first half of the semester students learned how to develop an open source client-based project through a variety of activities: 1) experimenting with the *Homebase* and *Homeroom* projects in the textbook (Tucker, Morelli, & de Silva, 2011) ; 2) meeting with the YWCA business director to get clarifications on the system requirements; 3) learning from three experienced FOSS developers who joined the class via Skype; 4) receiving feedback and having their work reviewed by the software engineers who were auditing the class; and 5) doing assignments that provided practice with: PHP and SQL, model-view-controller architectural pattern, techniques for specifying requirements and design decisions, XAMPP run-time environment configuration, and a comprehensive development toolkit (Eclipse PHP, Xdebug, Doxygen, SimpleTest, and Balsamiq). In addition, a project forum, wiki, and hosted project version control and issue tracking supported teamwork and collaboration. The code base became the "common denominator" for all design and implementation decisions. That is why the assignments were grounded in the code base and allowed students to learn first-hand about the developer roles needed for the project. By the time the six week-long project sprints occurred in the second half of the semester, students understood well the roles they would assume to maximize their contributions to the project.

At UNH-Manchester, the *Homebase* approach yielded several good outcomes. First, having continuing education engineers as observers allowed creation of the role of configuration manager, who was the most active and involved student during class meetings. These observers also participated in client meetings, reviewed and gave feedback on use cases, database and UI design, and code quality. Students also learned from one of the invited speakers about the Asana service for project management activities, which they ended up using for issue tracking.

CONCLUSION

Trying to incorporate real world software practices and project experience into the typical one semester undergraduate software engineering course is always challenging. The CO-FOSS approach, which occupies a middle ground between proprietary software development for individual customers and working within a large, existing FOSS project, offers a number of advantages. Projects developed in conjunction with a nonprofit client can be tailored to be student friendly and fit within the constraints of the academic semester. At the same time, following FOSS practices means that the architecture, codebase, and documentation can be reused and adapted for future projects. As shown in this chapter, a CO-FOSS project does not have to remain resident within one college but can be adapted by other schools in true open source fashion. Our future directions involve efforts to bring more schools to this approach, creating ways to support maintenance of these projects, and investigating ways to foster communication among students working on similar projects at different schools.

In conclusion, this chapter illustrates how a CO-FOSS project course can be designed and delivered. Such a course embodies a novel curriculum design that teaches communication skills, teamwork, and writing skills. It prepares students by engaging them in real-world problems, introducing them to fresh technologies (such as agile programming, layered architecture, and mobile computing), and leveraging all the advantages of open source development and tool use. We hope that our work will inspire others to rethink their courses in these ways, so that their students will become better prepared to enter the software industry with both feet on the ground.

REFERENCES

Bredeson, A. (2013, June). Second helpings volunteers get much-needed technological boost from app. built by students. *The Island Packet*.

Cheng, Y., & Lin, J. (2010). A constrained and guided approach for managing software engineering course projects. *IEEE Transactions on Education*, *53*(3), 430–436. doi:10.1109/TE.2009.2026738

Coppit, D., & Haddox-Schatz, J. (2005). Large team projects in software engineering courses. In *Proceedings of the 36th SIGCSE Technical Symposium on Computer Science Education* (pp. 137–141). ACM Computer Society.

Coyle, E. J., Jamieson, L. H., & Oakes, W. C. (2005). EPICS: Engineering projects in community service. *International Journal of Engineering Education*, *21*(1), 139–150.

Ellis, H. J., Morelli, R. A., De Lanerolle, T. R., & Hislop, G. W. (2007). Holistic software engineering education based on a humanitarian open source project. In *Proceedings of the 20th Conference on Software Engineering Education & Training* (pp. 327–335). IEEE.

Farrell, V., Ravalli, G., & Farrell, G. (2012). Capstone project: Fair, just and accountable assessment. In *Proceedings of the 17th ACM Annual Conference on Innovation and Technology in Computer Science Education* (pp. 168–173). ACM.

Fogel, K. (2005). *Producing open source software*. Sebastopol, CA: O'Reilly.

Grundy, J. (1996). A comparative analysis of design principles for project-based IT courses. In *Proceedings of the 2nd Australasian Conference on Computer Science Education* (pp. 170–177). Australian Computer Society.

Hauge, O., Ayala, C., & Conradi, R. (2010). Adoption of open source software in software-intensive organizations – A systematic literature review. *Information and Software Technology, 52*(11), 1133–1154. doi:10.1016/j.infsof.2010.05.008

Judith, W., Bair, B., & Börstler, J. (2003). Client sponsored projects in software engineering courses. In *Proceedings of the 34th SIGCSE Technical Symposium on Computer Science Education* (pp. 401–402). ACM.

Linos, P. K., Herman, S., & Lally, J. (2003). A service-learning program for computer science and software engineering. In *Proceedings of the 8th Annual Conference on Innovation and Technology in Computer Science Education* (pp. 30–34). ACM.

MacKellar, B. (2013). Analyzing coordination among students in a software engineering project course. In *Proceedings of the 26th IEEE Conference on Software Engineering Education and Training* (pp. 279–283). IEEE Press.

Mackellar, B. K. (2011). A software engineering course with a large-scale project and diverse roles for students. *Journal of Computing Sciences in Colleges, 26*(6), 93–100.

MacKellar, B. K., Sabin, M., & Tucker, A. (2013). Scaling a framework for client-driven open source software projects: A report from three schools. *Journal of Computing Sciences in Colleges, 28*(6), 140–147.

Madsen, F., & Nürnberg, P. (2005). Calliope: Supporting high-level documentation of open-source projects. In *Proceedings of the 2005 Symposia on Metainformatics* (pp. 10). ACM.

Marmorstein, R. (2011). Open source contribution as an effective software engineering class project. In *Proceedings of the 16th Annual Joint Conference on Innovation and Technology in Computer Science Education* (pp. 268–272). ACM.

McConnell, J. J. (2006). Active and cooperative learning. *ACM SIGCSE Bulletin, 38*(2), 24. doi:10.1145/1138403.1138426

Meneely, A., Williams, L., & Gehringer, E. (2008). ROSE: A repository of education-friendly open-source projects. In *Proceedings of the 13th Annual Conference on Innovation and Technology in Computer Science Education* (pp. 7–11). ACM.

Morelli, R., de Lanerolle, T., & Tucker, A. (2012). The HFOSS project: Engaging students in service learning through building software. In *Service learning in the computer and information sciences*. New York: Wiley and IEEE Press. doi:10.1002/9781118319130.ch5

Olsen, A. L. (2008). A service learning project for a software engineering course. *Journal of Computing Sciences in Colleges, 24*(2), 130–136.

Poger, S., & Bailie, F. (2006). Student perspectives on a real world project. *Journal of Computing Sciences in Colleges, 21*(6), 69–75.

Richards, D. (2009). Designing project-based courses with a focus on group formation and assessment. *Transactions on Computer Education, 9*(1), 1–40. doi:10.1145/1513593.1513595

Tadayon, N. (2004). Software engineering based on the team software process with a real world project. *Journal of Computing Sciences in Colleges, 19*(4), 133–142.

Tan, J., & Jones, M. (2008). A case study of classroom experience with client-based team projects. *Journal of Computing Sciences in Colleges, 23*(5), 150–159.

Tucker, A., Morelli, R., & de Silva, C. (2011). *Software development: An open source approach*. Boca Raton, FL: CRC Press.

Venkatagiri, S. (2006). Engineering the software requirements of nonprofits: A service-learning approach. In *Proceedings of the 28th IEEE International Conference on Software Engineering* (pp. 643–648). IEEE.

ADDITIONAL READING

Anewalt, K. (2009). Dynamic group management in a software projects course. *Journal of Computing Sciences in Colleges*, *25*(2), 146–151.

Beck, J. (2005). Using the CVS version management system in a software engineering course. *Journal of Computing Sciences in Colleges*, *20*(6), 57–65.

Beck, J., Almstrum, V. L., Ellis, H. J. C., & Towhidnejad, M. (2009). Best practices in software engineering project class management. *ACM SIGCSE Bulletin*, *41*(1), 201–202. doi:10.1145/1539024.1508939

Bernhart, M., Grechenig, T., Hetzl, J., & Zuser, W. (2006). Dimensions of software engineering course design. In *Proceedings of the 28th International Conference on Software Engineering* (pp. 667–672). ACM.

Ellis, H., Morelli, R., & Hislop, G. (2008). Work in progress-challenges to educating students within the Community of Open Source Software for Humanity. In *Proceedings of Frontiers in Education Conference* (pp. 7–8). IEEE Press.

Ellis, H. J. C., & Mitchell, R. (2004). Self-grading in a project-based software engineering course. In *Proceedings of the 17th Conference on Software Engineering Education and Training* (pp. 138–143). IEEE Press.

Ellis, H. J. C., Purcell, M., & Hislop, G. W. (2012). An approach for evaluating FOSS projects for student participation. In *Proceedings of the 43rd ACM Technical Symposium on Computer Science Education.* (pp. 415). ACM.

Filho, W. P. (2006). A software process for time-constrained course projects. In *Proceedings of the 28th International Conference on Software Engineering* (pp. 707–710). ACM.

Gehringer, E. F. (2011). From the manager's perspective: Classroom contributions to open-source projects. In *Proceedings of Frontiers in Education Conference* (pp. 1–5). IEEE Press.

Gehrke, M., Giese, H., Nickel, U. A., Niere, J., Tichy, M., Wadsack, J. P., & Zündorf, A. (2002). Reporting about industrial strength software engineering courses for undergraduates. In *Proceedings of the 24th International Conference on Software Engineering* (pp. 395–405). ACM.

Hart, D. (2009). A survey of source code management tools for programming courses. *Journal of Computing Sciences in Colleges*, *24*(6), 113–114.

Hayes, J. H., Lethbridge, T. C., & Port, D. (2003). Evaluating individual contribution toward group software engineering projects. In *Proceedings of the 25th International Conference on Software Engineering* (pp. 622–627). IEEE.

Ikonen, M., & Kurhila, J. (2009). Discovering high-impact success factors in capstone software projects. In *Proceedings of the 10th ACM Conference on Information Technology Education* (pp. 235–244). ACM.

Lancor, L. (2008). Collaboration tools in a one-semester software engineering course: What worked? what didn't? *Journal of Computing Sciences in Colleges*, *23*(5), 160–168.

Lancor, L., & Katha, S. (2013). Analyzing PHP frameworks for use in a project-based software engineering course. In *Proceedings of the 44th ACM Technical Symposium on Computer Science Education* (pp. 519–524), ACM.

Larman, C. (2004). *Applying UML and patterns: An introduction to object-oriented analysis and design and iterative development.* Prentice Hall PTR.

Liu, C. (2005). Using issue tracking tools to facilitate student learning of communication skills in software engineering courses. In *Proceedings of the 18th Conference on Software Engineering Education & Training* (pp. 61–68). IEEE.

Liu, C. (2005). Enriching software engineering courses with service-learning projects and the open-source approach. In *Proceedings of the 27th International Conference on Software Engineering* (pp. 613–614). ACM.

Morelli, R., Tucker, A., Danner, N., De Lanerolle, T., Ellis, H., & Izmirli, O. et al. (2009). Revitalizing computing education through free and open source software for humanity. *Communications of the ACM, 52*(8), 67–75. doi:10.1145/1536616.1536635

Nita-Rotaru, C., Dark, M., & Popescu, V. (2007). A Multi-Expertise Application-Driven Class. In *Proceedings of the 38th SIGCSE Technical Symposium on Computer Science Education* (pp. 119–123). ACM.

Pedroni, M., Bay, T., Oriol, M., & Pedroni, A. (2007). Open source projects in programming courses. *ACM SIGCSE Bulletin, 39*(1), 454. doi:10.1145/1227504.1227465

Reichlmayr, T. (2003). The agile approach in an undergraduate software engineering course project. In *Proceedings of Frontiers in Education Conference* (pp. 13–18). IEEE Press.

Sanders, D. (2007). Using Scrum to manage student projects. *Journal of Computing Sciences in Colleges, 23*(1), 79–79.

Sherrell, L. B., & Robertson, J. J. (2006). Pair programming and agile software development: Experiences in a college setting. *Journal of Computing Sciences in Colleges, 22*(2), 145–153.

Stein, M. V. (2002). Using large vs. small group projects in capstone and software engineering courses. *Journal of Computing Sciences in Colleges, 17*(4), 1–6.

Tan, J., & Jones, M. (2008). An evaluation of tools supporting enhanced student collaboration. In *Proceedings of the 38th Annual Frontiers in Education Conference* (pp. 7–12). IEEE Press.

Tucker, A. (2009). Teaching client-driven software development. *Journal of Computing Sciences in Colleges, 24*(4), 29–39.

Way, T. (2005). A company-based framework for a software engineering course. In *Proceedings of the 36th SIGCSE Technical Symposium on Computer Science Education* (pp. 132–136). ACM.

Wilkins, D. E., & Lawhead, P. B. (2000). Evaluating individuals in team projects. In *Proceedings of the 31st SIGCSE Technical Symposium on Computer Science Education* (Vol. 32, pp. 172–175). ACM.

KEY TERMS AND DEFINITIONS

Agile Development: Methodology to develop software in short iterations, called sprints, in which existing code is first refactored, tests for a new requirement are written, and test-driven code is developed.

Code Base: All the source code files stored in a source control repository that handles various versions and tracks implementation issues.

Domain Model: Representation of conceptual classes or real-situation objects in the domain of interest.

Free and Open Source Software: Software licensed and freely distributed along with its underlying code.

Layered Architecture: Organization of the software classes into layers such that higher layers call upon services of lower layers.

Model-View-Controller Architecture: A three-layer architecture in which domain knowledge is maintained by the domain model objects, displayed by the view (user interface) objects, and manipulated by control (application logic) objects.

Nonprofit Organization: Usually a charity or service organization that uses any surplus revenues to support its mission and achieve its goals.

Real-World Projects: Software engineering projects that are sponsored by real clients and result in production-grade software systems.

Requirements: Functional capabilities (what the system will do) to which the software system must confirm.

APPDENDIX: CO-FOSS RESOURCES

- **Examples of the *Homebase* and *Homeplate* requirements documents:**
  - http://code.google.com/p/rmh-homebase
  - http://code.google.com/p/sh-homeplate
- **The current version of *Homebase* and its requirements document:**
  - http://code.google.com/p/rmh-homebase
- **The *Homeplate* project:**
  - http://code.google.com/p/sh-homeplate.
- **Student oriented Mercurial and Eclipse tutorials:**
  - http://myopensoftware.org/content/supporting-materials
- **Example syllabi for CO-FOSS project courses:**
  - http://myopensoftware.org/content/extended-course-syllabus.
- **The *RMH-RoomReservationMaker* project:**
  - https://code.google.com/p/rmh-roomreservation-maker/
- **Examples of currently popular project repositories:**
  - SourceForge http://sourceforge.net/
  - Google Code https://code.google.com/
  - GitHub https://github.com/.
- **Examples of currently popular version control systems:**
  - CVS http://savannah.nongnu.org/projects/cvs
  - Subversion http://subversion.apache.org/
  - Mercurial http://mercurial.selenic.com/
  - Git http://git-scm.com/

Chapter 20
Teaching Software Architecture in Industrial and Academic Contexts:
Similarities and Differences

Paolo Ciancarini
University of Bologna, Italy

Stefano Russo
University of Naples Federico II, Italy

ABSTRACT

In this chapter, the authors describe their experiences in designing, developing, and teaching a course on Software Architecture that tested both in an academic context with their graduate Computer Science students and in an advanced context of professional updating and training with scores of system engineers in a number of different companies. The course has been taught in several editions in the last five years. The authors describe its rationale, the way in which they teach it differently in academia and in industry, and how they evaluate the students' learning in the different contexts. Finally, the authors discuss the lessons learnt and describe how this experience is inspiring for the future of this course.

INTRODUCTION

What is the role of the software architecture inside a large mission-critical system? How is it created? How is it managed? How can the concept foster software reuse and productivity? These questions are quite relevant for engineering companies,

which produce families of software intensive systems (Buschmann, 1996). As software systems become larger, more complex, and more expensive, companies - in particular system integrators - feel an increasing need for improving their productivity exploiting sound and effective techniques for the definition, analysis, and evaluation of software architectures. This is what we observed in a number of cooperations between academia and industry,

DOI: 10.4018/978-1-4666-5800-4.ch020

and that motivated our study of how Software Architecture can be taught.

The initial question from which we started our study was: "how can one introduce and teach software asset reuse and software architecture evaluation to engineers who have been designing systems for years without explicitly dealing with these concepts?" Then we added a related question: "how can our experience in teaching Software Architecture in an industrial context be imported in an academic context of Computer Science students?"

When the field of Software Architecture emerged, it was argued it should be considered a discipline in its own, separate from Computer Science and possibly encompassing Software Engineering (Clements, 2010). After almost twenty years the corpus of the scientific works in the field has developed consistently, but there is still a large gap between this body of knowledge and what is actually needed in academic and industrial settings. Many of the achievements in the field have not matured enough; an example are Architecture Definition Languages (ADLs), that have not replaced - and are not likely anymore to replace - standard modeling languages (Clements, 2012). Some others achievements are more mature, for instance some tools for architectural analysis: in (Bernardo, 2001) is described a tool for performance evaluation of a software architecture, whereas in (Sterling, 1996) is described a tool for architecture animations. However, these tools still need to be tailored to specific software systems, and even more to internal industrial practices.

The increase of the size and complexity of contemporary software-intensive systems raises critical challenges to the engineering companies which build them to be integrated into larger systems of systems. Production and management problems with software intensive systems are well known and related to requirements engineering, software design, systems' families management, and their continuous testing integration and evolution.

Thus, teaching software architectures in an industrial context requires to meet a company's expectations in terms of mature knowledge, special competences, and best practices transferred to practitioners, that they can subsequently turn into the engineering life cycle of complex systems.

This is not easy to achieve, as architecting large-scale complex software systems - having tens of thousands of requirements and millions of lines of code - requires very high abstraction and modeling skills. A number of methods and solutions to these problems are based on the introduction of software architecting in the industrial practice (Bass, 2012). However, to become effective an architecture-centric development approach needs to be assimilated and put in everyday practice by the company personnel, who need architectural education and training.

We have found that introducing Software Architecture principles, methods, and tools in an academic context poses different problems because Computer Science students are less expert and more interested in creativity and technology use rather than software reuse and architectural evaluations. Thus, in our classes we emphasize the relationship of Software Architecture with programming languages and formal methods for modeling and reasoning on software systems.

In this chapter we describe our experience in designing, developing, and teaching a course on Software Architecture, that we tested both with our graduate Computer Science students in an academic context and with several systems engineers in a number of different companies, in a context of professional updating and training. The course has been taught in several editions in the last five years. We describe its rationale, the way in which we teach it differently in academia and in industry, and how we evaluate the students in the different contexts. Finally, we discuss the lessons we learnt and describe how this experience is inspiring us for the future of this course.

The structure of this chapter is the following: in the next section we discuss the "Background" of our research; then in the following Section we discuss "Teaching Software Architecture" in industry first, where our experiences started, then in the academia, where we expanded and consolidated our approach. Then we present our "Solutions and Recommendations," and how we intend to proceed with the "Future Research" on Software Architecture education and training. The last section contains our "Conclusions."

BACKGROUND

Research on Software Architecture flourished in the last decade of the 20th century. When the Standard 1471 "Recommended Practice for Architecture Description of Software-Intensive Systems" was issued by IEEE in 2000, this event highlighted the relevance of the concept for software engineering practice. In this century initially the education in Software Architecture has been considered more an issue in training practitioners than as a part of a curriculum in Software Engineering.

Currently the typical academic Software Architecture course consists of lectures in which concepts and theories are presented, along with a small modeling project whose aim is to give learners the opportunity to put this knowledge into practice. A text reference for this kind of course is (Taylor, 2009).

In an industrial context these two activities, namely *ex-cathedra* lectures and toy projects, are not adequate because learners have already some or even strong practical experience of large software projects and, more important, there is scarce time available to present, compare, and put in practice a variety of architectural principles and best practices. A reference for a one-week course

in an industrial context - the Philips company - is (Muller, 2004). In that course the emphasis was on "non-technical context that plays a role in architecting, such as organization, process, people, market and business."

Although now several Universities offer courses in Software Architecture, papers reporting on the rationale of teaching a Software Architecture course in a post-graduate curriculum are still quite scarce: we have found the following ones.

In (Lago & vanVliet, 2005), the authors presented a course given at the Vrije Universiteit in Amsterdam where the emphasis was on "*social*" aspects of software architecture with respect to the concerns of a variety of stakeholders; the paper reports about two master-level courses in academia focusing on the communication mechanisms that enact the social development of the architecture.

In (Gast, 2008) the author described a course at the University of Tubingen whose rationale was to let the students to appreciate the importance of *traceability*, seen as a non-functional property of software systems which makes them more maintainable, adaptable, and extensible.

The course described in (Mannisto, 2008), given at the Helsinky University of Technology, aimed at teaching realistic software architecture issues, taken from the industry (Nokia), inside a curriculum for software engineers. The participating students were provided with a concrete understanding of the decision-making involved in the architectural process for a complex software intensive system. However, the paper does not report any test of the course in an industrial setting.

Finally, in (Wang & Wu, 2001) the authors describe how a software architecture course developed at the Norwegian University of Science and Technology in Trondheim was adapted to include a game development project. The paper shows how especially difficult is to adapt an existing Software Architecture course to a domain for which no architectural styles are known, and reusable components are scarcely available, as it is the videogame domain.

TEACHING SOFTWARE ARCHITECTURE

Teaching Software Architecture in the Industry

Our experience in teaching Software Architecture in the industry concerns some Finmeccanica companies in Italy. Finmeccanica is a large industrial corporate group operating worldwide in system engineering for the aerospace, defense, and security sectors. It is an industry leader in the fields of helicopters and defense electronics. It employs more than 70,000 people, has revenues of 15 Billion Euros, and invests 1.8 Billion Euros (12% of turnover) a year in R&D activities.

From Jan 2013 Selex Electronic Systems (in short Selex ES) is Finmeccanica's flagship company focusing on the design of systems of systems. Selex ES inherits the activities of SELEX Elsag, SELEX Galileo, and SELEX SI to create a unified company. The new company operates worldwide with a workforce of approximately 17,900 employers, with main operations in Italy and the UK and a customer base spanning five continents. As the company delivers mission critical systems in the above fields by utilizing its experience and proprietary sensors and hardware technologies - radar and electro-optic sensors, avionics, electronic warfare, communication, space instruments and sensors, UAV solutions - most of its engineers work on system integration projects and products. As software is becoming an increasing part of these systems, the need raised for updating and upgrading the employees' skills on architecting large-scale complex software systems.

SELEX SI (for the rest of this chapter we will use this old company name because the course we describe was commissioned by that company)

spends literally millions of hours in software development and integration inside its lines of products and systems. One of the main ways for minimizing development costs and maximizing the reuse of components consists of defining a reference software architecture for each product line (Clements & Northrop, 2002). In general, a product line reference architecture is a generic and somewhat abstract blueprint-type view of a system, that includes the systems major components, the relationships among them, and the externally visible properties of those components. To build this kind of software architecture, domain solutions have been generalized and structured for the depiction of one software system structure based on the harvesting of a set of patterns that have been observed in a number of successful implementations. In this way the basis for a number of software product lines has been established, aiming at defining a product line reference architecture for SELEX SI systems. A product line reference architecture is not usually designed for a highly specialized set of requirements. Rather, architects tend to use it as a starting point and specialize it for their own requirements.

We were required to define a course on Software Architecture strongly emphasizing the power of reuse product line reference architectures.

The company started some years ago a learning programme called *"SELEX SI Academy."* The Academy is an operational tool to enhance and to develop the corporate knowledge heritage, and improve it over time. Academy's activities are developed by means of three Schools: Radar and Systems School, System Engineering School, and Complex Software Systems School, overall accounting for more than 100 courses.

The academic institution which has contributed to designing the Complex Software System School is the Consortium of Italian Universities for Informatics (*Consorzio Interuniversitario Nazionale per l'Informatica*, in short: CINI). The Complex Software Systems School includes 22 courses. A number of Computer Science and

Engineering professors of different Italian universities affiliate to CINI cooperated to its set up, teaching, and evaluation.

Software Architecture was the first course of the School. The course was designed by a small committee including two university professors from CINI (the two first authors of this chapter) and a number of people from SELEX SI, including engineering staff members (among them the third author) and the human resources department.

The description of the software architecture of engineering products and systems is becoming an important topic for educating software engineers, as can be seen for instance in the latest version of the chapter on "Software design" of the Body of Knowledge of Software Engineering. The teaching of Software Architecture per se in an academic environment is yet not standardized, however there are several examples, as a Google enquiry can show.

Indeed, the engineering people of large companies expose a variety of experiences and training histories, and often produce very peculiar architectures for specific markets. Thus, in an industrial context any teaching of Software Architecture has to take into account the specific requirements of the company managers.

In our case the course committee mentioned above started from a set of learning requirements given by the company:

- The course had to target generically all software developers active in SELEX SI. These are a few hundreds, with a variety of ages and cultural backgrounds. The basic assumptions were some proficiency in object-oriented programming in languages like C++ and concurrent programming in languages like Ada;
- The contents of the course should be customized for such SELEX SI personnel, meaning that the examples and the exercises should focus on the actual software systems developed by the company. This

implied that the teachers should receive some information on the software of interest for the company. Moreover, a number of testimonials from actual projects would make short presentations showing how the main topics of the course were being applied in the company itself;

- The course, overall 35 hours, should last one week (five days) or two half-weeks, as chosen by the company according to the workload of the different periods.
- Each class would include 20 people on average. Prerequisites for attending the class were the knowledge and practice of basic notions of object-oriented programming using a language like C++ or Java.
- All material should be written in English, as customary for all Schools in the SELEX SI Academy.

The learning outcomes of the course on Software Architecture for SELEX SI were defined as follows: "a review course on software systems architectures and related software engineering methods for designing complex software intensive systems." The initial idea was to establish a coherent baseline of software development methods, modeling notations, techniques for software asset reuse based on architectural styles and design patterns. More specifically, the learners were expected to obtain a basic knowledge of the principles governing software architecture, and the relationship between requirements, system architectures and software components (Component-Based Software Engineering - CBSE).

In our case the syllabus was then refined as an introduction to Software Architecture with some reference to Software Product Lines, putting software architecting in a perspective of software reuse, and presenting the related technologies. More precisely, the course had to recall some advanced issues in software modeling and present the main ideas in architectural design for reuse.

What follows is a short description of the contents of each lecture.

1. **Introduction to software architecture and techniques for software reuse:** This lecture introduces the main topic: how software-intensive systems benefit form focusing all the development efforts on the software architecture, aiming at reusing the existing assets for the various product lines architectures. The reference texts for this introduction were the older editions of (Clements, 2010) and (Rozanski & Woods, 2012).

2. **Describing software architectures:** This lecture had the goal to establish a methodological and notational baseline describing an iterative, architecture-centric process able to develop and exploit architectural descriptions for software systems using UML. Thus its contents are the principles and the practices of modeling and documenting software systems with UML, aiming at the ability to describe, analyze and evaluate a software system using a modeling notation. The reference text for this lecture was (Gomaa, 2005).

3. **Reusable architectural design:** This lecture has the goal to discuss the reuse of architectural design ideas: after a short recall of the elementary object oriented design patterns GRASP (Larman, 2004) and the more complex and well known Gang-of-Four patterns (Gamma, 1995), the main architectural patterns are introduced and presented in a systematic way. An architectural style can be seen as providing the system with a high-level organization. A number of architectural styles have been identified, that are especially important and widely reusable for SELEX SI systems. The reference texts for this lecture were (Gorton, 2011) and (Qian, 2006).

4. **Understanding component and connector views in the context of pattern-oriented software architectures:** This lecture has the goal to present some advanced examples of software architectures presenting their components and behaviors. The reference texts for this lecture were (Buschmann, 1996) and (Schmidt, 2000).

5. **Software systems architecture:** This lecture has the goal to present some advanced issues of software architecture in the context of model driven engineering. The main topics are the Model Driven Architecture and SysML, as used in SELEX SI. The reference text on MDA was (Stahl, 2006); for SysML we used (Weilkiens, 2008) and (Friedenthal, 2008).

We started giving this course in 2008, and until 2013 there have been four editions. For each edition the same protocol has been followed. One month before the class week the slides prepared by the instructors were examined by personnel of the company, who if necessary could suggest some changes. Then after the modifications the slides were put on line in a reserved site, and access was granted to the students. Thus, the people selected for the class could have a copy of the teaching material about one week in advance.

We had two variants of the time table: class over one week and class over two half weeks. The one-week timetable is shown in Table 1.

When the course was split over two half weeks, each period had specific entrance and final tests. Each edition leveraged upon the results of the preceding one, meaning that some adjustments were required to the course contents, usually triggered by some specific need presented by the company. Some suggestions came from the students' comments we gathered in class, other from people in the company covering the role of training manager. The suggestions impacted mainly

Table 1. Structure of the class, version offered in one week

Time	Day1	Day2	Day3	Day4	Day5
09-11	Test-In&lecture	Lecture	Lecture	Lecture	Lecture
11-13	Lecture	Lecture	Lecture	Lecture	Lecture
14-15	Testimonial	Testimonial	Testimonial	Testimonial	Testimonial
15-17	Hands-on	Hands-on	Hands-on	Hands-on	Lecture&Test-Out

the topic covered by the testimonials. In fact, these varied across the different editions of the course; the recurring topics they covered were: i) The description of software architectures inside SELEX SI (Sabbatino, 2012); ii) Using UML for documenting SELEX SI systems; iii) Typical architectural styles used inside SELEX SI; iv) Experiences with MDA inside SELEX SI (Barbi, 2012); v) Using SysML for SELEX SI systems. All speakers were engineers of the company engaged in using the related technologies and methods in their projects.

The results were measured in two ways:

- Comparing the results of the entrance and final tests. A scoresheet with 15 questions with predefined answers (each question had four possible answers) was given at the beginning of the first day and at the end of last day. The same set of questions and answers were used in both tests - entrance and final. In the first two editions (for which we have completed the analysis) the entrance test had a 41% of correct answers, while the final test gave almost 80% correct answers.
- Gathering the answers of the students to a satisfaction questionnaire. The satisfaction was measured on seven different parameters, and averaged 82,7 on a scale 0-100. According to this figure, this course was one of the most appreciated in the School of Complex Software Systems.

Teaching Software Architecture in the Academia

Just after the first edition of the course for the industry described in the preceding section, we had the occasion of introducing a course of Software Architecture in a Computer Science (CS) master degree.

At the University of Bologna CS students at the bachelor level (a three years course) receive a standard education to the main CS topics. These include a basic course on Software Engineering, which introduces process models, requirements analysis, software design patterns, testing methods, and maintenance metrics. The UML is used especially for teaching requirement analysis and design topics: this is quite common, see for instance (Wrycza & B. Marcinkowski, 2005) for an example of a course using UML2 for teaching at post-graduate level.

We have described our activities in the basic course in Software Engineering at the University of Bologna in a number of papers (Bolognesi, 2006; Ciancarini, 2013; Zuppiroli, 2012).

At the Master level (a biennial course) we introduced in 2009 the Software Architecture course leveraging on the knowledge and basic skills developed during the Software Engineering bachelor's course. The course material is available in English at the address http://www.cs.unibo.it/~cianca/wwwpages/swarch.html.

The basic idea consisted of discussing some key and well known software-intensive systems from an architectural viewpoint. This has been obtained mainly presenting the principal architectural patterns and styles using UML2 as reference language for discussing some architectural views, like for instance those suggested in Clements (2010): the component-and-connectors (C&C) view, the module view, the allocation view. The C&C view is the most important, and also the most difficult to be appreciated by the students, because its understanding requires some experience in the development of large software-intensive systems.

Examples of well-known systems we have discussed from an architectural viewpoint are: Cloud computing architectures, like Google App Engine, Microsoft Azure, Eucalyptus, or Dropbox; software architectures for mobile systems, like instant messaging services or Android; frameworks for Web applications, like .NET, etc.

We have taught several editions of this course, of different lengths: one semester (40 hours, six credits) or two semesters (80 hours, 12 credits).

The Short Version

The syllabus of our Software Architecture course in the short version is shown in Table 2.

Table 2. The syllabus in the short version

Topic	Hours
Software products and software-intensive systems	2
Architecture-centric processes for software reuse	2
How to describe a software architecture: concerns and views	2
Basic architectural styles (e.g. Pipe & filter, repository, client-server, peer2peer)	6
Complex architectural styles (choice from MVC, SOA, cloud, POSA, etc.)	10
Advanced UML: packages, components, and profiles	6
DSSA and SW Product Lines	2
Meta-modeling a software architecture with UML	6
Evaluating a software architecture	4
Total	40

This syllabus is then complemented with the presentations by the students, on a specific software architecture of their choice. We give specific guidelines for this the presentation, aiming at standardizing the architectural descriptions given by the students. The suggested structure of the presentation is the following:

1. Description of the system or class of systems;
2. Architectural description;
3. Context (where this architecture is used?)
4. Structure (which components and connectors? which packages?);
5. Features (how requirements are mapped on components);
6. Behavior (how the architecture works?);
7. Rationale (why the architecture has this form? Which styles are reused?);
8. Analysis and evaluation (critical points? Possible or usual tests?);
9. Similar systems or derived architectural patterns;
10. Bibliography (at least five research papers published in archival journals or proceedings, not Web pages).

The Long Version

In the longer version (that we taught in the first two editions out of four) we added to the basic kernel described above the presentations of some technologies: in Table 3 we present the additional syllabus we used. Moreover, some topics from the short syllabus – *i.e.*, architectural styles and advanced UML - were presented more deeply and with more exercises.

When we used this long version of the syllabus the final exam was based on the evaluation of a project driven by a given architectural style and exploiting one or more among the technologies listed in Table3, and developed in teams using a wiki as shared collaborative platform (Al-Asmari & Yu, 2006). We have experimented both a project oriented approach, in which students had to

Table 3. The additions to the syllabus in the long version

Topic	Hours
Introduction to middleware	4
CORBA	4
Enterprise Java Beans (EJB)	6
Java Enterprise Edition (JEE)	6
SOAP Web services	6
The REST style and applications	4
Dynamic content in Web applications	6
Google Web Toolkit	4
Total	40

design and develop the software architecture of a simple system, and a report oriented approach, in which students study an existing well known system (e.g. Dropbox) and have to research, model, and discuss its software architecture.

The main topics in the academic version of the course are the architectural styles and the UML2 used as an architectural language.

The development of software-intensive systems is based on a sequence of design decisions about their elements and the way in which these elements are related and depend on each other. The behavior of a system is determined not just by the functionality of each individual element, but also by the features of the configuration in which these elements are related.

The idea of building software systems through reusable configurations of components and connectors is quite effective. The software architecture community has codified a collection of reusable architectural styles, which are descriptions of patterns and constraints on the configuration of architectural elements that exhibit desirable properties for particular types of systems or application domains. By introducing style-specific constraints on component types and how these components interconnect, architectural styles codify reusable sets of design decisions. While

they certainly provide a useful starting point for the design of specific systems, styles also act as a valuable reusable knowledge since they capture distilled architectural experience from past development efforts. Students usually appreciate quite easily the basic architectural style: pipe-and-filters, shared repository, client-server, peer-to-peer. In our experience some students have more difficulty with styles for interaction, namely the Model-View-Controller and the related family of styles.

The use of UML2 as an architectural language is especially important in our setting, because during the final exam we test the students on it. For instance a typical exam test we would ask the students to describe the architecture of a system in terms of a C&C diagram drawn in UML2. In Figure 1 we show an example of component diagram proposed by a student of ours for a cloud "Data-as-a-Service" system similar to Dropbox.

SOLUTIONS AND RECOMMENDATIONS

The teaching of Software Architecture in the context of a large system engineering company should stress the importance of reusing software assets. Another relevant issue is the analysis and evaluation of the software architecture before a system is built, possibly with the use of tools; for examples see for instance (Sterling, 1996; Bernardo, 2001).

However, in a large company like SELEX SI any effort would be probably useless without the strong support of management; and the management can be more easily convinced if there are a number of reusable assets that are actually reused in the products.

In our case the assets available were mostly some architectural styles chosen by the company; moreover, some CORBA-based middleware was also used: see (Barbi, 2012).

Figure 1. A UML component diagram for a system similar to Dropbox

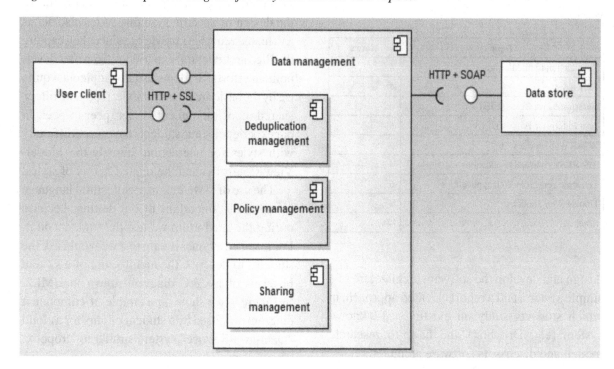

The prerequisites of a course on software architecture given in an industrial context must be kept quite basic, because in such a context the background of personnel is usually very varied, ranging from people having technical degrees ranging in Computer Science (usually a minority), Telecommunications, Electronic Engineering, Physics, and Mathematics to more non-technical degrees in Economics or even Philosophy.

Teaching software architectures in an industrial context requires to meet company's expectations in terms of mature knowledge and competences transferred to practitioners, that they can subsequently turn into the practice of engineering complex systems (Bass, 2008). This is not easy to achieve, as architecting large-scale complex software systems - having tens of thousands of requirements and millions of lines of code - requires very high abstraction and modeling skills.

In our experience, successful teaching software architectures in an industrial context is not a unidirectional activity, but it benefits a lot from a strong cooperation between company stakeholders for the courses and teaching staff, and it included:

- Selecting and teaching the real foundations of software architecture, which have their roots in the very basic principles of software engineering;
- Selecting and teaching those architectural styles and patterns that can meet the actual needs of practitioners in terms of design and reuse, with reference to their current target technologies; for instance, as modern middleware technologies are used in the design and development of large-scale software systems, this means defining and teaching proper guidelines that can help engineers to put architecture principles and patterns into their daily practice;
- Interleaving lectures and hands-on sessions with presentations by company testimonials, describing the practical problems, the current internal design standards,

and the needs for more modern yet mature techniques;

- Tailoring the theory of software architecture to the specific domain and existing industrial software processes;
- Defining and teaching practical guidelines that engineers can apply reasonably easily to assess and evaluate software architectures based on project/product stakeholders' needs and pre-defined criteria.

The teaching of Software Architecture in academic context should instead stress the "wicked" nature of software design, that is moreover the result of a social process based on consensus (de-Boer, 2009). The idea that there is no "absolutely good or absolutely bad" result when architecting is especially crucial to emphasize and discuss with students. It is important that they learn to understand the rationale behind architectural choices and to evaluate the consequences of those choices. Interestingly, this also applies when evaluating students' projects, meaning that also instructors should follow some architectural approach when grading post-mortem (Wang, 2005). The main

emphasis should be oriented to present and discuss case studies for the basic architectural styles. This is also the main suggestion given in a recent paper (Georgas, 2013).

Finally, we remark that the main reference book we suggested to our students has been (Taylor, 2009). This book is more suitable for an academic course than that we used for the industrial context, namely (Rozanski & Woods, 2012).

In Table 4 we summarize the main differences we experienced in the two different contexts.

FUTURE RESEARCH DIRECTIONS

The teaching of Software Architecture should evolve together with the new technologies for software-intensive systems. For instance, some recent architectural topics are a) the architectural styles for embedded systems, a) the architectural styles for systems based on Semantic Web technologies; b) the architectural styles for systems based on adaptive software technologies. All these research topics are quite hot in the respective engineering communities.

Table 4. Comparison of issues between the industrial and academic contexts

	Teaching SA in an Industrial Context	**Teaching SA in an Academic Context**
Goal	Software reuse	Increasing students' knowledge
Prerequisites	Low level, various background	Basic software engineering knowledge
Topics	Oriented to describing real systems	Architectural theories, then examples
Motivation	Given by management	Curiosity driven
Process	Planned, as defined by the company	Restricted to design the architecture
Activity	Describe an architecture and its requirements	Understand an architecture and be able to evaluate its properties
Exercises	Hands-on real systems of interest for the company	Modeling and meta-modeling simple systems
Use of UML	As a modeling language	As an architectural language, paying attention to non-functional requirements
Tools	Proprietary, as required by the company (mostly IBM)	Free educational license or Open source
Exam	Oriented to test new design knowledge	Oriented to test new design ability
Evaluation	Based on tests closed-answer	Architectural evaluation
Reference text	(Rozanski & Woods, 2012)	(Taylor, 2009)

An important topic that is related to Software Architecture is Enterprise Architecture, that is the discipline of modeling the organizations and their Information Technology infrastructures in order to efficiently support an organization's mission and activities. The relationships among Software Architecture, Enterprise Architecture and Knowledge Management are especially relevant for our students in Informatics for Management, a course described in (Bolognesi, 2006; Ciancarini, 2013). We intend to study such relationships in a future work.

Another issue that we intend to propose is the study and comparison of architectural theories. By *architectural theory* we mean a set of rules that govern architectural descriptions. For instance, the following diagram shows the relationships among the main architectural views according to (Clements, 2010). In this diagram the main de-

pendencies are UML stereotypes called <<implement>>, <<build>>, and <<manifest>>. The organization of the diagram shown in Figure 2 derives from the particular architectural theory described in (Clements, 2010).

Software Architecture is a concept both crucial and elusive when designing software-intensive systems. Theories of software architecture should describe how people think, build consensus, and work with software abstractions. There is a large body of work concerning the knowledge management of software architectures, a valuable resource of which the domain of research on software design has this far not taken full advantage. We intend to explore software architecture theories, aiming at studying how they are defined, how they are compared, and how they influence the practice of developing software intensive systems.

Figure 2. Dependencies among the main architectural views according to the architectural theory described in (Clements, 2010)

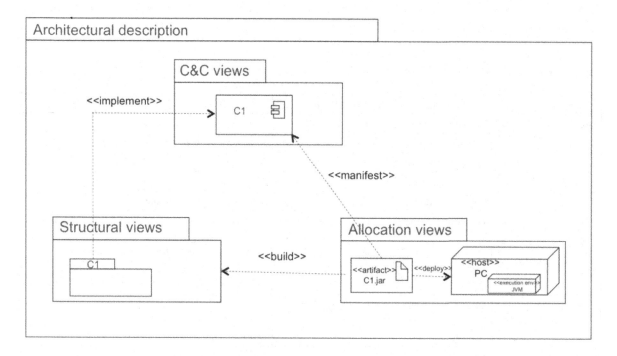

CONCLUSION

As mission-critical software systems become more complex, companies - in particular system integrators - feel an increasing need for improving their processes based on sound and effective techniques for software architectures definition and evaluation. This is what we observed in a number of public-private co-operations, which go even beyond the experiences described in this chapter.

When the field of Software Architecture emerged, it was argued it should be considered a discipline in its own (Shaw & Garlan, 1996). After almost two decades, the corpus of scientific work in the field has developed consistently, but there is still a large gap between this and what is actually needed in industrial and academic settings (Shaw & Clements, 2006). Many of the achievements in the field have not matured enough; an example are architecture definition languages (ADLs), that we believe have not replaced - and are not likely anymore to replace - (extensions to) standard modeling languages. Some others are more mature, but they need to be tailored to the specific software processes, internal industrial standards and practices.

ACKNOWLEDGMENT

The authors would like to thank prof. P. Prinetto, director of the CINI School on Design of Complex Software Systems, and E. Barbi, A. Galasso, F. Marcoz, V.Sabbatino, M. Scipioni, engineers from Finmeccanica companies, who in different stages and under different roles cooperated to the success of the various editions of the School.

REFERENCES

Al-Asmari, K., & Yu, L. (2006). Experiences in distributed software development with wiki. In *Proceedings of Software Engineering Research and Practice* (pp. 389–293). Academic Press.

Barbi, E., Cantone, G., Falessi, D., Morciano, F., Rizzuto, M., Sabbatino, V., & Scarrone, S. (2012). A model-driven approach for configuring and deploying systems of systems. In *Proceedings of the 7th International Conference on System of Systems* (pp. 214–218). IEEE.

Bass, L., Clements, P., Kazman, R., & Klein, M. (2008). *Models for evaluating and improving architecture competence (TR CMU/SEI-2008-006)*. Pittsburgh, PA: Software Engineering Institute.

Bass, L., Kazman, R., & Clements, P. (2012). *Software architecture in practice*. Reading, MA: Addison-Wesley.

Bernardo, M., Ciancarini, P., & Donatiello, L. (2001). Detecting architectural mismatches in process algebraic descriptions of software systems. In *Proceedings of working IEEE/IFIP Conference on Software Architecture* (pp. 77–86). IEEE.

Bolognesi, A., Ciancarini, P., & Moretti, R. (2006). On the education of future software engineers. In *Software engineering education in the modern age* (pp. 186–205). Berlin: Springer. doi:10.1007/11949374_12

Buschmann, F., Meunier, R., Rohnert, H., Sommerlad, P., & Stal, M. (1996). *Pattern-oriented software architecture: A system of patterns* (Vol. 1). New York: Wiley.

Ciancarini, P., Dos, C., & Zuppiroli, S. (2013). A double comparative study: Process models and student skills. In *Proceedings of the 26th IEEE Conference on Software Engineering Education and Training* (pp. 189–198). IEEE Computer Society.

Clements, P., Garlan, D., Bass, L., Stafford, J., Nord, R., Ivers, J., & Little, R. (2002). *Documenting software architectures: Views and beyond*. Upper Saddle River, NJ: Pearson Education.

Clements, P., & Northrop, L. (2002). *Software product lines: Practices and patterns*. Reading, MA: Addison-Wesley.

De Boer, R., Farenhorst, R., & van Vliet, H. (2009). A community of learners approach to software architecture education. In *Proceedings of the 22th IEEE Conference on Software Engineering Education and Training* (pp. 190–197). IEEE Computer Society.

Friedenthal, S. (2008). *A practical guide to SysML: The systems modeling language*. San Francisco: Morgan Kaufmann / The OMG Press.

Gamma, E., Helm, R., Johnson, R., & Vlissides, J. (1995). *Design patterns*. Reading, MA: Addison-Wesley.

Gast, H. (2008). Patterns and traceability in teaching software architecture. In *Proceedings of the 6th International Symposium on Principles and Practice of Programming in Java* (pp. 23–31). ACM.

Georgas, J. (2013). Supporting software architectural style education using active learning and role-playing. In *Proceedings of the 120th ASEE Conference*. ASEE.

Gomaa, H. (2005). *Designing software product lines with UML*. Reading, MA: Addison-Wesley.

Gorton, I. (2011). *Essential software architecture*. Berlin: Springer. doi:10.1007/978-3-642-19176-3

Lago, P., & van Vliet, H. (2005). Teaching a course on software architecture. In *Proceedings of the 18th IEEE Conference on Software Engineering Education and Training* (pp. 35–42). IEEE Computer Society.

Larman, C. (2004). *Applying UML and patterns: An introduction to object-oriented analysis and design and iterative development*. Upper Saddle River, NJ: Prentice-Hall.

Mannisto, T., Savolainen, J., & Myllarniemi, V. (2008). Teaching software architecture design. In *Proceedings of the 7th Working IEEE/IFIP Conference on Software Architecture* (pp. 117–124). IEEE.

Muller, G. (2004). Experiences of teaching systems architecting. In *Proceedings of International Council on Systems Engineering (INCOSE) Symposium*. INCOSE.

Qian, K., Fu, X., Tao, L., & Xu, C. (2006). *Software architecture and design illuminated*. Jones and Bartlett.

Rozanski, N., & Woods, E. (2012). *Software systems architecture*. Reading, MA: Addison-Wesley.

Sabbatino, V., Arecchi, A., Lanciotti, R., Leardi, A., & Tonni, V. (n.d.). Modeling the software architecture of large systems. In *Proceedings of International Conference on Systems of Systems Engineering*. IEEE Computer Society Press.

Schmidt, D. C., Stal, M., Rohnert, H., Buschmann, F., & Wiley, J. (2000). *Pattern-oriented software architecture: Patterns for concurrent and networked objects* (Vol. 2). New York: Wiley.

Shaw, M., & Clements, P. (2006). The golden age of software architecture. *IEEE Software*, *23*(2), 31–39. doi:10.1109/MS.2006.58

Shaw, M., & Garlan, D. (1996). *Software architecture: Perspectives on an emerging discipline*. Upper Saddle River, NJ: Prentice-Hall.

Stahl, T., Voelter, M., & Czarnecki, K. (2006). *Model-driven software development: Technology, engineering, management*. New York: Wiley.

Sterling, L., Ciancarini, P., & Turnidge, T. (1996). On the animation of not executable specifications by prolog. *International Journal of Software Engineering and Knowledge Engineering*, 6(1), 63–87. doi:10.1142/S0218194096000041

Taylor, R., Medvidovic, N., & Dashofy, E. (2009). *Software architecture: Foundations, theory, and practice*. New York: Wiley.

Wang, A. I., & Stalhane, T. (2005). Using post mortem analysis to evaluate software architecture student projects. In *Proceedings of the 18th International Conference on Software Engineering Education & Training* (pp. 43–50). IEEE.

Wang, A. I., & Wu, B. (2011). Using game development to teach software architecture. *International Journal of Computer Games Technology*, 4.

Weilkiens, T. (2008). *Systems engineering with SysML/UML: Modeling, analysis, design*. San Francisco: Morgan Kaufmann / The OMG Press.

Wrycza, S., & Marcinkowski, B. (2005). UML 2 teaching at postgraduate studies–Prerequisites and practice. In *Proceedings of ISECON* (Vol. 22). ISECON.

Zuppiroli, S., Ciancarini, P., & Gabbrielli, M. (2012). A role-playing game for a software engineering lab: Developing a product line. In *Proceedings of the 25th IEEE Conference on Software Engineering Education and Training* (pp. 13–22). IEEE.

ADDITIONAL READING

Albin, S. (2003). The art of software architecture: Design methods and techniques. Wiley.

Brown, A. (2000). *Large-scale, component-based development*. Prentice Hall.

Fowler, M. (2010). Domain specific languages. Addison Wesley.

Garland, J., & Anthony, R. (2003). Large scale software architecture: A practical guide using UML. Wiley.

Hohmann, L. (2003). Beyond software architecture: Creating and sustaining winning solutions. Addison-Wesley Longman Publishing.

Maier, M., & Rechtin, E. (2009). *The art of systems architecting*. CRC.

Malveau, R., & Mowbray, T. J. (2003). Software architect bootcamp. Prentice Hall Professional Technical Reference.

O'Reilly. Fairbanks, G. (2010). Just enough software architecture. Marhall & Brainerd.

Spinellis, D., & Gousios, G. (2009). Beautiful architecture leading thinkers reveal the hidden beauty in software design.

Vogel, O., Arnold, I., & Chughtai, A. (2011). *Software Architecture: A Comprehensive Framework and Guide for Practitioners*. Springer. doi:10.1007/978-3-642-19736-9

KEY TERMS AND DEFINITIONS

Architectural Description Language (ADL): Notation for the description of software architectures. Eg. UML2 can be used as an ADL.

Architectural Model: A diagram or set of diagrams which describes the structure or the behavior of a system assuming a given viewpoint related to some stakeholder.

Architectural Style: The abstract description of the structure of generic components and connectors, that can be instantiated to obtain a software architecture "compliant" with the style and inheriting its properties.

Architecture-Centric Software Process: Software development process exploiting architectural descriptions in all its phases.

Metamodeling an Architecture: Abstract description of the main concepts in a system. A metamodel is a usually simpler model of another model.

Model Driven Architecture: Both a method and a technology developed by OMG to develop systems described by models and metamodels.

Software Architecture: The basic structure of a software system, defined by its components, their relationships, and the set of decisions explaining the choice of both components and relationships.

Software Component: A part of a software architecture that encapsulates a set of related functions and is defined by an interface (e.g. an API, application programming interface) and by a behavior.

Software Connector: A part of a software architecture, which models the interaction among components. E.g. in an operating system the clipboard is a connector among different applications.

Section 9
Using Open-Source Tools

Chapter 21
Learning Software Industry Practices with Open Source and Free Software Tools

Jagadeesh Nandigam
Grand Valley State University, USA

Venkat N Gudivada
Marshall University, USA

ABSTRACT

This chapter describes a pragmatic approach to using open source and free software tools as valuable resources to affect learning of software industry practices using iterative and incremental development methods. The authors discuss how the above resources are used in teaching undergraduate Software Engineering (SE) courses. More specifically, they illustrate iterative and incremental development, documenting software requirements, version control and source code management, coding standards compliance, design visualization, software testing, software metrics, release deliverables, software engineering ethics, and professional practices. The authors also present how they positioned the activities of this course to qualify it for writing intensive designation. End of semester course evaluations and anecdotal evidence indicate that the proposed approach is effective in educating students in software industry practices.

INTRODUCTION

Software Engineering (SE) courses are some of the most challenging ones to teach in Computer Science (CS) curricula. Not only do students need to learn basic concepts, principles, and methods, but also master industry practices and tools in these courses. Lecture-based approaches to espousing software engineering principles hardly engage students' attention (Nandigam, Gudivada, & Hamou-Lhadj, 2008). Students often view software engineering principles as mere academic concepts and graduate without a clear understanding of how these principles can be used in practice. By the time students take their first SE

DOI: 10.4018/978-1-4666-5800-4.ch021

course, it is quite unlikely that they have written programs that are more than 500 lines long. It is also equally unlikely they had an opportunity to inspect large programs (> a few thousand lines of code) written by others.

One practice that seems to pervade across universities to bringing software engineering professional practices into the classroom is using a semester-long term project. In this project, students are expected to demonstrate their ability to apply software engineering practices and tools in solving a real-world problem in a team environment. However, there is no established approach to accomplishing the above goal due to various factors discussed below.

Selecting a right project with appropriate scope is in itself a challenge. In our experience with teaching SE courses, asking student teams to self-select a project rarely produces successful outcomes. Students typically overestimate or underestimate project scope and complexity. Overestimation leads to selecting a trivial project and embellishing it with superficial complexity. Underestimation results in switching to a trivial project halfway through the semester. In either case, the project scope and complexity are insufficient for students to fully experience professional software development practices.

The overarching goal of this chapter is to present our approach to teaching software engineering industry practices and tools in the backdrop of SE concepts, theory, methods, and principles. Using suitable software tools and team projects, we promote conceptual understanding and practical skills of the following topics: role of tools in the software development life cycle; iterative and incremental development as a means for timely project completion; requirements elicitation and specification; source code management with version control; importance of adhering to coding standards; design visualization; verification and validation through software testing; measuring and using software metrics as a means for improving software quality; software release management;

ethics and professional practice; and writing as a means to learning. We also discuss how Open Source code bases can be used in achieving the above learning goals.

SOFTWARE ENGINEERING COURSE

Our undergraduate SE course includes a semester-long (about 14 weeks) software development project to provide students hands-on experience with processes, methods, techniques, and tools of software development. The course first provides the necessary theoretical foundation for a broad range of topics – software engineering process models, project management, software requirements elicitation and specification, use case modeling, UML, object-oriented analysis and design, design patterns, test-driven development, version control, system building, software testing, mock object frameworks, software maintenance, software internationalization, SE ethics, and writing skills. Though the topics are quite a few, very focused and conceptually oriented lectures make this task possible. Students gain practical aspects of these topics by working on a realistic project in a team environment.

Students begin the course by writing a short formal paper on a SE ethics topic. The semester-long project involves development of a software product using an iterative and incremental development model. Students use Eclipse IDE (n.d.), and several free and open source tools and plugins available for the Eclipse IDE. The product is delivered incrementally in three releases with each release taking roughly 4 weeks of effort. The course also includes a midterm, a final exam, and several quizzes as part of formative and summative assessments. The weight distribution of various components in the course is: term paper (10%), ethics writing assignment (5%), term project (30%), midterm exam (20%), final exam (25%), and quizzes (10%).

SOFTWARE TOOLS USED

Though Computer-Aided Software Engineering (CASE) is not new, the range and ubiquity of both Open Source and free software tools have increased manifold in recent years. Eclipse IDE is used as the primary integrated development environment mainly because many of the Open Source and free software tools easily integrate with it. After careful exploration and experimentation, we chose the following open source tools and plug-ins for the Eclipse IDE to assist in various phases of software project development. The instructor provided students with detailed instructions for installing and using the tools selected for project development. The selection of tools for use in a software engineering course is an on-going activity as the instructor must be on the look out for better open source tools and free plug-ins that may become available.

Git (n.d.) is used for distributed source code management and version control. UCTool (Sourceforge, n.d.), CaseComplete (n.d.), and simple textual templates (provided by the instructor) are used for requirements documentation. UML Explorer from ObjectAid (n.d.) is used to visualize code and generate class diagrams from the code base.

Checkstyle (Eclipse CS, n.d.) is used for auditing adherence of source code with project-specific coding conventions and coding standards. Collection and analysis of source code based metrics for improving code quality is facilitated by EclipseMetrics (n.d.) plugin. JUnit (n.d.) is used for unit and integration testing, and CodeCover (n.d.) is used for system testing and coverage measurement. We used EasyMock (n.d.) tool for generating mock replacements for collaborators of a unit during unit testing.

ITERATIVE AND INCREMENTAL DEVELOPMENT

Given ever shrinking cycle times and increasing complexity of software projects, it is important that instructors of software engineering courses discuss briefly the traditional waterfall model and its origins, and disadvantages of using a strict waterfall model before introducing iterative and incremental development methods for building software.

It is important to provide students with background on the history of the waterfall model so they can appreciate the benefits of iterative and incremental software development. The waterfall model refers to a software development process that is strictly sequential where the process flows steadily downward (like a waterfall) through various phases of software development. Royce (1970) noted that this process is flawed, risky, and a non-working model that invites failure. He also discussed modified versions of this model that incorporate significant iteration and some degree of incremental approach to software development.

Iterative and incremental development (IID) techniques are not entirely new. Larman and Basili provide an excellent discussion of the history of iterative and incremental development and examples of large software systems that used these practices successfully since the mid-1950s (Larman & Basili, 2003). Cockburn (2008) advocates using both incremental and iterative development to improve software process and product quality, and manage changing requirements. He defines incremental development as a staging and scheduling strategy where various parts of the system are developed in stages and integrated as they are completed. Alternative to incremental integration is to build the entire system at once and use a big-bang integration at the end to build the system. Cockburn defines iterative development as a rework scheduling strategy where previously completed parts are revised for improvement. Alternative to iterative development is to get everything right the first time, which is unrealistic in software systems of significant size or complexity. Incremental development is used to improve development process and manage changing requirements, whereas iterative development is used to improve product quality. Martin (1999a, 1999b, 1999c) provides a detailed discussion of the waterfall model and its weaknesses, the benefits of using IID, and how to use IID even when a corporate culture is entrenched in the waterfall model.

Once the students are taught the waterfall model and IID, they are required to use IID for the team project. Students are also expected to read the articles mentioned in this section. The assigned software development project is divided into three increments (releases) where each release delivers a subset of a prioritized feature set. Each release takes about four weeks of effort in a 14-week semester. At the end of each release, each student team delivers a working software increment, provides a demo, and submits a document containing various artifacts (see Release Deliverables and Documentation Section).

REQUIREMENTS DOCUMENTATION

We employed use case driven approach for gathering and documenting product requirements. Cockburn (2001) defines a use case as a series of interactions between an outside entity (an actor) and the system to accomplish a specific task that provides business value to one or more stakeholders of the system. Complete description of a use case contains a set of scenarios that an actor and the system may go through to achieve a goal, either successfully or unsuccessfully. A scenario is a particular instance of a use case and represents

a single path through the use case. Scenarios are typically depicted using UML sequence diagrams.

Use cases are an excellent way to describe behavioral requirements of a system from the project stakeholders' perspective. Use cases are primarily textual descriptions. UML has support for use case diagrams, which are graphical representations of the relationships between an actor and a use case, between use cases, and between actors. These relationships are depicted using generalization, extend, and include associations. A use case is depicted as an oval in a use case diagram. It is important to have the students understand the difference between a use case and a use case diagram – the former as a textual description and the latter as a diagram showing relationships between and among actors and use cases. Figure 1 shows an example use case diagram for an online store.

Kulak presents four main advantages of using a use case driven approach to requirement specification (Kulak & Guiney, 2003). Use cases serve as an effective communication tool between system users and IT staff during requirements gathering. Use cases can be used to specify both functional and non-functional requirements (using use-case stereotypes) of a system. They serve as an effective aid in requirements traceability. Use cases can also discourage premature design from creeping into requirements.

Since use case descriptions are primarily textual, students are encouraged to use the following template (Cockburn, 2001) when writing use cases:

- Use case name
- Unique ID
- Description
- Primary and supporting actors
- Triggers
- Preconditions
- Primary flow (of steps/actions)
- Alternate flows (of steps/actions)
- Minimal guarantees
- Success guarantees
- Non-functional requirements

Figure 1. A use case diagram in UML

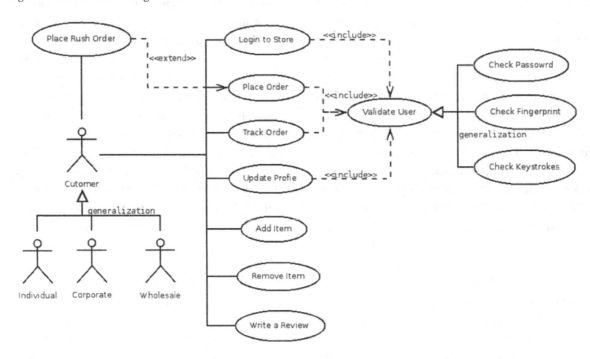

In our SE course, students may use one of three options for writing use cases – UCTool, CaseComplete, and a plain text-based template shown above. In UCTool, use cases are written in XML to a supplied XSD (XML Schema Definition) that adheres to the use case format described by Cockburn (2001). The tool then generates use case descriptions as a set of nicely formatted and hyper-linked HTML pages. UCTool works with the Eclipse IDE. CaseComplete is a commercial tool with a free 30-day trial option that provides a user-friendly GUI front-end with many features for writing use cases, creating use case diagrams, generating reports and test plans, modeling UI prototypes, and printing. Under the third option, students use a plain use case template document provided by the instructor where each use case is formatted as a table with two columns. The first column is populated with items in the template shown above. Many students liked working with XML and continued to use UCTool. Some chose to use the plain text template while few opted for the CaseComplete tool.

VERSION CONTROL AND SOURCE CODE MANAGEMENT

Developing version control and source code management skills is important for graduating computer science students. End of term course evaluations for several courses in our curricula consistently include comments related to skills in version control systems. In the SE course we require students to use a version control and source code management tool. Students are introduced to the basics of Git — a popular free and open source distributed version control system — through a class lecture. Student teams are encouraged to use Git by setting up their project code repositories on a Web-based Git hosting system.

Some student teams chose to use GitHub (n.d.), a Web-based Git hosting system that is free for storing and managing public code repositories. Since the free accounts on GitHub are public, some student teams expressed concern about their source code base privacy. They opted to use BitBucket (n.d.), another Web-based hosting service that uses Mercurial and Git version control systems. BitBucket offers a free account for hosting an unlimited number of private repositories that are accessed by up to five users. We also provided students with custom instructions for setting up their own Git server on campus Linux computers. A small percentage of student teams chose to use their accounts on Linux computers to set up a Git server.

CODING STANDARDS COMPLIANCE

It is important that students learn to produce source code that follows widely accepted coding standards in a consistent manner. Some benefits of coding standards are readability, program comprehension, maintainability, and increased inter-team communication. We used Checkstyle, an Eclipse plugin, for auditing adherence to a set of coding standards. Checkstyle automates the process of checking whether Java code produced by students conform to the Sun Microsystems coding conventions. Checkstyle is configurable and allows an instructor to alter the checks and ranges enforced to suit the needs of a given project. We configured Checkstyle to use a modified version of the Sun Microsystems checks suitable

for Eclipse and also provided students with a suppression filter file (an XML document). The latter is used to suppress certain checks (for example, missing javadoc comments, use of magic numbers or numeric literals) on JUnit source files only.

When Checkstyle is activated on a project code base, it runs immediately, applies configured checks and generates a violations report (Figure 2) as well as a violations chart. A row in a violations report or a segment in a violations pie chart can be selected to view details of all violations within that category.

We require student teams to run Checkstyle on their code base before making any changes that are related exclusively to coding conventions, and save the generated violations report and chart. Then they are encouraged to make any changes necessary to the source code to address all reported violations. Checkstyle is rerun and teams demonstrate adherence to coding standards through the subsequent violations report and chart. This is an iterative process. Checkstyle produces an empty list of violations if it fails to find any coding standards violations. The teams are required to submit their final violations report and chart so that the instructor can judge the effort put by the team in following the coding standards.

Figure 2. Checkstyle violations report in Eclipse IDE

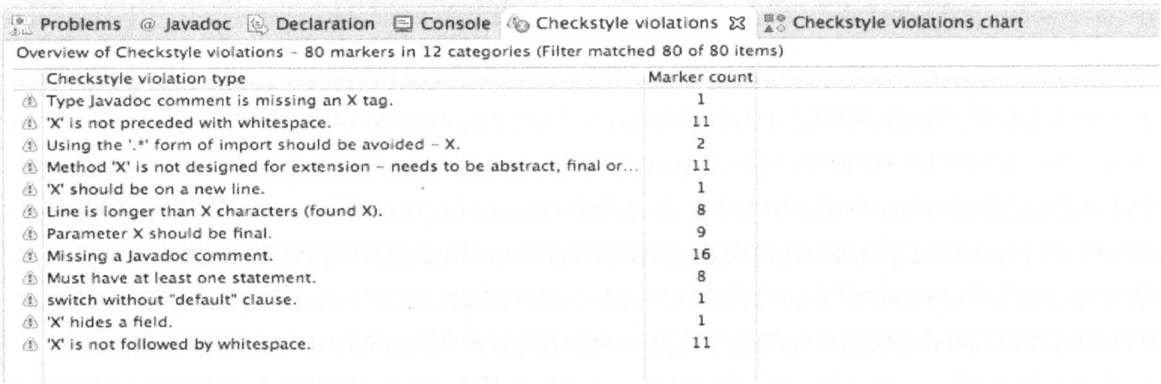

Checkstyle violation type	Marker count
Type Javadoc comment is missing an X tag.	1
'X' is not preceded with whitespace.	11
Using the '.*' form of import should be avoided – X.	2
Method 'X' is not designed for extension – needs to be abstract, final or...	11
'X' should be on a new line.	1
Line is longer than X characters (found X).	8
Parameter X should be final.	9
Missing a Javadoc comment.	16
Must have at least one statement.	8
switch without "default" clause.	1
'X' hides a field.	1
'X' is not followed by whitespace.	11

Problems | @ Javadoc | Declaration | Console | Checkstyle violations ⊠ | Checkstyle violations chart

Overview of Checkstyle violations – 80 markers in 12 categories (Filter matched 80 of 80 items)

DESIGN VISUALIZATION

During design and code development phases of the project, it is important to visualize the structure of the system under development. UML provides class diagrams to describe the static structure of a system. A class diagram contains classes and interfaces, their attributes and methods, and relationships among classes and interfaces (Figure 3). Relationships in a class diagram are primarily of four types: dependency, association (aggregation and composition as special forms of association), generalization, and realization.

Even though design artifacts (such as class diagrams) are generated prior to coding, getting the design right is an iterative process. Hence initial designs will likely change and may not stay in sync with the actual code developed. Therefore, it is essential to recreate or regenerate design artifacts frequently by reverse engineering the design from code. The resulting design artifacts can be used to visualize, conduct critical reviews of the current system architecture, and make possible changes to improve it as the development proceeds forward. It is helpful to have good automated tools in place to reverse engineer the design of either the entire system or selected portions from the current code base.

We used ObjectAid UML Explorer, another excellent Eclipse plugin, to generate class diagrams from source code. ObjectAid UML Explorer can be used to draw class diagrams as well as sequence diagrams. Though sequence diagrams can also be generated using this plugin, it requires a license. Figure 3 shows an example class diagram generated using the ObjectAid UML Explorer plugin. Student teams are required to generate one or more class diagrams from their code base.

Figure 3. A class diagram generated with ObjectAid UML Explorer

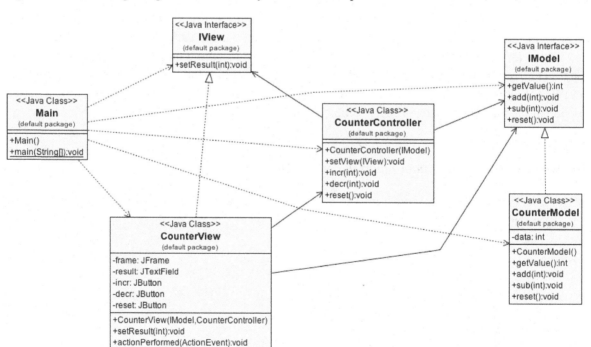

SOFTWARE TESTING

Verification and validation (V&V) is an important part of software development. Verification refers to the set of activities carried out to ensure that the software developed complies with its specifications. Verification activities help answer the question: *Are we building the product right*? (Boehm, 1981). Validation, on the other hand, refers to the set of activities carried out to make sure that the software built meets the needs and expectations of its end users and stakeholders. The main goal of validation activities is to answer the question: *Are we building the right product*? (Boehm, 1981).

Software testing can be viewed as a dynamic V&V activity. Myers defines software testing as the process of executing a program with the intent of finding errors (Myers, Sandler, & Bodgett, 2011). There are many levels in software testing: unit testing, integration testing, system testing, and acceptance testing. The first three levels focus on software verification part of V&V, where as the acceptance testing is mainly about software validation. The SWEBOK (n.d.) guide defines many other levels of testing that are based on various test objectives.

In the SE course, we focus on unit, integration, and system testing levels using the JUnit, EasyMock, and CodeCover tools, respectively. Students are required to use JUnit for unit testing. We also discuss the role of mock objects and how they facilitate unit testing in true isolation. A mock object is a test-oriented/test-aware replacement (test double) for a collaborator of the unit under test. EasyMock is a tool for dynamically generating mock objects for interfaces and classes to facilitate unit testing in isolation. We discuss EasyMock and encourage students to use it during unit testing.

The goal of integration testing is to ensure that various components that comprise a software system communicate as expected. We take an incremental approach to integration testing. We have attempted to do integration testing using JUnit for testing individual features of the system that required integration of one or more units. This

requires first setting up an environment/context that resembles the conditions in which a software feature under test is invoked. Each JUnit test case may end up containing significant setup code (context or preconditions) before an action can be triggered and action outcomes can be verified using assertions.

System testing is performed on a fully integrated system to verify that the system meets the specifications. CodeCover is used for both system testing and code coverage measurement. It is available in two versions: as a standalone and as an Eclipse plugin. CodeCover measures the following coverage metrics: statement, branch, term (subsumes modified condition/decision), loop, question mark (?) operator, and synchronized. It highlights the source code with a color scheme to show the coverage achieved in a test session. At the end of a test session, CodeCover presents the results of all enabled metrics for the test session. Figure 4 shows an example code coverage measurement report generated by CodeCover.

There are three ways to enable CodeCover to monitor the program execution and collect data used for code highlighting and coverage measurement. Under normal execution mode, CodeCover measures the system under test (SUT) in background. If a JUnit test suite is available for the SUT, CodeCover will measure the coverage resulting from executing that test suite. In Live Notification mode, a test case can be defined dynamically (for example, while the SUT is running, an end-to-end feature of the system can be tested). Figure 5 shows the Live Notification window of CodeCover. Measured coverage data from a test session can be exported as an HTML report.

Acceptance testing is simulated through tests conducted during product demonstrations that student teams make at the end of each release. The test cases used during acceptance testing is one of the key factors considered when assigning a score/grade to the product release made by a team.

Figure 4. CodeCover coverage results from a test session

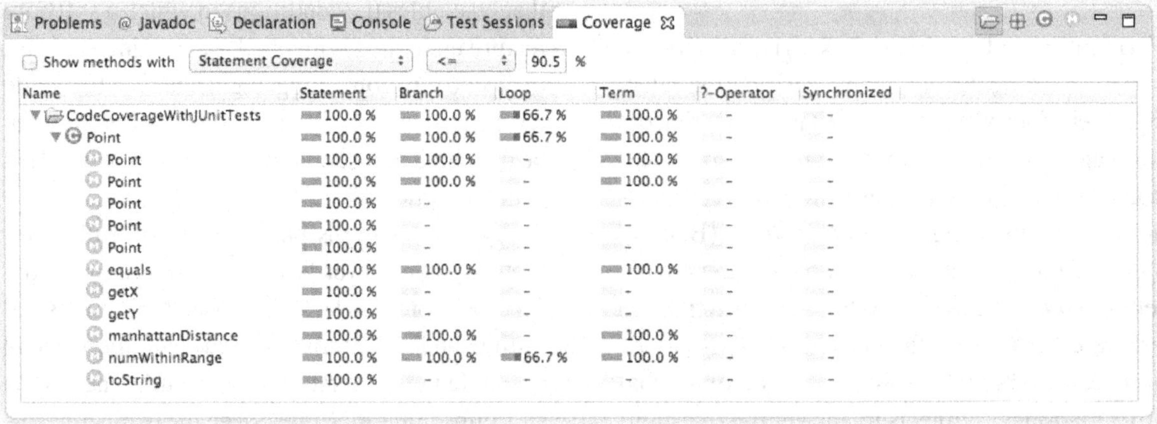

SOFTWARE METRICS

A software metric is a measurement of some attribute, characteristic, or property of a piece of software (Fenton, 1991). Metrics may be collected for a software product, process, project, or resources used. The main objective of using software metrics is to manage, control, or improve some aspect of software. Metrics support the software development philosophy: *you can't control what you can't measure* (DeMarco, 1986).

Software metrics can be used to improve software productivity and quality. Metrics can be helpful in determining when software is ready for production and for guidance at various stages in the development process. It is essential that students understand what software metrics are, how to measure frequently used product metrics, and how to analyze the resulting measurements to improve the health of the code base. It is also important that students realize that software metrics are more useful as indicators of unhealthy code

Figure 5. CodeCover live notification execution

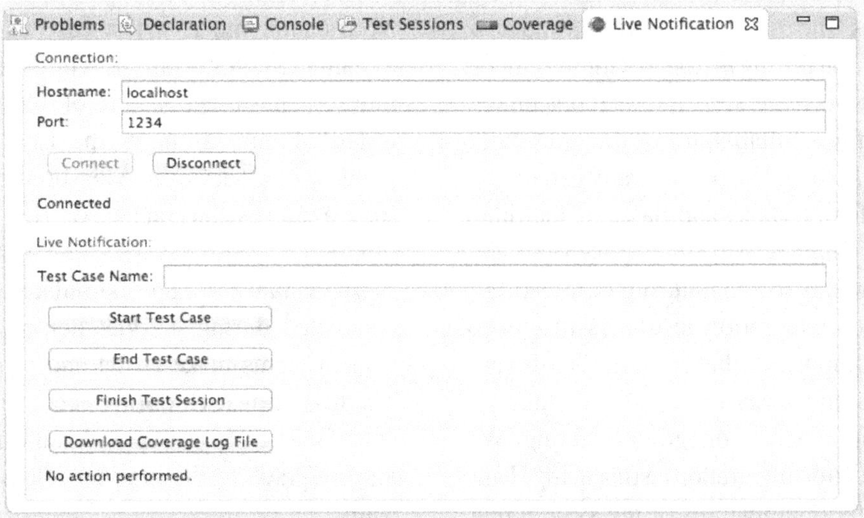

than as indicators of healthy code. Metrics can be used to determine the need for code refactoring.

We require student teams use a plugin called EclipseMetrics ("EclipseMetrics," n.d.) to measure various product metrics for the source code developed. Students are taught to pay close attention to frequently used metrics such as cyclomatic complexity, lines of code (LOC), coupling, and cohesion when using and applying the Checkstyle and EclipseMetrics tools to their code base. Teams are required to submit reports for a selected set of metrics including a rationale or critique for the measurements observed. The EclipseMetrics tool supports the following metrics:

- McCabe's cyclomatic complexity
- Efferent coupling of a class
- Feature envy of methods
- Lack of cohesion in methods
- Lines of code in methods
- Number of statements in methods
- Weighted methods per class
- Number of fields in a class
- Number of parameters in a method
- Number of levels in a method

RELEASE DELIVERABLES AND DOCUMENTATION

The semester-long project assigned in the SE course is carried out in teams of 3-4 students using an iterative and incremental development. Teams deliver the assigned project in three releases with each release taking roughly 4 weeks.

Successful completion of each release is marked by a working release and a demonstration of the features implemented in that release. For each release, student teams submit a professional document containing the following items, each in a separate section:

- Cover page with project title and team member names
- Project vision and scope
- Release objectives
- Role and Responsibilities of team members
- Use cases
- Class diagram(s) using ObjectAid UML Explorer
- Source code for the release
- Git log report that shows activity of team members with the code repository
- Checkstyle violations charts (before and after fixes)
- Eclipse metrics plugin reports
- JUnit test cases
- Code coverage reports from CodeCover tool
- Self-reflection by team
- User's guide for the release (optional)

ETHICS AND PROFESSIONAL PRACTICES

We believe that a software engineering course is an ideal place for discussing ethical and professional practices that one must observe in professional practice. We achieve this goal by discussing professional codes of ethics, samples of brief ethical scenarios, and a detailed case study on ethics in software engineering.

We introduce ethics with the discussion of the ACM Code of Ethics and Professional Conduct (ACM, n.d.) and the IEEE-CS/ACM Software Engineering Code of Ethics and Professional Practice (Gotterbarn, Miller, & Rogerson, 1999). Next, we present several brief ethical scenarios from (Anderson, Johnson, Gotterbarn, & Perrolle, 1993; Braude & Bernstein, 2010; Pfleeger, 1998) and solicit feedback and reactions from students in the classroom in an interactive mode.

Third, *The Case of the Killer Robot* (Epstein, 1994a, 1994b) articles are used as a case study on ethics in software engineering. The case study consists of nine articles and is about 70 pages long. These articles present a detailed and non-trivial scenario that combines specific elements of software engineering and computer ethics to acquaint students with the complexities of real-world software and ethical issues. The articles discuss topics including programmer psychology, team dynamics, project management, software process models, the nature of requirements, software prototyping, user interface design, software testing, software theft, and privacy. Students are required to read these articles and write reaction paper(s) addressing specific questions provided by the instructor. The questions are designed around ethical, social, and software engineering issues presented in the articles. Students are required to provide detailed responses to about 25 specific questions as a part of an ethics writing assignment. The questions we used were obtained from (Gerhardt, 2001, 2005) and through email communication with the author of these articles.

SUPPLEMENTAL WRITING SKILLS

The undergraduate software engineering course we offer is also designated as a SWS (Supplemental Writing Skills) course through the University SWS Program. The mission of the SWS Program is to enable students to write effectively for multiple purposes and audiences through the integration of writing across the curricula. Students turn in a total of at least 3,000 words of writing. A substantial amount of this writing is composed of finished essays, reports, and research papers. The instructor works with the students on revising drafts of papers. At least four hours of class time is devoted to writing instruction.

The SE course gives students an excellent opportunity to apply the writing process in the discipline. Students are given a list of topics in areas directly related to software engineering processes, methods, techniques, and tools. During the first week of the course, students either select a topic from the list of topics provided to them or propose a new topic with the consent of the instructor. During the third week of the course, students turn in a paper outline containing a thesis statement, topic outline, and a preliminary list of references. The instructor reviews the submitted paper outlines and returns them to students with comments so they can work on a rough draft based on the approved outline. Students turn in a draft paper mid-semester. The instructor reviews the draft and returns it to each student with comments. Students revise the draft and prepare the final version of the paper for submission to the instructor towards the end of the course. At any stage of this writing process, students can obtain assistance from the University Writing Center by making an appointment or by taking advantage of the Center's drop-in hours. Students are also provided with resources on writing a thesis statement, preparing a topic outline, preparing and revising drafts, and listing and citing references in an established style. The term paper that students write to satisfy the University SWS requirements is separate and distinct from the release documentation generated at the end of each release of the product. This is intentional because the SWS program requires that a written artifact must through multiple revisions.

USING OPEN SOURCE PROJECTS IN SE COURSES

Availability of numerous Open Source software code bases provides unprecedented and exciting opportunities for software engineering research and learning. We used the following source code bases for learning in software engineering courses: ImageJ (a Java image processing library with focus on biomedical imaging), Apache Derby (a small footprint, light-weight relational database suitable for embedding in Java applications),

Apache Lucene (a full-featured, high-performance search engine Java library), Hibernate (an object-relational mapping Java library), and JUnit (a unit testing framework for the Java programming language). Students in our software engineering courses have used these Open Source code bases for understanding the role of coding conventions and coding style, programming by intention to develop readable and maintainable code, assessing code quality using software metrics, refactoring, and reverse engineering to recover design elements (Nandigam et al., 2008).

FUTURE RESEARCH DIRECTIONS

Availability of numerous datasets under the *Open Data* initiative as well as *Open Source* software code bases will provide even more opportunities for teaching and learning software engineering practices in the classroom. Especially the advent of *Big Data* provides unprecedented opportunities for classroom learning. The 1000 Genomes project, an Open Data project, features a 200 TB dataset. This project aims to build the most detailed map of human genetic variation available from the genomes of more than 2,600 people from 26 populations around the world. In another Open Data project, astronomers are collecting more data than ever. Currently 1 petabyte (PB) of this data is electronically accessible to public, and this volume is growing at 0.5 PB per year. It is estimated that more than 60 PB of archived data will be accessible to astronomers in the near future. Challenging and interesting software engineering team projects can be developed using a trillion-word dataset (Brants & Franz, 2012). This dataset contains English word *n*-grams and their observed frequency counts. Data.gov is United States government Open Data initiative and provides access to over 70,892 datasets about various topics ranging from agriculture and business to safety and supply chain.

Downloading, installing, and configuring various tools discussed in this chapter is a tedious and time-consuming process. Given the operating system related issues, frequent releases of tools and their dependencies on code libraries, it would help to have a software tool whose functionality is similar to package management systems that come with various Linux distributions. This need and solution is best illustrated with the custom download of jQuery (n.d.) UI. Applying these principles for software engineering tools will feature classes of tools and choices within a class. For example, UML is a class and this class will feature various free UML tools; coding style compliance tools encompass another class. The user will simply indicate tool classes and choices within them, and a package management system will install the tools including their dependencies. This type of tool is essential to reap the benefits of Open Data and Open Source software code bases.

CONCLUSION

Students in software engineering courses often view the subject material as dry and assume no relevance to practice. This attitude of students poses several challenges for instructors who teach software engineering courses. However, availability of numerous datasets through the Open Data initiative and the Open Source code bases present unprecedented opportunities for making software engineering courses more engaging through realistic team projects. These two resources combined with free and Open Source software tools provide a sophisticated laboratory for exploring, experimenting, and learning software engineering industry practices in a classroom setting.

We have presented an approach to bringing realism to the teaching of software industry practices in university classrooms. Our approach has evolved over a period of five years and has been validated through many offerings of the course at two institutions.

We caution instructors who want to use this approach in their classrooms on two aspects. They should closely examine and determine which Open Data sources are going to be used well in advance. This helps them to determine scope of team projects and the type and extent of scaffolding to provide to students as they progress through execution of team projects. Second, all the tools need to be installed and tested in the lab before the beginning of the course to avoid project schedule slips.

Based on the course assessment data, we conclude that the approach presented in this chapter is effective for learning software industry practices in the classroom. But more importantly, students learn SE concepts, methods, and tools in an engaging environment and demonstrate their learning by solving a realistic problem.

REFERENCES

ACM. (n.d.). *Code of ethics.* Retrieved from http://www.acm.org/about/code-of-ethics

Anderson, R. E., Johnson, D. G., Gotterbarn, D., & Perrolle, J. (1993). Using the new ACM code of ethics in decision making. *Communications of the ACM, 36*(2), 98–107. doi:10.1145/151220.151231

Bitbucket. (n.d.). Retrieved from https://bitbucket.org/

Boehm, B. W. (1981). *Software engineering economics.* Englewood Cliffs, NJ: Prentice-Hall.

Brants, T., & Franz, A. (2012). Web 1T 5-gram version 1. *Linguistic Data Consortium Catalog.* Retrieved from http://www.ldc.upenn.edu/Catalog/CatalogEntry.jsp?catalogId=LDC2006T13

Braude, E. J., & Bernstein, M. E. (2011). *Software engineering: Modern approaches* (2nd ed.). Hoboken, NJ: J. Wiley & Sons.

Casecomplete. (n.d.). Retrieved from http://www.casecomplete.com/

Cockburn, A. (2001). *Writing effective use cases.* Boston: Addison-Wesley.

Cockburn, A. (2008). Using both incremental and iterative development. *CrossTalk – The Journal of Defense Software Engineering, 21*(2), 27–30.

Codecover. (n.d.). Retrieved from http://codecover.org/

Computer. (n.d.). Retrieved from http://www.computer.org/portal/Web/sWebok/html/contents

DeMarco, T. (1986). *Controlling software projects: Management, measurement & estimation.* New York, NY: Prentice Hall.

Easymock. (n.d.). Retrieved from http://easymock.org/

Eclipse CS. (n.d.). Retrieved from http://eclipse-cs.sourceforge.net/

Eclipse Metrics. (n.d.). Retrieved from http://eclipse-metrics.sourceforge.net/

Eclipse. (n.d.). Retrieved from http://eclipse.org/

Epstein, R. G. (1994a). The case of the killer robot (part 1). *SIGCAS Computers and Society, 24*(3), 20–28. doi:10.1145/191634.191640

Epstein, R. G. (1994b). The case of the killer robot (part 2). *SIGCAS Computers and Society, 24*(4), 12–32. doi:10.1145/190777.190778

Fenton, N. E. (1991). *Software metrics: A rigorous approach.* London: Chapman & Hall.

Gerhardt, J. (2001). Put ethics and fun into your computer course. *Journal of Computing Sciences in Colleges, 16*(4), 247–251.

Gerhardt, J. (2005). If you can teach it, you can measure it. *Journal of Computing Sciences in Colleges, 20*(3), 114–120.

GIT-SCM. (n.d.). Retrieved from http://git-scm.com/

Github. (n.d.). Retrieved from https://github.com/

Gotterbarn, D., Miller, K., & Rogerson, S. (1999). Computer society and ACM approve software engineering code of ethics. *Computer*, *32*(10), 84–88. doi:10.1109/MC.1999.796142

Jqueryi. (n.d.). Retrieved from http://jqueryui. com/

Junit. (n.d.). Retrieved from http://junit.org/

Kulak, D., & Guiney, E. (2003). *Use cases: Requirements in context*. London: Addison-Wesley.

Larman, C., & Basili, V. R. (2003). Iterative and incremental development. *IEEE Computer*, *36*(6), 47–56. doi:10.1109/MC.2003.1204375

Martin, R. C. (1999a). *Iterative and incremental development (part I)*. Retrieved July 11, 2013 from http://www.objectmentor.com/resources/articles/IIDI.pdf

Martin, R. C. (1999b). *Iterative and incremental development (part II)*. Retrieved July 11, 2013 from http://www.objectmentor.com/resources/articles/IIDII.pdf

Martin, R. C. (1999c). *Iterative and incremental development (part III)*. Retrieved July 11, 2013 from http://www.objectmentor.com/resources/articles/IIDIII.pdf

Myers, G. L., Sandler, C., & Badgett, T. (2011). *The art of software testing*. Hoboken, NJ: Wiley.

Nandigam, J., Gudivada, V. N., & Hamou-Lhadj, A. (2008). Learning software engineering principles using open source software. In *Proceedings of Frontiers in Education Conference*, (pp. S3H:18 – S3H:23). Academic Press.

Objectaid. (n.d.). Retrieved from http://www.objectaid.com/

Pfleeger, S. L. (1998). *Software engineering: Theory and practice*. Upper Saddle River, NJ: Prentice Hall.

Royce, W. W. (1970). Managing the development of large software systems. *IEEE WESCON*, *26*(8), 1–9.

Sourceforge. (n.d.). Retrieved from http://uctool.sourceforge.net/

ADDITIONAL READING

Biancuzzi, F., & Warden, S. (2009). *Masterminds of programming: Conversations with the creators of major programming languages*. O'Reilly.

Carter, J. D. (2011). *New programmer's survival manual: Navigate your workplace, cube farm, or startup*. Pragmatic Bookshelf.

Cauldwell, P. (2008). *Code leader: Using people, tools, and processes to build successful software*. Wrox.

Cheung, K. W. (2012). *The developer's code*. Pragmatic Bookshelf.

Davis, B., & Tucker, H. (2009). *97 things every project manager should know: Collective wisdom from the experts*. O'Reilly.

Doernhoefer, M. (2012). Surfing the net for software engineering notes. *SIGSOFT Software Engineering Notes*, *37*(2), 11–20. doi:10.1145/2108144.2108163

Ford, N. (2008). *The productive programmer*. O'Reilly.

Fowler, C. (2009). *The passionate programmer: Creating a remarkable career in software development*. Pragmatic Bookshelf.

Graham, P. (2010). *Hackers & painters: Big ideas from the computer age*. O'Reilly.

Henney, K. (2010). *97 things every programmer should know: Collective wisdom from the experts*. O'Reilly.

Humble, J., & Farley, D. (2010). *Continuous delivery: Reliable software releases through build, test, and deployment automation.* Addison-Wesley Professional.

Hunt, A. (2008). *Pragmatic thinking and learning: Refactor your wetware.* Pragmatic Bookshelf.

Hyde, R. (2004). Write great code: Vol. 1: Understanding the machine. No Starch Press.

Hyde, R. (2006). Write great code, Vol. 2: Thinking low-level, writing high-level. No Starch Press.

Martin, R. C. (2008). *Clean code: A handbook of agile software craftsmanship.* Prentice Hall.

McConnell, S. (2004). *Code complete: A practical handbook of software construction.* Apress.

McCuller, P. (2012). *How to recruit and hire great software engineers: Building a crack development team.* Apress. doi:10.1007/978-1-4302-4918-4

Mey, C. V. (2012). *Shipping greatness: Practical lessons on building and launching outstanding software, learned on the job at Google and Amazon.* O'Reilly.

Ryan, T. (2010). *Driving technical change.* Pragmatic Bookshelf.

Spraul, V. A. (2012). *Think like a programmer: An introduction to creative problem solving.* No Starch Press.

KEY TERMS AND DEFINITIONS

CASE: Computer-aided Software Engineering (CASE) is the application of a set of software tools and methods used in developing software systems.

Code Coverage: Refers to the percentage of the source code that has been tested using one or more test suites.

Coding Standards: These are a set of programming language-specific conventions and guidelines used for producing source code that is easy to read and maintain. Coding standards play a critical role especially in large projects by promoting consistency and uniformity. There may be one organization-level standard applicable across all projects or separate standard for each project.

Design Patterns: Generic solutions to commonly occurring problems in software design. They are proven best practices for producing software that is scalable, robust, and maintainable.

Design Visualization: Tools are used to visualize both static structure and dynamic behavior of a software system during the design and code development phases. Unified Modeling Language (UML) specifies various graphical notations such as class and sequence diagrams for design visualization.

Integrated Development Environment (IDE): An IDE is software application that provides comprehensive facilities for software development. IDEs may integrate several software tools such as compilers, interpreters, debuggers, build automation, version control, software refactoring, and profiling.

Integration Testing: A testing technique to insure that various components of a software system function as specified.

Iterative and Incremental Development (IID): A software development method for achieving high quality software in an environment that is characterized by incomplete and frequently changing requirements. Incremental development strategy advocates developing various parts of a system in stages and integrating the parts as they are completed. Iterative development specifies an approach for improving the quality of previously developed parts when they are integrated with new parts of a system.

LOC/SLOC: Source Lines of Code (SLOC) or simply Lines of Code (LOC) is a measure of the size of a software project. It is used to estimate software project budgets, measure productivity of software engineers, and characterize product quality in terms of defects per 1000 lines of code.

Object-Oriented Analysis and Design (OOAD): An analysis and design technique used in developing software. Object-Oriented Analysis (OOA) is used to capture functional requirements of a system. Object-Oriented Design (OOD) builds upon OOA and specifies components and interactions among the components needed to develop the system.

Project Scope: A high-level document which specifies what functionality is included in a software project as well as what functionality is not included.

Release Management: Managing the release cycle within a software project, which encompasses specification of various components (including their version numbers) that went into a software build, the build process itself, and infrastructure components of the operational environment.

Requirements Elicitation: Various techniques used to elicit requirements from stakeholders of a proposed software system.

Requirements Specification: Formal and semi-formal methods used to document the requirements of a proposed software system. Examples include state diagrams, petrinets, and use cases.

Software Design: Software design specifies major functional components, their interactions, and high-level implementation details of functional components. It serves as a blueprint for developing a software system.

Software Development Life Cycle: Specifies a process for planning, designing, building, testing, deploying, and operating a software application.

Software Ethics: A set of guidelines and practices for software engineers to conduct system development as a professional practice. For example, the ACM/IEEE-CS Software Engineering Code of Ethics and Professional Practice specifies such guidance and practices.

Software Metrics: A measure of some property of a software system or artifacts that are created to help build the system. Metrics are gathered to help improve software quality, cost, and project schedule. Lines of code, cost per line of code, defects per 1000 lines of code, average time to fix a defect are examples of software metrics.

Software Quality: Refers to non-functional attributes of a software system such as reliability, efficiency, security, scalability, and maintainability.

Software Testing: A suite of techniques used to verify that a software system performs according to its specifications. Test techniques include unit, integration, system, load, and stress.

System Building: Process of compiling, linking, and producing an executable code from source code and binaries. Examples of such tools include Apache Maven, Apache Ant, NAnt, GNU make, and CMake.

System Testing: A method used to test the entire software system in an environment that is identical or closely resembles the production environment. System is tested for validating functional requirements, software architecture, quality attributes, and operational environment factors.

Test Driven Development (TDD): A methodology for developing code where a test is written before writing the actual code for a feature. TDD cycle consists of adding a new test, running all the tests to see if the new test fails, writing new code, running all the tests, and refactoring the code.

UML: An ISO standard that specifies a modeling language to create visual models of software systems.

Unit Testing: Testing a unit of software in isolation. Units can be functions written in procedural programming languages, or classes written in object-oriented programming languages, among others.

Use Case: A requirements specification technique that describes a sequence of actions between a system user and software system to achieve a task. Use cases are effective in describing behavioral requirements of a system from end-users perspective.

Use Case Diagram: A pictorial representation of use cases, system users (aka actors), and association between use cases and system users.

Verification and Validation (V&V): Set of activities carried out to ensure that the software developed complies with its specifications. The goal of verification is to ensure that the product is developed using professional software engineering practices. Validation insures that the product being developed meets the needs of its stakeholders.

Version Control: A mechanism that is used to track the evolution of a piece of software over time. Version control tools facilitate multiple users concurrently working on a software project.

Waterfall Model: An approach to developing software systems, which employs a strict sequential process flowing from requirements elicitation, requirements specification, requirements analysis, design, construction, testing, and deployment.

Chapter 22
Incorporating Free/Open-Source Data and Tools in Software Engineering Education

Liguo Yu
Indiana University South Bend, USA

David R. Surma
Indiana University South Bend, USA

Hossein Hakimzadeh
Indiana University South Bend, USA

ABSTRACT

Software development is a fast-changing area. New methods and new technologies emerge all the time. As a result, the education of software engineering is generally considered not to be keeping pace with the development of software engineering in industry. Given the limited resources in academia, it is unrealistic to purchase all the latest software tools for classroom usage. In this chapter, the authors describe how free/open-source data and free/open-source tools are used in an upper-level software engineering class at Indiana University South Bend. Depending on different learning objectives, different free/open-source tools and free/open-source data are incorporated into different team projects. The approach has been applied for two semesters, where instructor's experiences are assembled and analyzed. The study suggests (1) incorporating both free/open-source tools and free/open-source data in a software engineering course so that students can better understand both development methods and development processes and (2) updating software engineering course regularly in order to keep up with the advance of development tools and development methods in industry.

DOI: 10.4018/978-1-4666-5800-4.ch022

1. INTRODUCTION

Software engineering is considered one of the most difficult topics in computer science program. Its difficulty is not like theory courses, such as algorithm analysis, nor programming courses, such as data structures. Software engineering is an empirical course. Students should learn software engineering methods through hands-on experience, which might include real-world software development, real-world customer interaction, real-world planning and estimation, and real-world decision-making and problem-solving.

However, given the limited resources in academia, it is hard for students to learn hands-on experience in a classroom environment. Software engineering educators have been working on this issue for years and various approaches have been adopted to overcome this hurdle. For example, in some programs, industry projects are introduced into the classroom (Hayes, 2002), where students practice software engineering principles through solving challenging and complicated real world-problems. In other programs, students are asked to participate in open-source software development (Lundell et al., 2007; Stamelos, 2008; Jaccheri & Osterlie, 2007), where the source code is available for analyzing and testing. In some cases, students could be assigned to tackle a reported bug. For example, Papadopoulos et al. (2012; 2013) have used free/libre open source software (FLOSS) projects to assist teaching software engineering for at least four years. Their experiences are well documented and analyzed.

The two methods described above are proven approaches that can better integrate software engineering education with software industry practices. They all can be classified as real-world project-based software engineering education.

The software engineering course offered at Indiana University South Bend is tool and data based, where students learn software engineering methods through using software tools and analyzing software data, more specifically, free/

open-source tools and free/open-source data. In this chapter, we describe how free/open-source tools and free/open-source data could be used in software engineering education to reduce the gap between industry expectations and what the academia can deliver.

The remaining of the chapter is organized as follows. In Section 2, we review related work and introduce our teaching approach. In Section 3, we describe our software engineering class, including the teaching method and the teaching experience. In Section 4, we summarize the analysis of our teaching approach. Conclusions and the improvement plan are presented in Section 5.

2. RELATED WORK AND OUR TEACHING APPROACH

Open-source software has been widely used in education (Lazic et al., 2011; Hoeppner & Boag, 2011), especially in computer science education. In software engineering field, open-source software has special usages. Because nowadays, software development largely depends on tools, which are computer software program that can facilitate the analysis, design, implementation, testing, and project management in software development. In other words, to be considered as a modern software engineer, one must know how to use various CASE (computer aided software engineering) tools.

Given the limited resources in academia, it is unrealistic to purchase all the latest commercial development tools for classroom usage. Therefore, open-source tools provide an opportunity for students to explore the latest technology development in software industry. Moreover, both the commercial software source code and commercial software development data are not accessible for most academic institutions. Without examining real-world source code and real-world development data, it is unlikely that the academia could deliver students the high standard knowledge and skills expected by the industry.

Fortunately, the thriving of open-source projects provides an avenue to study real-world software products and real-world software development process in a classroom environment. Accordingly, open-source software has been used in general computer science educations for years (Lakhan & Jhunjhunwala, 2008; Pedroni et al., 2007; O'Hara & Kay, 2003). For example, at Rochester Institute of Technology (Raj & Kazemian, 2006), open-source software has been used in teaching database systems, programming language, and language-based security. At Victoria University of Technology (Nelson & Ng, 2000), open-source software is used to teach computer networking.

Software engineering is a required subject in computer science program. Open-source software should play an important role in presenting students software engineering principles and methods (Kamthan, 2007). To the best of our knowledge, three papers that are most related to our work have been published. They are reviewed below.

Teel et al. (2012) used four different kinds of open-source tools in their software engineering courses. The tools include (1) Readmine, which is used for project management and project communication; (2) Turnkey applications, which provide application support such as Web service for course project; (3) Eclipse IDE for development; and (4) Apache Subversion for version control. In their approach, these software tools are integrated into course projects. In other words, the tools are mainly used for class project development and class project management.

Nandigam et al. (2008) used open-source Java tools in their software engineering class for source code browsing, measurement, and refactoring. The tools include (1) Eclipse IDE for java development; (2) Eclipse Checkstyle plugin; (3) Eclipse Metrics plugin; (4) Eclipse UML plugin; (5) Junit; (6) ImageJ for image processing and analysis; (7) Apache Derby, a relational database;

(8) Hibernate, a framework for mapping an object-oriented domain model to a traditional relational database; and (9) Apache Lucene, a text search engine library. Some of the aforementioned tools representing the latest development in industry are also used in our class, such as Junit.

Carrington and Kim (2003) used the following open-source tools in their software engineering classes: (1) ArgoUML, a UML modeling tool; (2) Eclipse IDE for Java development; (3) JEdit, a programmer's text editor; (4) JRefactory, a Java source code refactoring tool; (5) JUnit; (6) NetBeans IDE for Java development; (7) Process DashBoard, a personal software process support tool; (8) Apache Ant, a Java build tool; (9) DocWiz, a GUI tool to manage JavaDoc; and (9) PMD, a static Java source code checker. It can be seen that most of their tools are for Java development. This is also the case for most open-source tools.

To summarize, various open-source tools are reported being used in software engineering courses. Their major usage are for software development and project management with the ultimate goal of helping students understand software engineering principles and get familiar with the latest development of industry methods.

The difference between our teaching methods and the reported methods are twofold. First, both open-source tools and open-source data are used in our software engineering class to help students understand real world software development. Second, besides development tools, real-world project management, such cost estimation, is introduced in our class. To the best of our knowledge, there has been no report about using cost estimation as a project in software engineering class. Therefore, our approach is a combination of project-based education and tool-based education, where open-source data (such as open-source project source code), are used by students to understand software process, and open-source tools are used by students to get familiar with the latest development methods.

3. CLASS DESCRIPTION

At Indiana University South Bend, CSCI-P565 is a graduate level and upper-level undergraduate software engineering course. The objective of this course is to explore students with advanced topics in software engineering, such as project management, risk management, configuration management, cost estimation, and software testing. The assignments of this class include both individual homework and team projects. Because individual homework is unrelated to the topic of this chapter, it will not be discussed here. Therefore, we will only describe team projects that are relevant to the topic of this chapter.

In this class, there are five team projects in total. In each project, students will use some free/open-source tools to analyze data or study software systems. Each project is seamlessly integrated with a specific learning objective. The learning objective is achieved through using open-source tools and sometimes, analyzing open-source data. In the following of this section, we will describe each project in detail.

3.1 Project 1: Configuration Management

- **Objective**: To get students understand configuration management and acquire hands-on experience of using a configuration management tool.
- **Software tool**: Subversion (SVN) (n.d.). SVN is a version-control system. Developers use SVN to maintain current and historical versions of files such as source code, Web pages, and documentation. SVN has been widely used in industry, especially in open-source projects, such as Apache Software Foundation, FreeBSD, GCC, Ruby, and SourceForge.
- **Tool source**: SVN is a free and open-source tool. It is available on standard Linux installations. Students could use either the

Linux lab computers (remote logon or direct access) at school or their own Linux machines.

- **Tasks assigned**: Students are asked to work in a team so that they can explore all different features of SVN. These tasks include setting up the SVN server and practicing the SVN commands. These commands include both server operation commands and client operation commands.
- **Submission**: On the project due date, each team is asked to give a 30-35 minute demonstration about how to use SVN. This demonstration serves as a tutorial for those who are new to SVN. The demo should include (1) a brief introduction of version control system and SVN; (2) a demonstration of how to setup and manage SVN server; (3) a demonstration of how SVN manages different access permissions of clients: *none*, *read*, and *write*; and (4) a demonstration of how SVN manages concurrent editing by two or more clients under two different conditions: *without conflicts* and *with conflicts*. Students are also asked to demonstrate how to use some other commonly used SVN commands.
- **Experience**: In both semesters, about two out of five or six teams did very good in this project. Their demonstrations are really impressive. Some teams did not prepare well. However, through watching the demos given by other teams, they finally understood the mechanisms of version control systems. Version control is an abstract concept without hands-on experience. This project is successful as it endorses the learning through doing approaches.
- **Improvement**: Future plan for improving this project could be done through introducing different open-source version control tools, such as Git or GitHub. For example, some teams could be assigned with SVN and other teams could be assigned

with Git or GitHub. Accordingly, the demonstrations could be more meaningful and useful for teams using different tools.

3.2 Project 2: Cost Estimation

- **Objective**: To get students understand COCOMO cost estimation model and estimate the development cost of Linux kernel.
- **Background:** In October 2008, McPherson *et al.* (2008) published their work of estimating the development cost of *Fedora 9*, a Linux distribution. In their research, they repeated David Wheeler's estimation of the development cost of Linux (Wheeler, 2002).
- **Source tool**: LocMetrics (n.d.), which can be used to measure both the physical SLOC (source line of code) and logical SLOC (source line of code).
- **Data source:** Linux kernel (n.d.)
- **Tasks assigned**: Students are asked to (1) read the user manual of COCOMO 81, COCOMO 97, and COCOMO 2000; (2) measure the size of Linux kernel 2.6.25; (3) estimate the development cost of Linux kernel 2.6.25 using COCOMO model 81, post architecture model of COCOMO 97, and early design model and post architecture model of COCOMO 2000; and (4) compare and analyze the cost of different estimations.
- **Submission**: On project due date, each team (2 or 3 students) is asked to give a 20 minute presentation about their study.
- **Experience**: For students, the difficult part of this project is to assign cost drivers to each model. Different teams usually assigned dramatically different values for the same cost driver. Therefore, the final cost estimation of different teams ranges from 800 million US dollars to 1.9 billion US dollars for the development of Linux kernel 2.6.25. Generally, there are no single

correct assessments of the cost drivers. However, the students are asked to justify their assessment. This project is interesting to most students, because they learn some *magic* formulas that convert computer source code into US currency. This technique is certainly required for developing any real-world software product.

- **Improvement**: If we can find out the development cost of some commercial operating systems and compare them with the estimated development cost of Linux kernel, the project will be more interesting. That is one improvement we are going to try next time when this course is offered.

3.3 Project 3: Software Metrics

- **Objective**: Understand and measure the Chidamber and Kemerer's object-oriented metrics.
- **Software tool**: (1) ckjm (DMST, n.d.), which is used to measure the CK (Chidamber and Kemerer) metrics of Java classes; (2) LocMetrics, which is used to measure the McCabe's Cyclomatic complexity of Java source code.
- **Data source:** Apache Ant (n.d.), Apache Tomcat (n.d.), and Apache XML (n.d.).
- **Tasks assigned**: Students are asked to (1) read the original publication of Chidamber and Kemerer (1994) to understand the object-oriented metrics; (2) measure the six CK metrics and the McCabe's Cyclomatic complexity of Apache Ant, Apache Tomcat, and Apache XML; and (3) analyze the measurements and compare CK metrics with McCabe's Cyclomatic complexity.
- **Submission**: On project due date, each team (2 or 3 students) is asked to give a 20 minute presentation about their study. Statistical analysis tools, such as SPSS could be used to draw histograms and illustrate the results and the analysis.

- **Experience**: Software metrics are abstract concepts. Through measuring the most commonly used object-oriented metrics of real-world programs, students got hands-on understanding of the metrics. Through presenting, listening, and discussing the measurements, students had a better understanding of the relation between software metrics and software quality.

- **Improvement**: This project could be improved if we can mine the bug reports of these open-source projects and encourage students to study the relation between class metrics and the number of bugs reported to that class. In this sense, students could get a better understanding of the purposes of using software metrics. They will also get a hands-on experience of understanding the relation between software complexity and software quality.

3.4 Project 4: Unit Testing and Mutation Testing

- **Objective**: To get students understand and obtain hands-on experience with unit testing and mutation testing.

- **Software tool**: (1) Eclipse Java IDE (Eclipse, n.d.); (2) Junit (n.d.) for unit testing; (3) Jumble (Sourceforge, n.d.) for mutation testing.

- **Tasks assigned**: Students are asked to (1) write Junit test cases for a given program (provided by the instructor); (2) execute Junit test cases; (3) use Jumble to examine the effectiveness of the Junit test cases; and (4) modify the Junit test cases if needed so that Junit could detect 100% of the errors seeded by Jumble.

- **Submission**: On project due date, each team (2 or 3 students) is asked to explain

their Junit test cases and demonstrate how to use Junit and Jumble.

- **Experience**: This project involves programming, which is usually more interesting to students. Software testing is a topic that we usually talk more other than practice in software engineering courses. Through working on this project, students obtained hands-on experience of systematic software testing with tools. Moreover, test-driven development and mutation testing are relatively new technologies in software industry. They are rarely included in any textbooks. This project provides students an opportunity to experience software development methods in industry.

- **Improvement**: Currently, the program used in this project is written by the instructor for the purpose of simplicity. Real-world programs could be introduced in the future offerings of this class. On the other side, different testing tools could be used to allow students to practice different testing methods.

3.5 Project 5: Reverse Engineering, Automated Software Engineering, and Impact Analysis

- **Objective**: To get students obtain hands-on experience of reverse engineering, automated software engineering, and impact analysis.

- **Software tool**: (1) Eclipse Java IDE; (2) AgileJ, an Eclipse plugin (Agilej, n.d.); (3) Visual Paradigm for UML (a commercial product but with 30 days free trial) (VisualParadigm, n.d.).

- **Tasks assigned**: Students are asked to (1) use AgileJ to convert a given Java program (provided by the instructor) into a UML diagram; (2) use Visual Paradigm to generate source code based on the UML diagram created in Step 1; (3) compare the generated code with the original code given by the instructor; and (4) use Visual Paradigm to perform impact analysis.

- **Submission**: On project due date, each team (2 or 3 students) is asked to give a 20 minute demonstration about their study.
- **Experience**: Reverse engineering and automated software engineering are rarely covered in traditional software engineering courses. This project enables students to practice the latest development methods in software industry.
- **Improvement**: Currently, the program used in this project is written by the instructor for the purpose of simplicity. Real-world programs could be introduced in the future classes.

4. GENERAL ANYALSIS

The authors have offered this class (CSCI-P565) three times at Indiana University South Bend: Fall 2005, Spring 2010, and Spring 2012. In Fall 2005, it was taught with the traditional approach, where students learn software engineering principles through lectures and working on projects with limited commercial tools. In Spring 2010 and Spring 2012, this class was taught with the new approach described in this chapter, i.e, incorporating open-source tools and open-source data in team projects. In general, both the instructors' impression and the students' feedback are positive with the new teaching approach.

Based on our two semester's experiment with this combination of tool-based and project-based approach, we believe we achieved our objective of revamping this course: reducing the gap between industry expectation and what the academia can deliver. In other words, our students are equipped with some industry demanded-skills through taking this course. Below, we summarize the benefits of our combination of tool-based and project-based teaching approach.

- We can teach software engineering methods and principles through hands-on experience of working on real-world data and real-world project.
- Students could learn up-to-date software development techniques and use up-to-date software development tools.
- Students could learn project management skills, such as cost estimation, through analyzing large open-source projects.

In this course, students also practiced other soft skills that are expected by the industry. They are listed below.

- **Team working skills:** All five projects are assigned as teamwork. Students are encouraged to choose different team members for different projects so that they can practice team working skills through working with students of different working style and with different personality.
- **Communication and presentation skills:** Through working in teams, students need to communicate with their team members, schedule meetings, assign tasks, discuss problems, and offer solutions. Presentations are required for all five projects. Every team member is required to participate in the presentation.
- **Documentation skills:** Students are also asked to submit their document for evaluation by the instructor. Their study should be well documented, including the problem description, study method, results, and the analysis.

As with any other teaching approaches, there are certainly limitations to the approach presented in this chapter. They are discussed below.

- We have successfully applied this teaching approach in our graduate level software engineering course. However, this approach does not fit our undergraduate software engineering course syllabus, because in our undergraduate software engineering course, we have different learning objectives. Therefore, for instructors of different institutions, they should examine their learning objectives carefully before adopting this approach.

- This teaching approach brings great challenges to students. For students who have never worked on real-world problems, it means more work and rework. Some students have complained that the user manuals of some software tools are super long and difficult to understand. Requiring a prerequisite for this class might not be a good solution, because it will limit some potential self-motivating students.

- To achieve the best outcomes from this course, it also means the instructors should be ready for answering any technical questions at any time. Sometimes, the instructors need to work together with students to figure out certain issues of using a software tool. Therefore, this teaching approach requires the instructors to keep up with the latest development technology and be familiar with different kinds of development environment.

- The approach used in our software engineering course contains five independent projects and it works fine. However, to introduce more appropriate tools and more appropriate projects might not be an easy task, because these tools and projects should be integrated with the learning objectives. To adopt this approach, instructors of other institutions might need to consider different tools and different projects in order to meet their own learning objectives.

5. CONCLUSION AND IMPROVEMENT PLAN

In this chapter, we presented our experience of teaching a high-level software engineering course at Indiana University South Bend. The class is organized as a combination of project-based approach and tool-based approach. Students learn software engineering concepts and methods through hands-on experience of using the latest open-source software engineering tools and analyzing the latest open-source data and projects. This teaching approach is innovative and useful in that it is designed to reduce the gap between industry expectations and what the academia can deliver in a software engineering class. To improve this course in the near future offerings, the following approaches could be taken.

- Design and assign new projects that can allow students to learn new software development methods and new development tools that are introduced into the industry recently.

- The software tools should be evaluated every time this course is offered. This will make sure our students can practice using the latest software development techniques.

- Because this is a graduate level course, many students are part-time or full-time software professionals. We should encourage those students to bring real-world projects or real-world problems into the classroom. This could be another way to reduce the gap between academia offerings and industry expectations.

- The five projects might be able to be combined as one project that lasts the entire semester so that students could use different tools in different stages of this large project. This implementation will be more similar to an industry environment, where different skills, different methods, and different tools are used in the same project to solve different problems.

- This teaching approach is used in a graduate level software engineering course and it does not fulfill our undergraduate software engineering curriculum requirement. However, some of the lectures in this class can be organized as workshop/seminars, where graduate students give presentations or tutorials and interested undergraduate students could be the audience. This will allow undergraduates to learn some techniques that are not offered in other classes, such as version control tools.

Software engineering is a fast changing area. Our teaching approach should always be reexamined and reevaluated in order to keep up with the industry expectations. We wish our teaching experiences presented in this chapter are useful to educators of other instituations.

REFERENCES

Agilej. (n.d.). Retrieved from http://www.agilej. com/

Ant. (n.d.). Retrieved from http://ant.apache.org/

Apache. (n.d.). Retrieved from http://xml.apache. org/

Carrington, D., & Kim, S. K. (2003). Teaching software design with open source software. In *Proceedings of the 33rd ASEE/IEEE Frontiers in Education Conference* (Vol. 3, pp. S1C–9). Boulder, CO: IEEE.

Chidamber, S. R., & Kemerer, C. F. (1994). A metrics suite for object oriented design. *IEEE Transactions on Software Engineering*, *20*(6), 476–493. doi:10.1109/32.295895

DMST. (n.d.). Retrieved from http://www.dmst. aueb.gr/dds/sw/ckjm/

Eclipse. (n.d.). Retrieved from http://www.eclipse. org/

Hayes, J. H. (2002). Energizing software engineering education through real–world projects as experimental studies. In *Proceedings of the 15th Software Engineering Education and Training Conference* (pp. 192–206). Covington, KY: IEEE.

Hoeppner, K., & Boag, A. (2011). Open source software for education. *Global Learn*, *1*, 1658.

Jaccheri, L., & Osterlie, T. (2007). Open source software: A source of possibilities for software engineering education and empirical software engineering. In *Proceedings of the 1st International Workshop on Emerging Trends in FLOSS Research and Development* (pp. 1–5). Minneapolis, MN: IEEE.

Junit. (n.d.). Retrieved from http://www.junit.org/

Kamthan, P. (2007). A perspective on software engineering education with open source software. In K. St. Amant, & B. Still (Eds.), *Handbook of research on open source software: Technological, economic, and social perspectives* (pp. 690–702). Hershey, PA: Information Science Reference. doi:10.4018/978-1-59140-999-1.ch054

Kernel. (n.d.). Retrieved from https://www.kernel. org/

Lakhan, S. E., & Jhunjhunwala, K. (2008). Open source software in education. *EDUCAUSE Quarterly*, *31*(2), 32–40.

Lazic, N., Zorica, M. B., & Klindzic, J. (2011). Open source software in education. In *Proceedings of the 34th International Convention MIPRO* (pp. 1267–1270). Opatija, Croatia: IEEE.

Locmetrics. (n.d.). Retrieved from http://www.locmetrics.com/

Lundell, B., Persson, A., & Lings, B. (2007). Learning through practical involvement in the OSS ecosystem: Experiences from a masters assignment. In *Open source development, adoption and innovation* (pp. 289–294). Berlin: Springer. doi:10.1007/978-0-387-72486-7_30

McPherson, A., Proffitt, B., & Hale–Evans, R. (2008). *Estimating the total development cost of a Linux distribution*. The Linux Foundation.

Nandigam, J., Gudivada, V. N., & Hamou–Lhadj, A. (2008). Learning software engineering principles using open source software. In *Proceedings of the 38th Frontiers in Education Conference* (pp. S3H–18). Saratoga Springs, NY: IEEE.

Nelson, D., & Ng, Y. M. (2000). Teaching computer networking using open source software. *ACM SIGCSE Bulletin*, *32*(3), 13–16. doi:10.1145/353519.343056

O'Hara, K. J., & Kay, J. S. (2003). Open source software and computer science education. *Journal of Computing Sciences in Colleges*, *18*(3), 1–7.

Papadopoulos, P. M., Stamelos, I. G., & Meiszner, A. (2012). Students' perspectives on learning software engineering with open source projects: Lessons learnt after three years of program operation. In *Proceedings of the 4th International Conference on Computer Supported Education* (pp. 313–322). Porto, Portugal: SciTePress.

Papadopoulos, P. M., Stamelos, I. G., & Meiszner, A. (2013). Enhancing software engineering education through open source projects: Four years of students' perspectives. *Education and Information Technologies*, *18*(2), 381–397. doi:10.1007/s10639-012-9239-3

Pedroni, M., Bay, T., Oriol, M., & Pedroni, A. (2007). Open source projects in programming courses. *ACM SIGCSE Bulletin*, *39*(1), 454–458. doi:10.1145/1227504.1227465

Raj, R. K., & Kazemian, F. (2006). Using open source software in computer science courses. In *Proceedings of the 36th Annual Frontiers in Education Conference* (pp. 21–26). San Diego, CA: IEEE.

Sourceforge. (n.d.). Retrieved from http://jumble.sourceforge.net/index.html

Stamelos, I. (2008). Teaching software engineering with free/libre open source projects. *International Journal of Open Source Software and Processes*, *1*(1), 72–90. doi:10.4018/jossp.2009010105

Subversion. (n.d.). Retrieved from http://subversion.tigris.org/

Teel, S., Schweitzer, D., & Fulton, S. (2012). Teaching undergraduate software engineering using open source development tools. *Issues in Informing Science & Information Technology*, *9*, 62–73.

Tomcat. (n.d.). Retrieved from http://tomcat.apache.org/

Visual-Paradigm. (n.d.). Retrieved from http://www.visual-paradigm.com/product/vpuml/

Wheeler, D. A. (2001). *More than a gigabuck: Estimating GNU/Linux's size*. Retrieved November 25, 2013, from http://www.dwheeler.com/sloc/redhat71-v1/redhat71sloc.html

KEY TERMS AND DEFINITIONS

Open-Source Data: Data of an open-source project, including source code of programs, change logs, revision history, bug reports, etc.

Open-Source Project: A project to develop open-source software. It is managed by some open-source communities or foundations.

Open-Source Software: Computer software program with its source code available to the public and be freely to use under certain license.

Software Engineering Education: The discipline of educating professional software engineers who can design, develop, manage, and maintain a software product.

Tool-Based Teaching: Teach students theories, principles, and knowledge through encouraging students to use certain tool, especially, software tool.

Section 10
Adopting Digital Learning

Chapter 23
Improve Collaboration Skills Using Cyber–Enabled Learning Environment

Yujian Fu
Alabama A&M University, USA

ABSTRACT

Collaborative learning methods have been widely applied in online learning environments to increase the effectiveness of the STEM programs. However, simply grouping students and assigning them projects and homework does not guarantee that they will get effective learning outcomes and improve their collaboration skills. This chapter shows that students can improve their learning outcomes and non-technical skills (e.g. collaboration and communication skills) through the cyber-enabled learning environment. The data was collected mainly from software engineering and object-oriented design classes of both graduates and undergraduates. The authors apply a blended version of education techniques by taking advantage of online environment and classroom teaching. Based on the study, the authors show that students can improve their collaboration and communication skills as well as other learning outcomes through the blended version of learning environment.

INTRODUCTION

Collaborative learning methods have been widely applied in online learning environment to increase the effectiveness of STEM programs. However, purely application of the online learning environ-

ment may not work well for students to improve both technical and non-technical skills (such as collaboration and communication skills). First, simply grouping students and assigning projects and homework do not guarantee that they will get effective learning outcome. Second, existing instructor-centered learning environment in many online courses does not offer sufficient scope

DOI: 10.4018/978-1-4666-5800-4.ch023

for students to work collaboratively. To prepare students for their future information technology careers, it is necessary to foster collaboration and communication skills that are needed in the industry.

In addition, it is widely noticed that software engineering professionals working in industry are generally unsatisfied with the level of real-world preparedness possessed by recent university graduates entering the workforce (Cummings & Betsy, 2007; Callahan & Pedigo, 2002). Their frustration is understandable – in order for these graduates to be productive in an industrial setting, organizations that hire them must supplement their university education with extensive on-the-job training and preparation that provide them with the skills and knowledge they lack (Conn, 2002). The root of the problem seems to lie in the way software engineering is typically taught: theories and concepts are presented in a series of lectures, and students are required to complete a small, toy project in an attempt to put this newfound knowledge into practice. Thus, the student graduated has little demonstrated capability in solving problems of large scale systems or dealing with critical issues and is lack of adequate skills of collaboration, communication in the teamwork environment.

Having observed similar situations in Alabama A&M University in the past years, we have developed a framework using the blackboard learning system to encourage students to engage with online activities. This framework is intentionally designed to support student online activity so that they could actively interact with teaching content and collaborate and communicate with others. Another feature of this methodology is the development of reusable learning objects. A learning object is a learning unit that contains an objective, a learning activity and assessment, which represent a set of reusable and self-contained digital resources. The baseline data collection in

software engineering course at Alabama A&M University started in Spring 2008. The framework has been applied and validated since 2010 and has been improved in 2012. The measurement of the method was done in several ways – a pre-test and post-semester survey, a student interview, an alumni survey. The data was analyzed based on the satisfaction rate regarding to the course objectives. The survey questions are grouped by four categories regarding to the course objectives and program goals: background (including majors, minors, working experience), programming skills, project topics (information systems, embedded systems, security, government project, industrial project), and difficulty level (ranking from 1 to 5). In this chapter, we focus our analysis of the questions that are related to collaboration and communication skills.

In order to provide a meaningful context for students to learn and work collaboratively, this study is conducted in the software engineering and object oriented programming courses. We updated the current technology-based learning strategy as the background theory that supports this framework. Technology based teaching strategies utilizing Internet technology could provide remarkable educational opportunities for the 21st century learners. In our study, the upgraded technology-based learning (UTBL) includes communication devices other than just Internet based teaching. The communication devices used in this framework are robots, mobile devices (such as Android tablets, iPhone), social media and networks. These devices can be used for demo, example and project implementation, group discussion and peer communication during and after classroom time. Through synthesizing cyber enabled learning environment with technology-based teaching, the framework can dramatically motivate students and improve their learning outcomes.

THE DEFINITION OF COLLABORATION AND TEAMWORK

One of the main characteristics of our graduates considered by industry is the capability to work in team. Teamwork is not a new term and has been aware of by educators and employers for decades. In our study, we have considered the following capabilities and skills in teamwork: collaboration, communication, self-management, and leadership. Due to the limited space, in this chapter, we will focus on describing two skills – collaboration and communication.

In the dictionary, *collaboration* is defined as a style of working with each other to conduct a task and to achieve shared goals. In Wikipedia, it is defined as "a recursive process where two or more people or organizations work together to realize the shared goals." In education, *collaboration* means "two or more co-equal individual voluntarily brings their knowledge and experience together by interacting toward a common goal in the best interest of students for the betterment of their education success. Students achieve team building and communication skills meeting many curricular standards. Students have the ability to practice real-world communication experiences. Students gain leadership through collaboration and empowers peer to peer learning" (Wikipedia).

It is widely accepted that graduates should not only be technically competent but also be skilled in communication and teamwork. They should have certain social skills and be equipped with global awareness, self-managed and self-learning skills, which are important in their life time. However it is much less clear how these "soft skills" could be best developed in their undergraduate and even graduate studies. McLoughlin and Luca (2000) suggested that pedagogy needs to change from transmissive, didactic approaches to transformative, student-centered approaches. In this study, the framework that is developed focuses on stu-

dent's role instead of instructor's role. To achieve this goal, we implemented the framework in a context of project-based unit involving individual and group study format under an online learning environment.

This framework improves student achievement in two folds. First, it is a technology driven strategy. The power of digital instruction has been widely noticed. For example, digital instruction supports personalized learning through various vehicles. As instructions are required to align with college- and career-ready standards, digital learning can become increasingly student-centered and market-driven, individually tailored to provide the variety of paths and paces students need to achieve ambitious goals, and informed by adaptable technology and assessment data. The potential of digital instruction is enormous: In its next generation, it will likely become increasingly emotionally connective for students and provide them and their teachers with enhanced diagnostics and instructional roadmaps. These improvements will enable the consistent instructional differentiation and high standards for students' learning advancement that today typifies only the most excellent teachers and schools, while saving teachers' time so they could engage in other aspects of teaching.

The second benefit comes in the integrated approach's capacity to let institutions reach more students with excellent teachers who could ensure that students achieve their ambitious, personally fulfilling goals. This level of growth is essential for closing achievement gaps and helping average students leap ahead to higher standards. In the future, when technology makes the basics of learning available to all students globally, complex aspects of excellent teaching will become even more important: guiding students' selection of ambitious and engaging work, fostering student motivation, addressing the myriad learning barriers, and cultivating higher-order thinking.

BACKGROUND AND RATIONALE

Our country's success in the increasingly technology-driven and globalized economy will depend on how we prepare today's learners and students for tomorrow's job markets, for personal fulfillment and for civic engagement in an inter-connected world.

For this reason, it is important to set clear, ambitious goals for institutional education to generate high student growth and develop students' higher-order thinking skills. But goals alone will not set our students up for success. Students' learning needs are shaped by family supports and personal characteristics, such as past achievement, self-motivation, learning preferences, time management, and emotional stability. Even the best one-size-fits-all teaching methods do not meet the diverse needs that teachers encounter in classrooms. Our nation's educational challenge, then, is to maintain ambitious goals for all while helping each student find a path to meet them.

There is a moral and economic imperative to change the way instructors teach and students learn in the United States. All children graduated from high school should be ready for college and career, possessing the deeper learning skills they need in order to compete in today's rapidly changing economy. These skills include not only mastery of core content but also the ability to think critically, solve complex problems, work collaboratively, communicate effectively, be self-directed, and incorporate feedback (Alliance, 2011). Too many low-income students are still not developing the skills they need to succeed in modern life. Nationwide, only 72 percent of students earn a high school diploma. In the class of 2011, more than 1 million students dropped out before graduation. Among minority students, only 58 percent of Hispanic, 57 percent of African American, and 54 percent of American Indian and Alaska Native students in the United States graduate with a regular diploma, compared to 77 percent of white students and 83 percent of Asian Americans (Editorial, 2010).

Similarly, in Figure 1, data collected from ACT (2013) explicitly shows that college readiness rates for most of races are around or lower than 50%.

Figure 1. Graduation and college-ready rates: Data retrieved from ACT (2013)

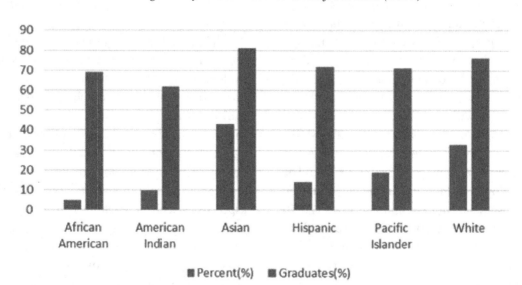

Specifically, African-American students have 5% college-readiness rate and 68% graduate rate. These numbers are astonishing to today's educators, especially for college education. It is highly desired for our researcher to find an efficient way to motivate students and encourage them to pursue and accomplish the college study and to prepare for the future workforce.

The rise of digital learning presents a unique opportunity to meet the aforementioned challenge. It has unprecedented potential to help underprepared students achieve ambitious goals by enabling personalized paths to learning success. Even in the digital age, the vast majority of U.S. students will probably attend brick-and-mortar schools. Many parents need to have their children stay at school while they work, because our schools act as connective fabric for the communities. As a result, most students will experience digital learning as part of blended learning: a combination of digital instruction and in-person teaching.

In this chapter, we explain how the proposed learning framework can succeed by bringing interesting projects with the best available technology to students under the guidance of instructor, motivating students and improving their collaboration and communication skills. When the instructors use the high quality project-based learning under cyber enabled learning environment, they can realize the great potential to dramatically improve student learning outcomes and non-technology skills.

DIMENSIONS OF TEAMWORK EDUCATION

This section discusses the key dimensions of teamwork education in the development, implementation and assessment of project-based, team-oriented framework in the cyber enabled learning environment. The dimensions of teamwork education include project-based learning (PBL),

technology based learning (TBL), participants, content, project, and assessment.

Project Based Learning (PBL)

The selected study is carried on in two courses: software engineering and object oriented programming and design. Software engineering class aims at conveying the fundamental concepts, design methodologies, validation and verification techniques for a software life cycle. Object oriented programming and design class aims at the advanced level of object oriented programming concepts, such as inheritance, method call, polymorphism and their relationship with design concepts. Both classes are concept oriented with a lot of design notations; students need to master and apply the design approaches to the real world applications. At the very beginning year of teaching both classes at Alabama A&M University, we found two dilemmas from students:

- **Group A:** Students struggle with the understanding of concepts. This situation will hinder the students to have further capability of applying the concepts.
- **Group B:** Students have hard time to implement and apply the design methodology to concrete problems, which usually are just some toy systems with minimum design. This situation happens to those students who even have a better understanding of notations and concepts and have demonstrated a very good grade in the quizzes and homework.

The author collected some student data with the consideration of age and background to make sure these students are at the similar level. The data includes the courses students registered, their current GPA, their final course grade, their background knowledge regarding to the software engineering, and the grades of some related courses. After analyzing the data, we have the following findings:

- In Group A, about 11% students maintain good GPA (above 3.0) and good grade in other courses;
- In Group B about 69% students maintain good GPA (above 3.0) and good grade in other courses.

To understand this situation, a simple survey with interview is offered to students in both classes. From student response, we have the following findings:

- Students are more interested in some new applications and realistic problems instead of the questions given in the classes.
- Students are looking for the connection between the design concepts and applications but have hard time to link them together.

Some student comments are "I feel it is interesting, but after a while, it is getting boring...," and "there are too many concepts, and I don't have time to study... I do not have notes..." From the data analysis and survey findings, we identified a key issue for student learning–motivation. Students need to be interested in the topic. Regarding to this issues, we started implementing project-based learning with cyber enabled environment in both classes to help students keep up and maintain their interests.

Project-based learning could improve and broaden the competence of computer science students. Through this learning approach, students could

- Understand the role of theoretical and real-world discipline-specific knowledge in a multi-disciplinary, collaborative, and project-centered platform;

- Recognize the relationship of the software engineering concepts to the enterprise context and the key role of this context in computer science decision;
- Learn how to work in a team manner and lead multiple roles to design, develop, manage, and maintain high quality large scale software intensive systems, effectively and economically.

Technology Based Learning Strategy

Internet is the most fluent technology in this century. It is one of the most popular applications in human history. Of course, several other technologies such as networking and protocol (WAN and LAN techniques), memory, Web page design, are the key supporting technologies for the development and booming of Internet. Besides Internet, mobile applications and robotics are all the latest emerging technology that currently flew into the classroom. Accordingly, today's technology based teaching is not limited to Internet. There are other new emerging technologies that have effects on the traditional teaching strategies. The definition of technology based learning (TBL) is changing from installing digital technology in classroom teaching to ad hoc multiple techniques and to a new cyber learning environment. One of the key ideas behind the new era of teaching is to motivate students and improve the teaching qualities of educators and instructors.

First and foremost, our technology-based learning as the education strategy includes various electronic technologies, such as Internet, intranets, satellite broadcasts, audio and video tape, video and audio conferencing, Internet conferencing, chat rooms, e-bulletin boards, Webcasts, computer-based instruction, and CD-ROM (ASTD, 2005). TBL also encompasses related terms, such as online learning and Web-based technology that occur via the Internet, and computer-based learning that is restricted to learning using computers. E-learning is synonymous with TBL and has largely replaced it in scholarship and industry as an alternative term. Therefore, this chapter uses these terms interchangeably. Distance learning sometimes is also referred as technology-delivered learning. However, it is worth to note that there is a difference between distance learning and TBL. TBL includes methodologies where instructors and learners are in the same room and instruction is computer-based, but there is no *distance* involved. On the other hand, TBL is more narrowly defined in that it does not include text-based learning and courses are conducted via written correspondence that would be covered by either distance learning or technology-delivered learning. Furthermore, technology-enhanced learning describes a methodology in which technology plays a subordinate role and serves to enrich a traditional face-to-face classroom. On earth, both TBL and digital learning aim at improving students' learning outcome through increasing their motivation and curiosity.

Participant

The students are required to self-group themselves based on willingness without instructor's interaction at the beginning of class. For the senior design class, this could be an easy process, because students are mostly acquainted with each other. Students have the choices to select one term project from the list given in the Blackboard. At the same time, students of the same group that have no interest in any of project in the list can propose a new project by their own. The proposal will be evaluated and approved by the instructor. By doing so, we want to ensure that the project students are working on is the one that are most interesting to them.

Content

The ultimate goal of the framework is to improve students' learning outcome through cyber enable learning environment with the context of project based learning and team oriented platform.

Through this type of learning, students will understand

- The fundamental concepts and design methodologies regarding to the context;
- How to apply design methodologies on real world systems;
- How information technology and collaboration skills could support their teamwork and help them achieve the learning outcomes.

Project

The projects students working on are multidisciplinary that require knowledge of software system design, programming language, system configuration and/or computer science theories. The students in a group are required to analyze the project, assign tasks and roles to team members, and setup timelines of the project. The project requires necessary documentations that contain design models, analysis, implementation, and validation through the software life cycle. During the project period, students will learn

- Software design and development methodologies and object oriented programming skills;
- Problem solving skills for both predictable issues and unexpected issues;
- Communication skills through face-to-face teamwork and multimedia social networking in the cyber learning environment;
- Collaboration skills through team meeting and multimedia social networking in the cyber learning environment;
- Using online materials provided in Internet and Blackboard learning systems;
- Leadership skills through collaborations and management.

Assessment

It is hard to evaluate non-technical skills using traditional assessment methods. In addition, teamwork in a digital learning environment is posing new assessment challenges. Current studies of university courses in which technology is a key component tend to focus on the technology part – specifically, on media selection and media effects. Neither of these issues addresses the individual learner (Walther, 1997). In software engineering education, project-based learning under the cyber enabled learning environment focus on determining how to design and build assessment within the perspective of cognitive and situated learning theory. If traditional assessment methods have limited value in evaluating a collaborative, multidisciplinary, geographically distributed team of students, they are ineffective in the measuring of student knowledge and collaboration skills.

All the aforementioned dimensions encompass various concepts, theories, strategies, and tactics that pertain to the motivation to learn (Keller, 1987a). They represent the major idea of the integration approach, which is the synthesis of PBL, TBL with the cyber enabled learning environment. They also provide the basis for the feature of this integration approach, which is the systematic design process that assists instructors in creating motivational tactics that match student characteristics and needs (Keller, 1987b).

As mentioned before, it is important for educators to prepare our college graduates with more skills than just knowledge. The proposed framework carries the following features – self-organized, simple, and easy-to-implement. In the following sections, we will discuss our framework upon the above dimensions.

DESIGN FRAMEWORK PROCESS

The overall idea of the framework is to increase student satisfaction to the content of both technology and nontechnology outcomes. Satisfaction of learners refers to the positive feelings about one's accomplishments and learning experiences. It means that students receive recognition and evidence of success that support their intrinsic feelings of satisfaction and they believe they have been treated fairly. This study is to show that students will demonstrate a high level of satisfaction through the learning process using this framework.

The framework contains a six-phase design process for the development of motivational systems in the work and learning settings (Figure 2). The first two phases – design & development and distribution – are the foundation of the tower, which include the basic components of the process and indicate the further evaluation and overall analysis of the process. These two phases produce information about the status quo and provide the basis for analyzing gaps and their causes. The middle two phases are engagement and implementation. These two phases focus on students' understanding of instructions. In other words, students should know what they need to do through the process of engagement. Implementation is the key component to apply the approach. This is the step to observe results. The last two phases are analysis and evaluation. Based on previous phases, the project's data is collected and analyzed, learning objectives are assessed, and future improvement is identified. These steps are more critical and analytical for the purpose of selecting solutions that best fit time, resources, and other constraining factors in the situation.

Figure 2. The design process

Design and Development

In our approach, it is best to work on specifically defined problems. Design starts with problem domain analysis. The domain will specify scope of the application. The content concentration will specify the concept inventory and related problems. At this initial phase, students should note the key concepts to learn and understand the problems to solve. The time frame and assignments are also necessary to be clarified in this phase. The development of problems and design of teaching strategies are well defined in this phase. The use of cyber enabled learning environment should also be clarified. Often, people will try to deal with other issues of how to improve motivation by adopting a global solution, such as a new set of curriculum materials or an entirely new approach to teaching. This approach may be successful for

a while, but after the novelty wears off, the old motivational problems tend to re-emerge.

After choosing a specific problem to solve, the primary task in the design & development phase is to brainstorm possible solutions. At this point, all potential solutions should be listed without considering their presumed feasibility. The goal, as in any brainstorming process, is to produce as many ideas as possible. In addition, students need to define the ideal solution without constraints. Each problem and ideal solution might be constructed from several specific suggestions utilizing the facilities of cyber enabled learning environment in various degrees. We also encourage students and instructors to use mobile and digital facilities if they have. At this stage, we do not worry about cost, organizational policies, or other constraints that might inhibit the discovery of an ideal solution.

Once the most feasible tactics are chosen and documented properly, the instructor needs to help students integrate them into the framework as an organizational structure with proper grading points. This step upgrades the previous step from encouraging restraintless envision to an ideal and possible solution. At this time, a best possible solution needs to be created by combining ideas and applying several selection criteria pertaining to expense, policy, acceptability, and proportionality.

Distribution

Design & Development provide support for all further activities. At this stage, learners will have a nutrition foundation that is full of digital and technical surroundings. The distributions mainly focus on the dissemination of the bipartite documents to all learners through two ways – digital and traditional ways. Digital distribution involves all possible synchronous and asynchronous methods, broadcasting style and multimedia style vehicles. For example, to distribute documents that do not need feedback, broadcast would be the most efficient way. To distribute documents that need to collect data, synchronous and asynchronous are necessary. In addition, in this phase, students can follow the document and employ the application. Other activities in this phase include documentation plan preparation, media development, developmental reviews, and implementation. As with any effective system development activity, it is important to have motivational tactics and strategies well integrated with other components. For example, tactics such as case studies at the beginning of a class can be a total waste of time if they do not meet specific needs of the audience.

These design & development and distribution phases are comprehensive, interleaving and time-consuming due to two limitations. First, it requires the motivational designer to have quite a bit of knowledge of the different factors represented by the three categories. Second, in reality, various situations where serious challenges always occur and unexpected cases appear frequently are hardly to handle. In some cases, it is highly critical to maximize the effectiveness and performance of a lesson or a course. The full six-phase can be the best approach to follow. But, in many situations these conditions are not met. With teachers or instructional designers, who have little or no formal knowledge of motivational concepts and principles, it would be good to have a simpler model. Such a model has been created and tested in several cyber-related learning environments.

Engagement

It is a challenge issue to engage students in the systematic design process. Two main factors are necessary to consider – level of students and degree of engagement. Engagement activities are varied with the class. Freshmen may be dramatically different from junior and seniors. Sophomore may be different from freshmen. Engagement is a typical issue and highly relevant to student motivations. Degree of engagement is related with the class, student motivation, and learning approach. In this framework, we use this as a separate phase because of our experience. We found that engagement of students can increase student curiosity and motivation, which will results in high quality of product. Otherwise, it is completely opposite. There are two difficulties in determining the degree of engagement since it is a fully subjective topic. First, there is no baseline for engagement. It is hard to find the symptom of non-engagement at the initial or on-going stage of the project. Second, there is no measurement for the degree of engagement. Right now, no research work has been published to evaluate and assess student engagement in an activity.

There are two main styles of integrations – classroom integration and cyber enabled engagement. In the digital learning era, how to engage students through current advanced technology remains a challenge. During lectures, materials and supplements are available online, shifting instructor-centered learning to student-centered learning. This requires students to be more reliable and sustainable during the learning process. Thus the engagement is between students and computers, students and learning environment. The question raised is how to design a cyber enabled learning environment that is interactive with and interesting to students. Many researchers have done some work to solve this problem. For example, WReSTT (Clarke et al., 2012) and its updated version WReSTT-CyLE are Web-based repository and cyber enabled learning environment to teach software testing.

"Students want to engage in technology, especially if it's socially based, whether it's with teachers, students, other schools, or experts around the world," says Julie Evans, CEO of Project Tomorrow. "But they want social interaction that is school-oriented, about serious topics and not the personal 'dramas' of Facebook." (Susan, 2011)

Implementation

After any systematic design process, the framework needs to be applied by the learning systems that involve students and instructors with multiple roles and responsibilities. In this case, motivation is resolved in the framework and combined with the instructors' development design from the beginning phase. Students are able to identify system characteristics and/or gaps which lead to their objectives. In this phase, there are two difficulties in determining the degree and nature of a problem. From our experience, the first and typical problem is that the applications students have chosen could not be realized, at least from the learner's side; and in some cases, the instructor is not aware of this situation. From the pedagogy point of view this is due to the insufficient documents of instruction. From the psychology point of view, this is due to the immature of learners' psycho-experience. The pre-seeing of this situation and tactically handling it is important for the framework to continue function. Students who

do not have and cannot get the skills required to perform satisfactorily will not be able to learn quickly thus cannot succeed to a satisfactory degree. They will develop low expectations for success, or even feelings of helplessness, and will be demotivated as evidenced by lowered levels of effort and performance.

Analysis

As with any systematic approach, the integration framework needs a process of analysis by collecting necessary data and information regarding to the project goal and objectives. The purpose of analysis lies in two aspects – for the assessment of the project, and for the resolution of the problems occurred during project execution.

There are several scientific methods that can be used to analyze the data once the project is evaluated. Without analysis, the evaluation results are meaningless. Usually, analysis is a little difficult and ad hoc. The method changes with the problems and domains. In addition, the problem lies in the nature of the project and is relevant to the project characteristics. Objectively, it is the data that carries the efficiency of content, domain, class, time, numbers of population, and ethnicity. Subjectively, it is varied with motivation, curiosity, maturity, performance and other psychology factors. It follows a curvilinear relationship with objective factors and subjective factors. As objective factors increase, subjective factors decrease to an optimal point. When one analyzes the data of a problem, it is desirable to include as many factors as possible in order to reduce the bias.

Comparable problems occur in other categories of the framework and require tactics to modify learner results into a more productive range. In conducting motivational analysis, it is important to identify the nature of technology gaps in these terms, and to realize that the problems might be different by subgroups or by individuals. It is also important to identify the presence of any positive factors.

Evaluation

Any approach needs a scientific evaluation to assess the results and effectiveness. There are two main groups of evaluation – summative and formative evaluation. This proposed framework has been used by the author in the past five years. A large amount of data has been collected including statistical data, surveys and assessment results. In the following section, we will present some non-technology activities that are implemented at Alabama A&M University using the proposed framework.

APPLICATION OF FRAMEWORK

Our application of the framework is implemented in the software engineering and object-oriented programming classes of Alabama A&M University starting 2008. The application of the framework does not have a comprehensive tool that includes all cyber learning environments. Thus, we utilized the existing available resources: Blackboard + Email + Yahoo Messenger/Google Talk. The Blackboard is updated with several integrated features to facilitate instructors to use. For example, students are automatically enrolled in the courses that are assigned to the instructor in the Blackboard. The Blackboard has included assessment process for quizzes, exam, homework and reports. In addition, it can realize social communications by using grouping and creating discussion forum.

The only weak point for Blackboard is video conference. Blackboard could support multimedia, such as video and audio. However, if the file size is getting bigger, it will be hard for learners to download the video. In addition, you cannot create videos using tools provided in Blackboard. The instructor has to find other ways to solve this problem. In the followings, we will describe how to implement the framework in our courses.

Step 1: Design and Development

Our approach was implemented in two classes at both undergraduate and graduate level. For the design & development, the difference is mainly in the courses, but not in the classification of students. The Software engineering courses offered at Alabama A&M University mainly focuses on software process, project management, design model, quality assurance. Object oriented programming & design course is offered to both undergraduates and graduates. The author implemented this approach at the graduate level course, which covers the principles of object oriented design and programming languages, object oriented design methodology (UML), advanced modeling and analysis.

We chose the application problems in the software intensive systems that include four categories – information systems (GUI design), mobile systems (iPhone, Android apps), robotics systems, and networking. In each category, one to three projects is carefully designed with short description of user requirement. The covered knowledge units are slightly different from category to category. The common knowledge for students is programming, software design and development. The projects aim at improving students' problem solving skills through sequence steps in a software development cycle.

Step 2: Distribution

All projects are distributed in three levels of documents – problem domain selection, project topics regarding to categories, and self-designed projects. The students are instructed to read through the documents in a top-down way from first level to third level. For each problem domain, a short description and sample project areas are listed. Several links of the existing student class projects are referred. After the first level of reading, students get the main ideas of each domain and they could connect them with their own background knowledge and skills. The tasks given to students are to evaluate themselves and find out their interest. After the first level of reading, most students can go through level 2 and select proper topics. To increase collaboration, the topic selection process must be done through a group discussion. The final topic will be reported online by the group coordinator.

Through this study, we found that the group discussion reduced a lot of collaboration problems that might occur later. Certainly, it cannot reduce all of problems, but we found that it completely reduce the problems regarding to project topics and domain areas, which we had in the first year.

Step 3: Engagement

Engagement, defined as "student-faculty interaction, peer-to-peer collaboration and active learning..." (Chen, Gonyea, & Kuh, 2008), has been positively related to the quality of the learning experience. Social learning or learning as part of a group is an important way to help students gain experience in collaboration and develop important skills of critical thinking, self-reflection, and co-construction of knowledge. Email has been a popular way for people to communicate. It has been used as one of the main communication techniques in author's classes. In this study, we adopted three ways of engagement: discussion board, social media, and announcement.

To encourage a broad discussion with participants and learners, the instructor created groups of discussion forums in Blackboard. Students can share information with colleagues and discuss some questions regarding to class contents and projects. The data collected from 2009 does not show a fully involvement of discussion forum in Blackboard. Since 2010, firmware is utilized and added to the class teaching. Several robotics and high level robots are bought and designed for student project. Some group members had problems with software and environment setup. Through the board discussion, some successful

stories were shared to help other students with similar problems.

Collaboration can be done in a specified group or within or outside of the class. Cyber enabled learning environment encourages students to share various types of information with their classmates, increases peer discussion and communication, and helps students in learning of the context in multiple aspects. Synchronous discussion is a key method of student collaboration and communication because of its efficiency in sharing ideas. Students are also encouraged to use social media to communicate, such as Google Talk, Yahoo Messenger, Twitter and Facebook. We expect students to provide the log file and archive for the class record.

Announcement provides a unidirectional way of communication. Only instructor is allowed to create and send announcements to students through Blackboard. The information passed will motivate students to participate in activities and get involved in projects. For example, once a local industrial conference info was passed to students and some students attended the conference, participated in discussions, presented their projects, and built network connections. One student came from the conference commented "it is a great experience... He (she talked to a person in the conference) is very interested in my project ..." Because of this experience, she had been diligently working on her robot project during the remaining four weeks and solved one hard problem by herself.

Step 4: Implementation

As discussed in the first step, this approach was implemented in two classes at both undergraduate and graduate level. Software engineering topics are offered for both graduates and undergraduates and object oriented programming & design is offer for graduates. During the implementation, to reduce the bias, we implemented one class as study group and another class as control group. The class content, homework, quizzes and exams are all the same. The participated student number is shown in Table 1.

Among all projects, robotics projects are the most favorite and were selected by 87% of students. Other projects are purely information system design.

Step 5: Analysis and Assessment

Data was gathered from two classes from Spring 2008 to Spring 2013. The collaborative work has demonstrated more efficient learning results in the concepts, such as UML diagrams, system design model, and analysis. However, it shows that students of study group are not good at understanding and mastering of definitions and terminologies. In contrast, the control group students are good at definition and concepts, but lack of the skills of system design. Students learning outcome is evaluated by homework, exams, and quizzes. We categorize the questions into several groups: definition, programming concepts, design concepts, programming skills, UML syntax, design model, and result analysis. The grading results of control group and study group are listed in Table 2 and Table 3, respectively

Table 1. Number of students participated in the study

	Control Group	Study Group	Number of Projects
Spring 2008	9	10	2
Fall 2008		9	3
Spring 2009	10	11	3
Fall 2009		10	3
Spring 2010	14	11	3
Fall 2010		10	3
Spring 2011	17	12	4
Fall 2011		7	3
Spring 2012	13	9	3
Fall 2012		5	2
Spring 2013	21		0

Table 2. Distribution of student grading results (control group)

Questions	A	B	C	Below C
Definition	75%	15%	10%	0%
Programming concepts	55%	35%	5%	5%
Design concepts	25%	50%	15%	10%
Programming skills	30%	40%	20%	10%
UML syntax	30%	50%	15%	5%
Design model	28%	44%	23%	5%
Result analysis	12%	21%	43%	24%

There is a slight difference in student grades between the control group and study group (shown in Tables 2 & 3). A factor could be that study group implemented collaborative team project and students in this group are evaluated by the items that do not reflect their behavior and contributions to the project. Regarding to this issue, we increased assessment for the collaboration evaluation.

Collaboration skills are evaluated based on learning outcomes and some other non-score features. The group projects were assigned a cumulative grade based on the collaborative group process (a series of documents), the final product (group paper, project demo and project presentation), and peer-and-self evaluations of the collaborative work.

Through the above assessment process of collaborative work, we saw some differences among students in the same project. After carefully analyzing the data and assessment results, we found that students who are involved in the group project but do not contribute well will have poor peer evaluations and perform poorly in the exam and quizzes. These groups of students are the main factor to lower the outcome of project-based learning.

In the end of semester survey, many students responded that they like this class. For example, one student commented "the project (robot) is very interesting, and it really helps me to un-

Table 3. Distribution of student grading results (study group)

Questions	A	B	C	Below C
Definition	85%	10%	5%	0%
Programming concepts	45%	35%	10%	10%
Design concepts	45%	35%	15%	5%
Programming skills	35%	35%	20%	10%
UML syntax	35%	50%	10%	5%
Design model	35%	45%	15%	5%
Result analysis	21%	32%	43%	4%

derstand programming concepts and software design model." Regarding to group project and collaboration skills, some students commented "Group project is fun, I like group discussion and meetings, a lot of ideas come (during meeting). … Leading a group is not simple but tricky, (I) like work with people."

Cyber enable learning environment provides a suitable platform for students to learn and study class content in digital style. Some student commented that "online discussion is good and helpful for our project. Information in blackboard is really helpful and saves us a lot of time." A student who disliked team work at the beginning of the semester commented "online learning helps the teamwork, with this I can get a lot of my work done (faster), now, I like work in a team."

Discussion

In Alabama A&M University, we have implemented this framework and found that cyber enabled learning environment combined with project based learning strategy can improve student collaboration skills, motivate student as well as improve their learning outcomes. Our data may be different from other institutions. However, because of the large amount of minority students (African American students) in Alabama A&M University, we believe our survey results are representatives of other institutions that have similar student populations.

Students were excited for every step of their improvement. In 2009, some students gave the following evaluations for this class: "I like the videos (in the blackboard), it helps me a lot, esp. when I missed some lectures." and "(Materials in blackboard) is helpful for our group discussion, whenever we have some issues, just go to the blackboard, either send message to discussion forum, or send an email to professor, or look at lectures and supplements. It helps our collaboration and team work."

In 2011, robotics project is introduced in this framework. More than 50% students chose the robot project. Before the end of the semester, two students have told the instructor that they got internships from a company after they presented their robot project during the interview. Some students commented "(using blackboard for class teaching is) very good for me, the project (robot) is very interesting, love it!"

CONCLUSION

It is the digital era now and it is the time to adopt cyber enabled learning environment technique to support student learning and improve student skills. Cyber enabled learning environment not only supports gaining knowledge, but is also an important platform for improving the non-technological skills. The world is shifting from hard copy documents to electronic versions. It is the time to make our classroom green – paperless. However, without well thought-out design of classes and context, skill training cannot be achieved in a passive electronic platform. A well design of teaching framework can widely extend the benefit of cyber enabled learning environment. Project-based learning framework presented in this chapter provides a solid foundation for student-centered learning. In addition, this framework can be easily extend to other courses in STEM areas especially for the courses that require more credit hours, because this framework focuses on improving student learning and student engagement in a collaborative, crosscutting process and can be implemented effectively.

REFERENCES

ACT. (2013). *The condition of college & career readiness*. ACT, Inc.

Alliance. (2011). *A time for deeper learning: Preparing students for a changing world*. Alliance for Excellent Education.

Ark, T. (2012). Blended learning can improve working conditions, teaching & learning. *Getting Smart*. Retrieved from http://gettingsmart.com/blog/2012/06/blended-learning-can-improve-working-conditions-teaching-learning/

ASTD. (2005). *ASTD's e-learning glossary*. Learning Circuits.

Callahan, D., & Pedigo, B. (2002). Educating experienced IT professionals by addressing industry's needs. *IEEE Software*, *19*(5), 57–62. doi:10.1109/MS.2002.1032855

Chen, P., Gonyea, R., & Kuh, G. (2008). Learning at a distance: Engaged or not?. *Innovate, 4*(3).

Clarke, P. J., Pava, J., Davis, D., Hernandez, F., & King, T. M. (2012). Using WReSTT in SE courses: An empirical study. In *Proceedings of the 43rd ACM Technical Symposium on Computer Science Education* (pp. 307–312). ACM.

Conn, R. (2002). Developing software engineers at the C-130J software factory. *IEEE Software*, *19*(5), 25–29. doi:10.1109/MS.2002.1032849

Cummings, B. (2007). *Real world careers: Why college is not the only path to becoming rich*. Business Plus.

Editorial. (2010). Graduating by the number: Putting data to work for student success, special issue. *Education Week, 29*(34).

Keller, J. M. (1987a). Strategies for stimulating the motivation to learn. *Performance & Instruction*, *26*(8), 1–7. doi:10.1002/pfi.4160260802

Keller, J. M. (1987b). The systematic process of motivational design. *Performance & Instruction*, *26*(9), 1–8. doi:10.1002/pfi.4160260902

McLester, S. (2011). Building a blended learning program. *District Administration*. Retrieved from http://www.districtadministration.com/article/building-blended-learning-program

McLoughlin, C., Baird, J., Pigdon, K., & Wooley, M. (2000). Fostering teacher inquiry and reflective learning processes through technology enhanced scaffolding in a multimedia environment. In *Proceedings of World Conference on Educational Multimedia, Hypermedia and Telecommunications* (pp. 185–190). Academic Press.

McLoughlin, C., & Luca, J. (2000). Developing professional skills and competencies in tertiary learners through on-line assessment and peer support. In *Proceedings of World Conference on Educational Multimedia, Hypermedia and Telecommunications* (pp. 668–673). Academic Press.

Staker, H. (2011). The rise of K-12 blended learning. *District Administration*. Retrieved from http://www.districtadministration.com/article/building-blended-learning-program

Visser, L. (1998). *The development of motivational communication in distance education support*. (Unpublished doctoral dissertation). Educational Technology Department, The University of Twente, Twente, The Netherlands.

Walther, J. B. (1997). Group and interpersonal effects in international computer-mediated collaboration. *Human Communication Research*, *23*(3), 342–369. doi:10.1111/j.1468-2958.1997.tb00400.x

KEY TERMS AND DEFINITIONS

Cyber Enabled Learning: A teaching environment, where information technology is used as the infrastructure.

Collaboration Skills: Skills that require players to work together to achieve certain goals.

Project Based Learning: A learning approach that involves students to solve a complicated problem.

STEM Programs: Science, Technology, engineering, and mathematics programs.

Technology Based Learning Strategy: Using technology to assist students to achieve their learning objectives.

Chapter 24
Applying Online Learning in Software Engineering Education

Zuhoor Abdullah Salim Al-Khanjari
Sultan Qaboos University, Oman

ABSTRACT

Software Engineering education involves two learning aspects: (1) teaching theoretical material and (2) conducting the practical labs. Currently, Software Engineering education faces a challenge, which comes from the new learning opportunities afforded by the Web technologies. Delivering a Software Engineering curriculum by online distance learning requires innovative and flexible approaches to present and manage the theoretical and practical learning materials. E-Learning could support Software Engineering education through utilizing special e-Learning concepts, techniques, and tools. E-Learning could also change the mode of teaching from knowledge-as-transmission to knowledge-as-construction. This is called "Software Engineering e-Learning." This chapter provides a review on Software Engineering education and e-Learning technology. It explores the need to adopt a Software Engineering e-Learning model to help the facilitators/instructors prepare and manage the online Software Engineering courses. This chapter also addresses how e-Learning environment could simplify the application of the constructivist learning model towards Software Engineering education.

1. INTRODUCTION

Software engineering as a computer science discipline can be deliberated from two perspectives: (1) "Software Engineering e-Learning" and (2) "e-Learning Software Engineering." The former perspective arises when e-Learning concepts, techniques, and tools are used to support software engineering education. However, the later perspective takes place when the software engineering development methodologies and techniques are used for the development of e-Learning Management Systems (LMS). Both aspects have been thoroughly investigated in a large joint project, termed MuSofT (Multimedia in Software Engineering) (Doberkat et al., 2005).

DOI: 10.4018/978-1-4666-5800-4.ch024

Software products are significant assets in daily life of all organizations. Therefore, providing high quality software represents the goal of any software development and maintenance activity. To achieve this objective, it is important to educate IT professionals with high standards of development techniques.

Software Engineering is the sub-discipline and the umbrella of computer science that incorporates various accepted methodologies to design software. It deals with concepts, techniques and tools for supporting the development of high quality software systems. Teaching Software Engineering courses involves the consideration of theoretical and practical parts. In traditional ways, instructors use lectures to discuss the theoretical aspects and use labs to show students how to use and apply the concepts and techniques practically. We consider this situation in the lower level Software Engineering courses. However, the concern here is teaching the high level Software Engineering courses, which involve putting hands on practical work. Although, this is seen as easy as the lower level courses, the fact is there is a need to develop good and professional Software Engineering development team members. Therefore, teaching Software Engineering is considered a difficult process. The reason for this is that the concepts of Software Engineering cannot be realized and understood without a good practice and training on real life problems. The instructors are facing challenges to present this component in normal classrooms because of the need for students to work on industrial projects.

E-Learning concepts and techniques represent a good compensation solution to the above mentioned problem. E-Learning technology empowers the educational institutions to achieve better learning outcomes in a cost-efficient way. Multimedia techniques in particular support teaching of complex data structures and algorithms. Also, animations demonstrate the usage of complex software development tools in an interactive and simple manner. Students can benefit from the e-Learning environment in that the demonstration is available online and can be repeated as needed.

This chapter explores the use of e-Learning technologies to support the delivery of Software Engineering courses online. New technological advances such as high-speed Internet connections, virtual classrooms and virtual labs have removed the barriers limiting the learning opportunities of distance education. This chapter addresses some of the important issues related to utilizing e-Learning in Software Engineering education. It provides a background on the topic. It describes strategic learning models, e-Learning and Learning Management Systems. Further it discusses the importance of employing Software Engineering e-Learning model in Software Engineering education. Finally, it provides concluding remarks and future suggestions.

2. BACKGROUND

Over the last decade, the nature of education has changed significantly following the advances of information technology. The importance of utilizing computers and technology in education was predicted by many researchers (Daniel 1996; Crossman, 1997), who have advocated enhancing student learning by using digital tools, i.e. the e-Learning. The biggest advantage of e-Learning is that it gives students active learning opportunities. For this purpose, the researchers and academics have used several terminologies in education. These technologies include Computer Assisted Instruction (CAI), Computer Based Training (CBT) and Computer Assisted Learning (CAL) (Grieve, 1992; McDonough et al., 1994; Serdiukov, 2000).

Considerable research has been done to study e-Learning. For example, Tsai and Tsai studied the importance of e-Learning (Tsai & Tsai, 2003). Barra & Usman (2013) stressed on the importance of using technology to enhance learning in Software Engineering education. Although e-Learning is promising, many researchers indicated that it needs to be supported with learning strategies (Tsai, 2009). For example, Hadjerrouit (2003) specifically indicated the importance of using the constructivism learning strategy to support e-Learning for Software Engineering education. Linn et al. (2004) and Lazarinis (2004) stated that technological, cognitive and meta-cognitive skills are critical for successful e-learning.

Learning Management System (LMS) is a software system for managing complex educational databases through digital frameworks prepared for managing curriculum, training materials, and evaluation tools. Before the advent of Web technology, LMSs have not been widely used in education in general. Researchers found that most if not all of LMSs deliver the educational material statically without considering the practical perspective of the courses. In their opinions, these systems failed to fulfill the aims of the academics (Golas, 1993; Carswell & Murpfy, 1994). However, due to the advances of technologies in recent years, it is possible to incorporate LMS in any practical-based courses, including Software Engineering.

3. STRATEGIC LEARNING MODELS

The strategic learning model focuses on the learner. It considers the learner to be an active individual, who can process information and construct knowledge (Weinstein, 1998). This model stresses the need of individuals to know how to manage their personal learning process. This could be elaborated further to enable the individuals to know how to plan, monitor, focus on and evaluate their own learning (Cano, 2006). Strategic learning is an approach aims to generate learning in support of future strategic initiatives that will help the organization to grow. In addition, the model uses evaluations to help organizations or groups learn quickly from their work and accordingly adjust their strategies. This could lead to differences in organizational performance. Researchers think that strategic learning has tremendous potential,

particularly for complex and dynamic change strategies (Tsai, 2009). Six important strategic learning models are discussed in this section. Before going into the details, Table 1 briefly summarizes the models.

3.1 Instructivist Learning Model

Instructivist learning model represents a teacher-centered model. In this model, the teacher as the main actor is transferring knowledge to the student. This model is based on the view of exhibiting information to students through the lecture format. It requires students to accept information and knowledge as presented by the instructor. Although, educators consider this model as the basis of education for centuries, it does not satisfy the needs of online class. To make the acquisition of knowledge and skill more efficient, effective and appealing to online education, instructional design is needed (Porcaro, 2011).

Instructional design is the practice of creating needs of the learner, defining the end goal of teaching, and creating some involvement to assist in the transition. The outcome of this kind of teaching may be directly observable and scientifically measurable or completely hidden. There are many instructional design models. They are based on five phases: analysis, design, development, implementation and evaluation. In order to prepare the instructional material, the educators should be aware of the requirements of each phase.

3.2 Objectivist Learning Model

Objectivist learning model states that learning should be reinforced to the learners so that they could produce accurate answers to questions related to the learned information. This model was derived from the works of psychologists such a B. F. Skinner and Robert Gagne, who believed that learning was shaped by reinforcement, where positive reinforcement increases desired behaviors. In education, the reinforcement of learning as a

Table 1. Comparison of the six strategic learning models

Strategic Learning Model	Objective of the Model
Instructivist Learning Model	It involves the student to accept information and knowledge as presented by the instructor.
Objectivist Learning Model	It reinforces learning to the learners to enable them to produce accurate answers to questions related to the learned information. This involves using learning objectives, activities and techniques to evaluate these activities.
Adaptivist Learning Model	It uses computers as interactive teaching devices.
Constructivist Learning Model	It directs the learner to know how to construct new knowledge based on analyzing new information and comparing it to previous knowledge.
Blended Learning Model	It combines face-to-face classroom methods with computer-mediated activities.
Personalized Learning Model	It is a unique, blended educational model that is tailored to the needs and interests of each individual learner.

process should be continued until the skills are mastered by the learners in a hierarchical order (Roblyer & Doering, 2010).

In the objectivist classroom, teachers should identify the objectives, select the proper activities, organize and present the activities in the best possible manner, and finally evaluate the learning. There should be a strong association between the objectives, the learning activities and the evaluation component.

E-Learning technology can help the objectivist educators set up their objectives, plan their activities, present their results, and find the best way of evaluations. This can be a level-based process. Students who perform well move to higher-difficulty level of activities. Others are given more practice on the activities (Vrasidas, 2000).

3.3 Adaptivist Learning Model

Adaptivist learning model is an educational model that uses computers as interactive teaching devices. Computers adapt the presentation of educational material according to students' learning needs, as indicated by their responses to questions and tasks. The motivation of this model is to allow electronic education to incorporate the value of the interactivity between students and an actual human teacher or tutor. The technology encompasses aspects derived from various fields of study including computer science, education, and psychology. In adaptive learning, the basic premise is that the tool or system will be able to adjust to the student/user's learning method, which results in a better and more effective learning experience for the user.

Adaptive learning implemented in the classroom environment using information technology is often referred to as an intelligent tutoring system. Adaptive learning systems have traditionally been divided into separate components as follows.

- **Expert model:** The model with the information which is to be taught
- **Student model:** The model which tracks and learns about the student
- **Instructional model:** The model which actually conveys the information
- **Instructional environment:** The user interface for interacting with the system

E-learning is now incorporating aspects of adaptive learning. Initial systems without adaptive learning were able to provide automated feedback to students who are presented questions from a preselected question bank. That approach however lacks the guidance which teachers in the classroom can provide. Current trends in e-learning recommend the use of adaptive learning to implement intelligent dynamic behavior in the learning environment.

3.4 Constructivist Learning Model

Constructivist learning model directs the learner to know how to construct new knowledge based on analyzing new information and compare it to previous knowledge. In this model, the instructors provide guidance to the learners in the lecture hall. The learner is the most important factor that determines whether or not the knowledge has been gained. The model is learner-centered.

Constructivism helps learners to know how they understand or learn a topic. Learners continually compare new knowledge with old knowledge to see which is better. This process is continued until the knowledge is gained. Examples of constructivist learning activities include building a model, designing a chart and completing a project.

Constructivist model has a set of guiding principles used to help designers and instructors create learner-centered, technology-supported collaborative environments. This kind of environment supports reflective and experiential processes. This model also supports online education.

3.5 Blended Learning Model

Blended learning model is a semi-automated learning model, which combines face-to-face classroom methods with computer-mediated activities. This model represents a formal education program through which a student learns partially online. In this case, the learner receives the knowledge partially through online material and instructions with some control over time, place, path and/or speed. In addition to providing online learning contents and instructions, this model still requires the learner to attend face-to-face classes.

3.6 Personalized Learning Model

The personalized learning model is a unique, blended educational model that is tailored to the needs and interests of each individual learner. In personalized learning, each individual's learning style and interest are identified and then a suitable learning session is offered. Ultimately the personalized learning approach guarantees that all individual learners will be successful in achieving their learning goals. Learning time may vary from one individual learner to another because learning sessions are adjusted according to students learning styles. Personalized learning should also consider several other factors such as the impact of changing technology on learners' psychology. The LMS must be able to provide user friendly personalized learning environment irrespective of any changes in the technology. The nature of personalized learning is that every time a learner personalizes the system, the system should be able to identify the learner's unique set of needs and adapt the environment appropriately.

4. E-LEARNING

Electronic learning (E-Learning) is simply online learning that involves using computers, laptops, mobile devices to access the Internet. E-Learning uses the computer and appropriate software systems to support learners during their learning process, such as understanding new concepts in a certain discipline. Despite of the educational discipline, supporting educational software systems have to be developed. These systems could be complete e-Learning Management Systems, over discipline-specific Internet portals or small dedicated supporting tools (Horton, 2011).

In e-Learning environment the nature of learning mode is basically *self-learning*. In this case, the learners during their learning process could interact with a Learning Management System (LMS) on the individual basis. Since the media of instruction in e-Learning is economical, convenient and disbursable to a larger audience, Web-based instruction is being rapidly embraced by most educational institutions across the world. In engineering education in general and in Software Engineering in particular, Web-mediated techniques are becoming important, since they combine both aspects of computer-assisted learning methodologies with the online access features of the e-Learning at the learners end. This type of delivery mode encourages researchers and instructors of engineering discipline to redefine the concept of distance learning and rethink the delivery of engineering education content.

4.1 E-Learning Strategic Models

In theory, there are two basic e-Learning strategic models, namely instructive e-Learning strategic model and constructivist e-Learning strategic model. The former e-Learning model follows the traditional class-based instruction model. This model provides instructions to the learners without considering their previous knowledge experience. In relation to achieving learning outcomes, this model has several drawbacks including: static nature of course content organization (i.e. one set of content for all learners), low level interaction between the learners and instructors. Since this model provides one general set of instructions in the whole learning path for all learners, the learning

outcome will tend to show large variations among the learners. In this model, the learner with insufficient background knowledge might not be able to reach the expected level of learning. However, with the constructive strategic e-Learning model, the case is different. The learners are allowed to build their own knowledge and use their own reasoning power. In this model, each learner is encouraged to follow a suitable learning path that fits to the learner's preferences, interests, reasoning skills and background knowledge (Tsai, 2009).

Online learning environment has several facets including the World Wide Web (WWW), email, asynchronous discussion forum and synchronous discussion forum. Asynchronous discussion forum by itself could be expanded to BBS, mailing list and newsgroup. On the other hand, synchronous discussion forum could be expanded to online chat room, video conference and online games (Linn, Davis & Bell, 2004). Figure1 shows the online learning environment and its features.

The features of online learning environment illustrated in Figure1 are explained as follows.

Figure 1. Online learning environment and its features

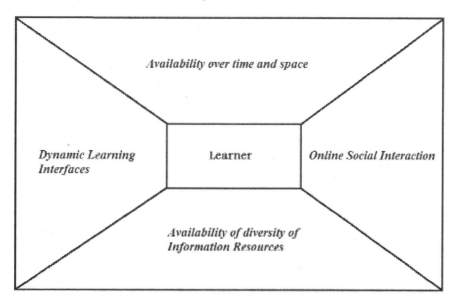

- **Availability over time and space:** Learners can perform the online learning tasks at any time or place provided that they have Internet access. Therefore, online learners become more free, flexible and convenient with respect to time and location. This requires learners to be aware of the environment and responsible for controlling and monitoring their time and learning.

- **Online social interaction:** Learners can benefit from the synchronous communication technologies significantly. Also, learners can use the asynchronous communication applications to elaborate their communication skills and freely interact with the instructor and other learners.

- **Availability of diversity of information resources:** Online learners need abundance and diversity of information resources in their Internet-based learning. They can utilize online information immediately from the resources.

- **Dynamic learning interfaces:** Online learners need to use interfaces that changes with time dynamically. This feature significantly influences learners' motivation, attitude and achievement of online learning.

4.2 Learning Management Systems

Learning Management System enables any organization to develop electronic coursework and manage its continued usage, reachability and flexibility over time. The LMS has become a powerful and popular tool for organizations that are specialized in staffing, training and enhancing the continuing education of its workforce. To achieve a successful individualized learning, LMS must be able to support each individual with a personalized instruction or personalized learning environment. This could be considered as the ultimate solution for successful self-learning. LMS continues to evolve and adapt to new learning challenges and technological capabilities. Current trends in technology and business are favoring migration of data storage to network-based storage known as the cloud.

4.3 Four Components for Successful E-Learning Course Design

Technology opens the door to electronic learning that is becoming more accessible and far reaching. E-learning design requires a succession of steps that is known as a critical path (Horton, 2011). Effective e-learning is the key to success for the course designers, technology providers, as well as the learners. To address learners' need, the course designers and the technology providers must focus on four pillars: instructional design, media design, software engineering, and economics. The following provides a brief discussion of each.

Instructional design takes into account of learning objectives and learning models. To increase learners' interest towards successful online learning, instructional design could use various learning technologies, including: simulations, discussion boards, peer editing, video conferencing, and online video lectures (Horton, 2011). The foundation of design is an understanding of how students learn, which varies by many factors. This influences instructional design since designers try to cover as many learning theories as possible. In fact, one theory could not address the needs of all learners. Therefore, a combination of theories must be utilized.

Media design is the second step in e-Learning design. This step considers the design of multimedia material delivered through computers such as audio, video, graphics, text, and anything deliverable through that format (Horton, 2011). Researchers found that e-learners were in favor of using e-Learning materials supported by the media. Information presented in text – audio – video format resulted in a higher perceived usefulness and concentration.

Using Software Engineering techniques and methodologies is core to a successful e-Learning platform. It has been realized that e-Learning does not mean scanning traditional classroom materials and transforming them into a digital format to be presented as an e-Learning course. The outcome of this process will not be same as those achieved through using quality e-Learning software. Preparing curriculum for e-Learning involves skilled designers and programmers (Ehlers, 2004). Teachers and educators should be trained on how to use technology to achieve a successful preparation of e-Learning course. Quality of an online course, such as usability, affects the learning outcome.

Economics is an important issue to be considered in a successful e-Learning design (Ehlers, 2004). The cost of delivering electronic learning includes (1) cost of organizations who provide the e-Learning environment, (2) cost of designers or course facilitators who design the e-Learning course, and (3) cost of E-Learning software providers that provide the technology and are responsible for all technical issues including maintenance.

5. SOFTWARE ENGINEERING E-LEARNING

To reach a successful implementation of e-Learning courses in reality, there is a need for careful consideration of the learning models and how to apply them online. This section discusses the issue of how e-Learning and the related technology can support Software Engineering education. E-Learning enables instructors and facilitators of Software Engineering courses to create dynamic educational courseware (Duggins & Walker, 2009). In these e-Learning courses, educational material can be categorized and thus be easily retrieved and associated with other available materials and components. E-Learning facilities allow instructors to update their course contents frequently. It also facilitates and controls interactions between the learners among themselves on one side and the learners and educators on the other side. Archi-

tecture of the Software Engineering e-Learning course should be prepared by the instructors or the facilitators. Once the architecture is setup, instructors can prepare the courseware, through which the material can be uploaded and used by the learners. For the future need, instructors can use the architecture of the Software Engineering e-Learning course and update the material as needed. We will discuss the strategic learning models, which could be used for e-Learning in Software Engineering education. It provides some hints on the required contents of the e-Learning Software Engineering course along with some e-Learning tools, which supports Software Engineering education.

5.1 Learning Models for E-Learning in Software Engineering

Traditional education depends on the objectivist learning model in which the learning process is transmitted from the teacher as information provider to the learner as information receiver. This model could not satisfy the new education vision, which states that the learners should construct their knowledge through the learning process under instructors' guidance. To achieve that goal, the instructor should provide the material to the learners and monitor the learning process rather than providing the knowledge to the learner in a spoon feeding manner. For this purpose, it was found that moving from the objectivist learning model towards the constructivist learning model represents an urgent need for Software Engineering education. Hadjerrouit (2003) discussed how to move towards a constructivist model to e-Learning in Software Engineering. Clearly, researchers and instructors need a clear understanding of the constructivist principles, constructivist characteristics, and guidelines that are useful for the development of Software Engineering course Web sites. Overall, the LMS needs to maintain a suitable learning content management for the purpose of presenting the constructivist model approach in Software Engineering courseware.

5.2 Content of E-Learning Courseware for Software Engineering Courses

E-learning tools are important components of the e-Learning courses. It is recommended that the following in e-Learning courseware should be included in online Software Engineering courses.

- Syllabus, question bank
- Assignments, quizzes
- Lectures notes
- Feedback form
- Important links to SEI (Software Engineering Institute)
- Important links to technical societies (e.g. IEEE Computer Society Chapter)
- Important link to QAI (Quality Assurance Institute), which offers a lot of e-learning courses.
- Books & their authors site
- Teaching resource
- Projects & instructions
- Lab manuals
- Important links to journals.

A virtual classroom represents an online learning environment. It is like a simulation of a real-world classroom. In this type of classrooms, learners and instructors participate in synchronous instruction. This requires both instructor and learner to log into the virtual learning environment at the same time. Virtual classroom employs software applications for synchronous technologies, including: Web conferencing, video conferencing, and live streaming. On the other hand, to enhance the educational process, virtual classroom can use software applications for asynchronous technologies, including: message boards and chat capabilities. Using asynchronous communication, the leaner can access the course material without actually being present virtually online.

Since in a constructivist classroom, the task of the instructor is to provide materials and guide the learners in ways that encourage students to build their own knowledge, the constructivist approach can be provided in a virtual learning environment. This could be realized to improve the outcomes of Software Engineering education.

Social collaboration helps learners to interact and share information. The learners can use this kind of collaboration to achieve their common goals. An important concept of social collaboration is that different learners can share their knowledge together through group communications. This helps the learners to understand the concepts based on the knowledge collected from the communication.

Constructivist approach which is proposed to be used in Software Engineering education requires the learners to acquire knowledge constructively. Actually, this approach views the learning process as individual learners obtaining different knowledge and experience. These learners make unique connections in building their knowledge by questioning each other's understanding and explaining their own perspectives. The use of group-based cooperative learning activities can be adopted in Software Engineering education.

5.3 Case Study

This section describes a case study, Moodle online Software Engineering course, which is developed by Department of Computer Science at Sultan Qaboos University (SQU) in the Sultanate of Oman. The course helped the students to overcome some of the difficulties with respect to time and pedagogical aspects.

Figure 2 shows a snapshot of the main page of the course. It is organized as sections and represented with icons. The sections include the course outline, course notes, evaluation tools, communication tools, and study tools. The main page also includes side screens that help instructors to manage the course and students to use the course. These side screens include: participants, activities, administrative aspects, upcoming events, and calendar.

Figure 2. The main page of the Software Engineering online course, SQU

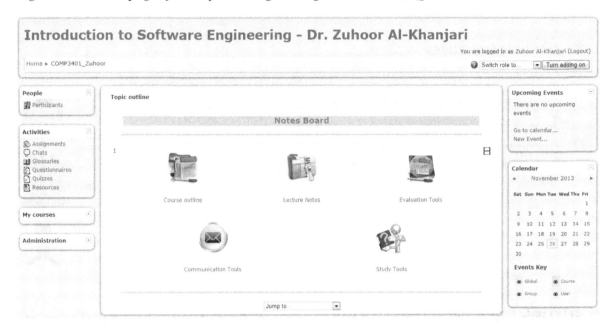

Students can use the provided link to view, download and/or print the course outline. The outline of the course is a PDF document uploaded by the instructor to the specified location in the Learning Management System—Moodle.

Although the lectures are delivered in the traditional lecture rooms, the lecture notes are kept in Moodle for online access. These lectures can be organized based on chapters, weeks, etc. The author organized the lectures based on the chapters. The online lecture notes are available anytime they are needed.

In an extra effort, the author requested the Center for Educational Technology (CET) in Sultan Qaboos University to help in trying to change the mode of the available lecture notes. The lecture notes were prepared based on an interactive mode so that they are attractive to students. This new mode demonstrates the lecture in a constructivism manner. Students can get a constructive list of topics so that they can manage their understanding of the chapter according to their needs. Also this mode of the lecture notes was supported with small exams to monitor the learning process. The students responded quite positively to the new introduced mode. Figure 3 illustrates the interactive

module of Chapter 1 of the Software Engineering online course. The topic could be an explanation, an example, or a sample exam.

Students can benefit from the online exam to test their understanding of certain knowledge. The tool used for the online exam could show immediate response on the answers whether they are correct or wrong and provide the marks immediately after the exam is finished.

Students in each team can download the assignment which originally was uploaded by the instructor. After solving it, the leader of the student team should upload the solution. Chatrooms are used for instructors to communicate with the students or for students to communicate between each other. The email tool could be used to send emails and notifications to the students if needed.

Our experience shows E-Learning technology is capable of enhancing Software Engineering education. The available tools for interactive communication and collaboration between the instructor and learners and among the learners themselves presented a new horizon in the Software Engineering education.

Figure 3. The interactive module of Chapter1 of the Software Engineering online course, SQU

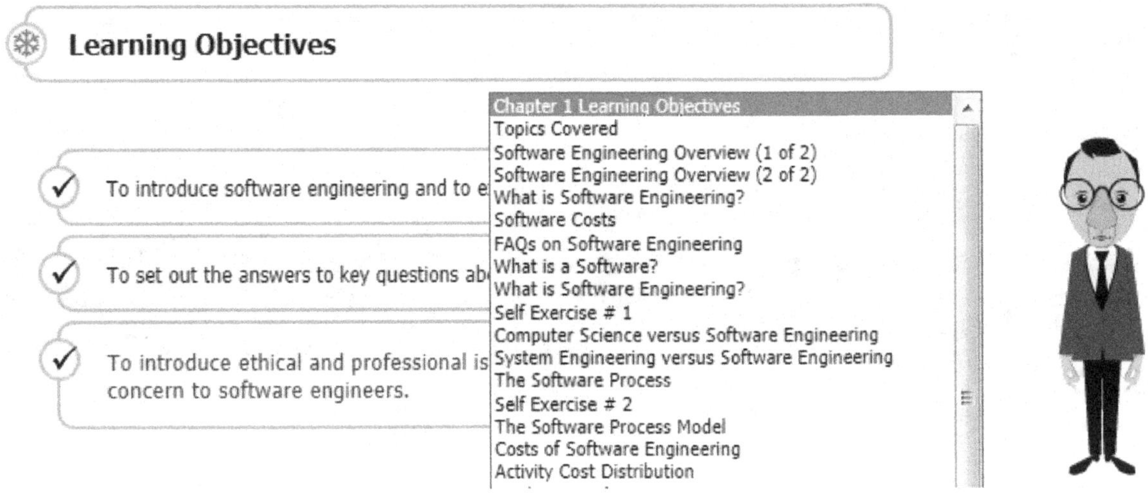

6. FUTURE RESEARCH DIRECTIONS

As for future research, we suggest the construction of a template e-Learning courseware for the Software Engineering education. This represents a good topic to be investigated. On the other hand the actual application of the environment needs to be further studied and explored. As a step forward in this direction, Figure 4 illustrates our plan to plug and use software tools in an online environment.

The purpose of plugging software tools in an online environment (e.g. Moodle) is to enhance the practical side of the online courses. This can be done in three ways

- **Software tool can be provided as a URL link in the Software Engineering online course:** Students can access the online course and use the URL link to download the software tool in their computers. Students should have the required hardware facilities in their computers to be able to use the software tool. More details about this can be found in (Al-Khanjari et al., 2006).

- **Software tool can be installed as an application in the Software Engineering online course:** The software tool should be installed and enabled to run under the Moodle environment. The students should be able to use the online software tool in the Learning Management Systems' environment.

- **Software tool can be called as a service from the services' provider in the World Wide Web:** Once students need the tool, they should click on the link. As a response, Moodle should take them to the service provider's environment and open the software tool application online. Students should be directed to the Moodle environment to complete their work if needed.

7. CONCLUSION

In this chapter, we described e-Learning technologies in Software Engineering courses. Online learning allows remote learners to interact with each other and with the representations of the

Figure 4. Plugging software tools to the Software Engineering online course, SQU

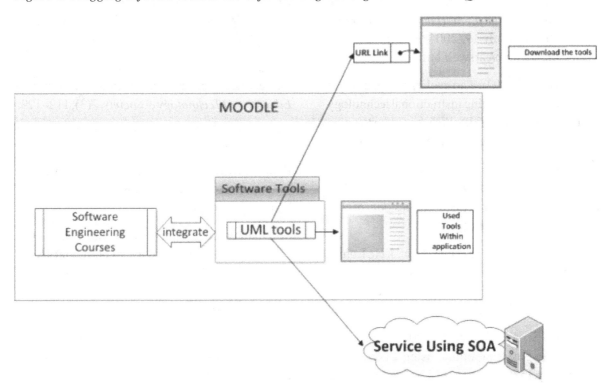

subject matter in a form that is different from the traditional classroom-based learning method. However, the role of the technology here is primarily to get remote learners into a position to learn as favorably as if they were campus-based, rather than offering a new teaching method.

In practice, it is important to understand the essential learning assumptions. In such a case the enhancement should be seen as realistic rather than pedagogic. Also the enhancement can be seen as achieving cost effective access to learning, rather than a new way to achieve deep understanding of a concept. The value of using educational tools and strategies may be enormous if it could be exploited through an educational infrastructure.

REFERENCES

Al-Khanjari, Z. A., Kutti, N. S., & Hatam, M. (2006). An extended e-learning architecture: Integrating software tools within the e-learning portal. *The International Arab Journal for Information Technology, 3*(1).

Barra, P., & Usman, S. (2013). *Technology enhanced learning and assessment.* Paper presented at the HEA Biosciences New-to-Teaching Workshop. Kingston, Jamaica.

Cano, F. (2006). An in-depth analysis of the learning and study strategies inventory (LASSI). *Educational and Psychological Measurement, 66*(6), 1023–1038. doi:10.1177/0013164406288167

Carswell, L., & Murphy, M. (1994). *Pragmatic methodology for educational courseware development*. Retrieved July 19, 2004 from http://www.ulst.ac.uk/cticomp/carswell.html

Crossman, D. (1997). The evolution of the world wide web as an emerging instructional technology tool. In B. H. Khan (Ed.), *Web-based instruction* (pp. 19–23). Educational Technology Publications.

Daniel, J. S. (1996). *Mega-universities and knowledge media: Technology strategies for higher education*. London: Keegan Press.

Doberkat, E., Engels, G., Hausmann, J. H., Lohmann, M., Pleumann, J., & Schröder, J. (2005). Software engineering and eLearning: The MuSofT project. *Language*, *1*, 2.

Duggins, S., & Walker, R. (2009). Teaching software engineering online using 21st century technology. In *Proceedings of ASEE Southeastern Section Annual Conference*. ASEE.

Ehlers, M. (2004). Deconstructing the axon: Wallerian degeneration and the ubiquitin-proteasome system. *Trends in Neurosciences*, *27*(1), 3–6. doi:10.1016/j.tins.2003.10.015 PMID:14698600

Golas, K. C. (1993). *Estimating time to develop interactive courseware in the 1990s*. Retrieved July 19, 2004 from http://www.coedu.usf.edu/inst_tech/resources/estimating.html

Grieve, C. (1992). Knowledge increment assessed for three methodologies of teaching physiology. *Medical Teacher*, *14*(1), 27–31. doi:10.3109/01421599209044011 PMID:1608324

Hadjerrouit, S. (2003). Toward a constructivist approach to e-learning in software engineering. In A. Rossett (Ed.), *Proceedings of World Conference on E-Learning in Corporate, Government, Healthcare, and Higher Education* (pp. 507–514). Academic Press.

Horton, W. (2011). *E-learning by design*. Hoboken, NJ: Wiley. doi:10.1002/9781118256039

Lazarinis, F. (2004). A template based system for automatic construction of online courseware for secondary educational institutes. *Journal of Educational Technology & Society*, *7*(3), 112–123.

Linn, M. C., Davis, E. A., & Bell, P. (2004). Inquiry and technology. In M. C. Linn, E. A. Davis, & P. Bell (Eds.), *Internet environments for science education* (pp. 3–28). Mahwah, NJ: Lawrence Erlbaum Associates.

McDonough, D., Strivens, J., & Rada, R. (1994). Current development and use of computer-based teaching at the University of Liverpool. *Computers & Education*, *22*(4), 335–343. doi:10.1016/0360-1315(94)90055-8

Porcaro, D. (2011). Applying constructivism in instructivist learning cultures. *Multicultural Education & Technology Journal*, *5*(1), 39–54. doi:10.1108/17504971111121919

Roblyer, M. D., & Doering, A. H. (2010). *Integrating technology into teaching*. Boston, MA: Pearson Education, Inc.

Serdiukov, P. (2000). Educational technology and its terminology: New developments in the end of the 20th century. In J. Bourdeau & R. Heller (Eds.), *Proceedings of ED-MEDIA* (pp. 1022–1025). AACE.

Tsai, M. J. (2009). The model of strategic e-learning: Understanding and evaluating student e-learning from metacognitive perspectives. *Journal of Educational Technology & Society*, *12*(1), 34–48.

Tsai, M. J., & Tsai, C. C. (2003). Information searching strategies in web-based science learning: The role of internet self-efficacy. *Innovations in Education and Teaching International*, *40*(1), 43–49. doi:10.1080/1355800032000038822

Vrasidas, C. (2000). Constructivism versus objectivism: Implications for interaction, course design, and evaluation in distance education. *International Journal of Educational Telecommunications*, 6(4), 339–362.

Weinstein, C. E., & McCombs, B. L. (1998). *Strategic learning: The merging of skill, will and self-regulation*. Hillsdale, NJ: Lawrence Erlbaum Associates. doi:10.1037/10296-001

KEY TERMS AND DEFINITIONS

Adaptivist Learning Model: Adaptivist learning model known as computer-based learning model is an educational model which uses computers as interactive teaching devices.

Constructivist Learning Model: Constructivist learning model directs the learner to know how to construct new knowledge based on analyzing new information and comparing it to previous knowledge.

E-Learning: Electronic learning (E-Learning) is simply online learning which involves using computers, laptops, mobile devices to access the Internet.

E-Learning Software Engineering: This perspective, takes place when the software engineering development methodologies and techniques are used for the development of e-Learning Learning Management Systems (LMSs).

Instructivist Learning Model: This model represents a teacher-centered model. In this model, the teacher as the main actor is transferring the knowledge to the student.

Learning Management Systems (LMS): Learning Management System (LMS) is a powerful software system for managing complex educational databases through digital frameworks prepared for managing curriculum, training materials, and evaluation tools.

Objectivist Learning Model: Objectivist learning model states that learning should be reinforced to the learners to enable them to produce accurate answers to questions related to the learned information.

Software Engineering: Software Engineering is defined as the sub-discipline of computer science that deals with concepts, techniques and tools for supporting the development of high quality software systems.

Software Engineering E-Learning: In this perspective, e-Learning concepts, techniques, and tools are used to support Software Engineering education.

Strategic Learning Models: Strategic learning is an approach aims to generate learning in support of future strategic initiatives that will help the organization to grow. In addition to this, the model uses evaluation to help organizations or groups learn quickly from their work and accordingly adjust their strategies.

Virtual Classroom: A virtual classroom represents an online learning environment. It is like a simulation of a real-world classroom.

Compilation of References

ABET. (2012-2013). *Criteria for accrediting engineering programs*. Retrieved November 2013, from http://www.abet.org/DisplayTemplates/DocsHandbook.aspx?id=3143

ABET. (2012a). *Criteria for accrediting computing programs, 2013 - 2014*. Retrieved from http://www.abet.org/DisplayTemplates/DocsHandbook.aspx?id=3148

ABET. (2012b). *Criteria for accrediting engineering programs, 2013-2014*. Retrieved from http://www.abet.org/DisplayTemplates/DocsHandbook.aspx?id=3149

Abeysekera, I. (2006). Issues relating to designing a work-integrated learning program in an undergraduate accounting degree program and its implications for the curriculum. *Asia-Pacific Journal of Cooperative Education*, 7(1), 7–15.

Abran, A., Bourque, P., Dupuis, R., Moore, J. W., & Tripp, L. (2004). *Guide to the software engineering body of knowledge (SWEBOK)*. Los Alamitos, CA: IEEE Computer Society Press.

Abran, A., & Moore, J. W. (2004). *Guide to the software engineering body of knowledge*. IEEE Computer Society.

ACCI. (2002) Employability skills - An employer perspective: Getting what employers want out of the too hard basket. *ACCI Review, 88*.

Accreditation Board for Engineering and Technology (ABET). (2013). *Criteria for accrediting engineering programs*. ABET Inc.

ACM. (n.d.). *Code of ethics*. Retrieved from http://www.acm.org/about/code-of-ethics

ACM/IEEE. (2008). *Computer science curriculum 2008: An interim revision of CS 2001*. ACM/IEEE Computer Society.

ACM-Curricula-IS. (2002). *IS 2002: Model curriculum and guidelines for undergraduate degree programs in information systems. Association for Computing Machinery (ACM), Association for Information Systems (AIS), Association of Information Technology Professionals*. AITP.

ACM-Curricula-IS. (2010). *IS 2010: Curriculum guidelines for undergraduate degree programs in information systems. Association for Computing Machinery (ACM), Association for Information Systems*. AIS.

ACM-Curricula-MSIS. (2006). *Model curriculum and guidelines graduate degree programs in information systems. Association for Information Systems*. AIS.

ACT. (2013). *The condition of college & career readiness*. ACT, Inc.

Agile Manifesto . (n.d.). Retrieved from http://www.agilemanifesto.org/

Agilej . (n.d.). Retrieved from http://www.agilej.com/

Ahmed, O. (2005). *Migrating from proprietary to open source learning content management systems*. (PhD thesis). Carleton University, Ottawa, Canada.

Al-Ajlan, A., & Zedan, H. (2008). Why Moodle. In *Proceedings of 12th International Workshop on Future Trends of Distributed Computing Systems* (pp. 58–64). IEEE.

Al-Asmari, K., & Yu, L. (2006). Experiences in distributed software development with wiki. In *Proceedings of Software Engineering Research and Practice* (pp. 389–293). Academic Press.

Albanese, R., & van Fleet, D. D. (1985). Rational behavior in groups: The free-riding tendency. *Academy of Management Review, 10*(2), 244–255.

Aldrich, C. (2005). *Learning by doing: A comprehensive guide to simulations computer games, and pedagogy in e–learning and other educational experiences.* Hoboken, NJ: Wiley.

Alegría, J. A. H., & Bastarrica, M. C. (2006). Implementing CMMI using a combination of agile methods. *CLEI Electronic Journal, 1*(1).

Alfonso, M. I., & Botia, A. (2005). An iterative and agile process model for teaching software engineering. In *Proceedings of 18ᵗʰ Conference on Software Engineering Education and Training.* Academic Press.

Ali, M. R. (2006). Imparting effective software engineering education. *ACM SIGSOFT Software Engineering Notes, 31*(4), 1–3. doi:10.1145/1142958.1142960

Al-Khanjari, Z. A., Kutti, N. S., & Hatam, M. (2006). An extended e-learning architecture: Integrating software tools within the e-learning portal. *The International Arab Journal for Information Technology, 3*(1).

Alliance. (2011). *A time for deeper learning: Preparing students for a changing world.* Alliance for Excellent Education.

Ally, M. (2004). Foundations of educational theory for online learning. In *Theory and practice of online learning* (pp. 3–31). Athabasca University, Canada's Open University.

Amabile, T. M. (1996). *Creativity in context.* Boulder, CO: Westview Press.

Ambler, S., & Lines, M. (2012). *Disciplined agile delivery: A practitioner's guide to agile software delivery in the enterprise.* IBM Press.

Anderson, D. J. (2010). *Kanban: Successful evolutionary change for your technology business.* Blue Hole Press.

Anderson, R. E., Johnson, D. G., Gotterbarn, D., & Perrolle, J. (1993). Using the new ACM code of ethics in decision making. *Communications of the ACM, 36*(2), 98–107. doi:10.1145/151220.151231

Anderson, T. (Ed.). (2008). *The theory and practice of online learning.* Athabasca University Press.

Andrews, J., & Higson, H. (2008). Graduate employability, 'soft skills' versus 'hard' business knowledge: A European study. *Higher Education in Europe, 33*(4), 411–422. doi:10.1080/03797720802522627

Android Tips. (n.d.). Retrieved from http://tools.android.com/tips/lint

Android. (n.d.). Retrieved from http://developer.android.com/tools/sdk/eclipse-adt.html

Ant. (n.d.). Retrieved from http://ant.apache.org/

Antunes, C. (2010). Anticipating student's failure as soon as possible. In C. Romero et al. (Eds.), *Handbook of educational data mining.* CRC Press. doi:10.1201/b10274-28

Apache. (n.d.). Retrieved from http://xml.apache.org/

Apelo, J. (2010). *Management 3.0: Leading agile developers, developing agile developers.* Addison-Wesley.

Appaloosa. (n.d.). Retrieved from http://www.appaloosa-store.com/

ARC. (n.d.). Retrieved from http://www.arc.gov.au/applicants/rib.htm

Argyris, C., & Schön, D. A. (1974). *Theory in practice: Increasing professional effectiveness.* San Francisco, CA: Jossey Bass.

Ark, T. (2012). Blended learning can improve working conditions, teaching & learning. *Getting Smart.* Retrieved from http://gettingsmart.com/blog/2012/06/blended-learning-can-improve-working-conditions-teaching-learning/

ASTD. (2005). *ASTD's e-learning glossary.* Learning Circuits.

Atchison, M., Pollock, S., Reeders, E., & Rizetti, J. (2002). *Work integrated learning paper.* Retrieved November 2013, from http://mams.rmit.edu.au/a0o4e48729rdz.pdf

Atlassian. (n.d.). Retrieved from https://www.atlassian.com/software/jira

Augar, N., Raitman, R., & Zhou, W. (2004). Teaching and learning online with wikis. In R. Atkinson, C. McBeath, D. Jonas-Dwyer, & R. Phillips (Eds.), *Beyond the comfort zone: Proceedings of the 21st Ascilite conference* (pp. 95–104). Perth, Australia: Australasian Society for Computers in Learning in Tertiary Education.

Avison, D., & Wilson, D. (2001). A viewpoint on software engineering and information systems: What we can learn from the construction industry? *Information and Software Technology*, *43*(13), 795–799. doi:10.1016/S0950-5849(01)00186-0

Baker, A. (2003). *Problems and programmers*. (Honors Thesis). Department of Informatics, School of Information and Computer Science, University of California, Irvine, CA.

Baker, A., Oh Navarro, E., & Van Der Hoek, A. (2005). An experimental card game for teaching software engineering processes. *Journal of Systems and Software*, *75*(1), 3–16. doi:10.1016/j.jss.2004.02.033

Banks, D. A. (2003). Belief, inquiry, argument and reflection as significant issues in learning about information systems development methodologies. In T. McGill (Ed.), *Current issues in IT education* (pp. 1–10). Hershey, PA: IRM Press.

Barbi, E., Cantone, G., Falessi, D., Morciano, F., Rizzuto, M., Sabbatino, V., & Scarrone, S. (2012). A model-driven approach for configuring and deploying systems of systems. In *Proceedings of the 7th International Conference on System of Systems* (pp. 214–218). IEEE.

Barbosa, E. F., & Maldonado, J. C. (2006). Towards the establishment of a standard process for developing educational modules. In *Proceedings of the 36th ASEE/IEEE Frontiers in Education Conference*. San Diego, CA: ASEE/IEEE.

Barra, P., & Usman, S. (2013). *Technology enhanced learning and assessment*. Paper presented at the HEA Biosciences New-to-Teaching Workshop. Kingston, Jamaica.

Barrie, S. C. (2004). A research-based approach to generic graduate attributes policy. *Higher Education Research & Development*, *23*(3), 261–275. doi:10.1080/0729436042000235391

Barros, M., & Araújo, R. (2008). Ensinando construção de software aplicada a sistemas de informação do mundo real. In *Proceedings of the 1st Forum on Education in Software Engineering*. Campinas, Brazil: Academic Press.

Barrows, H. S. (1984). A specific, problem-based, self-directed learning method designed to teach medical problem-solving skills, self-learning skills and enhance knowledge retention and recall. In H. G. Schmidt, & M. L. De Volder (Eds.), *Tutorials in problem-based learning* (pp. 16–32). Van Gorcum.

Barrows, H. S. (1986). A taxonomy of problem-based learning methods. *Medical Education*, *20*, 481–486. doi:10.1111/j.1365-2923.1986.tb01386.x PMID:3796328

Barrows, H. S. (1996). Problem-based learning in medicine and beyond: A brief overview. *New Directions for Teaching and Learning*, *68*, 3–12. doi:10.1002/tl.37219966804

Barrows, H. S. (1998). The essentials of problem-based learning. *Journal of Dental Education*, *62*(9), 630–633. PMID:9789484

Barrows, H. S. (2000). *Problem-based learning applied to medical education*. Southern Illinois University School of Medicine.

Barrows, H. S. (2002). Is it truly possible to have such a thing as DPBL? *Distance Education*, *23*(1). doi:10.1080/01587910220124026

Barrows, H. S., Myers, A. N. N., & Williams, R. G. (1986). Large group problem-based learning: A possible solution for the 2 sigma problem. *Learning*, *8*(4), 325–331.

Barrows, H. S., & Tamblyn, R. M. (1977). The portable patient problem pack: A problem-based learning unit. *Journal of Medical Education*, *52*(12), 1002–1004. PMID:926146

Barrows, H. S., & Tamblyn, R. M. (1980). *Problem-based learning: An approach to medical education*. New York: Springer.

Barrows, H. S., & Wee, K. N. L. (2010). *Principles and practice of a PBL*. Springfield, IL: Southern Illinois University School of Medicine.

Bass, L., Clements, P., Kazman, R., & Klein, M. (2008). *Models for evaluating and improving architecture compe-tence (TR CMU/SEI-2008-006)*. Pittsburgh, PA: Software Engineering Institute.

Bass, L., Kazman, R., & Clements, P. (2012). *Software architecture in practice*. Reading, MA: Addison-Wesley.

Bavota, G., De Lucia, A., Fasano, F., Oliveto, R., & Zottoli, C. (2012). Teaching software engineering and software project management: An integrated and practical approach. In *Proceedings of the 2012 International Conference on Software Engineering* (pp. 1155–1164). IEEE Press.

Baxter, G. P., & Shavelson, R. J. (1994). Science per-formance assessments: Benchmarks and surrogates. *International Journal of Educational Research, 21*(3), 279–298. doi:10.1016/S0883-0355(06)80020-0

Baxter-Magnola, M. B. (2001). A constructivist revision of the measure of epistemological reflection. *Journal of College Student Development, 42*(6), 520–534.

Beasley, R. E. (2003). Conducting a successful senior capstone course in computing. *Journal of Computing Sciences in Colleges, 19*(1), 122–131.

Beck, K. (2000). *Extreme programming eXplained: Em-brace change*. Reading, MA: Addison-Wesley.

Beck, K., & Andres, C. (2004). *Extreme programming explained: Embrace change*. Reading, MA: Addison-Wesley Longman.

Beck, K., Beedle, M., Van Bennekum, A., Cockburn, A., Cunningham, W., Fowler, M., & Thomas, D. (2001). *Mani-festo for agile software development*. Academic Press.

Belbin. (n.d.). Retrieved from http://www.belbin.com/rte.asp?id=8

Belbin, M. (2010). *Management teams: Why they succeed or fail*. Oxford, UK: Butterworth Heinemann.

Belbin, R. M. (2010b). *Team roles at work* (2nd ed.). Oxford, UK: Butterworth-Heinemann.

Belland, B. R. (2010). Portraits of middle school students constructing evidence-based arguments during problem-based learning: The impact of computer-based scaffolds. *Educational Technology Research and Development, 58*, 285–309. doi:10.1007/s11423-009-9139-4

Benamati, J. H., Zafer, D., Ozdemir, Z. D., & Smith, H. J. (2010). Aligning undergraduate IS curricula with industry needs. *Communications of the ACM, 53*(3), 152–156. doi:10.1145/1666420.1666458

Benarek, G., Zuser, W., & Grechenig, T. (2005). Functional group roles in software engineering teams. In *Proceedings of the ACM Workshop on Human and Social Factors of Software Engineering* (pp. 1–7). ACM.

Benne, K. D., & Sheats, P. (1948). Functional roles of group members. *The Journal of Social Issues, 4*(2), 41–49. doi:10.1111/j.1540-4560.1948.tb01783.x

Benson, S. (2003). Metacognition in information systems education. In T. McGill (Ed.), *Current issues in IT educa-tion* (pp. 213–222). Hershey, PA: IRM Press.

Bentley, J. F., Lowry, G. R., & Sandy, G. A. (1999). Towards the compleat information systems graduate: A problem based learning approach. In *Proceedings of the 10th Australasian Conference on Information Systems*.

Benyon, J. (2012). *Response to the senate inquiry: The shortage of engineering and related employment skills, on behalf of the Australian Council of Engineering Deans Inc.* Retrieved from http://www.aph.gov.au/DocumentStore.ashx?id=14789c30-d180-4e1e-973e-592350643e39&ei=FL4WUvSuEae4iAfq6IHQCA&usg=AFQjCNE3qHqBqR8dMFlZ2HKZvDJ6ocOijQ&sig2=XEIzLBZ_xLPdcs1ukes2QQ&bvm=bv.51156542,d.aGc&cad=rja

Bergandy, J. (2008). Software engineering capstone project with rational unified process (RUP). In *Proceedings of the 38th ASEE/IEEE Frontiers in Education Conference*. New York: ASEE/IEEE.

Bergin, J. (2006). Active learning and feedback pat-terns. In Proceedings of PloP 2006: Pattern Languages of Programs. PloP.

Berkenkotter, C., & Huckin, T. N. (1995). *Genre knowledge in disciplinary communication: Cognition/culture/power*. Hillsdale, NJ: Lawrence Erlbaum Associates.

Bernardo, M., Ciancarini, P., & Donatiello, L. (2001). Detecting architectural mismatches in process algebraic descriptions of software systems. In *Proceedings of work-ing IEEE/IFIP Conference on Software Architecture* (pp. 77–86). IEEE.

Bernsteiner, R., Ostermann, H., & Staudinger, R. (2008). Facilitating e-learning with social software: Attitudes and usage from the student's point of view. *International Journal of Web-Based Learning and Teaching Technologies*, *3*(3), 16–33. doi:10.4018/jwltt.2008070102

Beyer, H. (2010). User-centered agile methods. *Synthesis Lectures on Human-Centered Informatics*, *3*(1), 1–71. doi:10.2200/S00286ED1V01Y201002HCI010

BIE. (n.d.). *What is PBL?* Retrieved from http://www.bie.org/about/what_is_pbl

Biggs, J. (2003). Aligning teaching and assessing to course objectives. In *Teaching and learning in higher education: New trends and innovations*. University of Aveiro.

Biggs, J., & Tang, C. (2007). *Teaching for quality learning at university*. Berkshire, UK: Open University Press.

Birenbaum, M. (1996). Assessment 2000: Towards a pluralistic approach to assessment. In *Alternatives in assessment of achievements, learning processes and prior knowledge* (pp. 3–29). Berlin: Springer. doi:10.1007/978-94-011-0657-3_1

Birk, A., Heller, G., John, I., Schmid, K., von der Massen, T., & Muller, K. (2003). Product line engineering, the state of the practice. *IEEE Software*, *20*(6), 52–60. doi:10.1109/MS.2003.1241367

Birkhoelzer, T., Navarro, E., & van der Hoek, A. (2005). Teaching by modeling instead of by models. In *Proceedings of the 6th International Workshop on Software Process Simulation and Modeling*. St. Louis, MO: Academic Press.

Bitbucket . (n.d.). Retrieved from https://bitbucket.org/

Blake, M. B. (2005). Integrating large-scale group projects and software engineering approaches for early computer science courses. *IEEE Transactions on Education*, *48*(1), 63–72. doi:10.1109/TE.2004.832875

Bloom, B. S. (1956). *Taxonomy of educational objectives: The classification of educational goals Handbook 1: Cognitive domain*. New York: David Mackay.

Bloom, B. S., Engelhart, M., Furst, E., Hill, W., & Krathwohl, D. R. (1956). *Taxonomy of educational objectives – The classification of educational goals – Handbook 1: Cognitive domain*. London: Longmans, Green & Co. Ltd.

Boehm, B. W. (1981). *Software engineering economics*. Englewood Cliffs, NJ: Prentice-Hall.

Boehm, B., & Turner, R. (2003). *Balancing agility and discipline: A guide for the perplexed*. Reading, MA: Addison-Wesley. doi:10.1109/ADC.2003.1231450

Bollin, A., Hochmuller, E., & Mittermeir, R. T. (2011). Teaching software project management using simulations. In *Proceedings of the 24ᵗʰ Conference on Software Engineering Education and Training* (pp. 81–90). IEEE.

Bolognesi, A., Ciancarini, P., & Moretti, R. (2006). On the education of future software engineers. In *Software engineering education in the modern age* (pp. 186–205). Berlin: Springer. doi:10.1007/11949374_12

Borges, P., Monteiro, P., & Machado, R. J. (2011). Tailoring RUP to small software development teams. In *Proceedings of the 37th EUROMICRO Conference on Software Engineering and Advanced Applications* (pp. 306–309). IEEE.

Bornat, R., & Dehnadi, S. (2008). Mental models, consistency and programming aptitude. In *Proceedings of the 10ᵗʰ Conference on Australasian Computing Education* (vol. 78, pp. 53–61). Australian Computer Society, Inc.

Bothe, K., Budimac, Z., Cortazar, R., Ivanović, M., & Zedan, H. (2009). Development of a modern curriculum in software engineering at master level across countries. *Computer Science and Information Systems*, *6*(1), 1–21. doi:10.2298/CSIS0901001B

Boud, D. (1985). Problem-based learning in perspective. In D. Boud (Ed.), *Problem-based learning in education for the professions* (pp. 13–18). Sydney: Higher Education Research Society of Australasia.

Boud, D. J., & Feletti, G. (1998). *The challenge of problem-based learning*. New York: Routledge.

Boud, D., Cohen, R., & Sampson, J. (1999). Peer learning and assessment. *Assessment & Evaluation in Higher Education*, *24*(4), 413–426. doi:10.1080/0260293990240405

Boud, D., & Falchikov, N. (2006). Aligning assessment with long-term learning. *Assessment & Evaluation in Higher Education*, *31*(4), 399–413. doi:10.1080/02602930600679050

Boud, D., Keough, R., & Walker, D. (1985). *Reflection: Turning experience into learning.* London: Kogan Page.

Bourque, P., & Dupuis, R. (Eds.). (2004). *Guide to the software engineering body of knowledge - SWEBoK.* Los Alamitos, CA: IEEE Computer Society.

Bowden, J., Hart, G., King, B., Trigwell, K., & Watts, O. (2000). Generic capabilities of ATN university graduates. Sydney Teaching and Learning Committee, Australian Technology Network.

Bransford, J. D., Brown, A. L., & Cocking, R. R. (Eds.). (1999). *How people learn: Brain, mind, experience and school.* Washington, DC: National Academy Press.

Bransford, J. D., & Stein, B. S. (1993). *The ideal problem solver: A guide for improving thinking, learning, and creativity.* New York: W.H. Freeman.

Brants, T., & Franz, A. (2012). Web 1T 5-gram version 1. *Linguistic Data Consortium Catalog.* Retrieved from http://www.ldc.upenn.edu/Catalog/CatalogEntry.jsp?catalogId=LDC2006T13

Braude, E. J., & Bernstein, M. E. (2011). *Software engineering: Modern approaches* (2nd ed.). Hoboken, NJ: J. Wiley & Sons.

Brazier, P., Garcia, A., & Vaca, A. (2007). A software engineering senior design project inherited from a partially implemented software engineering class project. In *Proceedings of the 37th ASEE/IEEE Frontiers in Education Conference* (pp. F4D–7). IEEE.

Bredeson, A. (2013, June). Second helpings volunteers get much-needed technological boost from app. built by students. *The Island Packet.*

Brenner, L. A., Koehler, D. J., & Tversky, A. (1996). On the evaluation of one-sided evidence. *Journal of Behavioral Decision Making, 9*(1), 59–70. doi:10.1002/(SICI)1099-0771(199603)9:1<59::AID-BDM216>3.0.CO;2-V

Brickell, J. L., Porter, D. B., Reynolds, M. F., & Cosgrove, R. D. (1994). Assigning students to groups for engineering design projects: A comparison of five methods. *Journal of Engineering Education, 83*(3), 259–262. doi:10.1002/j.2168-9830.1994.tb01113.x

Bridges, M. E. (1992). *Problem-based leaning for administrators. ERIC Clearinghouse on Educational Management.* University of Oregon.

Broman, D. (2010). Should software engineering projects be the backbone or the tail of computing curricula? In *Proceedings of the 23rd IEEE Conference on Software Engineering Education and Training.* Pittsburgh, PA: IEEE.

Broman, D., Sandahl, K., & Abu Baker, M. (2012). The company approach to software engineering project courses. *IEEE Transactions on Education, 55*(4). doi:10.1109/TE.2012.2187208

Brown, J. S., Collins, A., & Duguid, P. (1989). Situated cognition and the culture of learning. *Educational Researcher, 18*, 32–42. doi:10.3102/0013189X018001032

Brown, P., & Hesketh, A. J. (2004). *The mismangement of talent: Employability and jobs in the knowledge-based economy.* Oxford, UK: Oxford University Press. doi:10.1093/acprof:oso/9780199269532.001.0001

Brown, S. (2004). Assessment for learning: The changing nature of assessment. *Learning and Teaching in Higher Education, 1*, 81–89.

Bruckman, A. (1998). Community support for constructionist learning. *Computer Supported Cooperative Work: The Journal of Collaborative Computing, 7*, 47–86. doi:10.1023/A:1008684120893

Bruegge, B., Reiss, M., & Schiller, J. (2009, April). Agile principles in academic education: A case study. In *Proceedings of the 6th International Conference Information on Technology: New Generations* (pp. 1684–1686). IEEE.

Budgen, D. (2003). *Software design.* Harlow, UK: Pearson Education Ltd.

Budimac, Z., Ivanović, M., Putnik, Z., & Bothe, K. (2011). Studies in wonderland – Sharing of courses, lectures, tasks, assignments, tests and pleasure. In *Proceedings of the 22nd EAEEIE Annual Conference* (pp. 213–219). EAEEIE.

Bunse, C., Feldmann, R. L., & Dörr, J. (2004). Agile methods in software engineering education. In *Extreme programming and agile processes in software engineering* (pp. 284–293). Berlin: Springer. doi:10.1007/978-3-540-24853-8_43

Burge, J., & Troy, D. (2006). Rising to the challenge: Using business-oriented case studies in software engineering education. In *Proceedings of the 19th Conference on Software Engineering Education and Training* (pp. 43–50). IEEE.

Burge, J., & Wallace, C. (2008). Teaching communication skills in the software engineering curriculum. In *Proceedings of the 21st Conference on Software Engineering Education and Training* (pp. 265–266). IEEE.

Burgstaller, B., & Egyed, A. (2010). Understanding where requirements are implemented. In *Proceedings of the 26th IEEE International Conference on Software Maintenance* (pp. 1–5). IEEE.

Buschmann, F., Meunier, R., Rohnert, H., Sommerlad, P., & Stal, M. (1996). *Pattern-oriented software architecture: A system of patterns* (Vol. 1). New York: Wiley.

Button, G., & Sharrock, W. (1996). Project work: The organisation of collaborative design and development in software engineering. *Computer Supported Cooperative Work, 5*(4), 369–386. doi:10.1007/BF00136711

Callaghan, M. J., McCusker, K., Losada, J. L., Harkin, J. G., & Wilson, S. (2009). Teaching engineering education using virtual worlds and virtual learning environment. In *Proceedings of 2009 Conference on Advances in Computing, Control and Telecommunications Technologies* (pp. 295–299). Kerela, India: Academic Press.

Callahan, D., & Pedigo, B. (2002). Educating experienced IT professionals by addressing industry's needs. *IEEE Software, 19*(5), 57–62. doi:10.1109/MS.2002.1032855

Callele, D., & Makaroff, D. (2006). Teaching requirements engineering to an unsuspecting audience. *ACM SIGCSE Bulletin, 38*(1), 433–437. doi:10.1145/1124706.1121475

Cano, F. (2006). An in-depth analysis of the learning and study strategies inventory (LASSI). *Educational and Psychological Measurement, 66*(6), 1023–1038. doi:10.1177/0013164406288167

Cano, M. D. (2011). Students' involvement in continuous assessment methodologies: A case study for a distributed information systems course. *IEEE Transactions on Education, 54*(3), 442–451. doi:10.1109/TE.2010.2073708

Cardón, A. (2003). Coaching d'equipe. Ed.s d'Organisation.

Carmel, E. (1999). *Global software engineering teams*. Upper Saddle River, NJ: Prentice Hall.

Carr, D. (2009). Textual analysis, digital games, zombies. In *Proceedings of DiGRA 2009 Conference: Breaking New Ground: Innovation in Games, Play, Practice and Theory*. DiGRA.

Carrington, D., & Kim, S. K. (2003). Teaching software design with open source software. In *Proceedings of the 33rd ASEE/IEEE Frontiers in Education Conference* (Vol. 3, pp. S1C–9). Boulder, CO: IEEE.

Carswell, L., & Murphy, M. (1994). *Pragmatic methodology for educational courseware development*. Retrieved July 19, 2004 from http://www.ulst.ac.uk/cticomp/carswell.html

Casecomplete. (n.d.). Retrieved from http://www.casecomplete.com/

Cassidy, S. (2006). Developing employability skills: Peer assessment in higher education. *Education + training, 48*(7), 508–517.

Catme. (n.d.). Retrieved from http://www.catme.org

Caulfield, C., Xia, J., Veal, D., & Maj, S. P. (2011). A systematic survey of games used for software engineering education. *Modern Applied Science, 5*(6), 28. doi:10.5539/mas.v5n6p28

Cavrak, I., Orlic, M., & Crnkovic, I. (2012). Collaboration patterns in distributed software development projects. In *Proceedings of the 34th International Conference on Software Engineering* (pp. 1235–1244). Academic Press.

CCSE. (n.d.). Retrieved from http://sites.computer.org/ccse/SE2004Volume.pdf

Chaczko, Z., Davis, D., & Mahadevan, V. (2004). New perspectives on teaching and learning software systems development in large groups. In *Proceedings of the 5th International Conference on Information Technology Based Higher Education and Training* (pp. 409–414). IEEE.

Chamillard, A. T., & Braun, K. A. (2002). The software engineering capstone: Structure and tradeoffs. *ACM SIGCSE Bulletin, 34*(1), 227–231. doi:10.1145/563517.563428

Chen, P., Gonyea, R., & Kuh, G. (2008). Learning at a distance: Engaged or not?. *Innovate, 4*(3).

Cheng, Y., & Lin, J. (2010). A constrained and guided approach for managing software engineering course projects. *IEEE Transactions on Education, 53*(3), 430–436. doi:10.1109/TE.2009.2026738

Chidamber, S. R., & Kemerer, C. F. (1994). A metrics suite for object oriented design. *IEEE Transactions on Software Engineering, 20*(6), 476–493. doi:10.1109/32.295895

Christel, M. G., & Kang, K. C. (1992). *Issues in requirements elicitation (No. CMU/SEI-92-TR-12)*. Pittsburgh, PA: Carnegie Mellon University.

Christensen, M. J., & Thayer, R. H. (2001). *The project manager's guide to software engineering's best practices*. Los Alamitos, CA: IEEE Computer Society Press.

Chung, J., & Chow, S. (2004). Promoting student learning through a student-centred problem-based learning subject curriculum. *Innovations in Education and Teaching International, 41*, 157–168. doi:10.1080/1470329042000208684

Ciancarini, P., Dos, C., & Zuppiroli, S. (2013). A double comparative study: Process models and student skills. In *Proceedings of the 26th IEEE Conference on Software Engineering Education and Training* (pp. 189–198). IEEE Computer Society.

Clark, N. (2005). Evaluating student teams developing unique industry projects. *Proceedings of the 7th Australasian Conference on Computer Education, 42*, 21–30.

Clark, N., Davies, P., & Skeers, R. (2005). Self and peer assessment in software engineering projects. In *Proceedings of the 7th Australasian Conference on Computing Education* (vol. 42, pp. 91–100). Australian Computer Society.

Clarke, P. J., Pava, J., Davis, D., Hernandez, F., & King, T. M. (2012). Using WReSTT in SE courses: An empirical study. In *Proceedings of the 43rd ACM Technical Symposium on Computer Science Education* (pp. 307–312). ACM.

Claxton, G. (1998). *Hare brain, tortoise mind*. London: Fourth Estate.

Clements, P., Garlan, D., Bass, L., Stafford, J., Nord, R., Ivers, J., & Little, R. (2002). *Documenting software architectures: Views and beyond*. Upper Saddle River, NJ: Pearson Education.

Clements, P., & Northrop, L. (2002). *Software product lines: Practices and patterns*. Reading, MA: Addison-Wesley.

Cockburn, A. (2008). Using both incremental and iterative development. *CrossTalk – The Journal of Defense Software Engineering, 21*(2), 27–30.

Cockburn, A. (2001). *Writing effective use cases*. Boston: Addison-Wesley.

Cockburn, A. (2006). *Agile software development: The cooperative game*. Reading, MA: Addison-Wesley.

Code . (n.d.). Retrieved from https://code.google.com

Codecover . (n.d.). Retrieved from http://codecover.org/

Coffield, F., Moseley, D., Hall, E., & Ecclestone, K. (2004). *Should we be using learning styles?* Learning and Skills Research Centre.

Cohn, M. (2013). *Agile & scrum software development*. Retrieved November 2013, from www.mountaingoatsoftware.com

Cohn, M. (2004). *User stories applied for agile software development*. Reading, MA: Addison-Wesley.

Cohn, M. (2004). *User stories applied: For agile software development*. Reading, MA: Addison-Wesley.

Cohn, M. (2005). *Agile estimating and planning*. Pearson Education.

Coleman, B., & Lang, M. (2012). Collaboration across the curriculum: A disciplined approach to developing team skills. In *Proceedings of the 2012 Special Interest Group on Computer Science Education Technical Symposium*. Academic Press.

Coleman, B., & Lang, M. (2012). Collaboration across the curriculum: A disciplined approach to developing team skills. In *Proceedings of the 43rd ACM Technical Symposium on Computer Science Education* (pp. 277–282). ACM.

Collins, A., Brown, J. S., & Newman, S. E. (1989). Cognitive apprenticeship: Teaching the crafts of reading, writing and mathematics. In L. Resnick (Ed.), *Knowing, learning and instruction: Essays in honour of Robert Glaser* (pp. 453–494). Hillsdale, NJ: Erlbaum.

Computer. (n.d.). Retrieved from http://www.computer.org/portal/Web/sWebok/html/contents

Connor, A. M., Buchan, J., & Petrova, K. (2009). Bridging the research-practice gap in requirements engineering through effective teaching and peer learning. In *Proceedings of the 6th International Conference on Information Technology: New Generations* (pp. 678–683). IEEE.

Connor, H., & Shaw, S. (2008). Graduate training and development: Current trends and issues. *Education + Training, 50*, 357–365.

Conn, R. (2002). Developing software engineers at the C-130J software factory. *IEEE Software, 19*(5), 25–29. doi:10.1109/MS.2002.1032849

Conway, M. E. (1968). How do committees invent? *Datamation, 14*(5), 28–31.

Cooper, L., Orrell, J., & Bowden, M. (2010). *Work integrated learning: A guide to effective practice.* Abingdon, UK: Routledge.

Coplien, J. O., & Harrison, N. B. (2004). *Organizational patterns of agile software development.* Upper Saddle River, NJ: Prentice-Hall.

Coppit, D., & Haddox-Schatz, J. (2005). Large team projects in software engineering courses. In *Proceedings of the 36th SIGCSE Technical Symposium on Computer Science Education* (pp. 137–141). ACM Computer Society.

Costa, A. C., Roe, R. A., & Taillieu, T. (2001). Trust within teams: The relation with performance effectiveness. *European Journal of Work and Organizational Psychology, 10*(3), 225–244. doi:10.1080/13594320143000654

Cotton, K. (1993). *Developing employability skills.* School Improvement Research Series. Retrieved from http://educationnorthwest.org/Webfm_send/524

Coupal, C., & Boechler, K. (2005). Introducing agile into a software development capstone project. In *Proceedings of Agile Conference* (pp. 289–297). Academic Press.

Coursera. (n.d.). Retrieved from https://www.coursera.org

Coyle, E. J., Jamieson, L. H., & Oakes, W. C. (2005). EPICS: Engineering projects in community service. *International Journal of Engineering Education, 21*(1), 139–150.

Crane, T., & Patrick, L. N. (1998). *The heart of coaching: Using transformational coaching to create a high-performance culture.* San Diego, CA: FTA Press.

Crossman, D. (1997). The evolution of the world wide web as an emerging instructional technology tool. In B. H. Khan (Ed.), *Web-based instruction* (pp. 19–23). Educational Technology Publications.

Csikszentmihalyi, M. (1996). *Creativity: Flow and the psychology of discovery and invention.* New York: HarperPerennial.

Cubric, M. (2007). Using wikis for summative and formative assessment. In *Reap international online conference on assessment design for learner responsibility.* Academic Press.

Cummings, B. (2007). *Real world careers: Why college is not the only path to becoming rich.* Business Plus.

Curtin, P. (2004). Employability skills for the future. In J. Gibb (Ed.), *Generic skills in vocational education and training: Research readings* (pp. 40–41). Adelaide, Australia: National Centre for Vocational Education Research.

Curtis, D. D., & Lawson, M. J. (2001). Exploring collaborative online learning. *Journal of Asynchronous Learning Networks, 5*(1).

Dahiya, D. (2010). Teaching software engineering: A practical approach. *ACM SIGSOFT Software Engineering Notes, 35*(2).

Dahlgren, M. A., & Dahlgren, L. O. (2002). Portraits of PBL : Students' experiences of the characteristics of problem-based learning in physiotherapy, computer engineering. *Instructional Science, 30*, 111–127. doi:10.1023/A:1014819418051

Damian, D. E., & Zowghi, D. (2003). RE challenges in multi-site software development organisations. *Requirements Engineering, 8*(3), 149–160. doi:10.1007/s00766-003-0173-1

Daniel, J. S. (1996). *Mega-universities and knowledge media: Technology strategies for higher education.* London: Keegan Press.

Danielsen, A. (2010). *Teaching requirements engineering: An experimental approach. Norsk Informatikkonferanse.* NIK.

Dantas, A. R., de Oliveira Barros, M., & Werner, C. M. L. (2004). A simulation-based game for project management experiential learning. In *Proceedings of the 16th International Conference on Software Engineering and Knowledge Engineering* (pp. 19–24). Academic Press.

Davies, L. (2000). Why kick the L out of learning? The development of students' employability skills through part-time working. *Education + Training, 42*, 436–445.

Davies, S. (2009). Appointing team leads for student software development projects. *Journal of Computing Sciences in Colleges, 25*(2), 92–99.

Davis, C., & Wilcock, E. (2013). Teaching materials using case studies. *UK Centre for Materials Education.* Retrieved from http://www.materials.ac.uk/guides/casestudies.asp

Davis, B. (1993). *Tools for teaching*. San Francisco: Jossey-Bass.

Davis, B. (2009). *97 things every project manager should know*. Sebastopol, CA: O'Reilly.

Dawson, R. (2000). *Twenty dirty tricks to train software engineers*. Paper presented at the 22nd International Conference on Software Engineering. New York, NY.

De Boer, R., Farenhorst, R., & van Vliet, H. (2009). A community of learners approach to software architecture education. In *Proceedings of the 22th IEEE Conference on Software Engineering Education and Training* (pp. 190–197). IEEE Computer Society.

De Corte, E. (2000). Marrying theory building and the improvement of school practice: A permanent challenge for instructional psychology. *Learning and Instruction, 10*, 249–266. doi:10.1016/S0959-4752(99)00029-8

De Freitas, S. (2006). *Learning in immersive worlds*. London: Joint Information Systems Committee.

de Graaff, E., & Kolmos, A. (2003). Characteristics of problem-based learning. *International Journal of Engineering Education, 19*(5), 657–662.

de Souza, C. S. (2005). *The semiotic engineering of human–computer interaction*. Cambridge, MA: MIT Press.

de Souza, C. S., Leitão, C. F., Prates, R. O., Bim, S. A., & da Silva, E. J. (2010). Can inspection methods generate valid new knowledge in HCI? The case of semiotic inspection. *International Journal of Human-Computer Studies, 68*(1/2), 22–40. doi:10.1016/j.ijhcs.2009.08.006

Dede, C. (2009). Immersive interfaces for engagement and learning. *Science, 323*(5910), 66–69. doi:10.1126/science.1167311 PMID:19119219

DEEWR. (2012). *Employability skills framework stage 1 – Final report*. Canberra, Australia: Australian Government. DEEWR Retrieved from http://foi.deewr.gov.au/system/files/doc/other/employability_skills_framework_stage_1_final_report.pdf

Deiters, C., Herrmann, C., Hildebrandt, R., Knauss, E., Kuhrmann, M., Rausch, A., et al. (2011). GloSE-lab: Teaching global software engineering. In *Proceedings of the 6th IEEE International Conference on Global Software Engineering*. IEEE.

Delaney, D., & Mitchell, G. G. (2002). PBL applied to software engineering group projects. In *Proceedings International Conference on Information and Communication in Education* (pp. 1093-1098). Government of Extremadura.

Delisle, S., Barker, K., & Biskri, I. (1999). Object-oriented analysis: Getting help from robust computational linguistic tools. *Application of Natural Language to Information Systems, Oesterreichische Computer Gesellschaft*, 167–172.

Delvin, M., & Phillips, C. (2010). Assessing competency in undergraduate software engineering teams. In *Proceedings of IEEE EDUCON Education Engineering 2010 – The Future of Global Learning Engineering Education* (pp. 271–278). IEEE Computer Society Press.

DeMarco, T. (1986). *Controlling software projects: Management, measurement & estimation*. New York, NY: Prentice Hall.

Deming, W. E. (1986). *Out of the crisis*. Cambridge, MA: MIT Center for Advanced Engineering Study.

Denning, P. J. (1992). Educating the new engineer. *Communications of the ACM, 35*(12), 83–97. doi:10.1145/138859.138870

Denning, P. J., & Yaholkovsky, P. (2008). Getting to we. *Communications of the ACM, 51*(4), 19–24. doi:10.1145/1330311.1330316

Derby, E., Larsen, D., & Schwaber, K. (2006). *Agile retrospectives: Making good teams great.* Pragmatic Bookshelf.

DEST. (2006). Employability skills - From framework to practice. Canberra, Australia: Commonwealth of Australia. Department of Education, Science and Training.

Deutsch, M. (1949). A theory of cooperation and competition. *Human Relations, 2*, 129–152. doi:10.1177/001872674900200204

Deutsch, M. (1962). Cooperation and trust: Some theoretical notes. In M. R. Jones (Ed.), *Nebraska symposium on motivation* (pp. 275–319). Lincoln, NE: University of Nebraska Press.

Devedzic, V., & Milenkovic, S. R. (2011). Teaching agile software development: A case study. *IEEE Transactions on Education, 54*(2), 273–278. doi:10.1109/TE.2010.2052104

Dewey, J. (1916). *Democracy and education.* New York: Macmillan.

Dewey, J. (1929). *The quest for certainty.* New York: Minton.

Dewey, J. (2009). *Democracy and education: An introduction to the philosophy of education.* New York: WLC Books.

Di Domenico, F., Panizzi, E., Sterbini, A., & Temperini, M. (2005). *Analysis of commercial and experimental e-learning systems. Quality, Interoperability and Standards in e-Learning Team.* TISIP Research Foundation.

Diaz, J., Garbajosa, J., & Calvo-Manzano, J. A. (2009). Mapping CMMI level 2 to scrum practices: An experience report. *Software Process Improvements, 42*, 93–104. doi:10.1007/978-3-642-04133-4_8

Dick, W. O., Carey, L., & Carey, J. O. (2005). *The systematic design of instruction.* Pearson/Allyn & Bacon.

DigitalHome Case Study Project. (2013). Retrieved from www.softwarecasestudy.org

DMST. (n.d.). Retrieved from http://www.dmst.aueb.gr/dds/sw/ckjm/

Doberkat, E., Engels, G., Hausmann, J. H., Lohmann, M., Pleumann, J., & Schröder, J. (2005). Software engineering and eLearning: The MuSofT project. *Language, 1*, 2.

Doke, E. R., & Williams, S. R. (1999). Knowledge and skill requirements for information systems professionals: An exploratory study. *Journal of Information Systems Education, 10*(1), 10–18.

Donald, J. G. (2002). *Learning to think.* San Francisco: Jossey-Bass.

Dos Santos, S. C., Da Conceicao Moraes Batista, M., Cavalcanti, A. P. C., Albuquerue, J. O., & Meira, S. R. L. (2009). Applying PBL in software engineering education. In *Proceedings of 22nd Conference on Software Engineering Education and Training* (pp. 182-189). IEEE.

Draper, S., & Cutts, C. (2006). Targeted remediation for a computer programming course using student facilitators. *Practice and Evidence of Scholarship of Teaching and Learning in Higher Education, 1*(2), 117–128.

Drappa, A., & Ludewig, J. (2001). Simulation in software engineering training. In *Proceedings of the 23rd International Conference on Software Engineering* (pp. 199 – 208). ACM.

Dreyfus, H. L. (2001). *On the internet.* London: Routledge.

Dreyfus, H. L., & Dreyfus, S. E. (1986). *Mind over machine.* New York: Free Press.

Drive . (n.d.). Retrieved from http://www.google.com/drive

Dröschel, W. (2000). *Das V-modell 97: Der standard für die entwicklung von IT-systemen mit anleitung für den praxiseinsatz.* München: Oldenbourg.

Dubinsky, Y., & Hazzan, O. (2003). Extreme programming as a framework for student-project coaching in computer science capstone courses. In *Proceedings of IEEE International Conference on Software: Science, Technology and Engineering* (pp. 53–59). IEEE.

Dubinsky, Y., & Hazzan, O. (2005). A framework for teaching software development methods. *Computer Science Education, 15*(4), 275–296. doi:10.1080/08993400500298538

Duggins, S., & Walker, R. (2009). Teaching software engineering online using 21st century technology. In *Proceedings of ASEE Southeastern Section Annual Conference*. ASEE.

Dukovska-Popovska, I., Hove-Madsen, V., & Nielsen, K. B. (2008). Teaching lean thinking through game: Some challenges. *In Proceedings of the 36th European Society for Engineering Education on Quality Assessment, Employability & Innovation*. Academic Press.

Dunlap, J. C. (2005). Problem-based learning and self-efficacy: How a capstone course prepares students for a profession. *Educational Technology Research and Development, 53*(1), 65–83. doi:10.1007/BF02504858

Dutoit, A. H., McCall, R., Mistrik, I., & Paech, B. (2006). Rationale management in software engineering: Concepts and techniques. In A. H. Dutoit, R. McCall, I. Mistrik, & B. Paech (Eds.), *Rationale management in software engineering*. Springer. doi:10.1007/978-3-540-30998-7_1

EA. (2011). *Stage 1 competency standard for professional engineer*. Retrieved from http://www.engineersaustralia.org.au/sites/default/files/shado/Education/Program%20Accreditation/110318%20Stage%201%20Professional%20Engineer.pdf

EA. (n.d.). *Stage 1 competency standard for professional engineer*. Canberra, Australia: Engineers Australia (EA). Retrieved November 2013, from https://www.engineersaustralia.org.au/sites/default/files/shado/Education/Program%20Accreditation/130607_stage_1_pe_2013_approved.pdf

Easymock . (n.d.). Retrieved from http://easymock.org/

Ebner, M., & Holzinger, A. (2007). Successful implementation of user–centered game based learning in higher education: An example from civil engineering. *Computers & Education, 49*(3), 873–890. doi:10.1016/j.compedu.2005.11.026

Echevarría, R. (2008). *Ontología del lenguaje*. Buenos Aires, Argentina: Granica.

Eclipse CS . (n.d.). Retrieved from http://eclipse-cs.sourceforge.net/

Eclipse Metrics . (n.d.). Retrieved from http://eclipse-metrics.sourceforge.net/

Eclipse . (n.d.). Retrieved from http://www.eclipse.org/

Editorial. (2010). Graduating by the number: Putting data to work for student success, special issue. *Education Week, 29*(34).

Edmonds, E., & Candy, L. (2002). Creativity, art practice and knowledge. *Communications of the ACM, 45*(10), 91–95. doi:10.1145/570907.570939

Educause. (2013). *7 things you should know about… Wikis*. Retrieved March 5, 2013, from http://Net.Educause.Edu/Ir/Library/Pdf/Eli7004.pdf

Ehlers, M. (2004). Deconstructing the axon: Wallerian degeneration and the ubiquitin-proteasome system. *Trends in Neurosciences, 27*(1), 3–6. doi:10.1016/j.tins.2003.10.015 PMID:14698600

Ellis, H. J., Morelli, R. A., De Lanerolle, T. R., & Hislop, G. W. (2007). Holistic software engineering education based on a humanitarian open source project. In *Proceedings of the 20ᵗʰ Conference on Software Engineering Education & Training* (pp. 327–335). IEEE.

Ellis, T. J., & Hafner, W. (2007). Assessing collaborative, project-based learning experiences: Drawing from three data sources. In *Proceedings of the 37th Annual Frontiers In Education Conference-Global Engineering: Knowledge Without Borders, Opportunities Without Passports* (pp. T2G–13). IEEE.

Engle, C. B. (1989). Software engineering is not computer science. In N. Gibbs (Ed.), *Software engineering education* (Vol. 376, pp. 257–262). New York: Springer. doi:10.1007/BFb0042363

Engstrom, M. E., & Jewett, D. (2005). Collaborative learning the wiki way. *TechTrends, 49*(6), 12–15. doi:10.1007/BF02763725

Entwistle, N. (2000). *Promoting deep learning through teaching and assessment: Conceptual frameworks and educational contexts*. Paper presented at TLRP Conference. Leicester, UK.

Entwistle, N. J., & Ramsden, P. (1983). *Understanding student learning*. London: Croom Helm.

Epstein, R. G. (1994a). The case of the killer robot (part 1). *SIGCAS Computers and Society, 24*(3), 20–28. doi:10.1145/191634.191640

Epstein, R. G. (1994b). The case of the killer robot (part 2). *SIGCAS Computers and Society*, *24*(4), 12–32. doi:10.1145/190777.190778

European Commission. (2010). *The Bologna process - Towards the european higher education area*. Retrieved 08-07-2013, 2013, from http://ec.europa.eu/education/higher-education/bologna_en.htm

Fagan, M. E. (1976). Design and code inspections to reduce errors in program development. *IBM Systems Journal*, *15*(3), 258–287. doi:10.1147/sj.153.0182

Fagan, M. E. (1986). Advances in software inspections. *IEEE Transactions on Software Engineering*, *12*(7), 744–751. doi:10.1109/TSE.1986.6312976

Fairly, D. (Ed.). (2013). Guide to the software engineering body of knowledge (SWEBOK V3). ACM/IEEE Computer Society.

Falchikov, N. (2005). *Improving assessment through student involvement*. London: Routledge Falmer.

Fancott, T., Kamthan, P., & Shahmir, N. (2011). Using the social web for teaching and learning user stories. In *Proceedings of the 6ᵗʰ International Conference on e-Learning*. Academic Press.

Farrell, V., Ravalli, G., & Farrell, G. (2012). Capstone project: Fair, just and accountable assessment. In *Proceedings of the 17th ACM Annual Conference on Innovation and Technology in Computer Science Education* (pp. 168–173). ACM.

Farrell, V., Ravalli, G., Farrell, G., Kindler, P., & Hall, D. (2012). Capstone project: Fair, just and accountable assessment. In *Proceedings of the 17th Conference on Innovation and Technology in Computer Science Education* (pp. 168–173). Academic Press.

Felder, R. M., & Brent, R. (2004). *The ABC's of engineering education: ABET, Bloom's taxonomy, cooperative learning, and so on*. Paper presented at the 2004 American Society for Engineering Education Annual Conference & Exposition. New York, NY.

Felder, R., & Brent, R. (2009). Active learning: An introduction. *ASQ Higher Education Brief*, *2*(4).

Felder, R. M., & Silverman, L. K. (1988). Learning and teaching styles in engineering education. *English Education*, *78*(7), 674–681.

Felder, R. M., & Spurlin, J. E. (2005). Applications, reliability, and validity of the index of learning styles. *International Journal of Engineering Education*, *21*(1), 103–112.

Fenton, N. E. (1991). *Software metrics: A rigorous approach*. London: Chapman & Hall.

Figl, K., & Motschnig, R. (2008). Researching the development of team competencies in computer science courses. In *Proceedings of the 38th Annual Frontiers in Education Conference* (pp. T1A 1–6). Piscataway, NJ: IEEE.

Figueiredo, E., Lobato, C., Dias, K., Leite, J., & Lucena, C. (2007). Um Jogo para o ensino de engenharia de software centrado na perspectiva de evolução. In *Proceedings of Workshop on Education in Computer (WEI – 2007)* (pp. 37–46). Rio de Janeiro, Brazil: WEI.

Finkelstein, A. (2011). Ten open challenges at the boundaries of software engineering and information systems. In H. Mouratidis, & C. Rolland (Eds.), *Advanced information systems engineering* (Vol. 6741, pp. 1–1). New York: Springer Berlin Heidelberg. doi:10.1007/978-3-642-21640-4_1

Fiske, J., & Jenkins, H. (2011). *Introduction to communication studies*. London: Routledge.

Fogel, K. (2005). *Producing open source software*. Sebastopol, CA: O'Reilly.

Foppa, K. (1975). *Lernen, gedächtnis, verhalten: Ergebnisse und probleme der lernpsychologie*. Köln: Kiepenheuer & Witsch.

Ford, C. W., & Minsker, S. (2003). TREEZ – An educational data structures game. *Journal of Computing Sciences in Colleges*, *18*(6), 180–185.

Ford, M., & Venema, S. (2010). Assessing the success of an introductory programming course. *Journal of Information Technology Education*, *9*, 133–145.

Forio . (n.d.). Retrieved from http://forio.com

Fowler, M., & Foemmel, M. (2006). *Continuous integration*. Retrieved from http://www.martinfowler.com/articles/continuousIntegration.html

Freudenberg, B., Brimble, M., & Vyvyan, V. (2010). The penny drops: Can work integrated learning improve students' learning?. *e-Journal of Business Education & Scholarship of Teaching, 4*(1), 42–61.

Freudenberg, B., Brimble, M., & Cameron, C. (2010). Where there is a WIL there is a way. *Higher Education Research & Development, 29*(5), 575–588. doi:10.1080/07294360.2010.502291

Frey, L. R., Botan, C. H., & Kreps, G. L. (2000). *Investigating communication*. New York: Allyn & Bacon.

Friedenthal, S. (2008). *A practical guide to SysML: The systems modeling language*. San Francisco: Morgan Kaufmann / The OMG Press.

Friedman, K. (2001). *Creating design knowledge: From research into practice*. Paper presented at the IDATER 2000 Conference. Loughborough, UK. https://dspace.lboro.ac.uk/2134/1360

Friedman, G., & Sage, A. (2004). Case studies of systems engineering and management in systems acquisition. *Systems Engineering, 7*(1), 84–96. doi:10.1002/sys.10057

Fritzsche, M., & Keil, P. (2007). Agile methods and CMMI: Compatibility or conflict? *e-Informatica. Software Engineering Journal, 1*(1), 9–26.

Froyd, J. E., Wankat, P. C., & Smith, K. A. (2012). Five major shifts in 100 years of engineering education. *Proceedings of the IEEE, 100*(13), 1344–1360. doi:10.1109/JPROC.2012.2190167

Fung, R. Y. K., Tam, W. T., Ip, A. W. H., & Lau, H. C. W. (2002). Software process improvement strategy for enterprise information systems development. *International Journal of Information Technology and Management, 1*(2-3), 225–241. doi:10.1504/IJITM.2002.001198

Gabriel, G. (2012). *Coaching scolaire: Augmenter le potentiel des élèves en difficulté*. Brussels, Belgium: De Boeck.

Gamma, E., Helm, R., Johnson, R., & Vlissides, J. (1995). *Design patterns*. Reading, MA: Addison-Wesley.

Garris, R., Ahlers, R., & Driskell, J. E. (2002). Games, motivation, and learning: A research and practice model. *Simulation & Gaming, 33*(4), 441–467. doi:10.1177/1046878102238607

Gast, H. (2008). Patterns and traceability in teaching software architecture. In *Proceedings of the 6th International Symposium on Principles and Practice of Programming in Java* (pp. 23–31). ACM.

Georgas, J. (2013). Supporting software architectural style education using active learning and role-playing. In *Proceedings of the 120th ASEE Conference*. ASEE.

Gerhardt, J. (2001). Put ethics and fun into your computer course. *Journal of Computing Sciences in Colleges, 16*(4), 247–251.

Gerhardt, J. (2005). If you can teach it, you can measure it. *Journal of Computing Sciences in Colleges, 20*(3), 114–120.

Ghezzi, C., Jazayeri, M., & Mandrioli, D. (2002). *Fundamentals of software engineering*. Upper Saddle River, NJ: Prentice-Hall.

Gijselaers, W. H. (2011). Problem-based learning (PBL) today and tomorrow. In *Proceedings of FACILITATE Conference: PBL 3.0 in Transformative Times*. FACILITATE.

Gijselaers, W. H. (1995). Perspectives on problem-based Learning. In W. H. Gijselaers, D. T. Tempelaar, P. K. Keizer, J. M. Blommaert, E. M. Bernard, & H. Kasper (Eds.), *Educational innovation in economics and business administration: The case of problem-based learning*. Kluwer. doi:10.1007/978-94-015-8545-3_5

Gijselaers, W. H. (1996). Connecting problem-based practices with educational theory. *New Directions for Teaching and Learning, 199*(68), 13–21. doi:10.1002/tl.37219966805

Gijselaers, W., & Schmidt, H. (1990). Towards a causal model of student learning within the context of problem-based curriculum. In Z. Nooman, H. Schmidt, & E. Ezzat (Eds.), *Innovations in medical education: An evaluation of its present status* (pp. 95–114). Springer.

Github . (n.d.). Retrieved from https://github.com

GIT-SCM . (n.d.). Retrieved from http://git-scm.com/

Glass, R. L. (2007). Is software engineering fun? Part 2. *IEEE Software, 24*(2), 104–103. doi:10.1109/MS.2007.46

Glass, R. L. (2007). Is software engineering fun? *IEEE Software, 24*(1), 96–95. doi:10.1109/MS.2007.18

Glazer, H. (2008). *CMMI or agile: Why not embrace both! (Technical Note, CMU/SEI-2008-TN-003)*. Pittsburgh, PA: Software Engineering Process Management, Carnegie Mellon.

Glazer, H. (2010). Love and marriage: CMMI and agile need each other. *CrossTalk, 23*(1), 29–34.

Gobet, F., & Wood, D. (1999). Expertise, models of learning and computer-based tutoring. *Computers & Education, 33*, 189–207. doi:10.1016/S0360-1315(99)00032-9

Golas, K. C. (1993). *Estimating time to develop interactive courseware in the 1990s*. Retrieved July 19, 2004 from http://www.coedu.usf.edu/inst_tech/resources/estimating.html

Golbeck, J. (2013). *Analyzing the social web*. London: Elsevier.

Goldberg, A. (2002). Collaborative software engineering. *Journal of Object Technology, 1*(1), 1–19. doi:10.5381/jot.2002.1.1.c1

Goleman, D. (1996). *Emotional intelligence: Why it can matter more than IQ*. Bantam Books. doi:10.1037/e538982004-001

Goleman, D. (1998). *Working with emotional intelligence*. Bantam Books.

Gomaa, H. (2005). *Designing software product lines with UML*. Reading, MA: Addison-Wesley.

González-Morales, D., de Antonio, L. M. M., & García, J. L. R. (2011). Teaching soft skills in software engineering. In *Proceedings of 2011 Global Engineering Education Conference* (pp. 630–637). IEEE.

Gordon, N. A. (2010). Group working and peer assessment—Using WebPA to encourage student engagement and participation. *Innovation in Teaching and Learning in Information and Computer Sciences, 9*(1), 20–31. doi:10.11120/ital.2010.09010020

Gorton, I. (2011). *Essential software architecture*. Berlin: Springer. doi:10.1007/978-3-642-19176-3

Gothelf, J., & Seiden, J. (2013). *Lean UX: Applying lean principles to improve user experience*. Sebastopol, CA: O'Reilly.

Gotterbarn, D., Miller, K., & Rogerson, S. (1999). Computer society and ACM approve software engineering code of ethics. *Computer, 32*(10), 84–88. doi:10.1109/MC.1999.796142

Gott, S. P., Hall, E. P., Pokorny, R. A., Dibble, E., & Glaser, R. (1993). A naturalistic study of transfer: Adaptive expertise in technical domains. In D. K. Detterman, & R. J. Sternberg (Eds.), *Transfer on trial: Intelligence, cognition and instruction* (pp. 258–288). Norwood, NJ: Ablex.

Graf, S., & List, B. (2005). An evaluation of open source e-learning platforms stressing adaptation issues. In *Proceedings of the 5th IEEE International Conference On Advanced Learning Technologies* (pp. 163–165). IEEE.

Gregg, D., Kulkarni, U., & Vinzé, A. (2001). Understanding the philosophical underpinnings of software engineering research in information systems. *Information Systems Frontiers, 3*(2), 169–183. doi:10.1023/A:1011491322406

Grenning, J. (2002). *Planning poker or how to avoid analysis paralysis while release planning*. Retrieved from http://renaissancesoftware.net/files/articles/PlanningPoker-v1.1.pdf

Grieve, C. (1992). Knowledge increment assessed for three methodologies of teaching physiology. *Medical Teacher, 14*(1), 27–31. doi:10.3109/01421599209044011 PMID:1608324

Grow, G. O. (1996). Teaching learners to be self-directed. *Adult Education Quarterly, 41*(3), 125–149. doi:10.1177/0001848191041003001

Grundy, J. (1996). A comparative analysis of design principles for project-based IT courses. In *Proceedings of the 2nd Australasian Conference on Computer Science Education* (pp. 170–177). Australian Computer Society.

Grupe, F. H., & Jay, J. K. (2000). Incremental cases: Real-life, real-time problem solving. *College Teaching, 48*(4), 123–128. doi:10.1080/87567550009595828

GSwE. (2009). *Graduate software engineering 2009 (GSwE2009), curriculum guidelines for graduate degree programs in software engineering*. Hoboken, NJ: Stevens Institute of Technology.

GSwE. (2009). *Graduate software engineering 2009: Curriculum guidelines for graduate degree programs in software engineering, version 1.0*. Retrieved in July 2013 from http://www.gswe2009.org/fileadmin/files/GSwE2009_Curriculum_Docs/GSwE2009_version_1.0.pdf

Guarnieri, S., & Ortiz de Zarate, M. (2010). *No es lo mismo*. Madrid, Spain: LID Editorial Empresarial.

Guindon, R. (1989). The process of knowledge discovery in system design. In G. Salvendy, & M. J. Smith (Eds.), *Designing and using human-computer interfaces and knowledge based systems* (pp. 727–734). Amsterdam: Elsevier.

Guindon, R. (1990). Knowledge exploited by experts during software systems design. *International Journal of Man-Machine Studies*, *33*, 279–304. doi:10.1016/S0020-7373(05)80120-8

Hadjerrouit, S. (2003). Toward a constructivist approach to e-learning in software engineering. In A. Rossett (Ed.), *Proceedings of World Conference on E-Learning in Corporate, Government, Healthcare, and Higher Education* (pp. 507–514). Academic Press.

Hadjerrouit, S. (2005). Constructivism as guiding philosophy for software engineering education. *ACM SIGCSE Bulletin*, *37*(4), 45–49. doi:10.1145/1113847.1113875

Hagan, D. (2004). *Employer satisfaction with ICT graduates*. Paper presented at the 6th Conference on Australasian Computing Education. Canberra, Australia.

Hainey, T. (2009). Games-based learning in computer science, software engineering and information systems education. *Computing and Information Systems Journal*, *13*(3), 1–13.

Hainey, T., Connolly, T. M., Stansfield, M., & Boyle, E. A. (2011). Evaluation of a game to teach requirements collection and analysis in software engineering at tertiary education level. *Computers & Education*, *56*(1), 21–35. doi:10.1016/j.compedu.2010.09.008

Hales, C. (2007). Moving down the line? The shifting boundary between middle and first-line management. *Journal of General Management*, *32*(2), 31–56.

Hallinger, P., & Bridges, M. E. (2007). *A problem-based approach for management education: Preparing managers for action*. Berlin: Springer. doi:10.1007/978-1-4020-5756-4_2

Hanisch, J., & Corbitt, B. J. (2004). Requirements engineering during global software development: Some impediments to the requirements engineering process – A case study. In *Proceedings of the 13th European Conference on Information Systems* (pp. 628–640). Academic Press.

Hanks, K. S., Knight, J. C., & Strunk, E. A. (2001). *A linguistic analysis of requirements errors and its application* (Technical Report CS-2001-30). University of Virginia Department of Computer Science.

Harland, T. (2003). Vygotsky's zone of proximal development and problem-based learning: Linking a theoretical concept with practice through action research. *Higher Education*, *8*(2), 263–272.

Hart, G., & Stone, T. (2002). Conversations with students: The outcomes of focus groups with QUT students. In *Proceedings of the 2002 Annual International Conference of the Higher Education Research and Development Society of Australasia (HERDSA)*. HERDSA.

Hart, D. (1994). *Authentic assessment: A handbook for educators*. Reading, MA: Addison-Wesley.

Harvey, L., & Bowers-Brown, T. (2004, Winter). Employability cross-country comparisons. *Graduate Market Trends*.

Hauge, O., Ayala, C., & Conradi, R. (2010). Adoption of open source software in software-intensive organizations – A systematic literature review. *Information and Software Technology*, *52*(11), 1133–1154. doi:10.1016/j.infsof.2010.05.008

Hawk, T. F., & Shah, A. J. (2007). Using learning style instruments to enhance student learning. *Decision Sciences Journal of Innovative Education*, *5*(1), 1–19. doi:10.1111/j.1540-4609.2007.00125.x

Hayes, D., Hill, J., Mannette-Wright, A., & Wong, H. (2006). Team project patterns for college students. In *Proceedings of the 2006 Conference on Pattern Languages of Programs*. ACM.

Hayes, J. H. (2002). Energizing software engineering education through real-world projects as experimental studies. In *Proceedings of the 15th Conference on Software Engineering Education and Training* (pp. 192–206). IEEE.

Hayes, J. H. (2002). Energizing software engineering education through real–world projects as experimental studies. In *Proceedings of the 15th Software Engineering Education and Training Conference* (pp. 192–206). Covington, KY: IEEE.

Hayes, J. H., Lethbridge, T. C., & Port, D. (2003). Evaluating individual contribution toward group software engineering projects. In *Proceedings of the 25th International Conference on Software Engineering* (pp. 622–627). IEEE.

Hazzan, O., & Dubinsky, Y. (2006). Teaching framework for software development methods. In *Proceedings of the 28th International Conference on Software Engineering* (pp. 703–706). ACM.

Hazzan, O., & Kramer, J. (2007). Abstraction in computer science & software engineering: A pedagogical perspective. *Frontier Journal, 4*(1), 6–14.

Hazzan, O., Lapidot, T., & Ragonis, N. (2011). *Guide to teaching computer science: An activity-based approach*. Berlin: Springer. doi:10.1007/978-0-85729-443-2

Hear-See-Do . (n.d.). Retrieved from http://hear-see-do.com/

Hedin, G., Bendix, L., & Magnusson, B. (2003). Introducing software engineering by extreme programming. In *Proceedings of the 25th International Conference on Software Engineering* (pp. 586–593). IEEE Computer Society Press.

Hedin, G., Bendix, L., & Magnusson, B. (2005). Teaching extreme programming to large groups of students. *Journal of Systems and Software, 74*(2), 133–146. doi:10.1016/j.jss.2003.09.026

Hellens, L. A. (1997). Information systems quality versus software quality a discussion from a managerial, an organisational and an engineering viewpoint. *Information and Software Technology, 39*(12), 801–808. doi:10.1016/S0950-5849(97)00038-4

Helmo-Silver, C. E. (2004). Problem-based learning: What and how do students learn? *Educational Psychology Review, 16*(3), 235–266. doi:10.1023/B:EDPR.0000034022.16470.f3

Hennessey, B. A., & Amabile, T. M. (2010). Creativity. *Annual Review of Psychology, 61*, 561–598. doi:10.1146/annurev.psych.093008.100416

Herbsleb, J. D. (2007). Global software engineering: The future of socio-technical coordination. In *Proceedings of Future of Software Engineering* (pp. 188–198). IEEE. doi:10.1109/FOSE.2007.11

Herbsleb, J. D., & Moitra, D. (2001). Global software development. *IEEE Software, 18*(2), 16–20. doi:10.1109/52.914732

Herreid, C. F. (1994). Case studies in science: A novel method of science education. *Journal of College Science Teaching, 23*(4), 221–229.

Herrington, J., & Oliver, R. (2000). An instructional design framework for authentic learning environments. *Educational Technology Research and Development, 48*(3), 23–48. doi:10.1007/BF02319856

Herrington, J., Reeves, T., & Oliver, R. (2003). Patterns of engagement in authentic online learning environments. *Australian Journal of Educational Technology, 19*(1), 59–71.

Hew, K., & Cheung, W. (2009). Use of wikis in K-12 and higher education: A review of the research. *International Journal of Continuing Engineering Education and Lifelong Learning, 19*(2/3), 141–165. doi:10.1504/IJCEELL.2009.025024

Higher Education Academy. (2003). *PBLE (project based learning in engineering)*. Retrieved 09-07-2013, 2013, from http://www.heacademy.ac.uk/resources/detail/resource_database/SNAS/PBLE_Project_Based_Learning_in_Engineering

Highsmith, J. (2004). *Agile project management-creating innovative products*. Boston: Addison-Wesley.

Highsmith, J. (2009). *Agile project management: Creating innovative products*. Upper Saddle River, NJ: Pearson Education.

Hilburn, T., & Towhidnejad, M. (2007). A case for software engineering. In *Proceedings of the 20ᵗʰ Conference on Software Engineering Education and Training* (pp. 107–114). IEEE.

Hilburn, T., Towhidnejad, M., & Salamah, S. (2008). The DigitalHome case study material. In *Proceedings of the 21st Conference on Software Engineering Education and Training* (pp. 279–280). IEEE.

Hmelo-Silver, C. E., & Barrows, H. S. (2006). Goals and strategies of a problem-based learning facilitator. *Learning, 1*(1), 21–39.

Hoeppner, K., & Boag, A. (2011). Open source software for education. *Global Learn, 1*, 1658.

Hogan, J. M., & Thomas, R. (2005). Developing the software engineering team. In *Proceedings of 2005 Australasia Computing Education Conference*. Newcastle, Australia: Academic Press.

Hon, A., & Chun, W. (2004). Teaching agile teaching/learning methodology and its e-learning platform. *LNCS-Advances in Web-Based Learning, 3143*, 11–18.

Horton, W. (2011). *E-learning by design*. Hoboken, NJ: Wiley. doi:10.1002/9781118256039

Horvath, I., Wiersma, M., Duhovnik, J., & Stroud, I. (2004). Navigated active learning in an international academic virtual enterprise. *European Journal of Engineering Education, 29*(4), 505–519. doi:10.1080/03043 790410001716275

Huang, L., Dai, L., Guo, B., & Lei, G. (2008). Project-driven teaching model for software project management course. In *Proceedings of 2008 International Conference on Computer Science and Software Engineering* (Vol. 5, pp. 503–506). IEEE.

Huba, M. E., & Freed, J. E. (2000). Learner centered assessment on college campuses: Shifting the focus from teaching to learning. *Community College Journal of Research and Practice, 24*(9), 759–766.

Humphrey, W. S. (1997). *Introduction to the personal software process*. Reading, MA: Addison-Wesley Longman, Inc.

Humphrey, W. S. (2000). *Introduction to the team software process*. Reading, MA: Addison-Wesley Longman, Inc.

Humphrey, W. S. (2005). *PSP(SM) - A self-improvement process for software engineers*. Reading, MA: Addison-Wesley.

IBM. (n.d.). *Developerworks*. Retrieved from http://www.ibm.com/developerworks/rational/library/apr05/crain/

IBM. (n.d.). *Solutions*. Retrieved from http://www-01.ibm.com/software/solutions/soa/innov8/index.html

IEEE. Computer Society & Association for Computing Machinery ACM. (2004). *Software engineering 2004 curriculum guidelines for undergraduate degree programs in software engineering: A volume of the computing curricula series*. Retrieved September 24, 2013, from http://sites.computer.org/ccse/SE2004Volume.pdf

Ifversen, J. (2003). Text, discourse, concept: Approaches to textual analysis. *Kontur, 7*, 60–69.

iPeer . (n.d.). Retrieved from http://sourceforge.net/projects/ipeer/

Isistan . (n.d.). Retrieved from http://isistan.exa.unicen.edu.ar/u3d/

Isomoettoenen, V., & Kaerkkaeinen, T. (2008). The value of a real customer in a capstone project. In *Proceedings of the 21st Conference on Software Engineering Education & Training*, (pp. 85–92). Academic Press.

Ivanović, M., Welzer, T., Putnik, Z., Hölbl, M., Komlenov, Ž., Pribela, I., & Schweighofer, T. (2009). Experiences and privacy issues – Usage of Moodle in Serbia and Slovenia. In *Proceedings of the Interactive Computer Aided Learning* (pp. 416–423). Academic Press.

Ivanović, M., Putnik, Z., Komlenov, Ž., Welzer, T., Hölbl, M., & Schweighofer, T. (2013). *Usability and privacy aspects of Moodle – Students' and teachers' perspective*. Informatica – An International Journal of Computing and Informatics.

Ivanović, M., Xinogalos, S., & Komlenov, Ž. (2011). Usage of technology enhanced educational tools for delivering programming courses. *International Journal of Emerging Technologies in Learning, 6*(4), 23–30.

Jaccheri, L., & Osterlie, T. (2007). Open source software: A source of possibilities for software engineering education and empirical software engineering. In *Proceedings of the 1ˢᵗ International Workshop on Emerging Trends in FLOSS Research and Development* (pp. 1–5). Minneapolis, MN: IEEE.

Jain, A., & Boehm, B. (2006). SimVBSE: Developing a game for value–based software engineering. In *Proceedings of 19th Conference on Software Engineering Education and Training* (pp. 103–114). IEEE.

Jalote, P. (2002). *Software project management in practice.* Boston, MA: Addison Wesley.

Jayaratna, N., & Sommerville, I. (1998). The role of information systems methodology in software engineering. *IEE Proceedings. Software, 145*(4), 93–94. doi:10.1049/ip-sen:19982193

Jazayeri, M. (2004). The education of a software engineer. In *Proceedings of the 19th IEEE International Conference on Automated Software Engineering* (pp. 18-27). IEEE Computer Society.

Jenkins. (n.d.). Retrieved from http://jenkins-ci.org/

Johns-Boast, L. F. (2010). *Group work and individual assessment.* Paper presented at the Past, Present, Future: 21st Annual Conference of the Australasian Association of Engineering Education. Sydney, Australia.

Johns-Boast, L. F., & Flint, S. (2009). *Providing students with 'real-world' experience through university group projects.* Paper presented at the 20th Annual Conference of the Australasian Association for Engineering Education. Adelaide, Australia.

Johns-Boast, L. F., & Flint, S. (2013). *Simulating Industry: An innovative software engineering capstone design course.* Paper presented at the 43rd Annual Frontiers in Education (FIE) Conference. Oklahoma City, OK.

Johns-Boast, L. F., & Patch, G. (2010). *A win-win situation: Benefits of industry-based group projects.* Paper presented at the Past, Present, Future: 21st Annual Conference of the Australasian Association of Engineering Education. Sydney, Australia.

Johnson, D. W. (1974). Communication and the inducement of cooperative behavior in conflicts. *Speech Monographs, 41*, 64–78. doi:10.1080/03637757409384402

Johnson, D. W. (1975). Cooperativeness and social perspective taking. *Journal of Personality and Social Psychology, 31*, 241–244. doi:10.1037/h0076285

Johnson, D. W., & Johnson, F. P. (2002). *Joining together: Group theory and group skills.* Boston: Allyn & Bacon.

Johnson, D. W., & Johnson, R. (1975/1994). *Learning together and alone: Cooperative, competitive, and individualistic learning.* Englewood Cliffs, NJ: Prentice-Hall.

Johnson, D. W., & Johnson, R. T. (1989). *Cooperation and competition: Theory and research.* Edina, MN: Interaction Book Company.

Johnson, D., Johnson, R., & Holubec, E. (1992). *Advanced cooperative learning.* Edina, MN: Interaction Book Company.

Johnson, D., Johnson, R., & Smith, K. (1998). *Active learning: Cooperation in the college classroom.* Edina, MN: Interaction Book Company.

Jollands, M., Jolly, L., & Molyneaux, T. (2012). Project-based learning as a contributing factor to graduates' work readiness. *European Journal of Engineering Education, 37*(2), 143–154. doi:10.1080/03043797.2012.665848

Jonassen, D. H. (1992). Semantic networking as cognitive tools. In P. A. M. Kommers, D. H. Jonassen, & J. T. Mayes (Eds.), *Cognitive tools for learning* (pp. 19–21). Heidelberg, Germany: Springer-Verlag.

Jonassen, D. H. (2002). Learning to solve problems online. In C. Vrasidas, & V. Glass (Eds.), *Distance education and distance learning* (pp. 75–98). Greenwich, CT: Information Age Publishing.

Jonassen, D. H., & Grabowski, B. L. (1993). *Handbook of individual differences, learning and instruction.* New York: Allyn & Bacon.

Jong, T. (2006). Technological advances in inquiry learning. *Science, 312*(5773), 532–533. doi:10.1126/science.1127750 PMID:16645080

Jorgensen, M. (2004). A review of studies on expert estimation of software development effort. *Journal of Systems and Software, 70*, 37–60. doi:10.1016/S0164-1212(02)00156-5

Jqueryi. (n.d.). Retrieved from http://jqueryui.com/

Judd, T., Kennedy, G., & Cropper, S. (2010). Using wikis for collaborative learning: Assessing collaboration through contribution. *Australasian Journal of Educational Technology*, *26*(3), 341–354.

Judith, W., Bair, B., & Börstler, J. (2003). Client sponsored projects in software engineering courses. In *Proceedings of the 34th SIGCSE Technical Symposium on Computer Science Education* (pp. 401–402). ACM.

Jung, E. (2002). *Projektunterricht - Projektstudium - Projektmanagement*. Retrieved July 25, 2013, from http://www.sowi-online.de/praxis/methode/projektunterricht_projektstudium_projektmanagement.html

Jung, E. (2010). *Kompetenzenerwerb: Grundlagen, didaktik, uberprüfbarkeit*. München: Oldenbourg, R/CVK.

Junit. (n.d.). Retrieved from http://junit.org/

Kabicher, S., Motschnig-Pitrik, R., & Figl, K. (2009). What competences do employers, staff and students expect of a computer science graduate? In *Proceedings of the 39th Frontiers in Education Conference* (pp. 1–6). IEEE.

Kafai, Y., & Resnick, M. (1996). *Constructionism in practice: Designing, thinking, and learning in a digital world*. Mahwah, NJ: Lawrence Erlbaum.

Kampa-Kokesch, S., & Anderson, M. Z. (2001). Executive coaching: A comprehensive review of the literature. *Consulting Psychology Journal: Practice and Research*, *53*(4), 205–228. doi:10.1037/1061-4087.53.4.205

Kamsties, E., & Lott, C. (1995). An empirical evaluation of three defect-detection techniques. In *Proceedings of the 5th European Software Engineering Conference*. Academic Press.

Kamthan, P. (2007). A perspective on software engineering education with open source software. In K. St. Amant, & B. Still (Eds.), *Handbook of research on open source software: Technological, economic, and social perspectives* (pp. 690–702). Hershey, PA: Information Science Reference. doi:10.4018/978-1-59140-999-1.ch054

Kamthan, P. (2009). A methodology for integrating the social Web environment in software engineering education. *International Journal of Information and Communication Technology Education*, *5*(2), 21–35. doi:10.4018/jicte.2009040103

Kane, G. C., & Fichman, R. G. (2009). The shoemaker's children: Using wikis for information systems teaching, research, and publication. *Management Information Systems Quarterly*, *33*(1), 1–17.

Kao, G. Y. M. (2013). Enhancing the quality of peer review by reducing student free riding: Peer assessment with positive interdependence. *British Journal of Educational Technology*, *44*(1), 112–124. doi:10.1111/j.1467-8535.2011.01278.x

Karunasekera, S., & Bedse, K. (2007). Preparing software engineering graduates for an industry career. In *Proceedings of the 20th Conference on Software Engineering Education & Training* (pp. 97–106). IEEE.

Katzenbach, J. R., & Smith, D. K. (1992). *The wisdom of teams: Creating the high-performance organization*. Boston: Harvard Business Press.

Kaufman, D. B., Felder, R. M., & Fuller, H. (1999). Peer ratings in cooperative learning teams. In *Proceedings of the 1999 Annual ASEE Meeting*. ASEE.

Kaufman, D. B., Felder, R. M., & Fuller, H. (2000). Accounting for individual effort in cooperative learning teams. *Journal of Engineering Education*, *89*(2), 133–140. doi:10.1002/j.2168-9830.2000.tb00507.x

Kayes, A. B., Kayes, D. C., & Kolb, D. A. (2005). Experiential learning in teams. *Simulation & Gaming*, *36*(3), 330–354. doi:10.1177/1046878105279012

Keenan, E., Steele, A., & Jia, X. (2010). Simulating global software development in a course environment. In *Proceedings of the 5th IEEE International Conference on Global Software Engineering* (pp. 201–205). IEEE.

Keirsey, D., & Bates, M. (1984). *Please understand me*. Prometheus Nemesis Book Company.

Keller, J. M. (1987a). Strategies for stimulating the motivation to learn. *Performance & Instruction*, *26*(8), 1–7. doi:10.1002/pfi.4160260802

Keller, J. M. (1987b). The systematic process of motivational design. *Performance & Instruction*, *26*(9), 1–8. doi:10.1002/pfi.4160260902

Kemp, S. (2011). *Constructivism and problem-based learning*. Retrieved October 09, 2013, from http://www.tp.edu.sg/pbl_sandra_joy_kemp.pdf

Kennan, M. A., Willard, P., Cecez-Kecmanovic, D., & Wilson, C. S. (2009). IS knowledge and skills sought by employers: A content analysis of Australian IS early career online job advertisements. *The Australasian Journal of Information Systems, 15*(2), 169–190.

Kernel. (n.d.). Retrieved from https://www.kernel.org/

Khan, F. A., Graf, S., Weippl, E. R., & Tjoa, A. M. (2010). Implementation of affective states and learning styles tactics in web-based learning management systems. In *Proceedings of the 10ᵗʰ IEEE International Conference on Advanced Learning Technologies* (pp. 734–735). IEEE.

King, A. (1993). From sage on the stage to guide on the side. *College Teaching, 41*(1), 30–35. doi:10.1080/8756 7555.1993.9926781

Kniberg, H. (2008). *The manager's role in scrum.* Retrieved from http://www.scrumalliance.org/resources/293

Knight, P. (1995). *Assessment for learning in higher education.* Psychology Press.

Knowles, M. S., Holton, E. F., & Swanson, R. A. (2011). *The adult learner: The definitive classic in adult education and human resource development.* Amsterdam: Elsevier.

Kolb, A. Y., & Kolb, D. A. (2009). The learning way: Meta-cognitive aspects of experiential learning. *Simulation & Gaming, 40*(3), 297–327. doi:10.1177/1046878108325713

Kolb, D. A. (1984). *Experiential learning experience as the source of learning and development.* Upper Saddle River, NJ: Prentice-Hall.

Kolb, D. A. (1984). *Experiential learning: Experience as the source of learning and development* (Vol. 1). Englewood Cliffs, NJ: Prentice-Hall.

Kolb, D. A. (1995). *Learning style inventory: Technical specifications.* Boston: McBer & Company.

Kolodner, J. L., Camp, P. J., Crismond, D., Fasse, B., Gray, J., & Holbrook, J. et al. (2003). Problem-based learning meets case-based reasoning in the middle-school science classroom: Putting learning by design(tm) into practice. *Journal of the Learning Sciences, 12*, 495–547. doi:10.1207/S15327809JLS1204_2

Komlenov, Ž., Budimac, Z., Putnik, Z., & Ivanović, M. (2013). Wiki as a tool of choice for students' team assignments. *International Journal of Information Systems and Social Change, 4*(3), 1–16. doi:10.4018/jissc.2013070101

Koolmanojwang, S., & Boehm, B. (2011). Educating software engineers to become systems engineers. In *Proceedings of the 24ᵗʰ Conference on Software Engineering Education & Training* (pp. 209–218). IEEE Computer Society Press.

Koppi, T., & Naghdy, F. (2009). *Discipline-based initiative: Managing educational change in the ICT discipline at the tertiary education level.* Wollongong, Australia: University of Wollongong.

Koranne, S. (2011). *Handbook of open source tools.* Berlin: Springer. doi:10.1007/978-1-4419-7719-9

Kornecki, A. J., Khajenoori, S., Gluch, D., & Kameli, N. (2003). On a partnership between software industry and academia. In *Proceedings of the 16th Conference on Software Engineering Education and Training* (pp. 60–69). IEEE.

Koschmann, T. D., Myers, A. C., Barrows, H. S., & Feltovich, P. J. (1994). Using technology to assist in realising effective learning and instruction: A principled approach to the use of computers in collaborative learning. *Journal of the Learning Sciences, 3*(3), 227–264. doi:10.1207/s15327809jls0303_2

Kovitz, B. (2003). Hidden skills that support phased and agile requirements engineering. *Requirements Engineering, 8*(2), 135–141. doi:10.1007/s00766-002-0162-9

Kroll, P., & Kruchten, P. (2003). *The rational unified process made easy: A practitioner's guide to the RUP.* Boston: Addison-Wesley.

Kruchten, P. (2011). Experience teaching software project management in both industrial and academic settings. In *Proceedings of the 24ᵗʰ Conference on Software Engineering Education & Training* (pp. 199–208). IEEE Computer Society Press.

Kruchten, P. (2003). *The rational unified process: An introduction.* Reading, MA: Addison-Wesley.

Kuhn, S. (2001). Learning from the architecture studio: Implications for project-based pedagogy. *International Journal of Engineering Education, 17*(4-5), 349–352.

Kuhrmann, M. (2012). A practical approach to align research with master's level courses. In *Proceedings of the 15th International Conference on Computational Science and Engineering* (pp.202-208). IEEE.

Kuhrmann, M., Fernández, D. M., & Münch, J. (2013). Teaching software process modeling. In *Proceedings of the 2013 International Conference on Software Engineering* (pp. 1138-1147). IEEE Press.

Kulak, D., & Guiney, E. (2003). *Use cases: Requirements in context*. London: Addison-Wesley.

Kulpa, M., & Johnson, K. (2003). *Interpreting the CMMI*. CRC Press. doi:10.1201/9780203504611

Kurbel, K., Krybus, I., & Nowakowski, K. (2013). *Lehrstuhl für wirtschaftsinformatik*. Retrieved 2013-07-05, 2013, from http://www.enzyklopaedie-der-wirtschaftsinformatik.de/wi-enzyklopaedie/lexikon/uebergreifendes/

Kurbel, K. E. (2008). *The making of information systems: Software engineering and management in a globalized world*. Berlin, Germany: Springer. doi:10.1007/978-3-540-79261-1

Lago, P., & vanVliet, H. (2005). Teaching a course on software architecture. In *Proceedings of the 18th IEEE Conference on Software Engineering Education and Training* (pp. 35–42). IEEE Computer Society.

Lähteenmäki, M. L., & Uhlin, L. (2011). Developing reflective practitioners through PBL in academic and practice environments. In *New approaches to problem-based learning: Revitalising your practice in higher education* (pp. 144–157). Academic Press.

Lakhan, S. E., & Jhunjhunwala, K. (2008). Open source software in education. *EDUCAUSE Quarterly, 31*(2), 32–40.

Larman, C. (2004). *Applying UML and patterns: An introduction to object-oriented analysis and design and iterative development*. Upper Saddle River, NJ: Prentice-Hall.

Larman, C., & Basili, V. R. (2003). Iterative and incremental development. *IEEE Computer, 36*(6), 47–56. doi:10.1109/MC.2003.1204375

Laurillard, D. (1993). *Rethinking university teaching: A framework for the effective use of educational technology*. London: Routledge.

Lave, J., Smith, S., & Butler, M. (1988). Problem solving as an everyday practice. In *Learning mathematical problem solving*. Palo Alto, CA: Institute for Research on Learning.

Lave, J., & Wenger, E. (1991). *Situated learning: Legitimate peripheral participation*. New York: Cambridge University Press. doi:10.1017/CBO9780511815355

Layman, L., Cornwell, T., & Williams, L. (2006). Personality types, learning styles, and an agile approach to software engineering education. *ACM SIGCSE Bulletin, 38*(1), 428–432. doi:10.1145/1124706.1121474

Lazarinis, F. (2004). A template based system for automatic construction of online courseware for secondary educational institutes. *Journal of Educational Technology & Society, 7*(3), 112–123.

Lazic, N., Zorica, M. B., & Klindzic, J. (2011). Open source software in education. In *Proceedings of the 34th International Convention MIPRO* (pp. 1267–1270). Opatija, Croatia: IEEE.

LeBlanc, R., & Sobel, A. E. K. (Eds.). (2004). *Software engineering 2004: Curriculum guidelines for undergraduate degree programs in software engineering*. Los Alamitos, CA: IEEE Computer Society Press.

Lee, D. M. S. (1999). *Information seeking and knowledge acquisition behaviors of young information systems workers: Preliminary analysis*. Paper presented at the 1999 Americas Conference on Information Systems. New York, NY.

Lee, D. M. S. (2004). Organizational entry and transition from academic study: Examining a critical step in the professional development of young IS workers. In M. Igbaria, & C. Shayo (Eds.), *Strategies for managing IS/IT personnel* (pp. 113–141). Hershey, PA: Idea Group.

Leffingwell, D. (2010). *Agile software requirements: Lean requirements practices for teams, programs, and the enterprise*. Reading, MA: Addison-Wesley.

Lefrançois, G. R. (2006). *Theories of human learning: What the old woman said* (5th ed.). Belmont, CA: Thomson/Wadsworth.

Lethbridge, T. C. (1999). *The relevance of education to software practitioners: Data from the 1998 survey.* Ottowa, Canada: School of Information Technology and Engineering, University of Ottowa.

Lethbridge, T. C. (2000). What knowledge is important to a software professional? *IEEE Computer, 33*(5), 44–50. doi:10.1109/2.841783

Lethbridge, T. C., & Laganière, R. (2001). *Object-oriented software engineering: Practical software development using UML and Java.* New York: McGraw Hill.

Leuf, B., & Cunningham, W. (2001). *The wiki way: Quick collaboration on the web.* Reading, MA: Addison-Wesley.

Levia, D. F., & Quiring, S. M. (2008). Assessment of student learning in a hybrid PBL capstone seminar. *The Journal of Geography, 32*(2), 217–231.

Levitin, A., & Papalaskari, M. A. (2002). Using puzzles in teaching algorithms. *ACM SIGCSE Bulletin, 34*(1), 292–296. doi:10.1145/563517.563456

Levy, A. (1986). Second order planned change: Definition and conceptualisation. *Organizational Dynamics, 15,* 5–20. doi:10.1016/0090-2616(86)90022-7

Lewin, K. (1935). *A dynamic theory of personality.* New York: McGraw-Hill.

Lewin, K. (1948). *Resolving social conflicts.* New York: Harper.

Limongelli, C., Sciarrone, F., & Vaste, G. (2008). Ls-plan: An effective combination of dynamic courseware generation and learning styles in web-based education. In *Proceedings of the 5th International Conference on Adaptive Hypermedia and Adaptive Web-Based Systems* (pp. 133–142). Springer-Verlag.

Lingard, R., & Berry, E. (2002). Teaching teamwork skills in software engineering based on understanding of factors affecting group performance. In *Proceedings of the 32nd Frontiers in Education Conference* (Vol. 3, pp. S3G-1). IEEE.

Linn, M. C., Davis, E. A., & Bell, P. (2004). Inquiry and technology. In M. C. Linn, E. A. Davis, & P. Bell (Eds.), *Internet environments for science education* (pp. 3–28). Mahwah, NJ: Lawrence Erlbaum Associates.

Linos, P. K., Herman, S., & Lally, J. (2003). A service-learning program for computer science and software engineering. In *Proceedings of the 8th Annual Conference on Innovation and Technology in Computer Science Education* (pp. 30–34). ACM.

Litchfield, A., Frawley, J., & Nettleton, S. (2010). Contextualising and integrating into the curriculum the learning and teaching of work-ready professional graduate attributes. *Higher Education Research & Development, 29*(5), 519–534. doi:10.1080/07294360.2010.502220

Litecky, C. R., Arnett, K. P., & Prabhakar, B. (2004). The paradox of soft skills versus technical skills in IS hiring. *Journal of Computer Information Systems, 45*(1), 69–76.

Litzinger, T. A., Lattuca, L. R., Hadgraft, R. G., & Newstetter, W. C. (2011). Engineering education and the development of expertise. *Journal of Engineering Education, 100*(1), 123–150. doi:10.1002/j.2168-9830.2011.tb00006.x

Liu, D. K., Huang, S. D., & Brown, T. A. (2007). Supporting teaching and learning of optimisation algorithms with visualisation techniques. In *Proceedings of 2007 Australian Association of Engineering Education Conference.* Academic Press.

Liu, E. Z. F., & Lin, S. S. (2007). Relationship between peer feedback, cognitive and metacognitive strategies and achievement in networked peer assessment. *British Journal of Educational Technology, 38*(6), 1122–1125. doi:10.1111/j.1467-8535.2007.00702.x

Liu, N. F., & Carless, D. (2006). Peer feedback: The learning element of peer assessment. *Teaching in Higher Education, 11*(3), 279–290. doi:10.1080/13562510600680582

Livingstone, D., & Lynch, K. (2002). Group project work and student-centred active learning: Two different experiences. *Journal of Geography in Higher Education, 26*(2), 217–237. doi:10.1080/03098260220144748

Locmetrics. (n.d.). Retrieved from http://www.locmetrics.com/

Lowry, G., & Turner, R. (2005). Information systems education for the 21st century: Aligning curriculum content & delivery with the professional workplace. In D. Carbonara (Ed.), *Technology literacy applications in learning environments* (pp. 171–202). Hershey, PA: IRM Press. doi:10.4018/978-1-59140-479-8.ch013

LTSN. (2002). *Constructive alignment and why it is important to the learner.* Retrieved from http://www.ltsneng. ac.uk/er/theory/constructivealignment.asp

Lu, H. K., Lo, C. H., & Lin, P. C. (2011). Competence analysis of IT professionals involved in business services — Using a qualitative method. In *Proceedings of the 24th Conference on Software Engineering Education and Training* (pp. 61–70). IEEE.

Lui, C., Sandell, K., & Welch, L. (2005). Teaching communication skills in software engineering courses. In *Proceedings of the 2005 American Society for Engineering Education Annual Conference & Exposition.* ASEE.

Lundell, B., Persson, A., & Lings, B. (2007). Learning through practical involvement in the OSS ecosystem: Experiences from a masters assignment. In *Open source development, adoption and innovation* (pp. 289–294). Berlin: Springer. doi:10.1007/978-0-387-72486-7_30

Lutteroth, C., Luxton-Reilly, A., Dobbie, G., & Hamer, J. (2007). A maturity model for computing education. In *Proceedings of the Ninth Australasian Conference on Computing Education* (vol. 6, pp. 107–114). Australian Computer Society.

Lynch, T. D., Herold, M., Bolinger, J., Deshpande, S., Bihari, T., Ramanathan, J., & Ramnath, R. (2011). An agile boot camp: Using a LEGO-based active game to ground agile development principles. In *Proceedings of Frontiers in Education Conference* (pp. F1H-1). IEEE.

Macaulay, L., & Mylopoulos, J. (1995). Requirements engineering: An educational dilemma. *Automated Software Engineering*, 2(4), 343–351. doi:10.1007/BF00871804

Machanick, P. (1998). The skills hierarchy and curriculum. In *Proceedings of 1998 Conference of the South African Institute of Computer Scientists and Information Technologists* (pp. 54–62). Academic Press.

MacKellar, B. (2013). Analyzing coordination among students in a software engineering project course. In *Proceedings of the 26th IEEE Conference on Software Engineering Education and Training* (pp. 279–283). IEEE Press.

Mackellar, B. K. (2011). A software engineering course with a large-scale project and diverse roles for students. *Journal of Computing Sciences in Colleges*, 26(6), 93–100.

MacKellar, B. K., Sabin, M., & Tucker, A. (2013). Scaling a framework for client-driven open source software projects: A report from three schools. *Journal of Computing Sciences in Colleges*, 28(6), 140–147.

Madsen, F., & Nürnberg, P. (2005). Calliope: Supporting high-level documentation of open-source projects. In *Proceedings of the 2005 Symposia on Metainformatics* (pp. 10). ACM.

Mager, R. F. (1992). *Preparing instructional objectives.* London: Kogan Page.

Maher, P. (2009). Weaving agile software development techniques into a traditional computer science curriculum. In *Proceeding on 6th Conference on Information Technology: New Generations.* Academic Press.

Mahnic, V. (2010). Teaching scrum through team-project work: Students' perceptions and teachers' observations. *The International Journal of Engineering Education.*

Mahnic, V. (2011). A case study on agile estimating and planning using scrum. *Electronics and Electrical Engineering*, 5.

Mahnic, V., & Rozanc, I. (2012). Students' perceptions of scrum practices. In *Proceedings of the 35th International Convention - Microelectronics, Electronics and Electronic Technology* (pp. 1178–1183). IEEE.

Mahnic, V. (2012). A capstone course on agile software development using Scrum. *IEEE Transactions on Education*, 55(1), 99–106. doi:10.1109/TE.2011.2142311

Mahnic, V., & Hovelja, T. (2012b). On using planning poker for estimating user stories. *Journal of Systems and Software*, 85(9), 2086–2095. doi:10.1016/j. jss.2012.04.005

Ma, L., Ferguson, J., Roper, M., & Wood, M. (2011). Investigating and improving the models of programming concepts held by novice programmers. *Computer Science Education*, 21(1), 57–80. doi:10.1080/0899340 8.2011.554722

Male, S. A. (2010). Generic engineering competencies: A review and modelling approach. *Education Research and Perspectives*, 37(1), 25–51.

Mandl-Strieglitz, P. (2001). How to successfully use software project simulation for educating software project managers. In *Proceedings of the 31ˢᵗ Frontiers in Education Conference* (Vol. 1, pp. T2D–19). IEEE.

Mannisto, T., Savolainen, J., & Myllarniemi, V. (2008). Teaching software architecture design. In *Proceedings of the 7ᵗʰ Working IEEE/IFIP Conference on Software Architecture* (pp. 117–124). IEEE.

Marçal, A. S. C., de Freitas, B. C. C., Soares, F. S. F., Furtado, M. E. S., Maciel, T. M., & Belchior, A. D. (2008). Blending scrum practices and CMMI project management process areas. *Innovations in Systems and Software Engineering, 4*(1), 17–29. doi:10.1007/s11334-007-0040-1

Marmorstein, R. (2011). Open source contribution as an effective software engineering class project. In *Proceedings of the 16th Annual Joint Conference on Innovation and Technology in Computer Science Education* (pp. 268–272). ACM.

Martin, R. C. (1999a). *Iterative and incremental development (part I)*. Retrieved July 11, 2013 from http://www.objectmentor.com/resources/articles/IIDI.pdf

Martin, R. C. (1999b). *Iterative and incremental development (part II)*. Retrieved July 11, 2013 from http://www.objectmentor.com/resources/articles/IIDII.pdf

Martin, R. C. (1999c). *Iterative and incremental development (part III)*. Retrieved July 11, 2013 from http://www.objectmentor.com/resources/articles/IIDIII.pdf

Mathiassen, L., & Pries-Heje, J. (2006). Business agility and diffusion of information technology. *European Journal of Information Systems*, 116–119. doi:10.1057/palgrave.ejis.3000610

McBride, N. (2003). A viewpoint on software engineering and information systems: Integrating the disciplines. *Information and Software Technology, 45*(5), 281–287. doi:10.1016/S0950-5849(02)00213-6

McConnell, J. J. (2006). Active and cooperative learning. *ACM SIGCSE Bulletin, 38*(2), 24. doi:10.1145/1138403.1138426

McConnell, S. (1997). *Software project survival guide*. Microsoft Press.

McConnell, S., & Tripp, L. (1999). Professional software engineering: Fact or fiction? *IEEE Software, 16*(6), 13–17. doi:10.1109/MS.1999.805468

McCracken, W. M. (1997). SE education: What academia can do. *IEEE Software, 14*(6), 27, 29.

McCracken, M., Almstrum, V., Diaz, D., Guzdial, M., Hagan, D., Kolikant, Y. B. D., & Wilusz, T. (2001). A multi-national, multi-institutional study of assessment of programming skills of first-year CS students. *ACM SIGCSE Bulletin, 33*(4), 125–180. doi:10.1145/572139.572181

McDonald, J. (2000). Teaching software project management in industrial and academic environments. In *Proceedings of the 13ᵗʰ Conference on Software Engineering Education & Training* (pp. 151–160). IEEE Computer Society Press.

McDonough, D., Strivens, J., & Rada, R. (1994). Current development and use of computer-based teaching at the University of Liverpool. *Computers & Education, 22*(4), 335–343. doi:10.1016/0360-1315(94)90055-8

McDowell, C., Werner, L., Bullock, H., & Fernald, J. (2002). The effects of pair-programming on performance in an introductory programming course. *ACM SIGCSE Bulletin, 34*(1), 38–42. doi:10.1145/563517.563353

McGinnes, S. (1995). Communication and collaboration: Skills for the new IT professional. In *Proceedings of the 2nd All-Ireland Conference on the Teaching of Computing*. Academic Press.

McLarty, R. (1998). *Using graduate skills in small and medium sized enterprises*. Ipswich, UK: University College Suffolk Press.

McLaughlan, R. G., & Kirkpatrick, D. (1999). A decision making simulation using computer mediated communication. *Australian Journal of Educational Technology, 15*, 242–256.

McLaughlan, R. G., & Kirkpatrick, D. (2004). Online roleplay: Design for active learning. *European Journal of Engineering Education, 29*(4), 477–490. doi:10.1080/03043790410001716293

McLean, P., Perkins, K., Tout, D., Brewer, K., & Wyse, L. (2012). *Australian core skills framework*. Canberra, Australia: Government of Australia.

McLennan, B., & Keating, S. (2008). *Work-integrated learning (WIL) in Australian universities: The challenges of mainstreaming WIL.* Paper presented at the ALTC NAGCAS National Symposium. Melbourne, Australia.

McLester, S. (2011). Building a blended learning program. *District Administration.* Retrieved from http://www.districtadministration.com/article/building-blended-learning-program

McLoughlin, C., & Luca, J. (2000). Developing professional skills and competencies in tertiary learners through on-line assessment and peer support. In *Proceedings of World Conference on Educational Multimedia, Hypermedia and Telecommunications* (pp. 668–673). Academic Press.

McLoughlin, C., Baird, J., Pigdon, K., & Wooley, M. (2000). Fostering teacher inquiry and reflective learning processes through technology enhanced scaffolding in a multimedia environment. In *Proceedings of World Conference on Educational Multimedia, Hypermedia and Telecommunications* (pp. 185–190). Academic Press.

McMeekin, D. A., von Konsky, B. R., Chang, E., & Cooper, D. J. (2009). Evaluating software inspection cognition levels using bloom's taxonomy. In *Proceedings of the 22nd Conference on Software Engineering Education and Training* (pp. 232–239). IEEE.

McPherson, A., Proffitt, B., & Hale–Evans, R. (2008). *Estimating the total development cost of a Linux distribution.* The Linux Foundation.

McQuaid, R. C., & Lindsay, C. (2005). The concept of employability. *Urban Studies (Edinburgh, Scotland), 42,* 197–219. doi:10.1080/0042098042000316100

McTavish, D. G., & Pirro, E. B. (1990). Contextual content analysis. *Quality & Quantity, 24*(3), 245–265. doi:10.1007/BF00139259

Meneely, A., Williams, L., & Gehringer, E. (2008). ROSE: A repository of education-friendly open-source projects. In *Proceedings of the 13th Annual Conference on Innovation and Technology in Computer Science Education* (pp. 7–11). ACM.

Merriam, S. B. (2001). Andragogy and self-directed learning: Pillars of adult learning theory. *New Directions for Adult and Continuing Education,* (89): 3. doi:10.1002/ace.3

Merriam, S. B., Caffarella, R. S., & Baumgartner, L. (2007). *Learning in adulthood: A comprehensive guide* (3rd ed.). San Francisco: Jossey-Bass.

Mervis, C. B., & Rosch, E. (1981). Categorization of natural objects. *Annual Review of Psychology, 32*(1), 89–115. doi:10.1146/annurev.ps.32.020181.000513

Metseker, S. (2006). *Design patterns in java.* Reading, MA: Addison-Wesley Professional.

Meyer, J. H. F., & Shanahan, M. P. (2000). Making teaching responsive to variation in student learning. In *Proceedings of the 7th Improving Student Learning Symposium.* Manchester, UK: Academic Press.

Michaelsen, L. K., & Schultheiss, E. E. (1988). Making feedback helpful. *Organizational Behavior Teaching Review, 13*(1), 109–113.

Milter, R. G., & Stinson, J. E. (1995). Educating leaders for the new competitive environment. In Educational innovation in economics and business administration: The case of problem-based learning administration. Boston: Kluwer.

Miranda, E. (2001). Improving subjective estimates using paired comparisons. *IEEE Software, 18*(1), 87–91. doi:10.1109/52.903173

MIT. (n.d.). *Beer game.* Retrieved from http://supplychain.mit.edu/games/beer-game

Mitchell, G. G., & Delaney, J. D. (2004). An assessment strategy to determine learning outcomes in a software engineering problem-based learning course. *International Journal of Engineering Education, 20*(3), 494–502.

Mohan, A., Merle, D., Jackson, C., Lannin, J., & Nair, S. S. (2010). Professional skills in the engineering curriculum. *IEEE Transactions on Education, 53*(4), 562–571. doi:10.1109/TE.2009.2033041

Monsalve, E. S. (2010). *Construindo um jogo educacional com modelagem intencional apoiado em princípios de transparência.* (Master Thesis). PUC–Rio.

Monsalve, E., Werneck, V., & Leite, J. C. S. P. (2010). Evolución de un juego educacional de ingeniería de software a través de técnicas de elicitación de requisitos. In *Proceedings of the 13th Workshop on Requirements Engineering* (pp. 12–23). Cuenca, Ecuador: Academic Press.

Monsalve, E., Werneck, V., & Leite, J. C. S. P. (2011). Teaching software engineering with simulESW. In *Proceedings of the 24th Conference on Software Engineering Education and Training* (pp. 31–40). IEEE.

Monteiro, M. P. (2011). On the cognitive foundations of modularity. In *Proceedings of 2011 Psychology of Programming Interest Group Conference*. Academic Press. González-Morales, D., de Antonio, L. M. M., & García, J. L. R. (2011). Teaching soft skills in software engineering. In *Proceedings of 2011 Global Engineering Education Conference* (pp. 630–637). IEEE.

Monteiro, P., Borges, P., Machado, R. J., & Ribeiro, P. (2012). A reduced set of RUP roles to small software development teams. In *Proceedings of International Conference on Software and System Process* (pp. 190–199). IEEE.

Monteiro, P., Machado, R., Kazman, R., Lima, A., Simões, C., & Ribeiro, P. (2013). Mapping CMMI and RUP process frameworks for the context of elaborating software project proposals. In *Software quality: Increasing value in software and systems development* (Vol. 133, pp. 191–214). Berlin: Springer. doi:10.1007/978-3-642-35702-2_12

Morelli, R., de Lanerolle, T., & Tucker, A. (2012). The HFOSS project: Engaging students in service learning through building software. In *Service learning in the computer and information sciences*. New York: Wiley and IEEE Press. doi:10.1002/9781118319130.ch5

Morrow, L. I. (2006). An application of peer feedback to undergraduates' writing of critical literature reviews. *Practice and Evidence of Scholarship of Teaching and Learning in Higher Education, 1*(2), 61–72.

Muller, G. (2004). Experiences of teaching systems architecting. In *Proceedings of International Council on Systems Engineering (INCOSE) Symposium*. INCOSE.

Münch, J., Pfahl, D., & Rus, I. (2005). Virtual software engineering laboratories in support of trade-off analyses. *International Software Quality Journal, 13*(4).

Myers Briggs. (n.d.). *MBTI personality type.* Retrieved from http://www.myersbriggs.org/my-mbti-personality-type/mbti-basics/

Myers, G. L., Sandler, C., & Badgett, T. (2011). *The art of software testing.* Hoboken, NJ: Wiley.

Myers, I. B., & Myers, P. B. (1995). *Gifts differing: Understanding personality type.* Nicholas Brealey Publishing.

Nandigam, J., Gudivada, V. N., & Hamou-Lhadj, A. (2008). Learning software engineering principles using open source software. In *Proceedings of Frontiers in Education Conference,* (pp. S3H:18 – S3H:23). Academic Press.

Nandigam, J., Gudivada, V. N., & Hamou–Lhadj, A. (2008). Learning software engineering principles using open source software. In *Proceedings of the 38th Frontiers in Education Conference* (pp. S3H–18). Saratoga Springs, NY: IEEE.

Narayan, A. (2010). Rote and algorithmic techniques in primary level mathematics teaching in the light of Gagne's hierarchy. In *Proceeding of the 3rd International Conference to Review Research on Science, Technology and Mathematics Education*. Academic Press.

Narayanan, N. H., Hundhausen, C., Hendrix, D., & Crosby, M. (2012). *Transforming the CS classroom with studio-based learning.* Paper presented at the SIGCSE'12. Raleigh, NC.

Navarro, E. O., & Van Der Hoek, A. (2007). Comprehensive evaluation of an educational software engineering simulation environment. In *Proceedings of the 20th Software Engineering Education & Training Conference* (pp. 195–202). IEEE.

Naveda, J. F. (1999). Teaching architectural design in an undergraduate software engineering curriculum. In *Proceedings of the 29th ASEE/IEEE Frontiers in Education Conference* (Vol. 2, pp. 12B1–1). IEEE.

Neenan, M., & Dryden, W. (2002). *Life coaching: A cognitive-behavioural approach.* Brunner-Routledge. doi:10.4324/9780203362853

Nelson, D., & Ng, Y. M. (2000). Teaching computer networking using open source software. *ACM SIGCSE Bulletin, 32*(3), 13–16. doi:10.1145/353519.343056

Nelson, W. A. (2003). Problem solving through design. *New Directions for Teaching and Learning*, (95): 39–44. doi:10.1002/tl.111

Nerur, S., & Balijepally, V. (2007). Theoretical reflections on agile development methodologies. *Communications of the ACM*, *50*, 79–83. doi:10.1145/1226736.1226739

Nicol, D. J., & Macfarlane-Dick, D. (2006). Formative assessment and self-regulated learning: A model and seven principles of good feedback practice. *Studies in Higher Education*, *31*(2), 199–218. doi:10.1080/03075070600572090

Nielsen, J. (1994). *Usability engineering*. London: Elsevier.

Noll, C. L., & Wilikens, M. (2002). Critical skills of IS professionals: A model for curriculum development. *Journal of Information Technology Education*, *1*(3), 143–154.

Nunan, T. (1999). *Graduate qualities, employment and mass higher education*. Paper presented at the HERDSA Annual International Conference. Melbourne, Australia.

Nuseibeh, B., & Easterbrook, S. (2000). Requirements engineering: A roadmap. In *Proceedings of the Conference on the Future of Software Engineering* (pp. 35–46). ACM.

O'Reilly, T. (2007). What is web 2.0: Design patterns and business models for the next generation of software. *Communications & Strategies*, (1), 17.

Oakley, B., Felder, R. M., Brent, R., & Elhajj, I. (2004). Turning student groups into effective teams. *Journal of Student Centered Learning*, *2*(1), 9–34.

Objectaid. (n.d.). Retrieved from http://www.objectaid.com/

O'Connor, J., & Lages, A. (2004). *Coaching with NLP: How to be a master coach*. Element.

OECD. (1996). *Lifelong learning for all*. Paris: OECD.

O'Grady, M., Barrett, G., Barrett, T., Delaney, Y., Hunt, N., Kador, T., & O'Brien, V. (2013). Reflecting on the need for problem triggers in multidisciplinary PBL. *AISHE-J: The All Ireland Journal of Teaching & Learning in Higher Education, 5*(2).

Oh Navarro, E., & Van der Hoek, A. (2005). Design and evaluation of an educational software process simulation environment and associated model. In *Proceedings of the 18th Conference on Software Engineering Education and Training* (pp. 25–32). IEEE.

O'Hara, K. J., & Kay, J. S. (2003). Open source software and computer science education. *Journal of Computing Sciences in Colleges*, *18*(3), 1–7.

Oliveira, E., & Lima, R. (2011). State of the art on the use of scrum in distributed development software. *Revista de Sistemas e Computação*, *1*(2), 106–119.

Oliver, R., & McLoughlin, C. (1999). *Using web and problem-based learning environments to support the development of key skills*. Paper presented at the Responding to Diversity. Brisbane, Australia.

Olsen, A. L. (2008). A service learning project for a software engineering course. *Journal of Computing Sciences in Colleges*, *24*(2), 130–136.

Orrell, J. (2004). *Work-integrated learning programmes: Management and educational quality*. Paper presented at the Australian University Quality Forum 2004. Canberra, Australia.

Osorio, J. A., Chaudron, M. R., & Heijstek, W. (2011). Moving from waterfall to iterative development – An empical evaluation of advantages, disadvantages and risks of RUP. In *Proceedings of 37th Conference on Software Engineering and Advanced Application* (pp. 453–460). IEEE.

Paasivaara, M., Durasiewicz, S., & Lassenius, C. (2009). Using scrum in distributed agile development: A multiple case study. In *Proceedings of the 4th IEEE International Conference on Global Software Engineering* (pp. 195–204). IEEE.

Padua, A. O., & Cysneiros, L. (2006). Defining strategic dependency situations in requirements elicitation. In *Proceedings of the 9th Workshop on Requirements Engineering* (pp. 12–23). Academic Press.

Pádua, W. (2010). Measuring complexity, effectiveness and efficiency in software course projects. In *Proceedings of the 32th International Conference on Software Engineering* (vol. 1, pp. 545–554). ACM.

Paechter, M., Maier, B., & Macher, D. (2010). Students' expectations of, and experiences in e-learning: Their relation to learning achievements and course satisfaction. *Computers & Education*, *54*(1), 222–229. doi:10.1016/j.compedu.2009.08.005

Papadopoulos, P. M., Stamelos, I. G., & Meiszner, A. (2012). Students' perspectives on learning software engineering with open source projects: Lessons learnt after three years of program operation. In *Proceedings of the 4th International Conference on Computer Supported Education* (pp. 313–322). Porto, Portugal: SciTePress.

Papadopoulos, P. M., Stamelos, I. G., & Meiszner, A. (2013). Enhancing software engineering education through open source projects: Four years of students' perspectives. *Education and Information Technologies*, *18*(2), 381–397. doi:10.1007/s10639-012-9239-3

Parker, K. R., & Chao, J. T. (2007). Wiki as a teaching tool. *Interdisciplinary Journal of Knowledge and Learning Objects*, *3*, 358–372.

Park, H. (2012). Relationship between motivation and student's activity on educational game. *International Journal of Grid and Distributed Computing*, *5*(1).

Parnas, D. L. (1999). Software engineering programs are not computer science programs. *IEEE Software*, *16*(6), 19–30. doi:10.1109/52.805469

Parnas, D. L., & Lawford, M. (2003). The role of inspection in software quality assurance. *IEEE Transactions on Software Engineering*, *29*(8), 674–676. doi:10.1109/TSE.2003.1223642

Parsons, D., & Stockdale, R. (2010). Cloud as context: Virtual world learning with open wonderland. In *Proceedings of the 9th World Conference on Mobile and Contextual Learning* (mLearn 2010). Valetta, Malta: mLearn.

Passow, H. J. (2007). *What competencies should engineering programs emphasize? A meta-analysis of practitioners' opinions informs curricular design*. Paper presented at the 3rd International CDIO Conference. New York, NY.

Passow, H. (2012). Which ABET competencies do engineering graduates find most important in their work? *Journal of Engineering Education*, *101*, 95–118. doi:10.1002/j.2168-9830.2012.tb00043.x

Patel, A., Kinshuk, R., & Russell, D. (2000). Intelligent tutoring tools for cognitive skill acquisition in life long learning. *Journal of Educational Technology & Society*, *3*(1), 32–40.

Patrick, C. J., Peach, D., Pocknee, C., Webb, F., Fletcher, M., & Pretto, G. (2008). *The WIL (work integrated learning) report: A national scoping study*. Queensland University of Technology.

Patry, J. L. (1998). *Transfer evaluation in educational processes: Models, results and problems - A theoretical approach*.

Paulus, P. B., & Nijstad, B. A. (2003). *Group creativity: Innovation through collaboration*. Oxford, UK: Oxford University Press. doi:10.1093/acprof:oso/9780195147308.001.0001

Pedroni, M., Bay, T., Oriol, M., & Pedroni, A. (2007). Open source projects in programming courses. *ACM SIGCSE Bulletin*, *39*(1), 454–458. doi:10.1145/1227504.1227465

Peñalver, O. (2010). *Emociones colectivas: La inteligencia emocional de los equipos*. Barcelona, Spain: Alienta.

Pérez-Martínez, J. E., & Sierra-Alonso, A. (2003). A coordinated plan for teaching software engineering in the Rey Juan Carlos University. In *Proceedings of the 16th Conference on Software Engineering Education and Training* (pp. 107–118). IEEE.

Periyasamy, K., & Garbers, B. (2006). A light weight tool for teaching the development and evaluation of requirements documents. In *Proceedings of the Annual Conference of American Society of Engineering Education*. American Society of Engineering Education.

Perrenet, J. C., Bouhuijs, P. A. J., & Smits, J. G. M. M. (2000). The suitability of problem-based learning for engineering education: Theory and practice. *Teaching in Higher Education*, *5*, 345–358. doi:10.1080/713699144

Peterson, D. (2009). *What is kanban?* Retrieved November, 2013, from http://www.kanbanblog.com/explained/

Petrides, K. V., Sangareau, Y., Furnham, A., & Frederickson, N. (2006). Trait emotional intelligence and children's peer relations at school. *Social Development*, *15*(3), 537–547. doi:10.1111/j.1467-9507.2006.00355.x

Pfleeger, S. L. (1998). *Software engineering: Theory and practice*. Upper Saddle River, NJ: Prentice Hall.

Pfleeger, S. L. (1999). Albert Einstein and empirical software engineering. *IEEE Computer, 32*(10), 32–37. doi:10.1109/2.796106

Phillips, L. (2005). *Authentic assessment: A briefing.* Retrieved from http://www.clubWebcanada.ca/l-pphillips/edarticles/assessment.htm

Piaget, J. (1968). *La structuralisme* (T. I. B. C. Maschler, Trans.). London: Routledge & Kegan Paul.

Picciano, A. (2002). Beyond student perceptions: Issues of interaction, presence, and performance in an online course. *Journal of Asynchronous Learning Networks, 6*(1), 21–40.

Pikkarainen, M., & Mantyniemi, A. (2006). *An approach for using CMMI in agile software development assessments: Experiences from three case studies.* University of Limerick Institutional Repository.

Piri, A., Niinimäki, T., & Lassenius, C. (2010). Fear and distrust in global software engineering projects. *Journal of Software Maintenance and Evolution: Research and Practice, 24*(2).

Platt, J. R. (2011). Career focus: Software engineering. *IEEE-USA Today's Engineer.* Retrieved from http://www.todaysengineer.org/2011/Mar/career-focus.asp

PMBOK. (2011). *A guide to the project management body of knowledge* (4th ed.). IEEE Computer Society.

Poger, S., & Bailie, F. (2006). Student perspectives on a real world project. *Journal of Computing Sciences in Colleges, 21*(6), 69–75.

Popescu, E. (2009). Evaluating the impact of adaptation to learning styles in a web-based educational system. In *Proceedings of the 8th International Conference on Advances in Web Based Learning* (pp. 343–352). Berlin: Springer-Verlag.

Porcaro, D. (2011). Applying constructivism in instructivist learning cultures. *Multicultural Education & Technology Journal, 5*(1), 39–54. doi:10.1108/17504971111121919

Port, D., & Boehm, B. (2001). Using a model framework in developing and delivering a family of software engineering project courses. In *Proceedings of the 14th Conference on Software Engineering Education and Training* (pp. 44–55). IEEE.

Porter, T. W., & Lilly, B. S. (1996). The effects of conflict, trust, and task commitment on project team performance. *The International Journal of Conflict Management, 7*(4), 361–376. doi:10.1108/eb022787

Powell, W., Powell, P. C., & Weenk, W. (2003). *Project-led engineering education.* Lemma Publishers.

Power, K. (2011). Using silent grouping to size user stories. In *Agile processes in software engineering and extreme programming* (pp. 60–72). Berlin: Springer. doi:10.1007/978-3-642-20677-1_5

Precision_Consultancy. (2007). *Graduate employability skills.* Canberra, Australia: Canberra Commonwealth of Australia.

Pressman, R. S. (2010). *Software engineering: A practitioner's approach.* New York: McGraw-Hill.

Prince, M. (2004). Does active learning work? A review of the research. *Journal of Engineering Education, 93*(3), 223–231. doi:10.1002/j.2168-9830.2004.tb00809.x

Prince, M. J., & Felder, R. M. (2006). Inductive teaching and learning methods: Definitions, comparisons, and research bases. *Journal of Engineering Education, 95*(2), 123–138. doi:10.1002/j.2168-9830.2006.tb00884.x

Prosser, M., & Trigwell, K. (1999). *Understanding learning and teaching: The experience in higher education.* Buckingham, UK: Oxford University Press.

Putnik, G., & Cunha, M. M. (2008). *Encyclopedia of networked and virtual organizations* (Vol. 2). Hershey, PA: Information Science Reference. doi:10.4018/978-1-59904-885-7

Putnik, Z., Ivanović, M., Budimac, Z., & Samuelis, L. (2012). Wiki – A useful tool to fight classroom cheating? In *Advances in web-based learning - ICWL 2012 (LNCS)* (Vol. 7558, pp. 31–40). Berlin: Springer. doi:10.1007/978-3-642-33642-3_4

Qian, K., Fu, X., Tao, L., & Xu, C. (2006). *Software architecture and design illuminated*. Jones and Bartlett.

Qin, S., & Mooney, C. H. (2009). Using game–oriented projects for teaching and learning software engineering. In *Proceedings* of *20th Annual Conference for the Australasian Association for Engineering Education* (pp. 49–54). Engineers Australia.

Quackback Peer Review . (n.d.). Retrieved from http://goldapplesoftware.ca/portfolio/quackback-peer-review

Queensu. (n.d.). *Learning strategies.* Retrieved from http://www.queensu.ca/ctl/resources/topicspecific/casebased/learningstrategies.html

Radermacher, A., & Walia, G. (2013). Gaps between industry expectations and the abilities of graduates: Systematic literature review findings. In *Proceedings of the 2013 Special Interest Group on Computer Science Education Technical Symposium*. Academic Press.

Rahman, A. N., & Sahibuddin, S. (2011). Extracting soft issues during requirements elicitation: A preliminary study. *International Journal of Information and Electronics Engineering*, *1*(2), 126–132.

Raj, R. K., & Kazemian, F. (2006). Using open source software in computer science courses. In *Proceedings of the 36th Annual Frontiers in Education Conference* (pp. 21–26). San Diego, CA: IEEE.

Ramakrishan, S. (2003). MUSE studio lab and innovative software engineering capstone project experience. *SIGCSE Bulletin*, *35*(3), 21–25. doi:10.1145/961290.961521

Ramsden, P. (2003). *Learning to teach in higher education*. London: Routledge Falmer.

Ratcliffe, L., & McNeill, M. (2011). *Agile experience design: A digital designer's guide to agile, lean, and continuous*. New Riders.

Reddy, B. B., & Gopi, M. M. (2013). The role of English language teacher in developing communication skills among the students of engineering and technology. *International Journal of Humanities and Social Science Invention*, *2*(4), 29–31.

Reischmann, J. (1986). Learning en passant: The forgotten dimension. In *Proceedings of Conference of the American Association of Adult and Continuing Education*. Columbus, OH: ERIC Clearinghouse on Adult, Career, and Vocational Education.

Retzer, S., Fisher, J., & Lamp, J. (2003). Information systems and business informatics: An Australian German comparison. In *Proceedings of the 14th Australasian Conference on Information Systems* (pp. 1–9). Edith Cowan University.

Review Process . (n.d.). Retrieved from http://www.finance.gov.au/gateway/review-process.html

Richards, D. (2009). Designing project-based courses with a focus on group formation and assessment. *Transactions on Computer Education*, *9*(1), 1–40. doi:10.1145/1513593.1513595

Richardson, I., & Delaney, Y. (2009). Problem based learning in the software engineering classroom. In *Proceedings of the 22nd Conference on Software Engineering Education and Training* (pp. 174–181). IEEE.

Richardson, I., & Delaney, Y. (2010). Software quality: From theory to practice. In *Proceedings of the 7th International Conference on the Quality of Information and Communications Technology* (pp. 150–155). Academic Press.

Richardson, I., & Hynes, B. (2008). Entrepreneurship education: Towards an industry sector approach. *Education + Training*, *50*(3), 188–198.

Richardson, I., Milewski, A. E., Mullick, N., & Keil, P. (2006). Distributed development: An education perspective on the global studio project. In *Proceedings of the 28th International Conference on Software Engineering* (pp. 679–684). ACM.

Richardson, I., Reid, L., Seidman, S. B., Pattinson, B., & Delaney, Y. (2011). Educating software engineers of the future: Software quality research through problem-based learning. In *Proceedings of the 24th Conference on Software Engineering Education and Training* (pp. 91–100). Academic Press.

Richardson, I., Reid, L., Seidman, S. B., Pattinson, B., & Delaney, Y. (2011). Educating software engineers of the future: Software quality research through problem-based learning. In *Proceedings of the 24th Conference on Software Engineering Education and Training* (pp. 91–100). IEEE.

Richardson, J. C., & Swan, K. (2003). Examining social presence in online courses in relation to students' perceived learning and satisfaction. *Journal of Asynchronous Learning Networks, 7*(1), 68–88.

Richardson, W. (2010). *Blogs, wikis, podcasts and other powerful web tools for the classroom.* Thousand Oaks, CA: Corwin Press.

Rico, D. F., & Sayani, H. H. (2009). Use of agile methods in software engineering education. In *Proceedings of the 2007 Agile Conference* (pp. 174–179). IEEE.

Rivera-Ibarra, J. G., Rodríguez-Jacobo, J., & Serrano-Vargas, M. A. (2010). Competency framework for software engineers. In *Proceedings of the 23rd Conference on Software Engineering Education and Training* (pp. 33–40). Academic Press.

Robillard, P. N. (2005). Opportunistic problem solving in software engineering. *IEEE Software, 22*(6), 60–67. doi:10.1109/MS.2005.161

Robillard, P. N., & Robillard, M. P. (2000). Types of collaborative work in software engineering. *Journal of Systems and Software, 53*(3), 219–224. doi:10.1016/S0164-1212(00)00013-3

Roblyer, M. D., & Doering, A. H. (2010). *Integrating technology into teaching.* Boston, MA: Pearson Education, Inc.

Rodríguez, G., Soria, A., & Campo, M. (2012a). Teaching scrum to software engineering students with virtual reality support. In *Advances in new technologies, interactive interfaces and communicability* (pp. 140–150). Berlin: Springer. doi:10.1007/978-3-642-34010-9_14

Rodríguez, G., Soria, A., & Campo, M. (2012b). Supporting virtual meetings in distributed scrum teams. *IEEE Latin America Transactions, 10*(6), 2316–2323. doi:10.1109/TLA.2012.6418138

Rombach, D., Münch, J., Ocampo, A., Humphrey, W. S., & Burton, D. (2008). Teaching disciplined software development. *International Journal of Systems and Software, 81*(5).

Royce, W. W. (1970). Managing the development of large software systems. *IEEE WESCON, 26*(8), 1–9.

Rozanski, N., & Woods, E. (2012). *Software systems architecture.* Reading, MA: Addison-Wesley.

Rummler, M. (2011). *Crashkurs hochschuldidaktik: Grundlagen und methoden guter lehre.* Weinheim: Beltz.

Rummler, M. (Ed.). (2012). *Innovative lehrformen: Projektarbeit in der hochschule: Projektbasiertes und problemorientiertes lehren und lernen.* Weinheim: Beltz.

Russell, D. R. (1998). *The limits of the apprenticeship models in WAC/WID research.* Paper presented at the Conference of College Composition and Communication. New York, NY.

Rust, C. (2002). The impact of assessment on student learning: How can the research literature practically help to inform the development of departmental assessment strategies and learner-centred assessment practices? *Active Learning in Higher Education, 3*(2), 145–158. doi:10.1177/1469787402003002004

Sabbatino, V., Arecchi, A., Lanciotti, R., Leardi, A., & Tonni, V. (n.d.). Modeling the software architecture of large systems. In *Proceedings of International Conference on Systems of Systems Engineering.* IEEE Computer Society Press.

Saddington, T. (n.d.). *What is experiential learning?* Retrieved from http://www.edb.gov.hk/attachment/sc/curriculum-development/kla/pshe/references-and-resources/ethics-and-religious-studies/experiential_learning_2.pdf

Saddington, T. (2000). The roots and branches of experiential learning. *NSEE Quarterly, 26*(1), 2–6.

Salamah, S., Towhidnejad, M., & Hilburn, T. (2011). Developing case modules for teaching software engineering and computer science concepts. In *Proceedings of Frontiers in Education Conference (FIE), 2011* (pp. T1H-1). IEEE.

Sangwan, R., Mullick, N., & Paulish, D. J. (2006). *Global software development handbook*. Auerbach Publishers Inc. doi:10.1201/9781420013856

Sanker, D. (2011). *Collaborate: The art of we*. New York: Wiley.

Savery, J. R. (2006). Overview of problem-based learning: Definitions and distinctions. *Interdisciplinary Journal of Problem-Based Learning and Teaching in Higher Education*, *1*(1), 9–20.

Savery, J. R., & Duffy, T. M. (1998). Problem based learning: An instructional model and its constructivist framework. In B. G. Wilson (Ed.), *Constructivist learning environments: Case studies in instructional design* (pp. 135–148). Englewood Cliffs, NJ: Educational Technology Publications.

Savery, J. R., & Duffy, T. M. (2001). *Problem based learning: An instructional model and its constructivist framework*. Bloomington, IN: Indian University.

Savin-Baden, M. (2001). The problem-based learning landscape. *Planet,* (4), 4–6.

Savin-Baden, M. (2000). *Problem-based learning in higher education: Untold stories*. Buckingham, UK: Society for Research into Higher Education and Open University Press.

Schmidt, D. C., Stal, M., Rohnert, H., Buschmann, F., & Wiley, J. (2000). *Pattern-oriented software architecture: Patterns for concurrent and networked objects* (Vol. 2). New York: Wiley.

Schmidt, H. G. (1993). Foundations of problem-based learning some explanatory notes. *Medical Education*, *27*, 422–432. doi:10.1111/j.1365-2923.1993.tb00296.x PMID:8208146

Schmidt, H. G., & De Volder, M. L. (1984). *Tutorial in problem-based learning*. Van Gorcum.

Schmidt, R., Lyytinen, K., Keil, M., & Cule, P. (2001). Identifying software project risks: An international Delphi study. *Journal of Management Information Systems*, *17*(4), 5–36.

Scholarios, D., van der Schoot, E., & van der Heijden, B. (2004). *The employability of ICT professional: A study of European SMEs*. Paper presented at the e-Challenges Conference. Vienna, Austria.

Scholtes, P. R. (1995). *The team handbook: How to use teams to improve quality*. Madison, WI: Joiner Associates, Inc.

Schön, D. A. (1983). *The reflective practitioner: How professionals think in action*. New York: Basic Books.

Schön, D. A. (1985). *The design studio*. London: RIBA Publications.

Schön, D. A. (1987). *Educating the reflective practitioner: Towards a new design for teaching in the professions. San Fransisco*. Jossey-Bass Inc.

Schroeder, A., & Klarl, A. (2012). Teaching agile software development through lab courses. In *Proceedings of IEEE Global Engineering Education Conference* (EDUCON) (pp. 1177–1186). IEEE.

Schulz von Thun, F. (1996). *Miteinander reden: Störungen und klärungen*. Reinbek: Rowohlt.

Schulz von Thun, F. (2008). *Six tools for clear communication: The hamburg approach in Englisch language*. Schulz von Thun, Institut für Kommunikation.

Schwaber, K. (2004). *Agile project management with Scrum*. Microsoft Press.

Schwaber, K., & Beedle, M. (2002). *Agile software development wit scrum*. Upper Saddle River, NJ: Prentice Hall.

Scott, E., Rodriguez, G., Soria, A., & Campo, M. (2013). El rol del estilo de aprendizaje en la enseñanza de prácticas de scrum: Un enfoque estadístico. In *Proceedings of the 14th Argentine Symposium on Software Engineering*. Academic Press.

Scott, G., & Wilson, D. (2002). *Tracking and profiling successful IT graduates: An exploratory study*. Paper presented at the 13th Australasian Conference on Information Systems. Canberra, Australia.

Scott, G., & Yates, W. (2002). Using successful graduates to improve the quality of undergraduate engineering programs. *European Journal of Engineering Education, 27*(4), 60–67. doi:10.1080/03043790210166666

Scraper, R. L. (2000). The art and science of Maieutic questioning within the Socratic method. *International Forum for Logotherapy, 23*(1), 14–16.

Scrumalliance. (n.d.). *Why scrum.* Retrieved from http://www.scrumalliance.org/why-scrum

Sedelmaier, Y., & Landes, D. (2013b). *Software engineering body of skills.* Unpublished manuscript.

Sedelmaier, Y., Claren, S., & Landes, D. (2013). Welche kompetenzen benötigt ein software ingenieur? In A. Spillner, & H. Lichter (Eds.), *Software engineering im unterricht der hochschulen 2013* (pp. 117–128). Academic Press.

Sedelmaier, Y., & Landes, D. (2013). A research agenda for identifying and developing competencies in software engineering. *International Journal of Engineering Pedagogy, 3*(2).

Sedelmaier, Y., & Landes, D. (2013a). A research agenda for identifying and developing required competencies in software engineering. *International Journal of Engineering Pedagogy, 3*(2), 30–35.

Séguin, N., Tremblay, G., & Bagane, H. (2012). Agile principles as software engineering principles: An analysis. In *Agile processes in software engineering and extreme programming* (pp. 1–15). Berlin: Springer. doi:10.1007/978-3-642-30350-0_1

Selenic. (n.d.). Retrieved from http://mercurial.selenic.com/

SEMAG. (2001). *Software engineering and management group.* Retrieved 01-06-2013, 2013, from https://sites.google.com/a/dsi.uminho.pt/semag/

Serdiukov, P. (2000). Educational technology and its terminology: New developments in the end of the 20th century. In J. Bourdeau & R. Heller (Eds.), *Proceedings of ED-MEDIA* (pp. 1022–1025). AACE.

Shams-Ul-Arif, M. R., Khan, M. Q., & Gahyyur, S. (2009). Requirements engineering processes, tools/technologies, & methodologies. *International Journal of Reviews in Computing.*

Shaw, K., & Dermoudy, J. (2005). Engendering an empathy for software engineering. In *Proceedings of the 7th Australasian Computing Education Conference* (Vol. 42, pp. 135–144). Newcastle, Australia: Academic Press.

Shaw, M. (2000, May). Software engineering education: A roadmap. In *Proceedings of the 2000 Conference on Future of Software Engineering* (pp. 371–380). ACM.

Shaw, M. (1992). We can teach software better. *Computing Research News, 4*, 2–4, 12.

Shaw, M., & Clements, P. (2006). The golden age of software architecture. *IEEE Software, 23*(2), 31–39. doi:10.1109/MS.2006.58

Shaw, M., & Garlan, D. (1996). *Software architecture: Perspectives on an emerging discipline.* Upper Saddle River, NJ: Prentice-Hall.

Sheppard, S. D., Macatangay, K., Colby, A., & Sullivan, W. M. (2008). *Educating engineers: Designing for the future of the field.* The Carnegie Foundation for the Advancement of Teaching.

Shere, K. D. (1988). *Software engineering management.* Upper Saddle River, NJ: Prentice Hall.

Sherer, M., & Eadie, R. (1987). Employability skills: Key to success. *Thrust, 17*, 16–17.

Shim, C. Y., Choi, M., & Kim, J. Y. (2009). Promoting collaborative learning in software engineering by adapting the PBL strategy. *World Academy of Science. Engineering and Technology, 53*, 1157–1160.

Shrivastava, S. V., & Date, H. (2010). Distributed agile software development: A review. *Journal of Computer Science and Engineering, 1*(1), 10–16.

Siebert, H. (2010). *Methoden für die bildungsarbeit: Leitfaden für aktivierendes Lehren.* Bielefeld: Bertelsmann.

Simul, E. S.-W. PES (2011). *Simuleswpes's blog.* Retrieved from http://simuleswpes.wordpress.com/

Siqueira, F. L., Barbaran, G. M. C., & Becerra, J. L. R. (2008). A software factory for education in software engineering. In *Proceedings of the 21st Conference on Software Engineering Education & Training* (pp. 215–222). IEEE.

Slaten, K. M., Droujkova, M., Berenson, S. B., Williams, L., & Layman, L. (2005). Undergraduate student perceptions of pair programming and agile software methodologies: Verifying a model of social interaction. In *Proceedings of the Agile Development Conference* (pp. 323–330). IEEE.

Sloan, M. I. T. (n.d.). *Simulations*. Retrieved from https://mitsloan.mit.edu/LearningEdge/simulations/platform-wars/Pages/default.aspx

Smith, K. A. (2000). Strategies for developing engineering student's teamwork and project management skills. In *Proceedings of the 2000 American Society for Engineering Education Annual Conference*. ASEE.

Smith, J., May, S., & Burke, L. (2007). Peer assisted learning: A case study into the value to student mentors and mentees. *Practice and Evidence of the Scholarship of Teaching and Learning in Higher Education, 2*(2), 80–109.

Smith, W. J., Belanger, F., Lewis, T. L., & Honaker, K. (2007). Training to persist in computing careers. *Inroads -. SIGCSE Bulletin, 39*(4), 119–120. doi:10.1145/1345375.1345429

Snoke, R., & Underwood, A. (1999). *Generic attributes of IS graduates - A Queensland study*. Paper presented at the 10th Australasian Conference on Information Systems. Wellington, NZ.

Soloman, B., & Felder, R. (1999). *Index of learning styles (ILS)*. Retrieved from http://www2.ncsu.edu/unity/lockers/users/f/felder/public/ILSpage.html

Sommerville, I. (2011). *Software engineering*. Boston: Pearson.

Song, L., Singleton, E., Hill, J., & Koh, M. (2004). Improving online learning: Student perceptions and challenging characteristics. *The Internet and Higher Education, 7*, 59–70. doi:10.1016/j.iheduc.2003.11.003

Soria, A., Rodriguez, G., & Campo, M. (2012). Improving software engineering teaching by introducing agile management. In *Proceedings of the 13th Argentine Symposium on Software Engineering* (pp. 215–229). Academic Press.

Soundararajan, S., Chigani, A., & Arthur, J. D. (2012, February). Understanding the tenets of agile software engineering: Lecturing, exploration and critical thinking. In *Proceedings of the 43rd ACM Technical Symposium on Computer Science Education* (pp. 313–318). ACM.

Sourceforge. (n.d.). Retrieved from http://sourceforge.net

Spiro, R. J., Feltovich, P. J., Jacobson, M., & Coulson, R. (1991). Cognitive flexibility, constructivism and hypertext: Random access instruction for advanced knowledge acquisition in ill-structured domains. *Educational Technology, 31*, 24–33.

Stahl, T., Voelter, M., & Czarnecki, K. (2006). *Model-driven software development: Technology, engineering, management*. New York: Wiley.

Staker, H. (2011). The rise of K-12 blended learning. *District Administration*. Retrieved from http://www.districtadministration.com/article/building-blended-learning-program

Stamelos, I. (2008). Teaching software engineering with free/libre open source projects. *International Journal of Open Source Software and Processes, 1*(1), 72–90. doi:10.4018/jossp.2009010105

Stemp-Morlock, G. (2009). Learning more about active learning. *Communications of the ACM, 52*(4), 11–13. doi:10.1145/1498765.1498771

Stepanyan, K., Mather, R., & Payne, J. (2007). Integrating social software into course design and tracking student engagement: Early results and research perspectives. In T. Bastiaens & S. Carliner (Eds.), *Proceedings of the World Conference on E-Learning in Corporate, Government, Healthcare, and Higher Education* (pp. 7386–7395). Academic Press.

Sterling, L., Ciancarini, P., & Turnidge, T. (1996). On the animation of not executable specifications by prolog. *International Journal of Software Engineering and Knowledge Engineering, 6*(1), 63–87. doi:10.1142/S0218194096000041

Stevens, D. D., & Levi, A. J. (2005). *Introduction to rubrics: An assessment tool to save grading time, convey effective feedback and promote student learning*. Stylus Publishing.

Stewart, B., Briton, D., Gismondi, M., Heller, B., Kennepohl, D., McGreal, R., & Nelson, C. (2007). Choosing Moodle: An evaluation of learning management systems at Athabasca University. *International Journal of Distance Education Technologies, 5*(3), 1–7. doi:10.4018/jdet.2007070101

Stiller, E., & LeBlanc, C. (2002). Effective software engineering pedagogy. *Journal of Computing Sciences in Colleges, 17*(6), 124–134.

Subversion. (n.d.). Retrieved from http://subversion.tigris.org/

Surakka, S. (2007). What subjects and skills are important for software developers? *Communications of the ACM, 50*(1), 73–78. doi:10.1145/1188913.1188920

Suri, D., & Sebern, M. J. (2004). Incorporating software process in an undergraduate software engineering curriculum: Challenges and rewards. In *Proceedings of the 17th Conference onSoftware Engineering Education and Training* (pp. 18–23). IEEE.

Surveymonkey . (n.d.). Retrieved from http://www.surveymonkey.com

Sutherland, J., Ruseng Jakobsen, C., & Johnson, K. (2008). Scrum and CMMI level 5: The magic potion for code warriors. In *Proceedings of Hawaii International Conference on System Sciences, Proceedings of the 41st Annual* (pp. 466–466). IEEE.

Sutton, S. M., Jr., & Rouvellou, I. (2001). Applicability of categorization theory to multidimensional separation of concerns. In *Proceedings of the Workshop on Advanced Separation of Concerns*. OOPSLA.

Svinicki, M., & McKeachie, W. J. (2011). *McKeachie's teaching tips: Strategies, research, and theory for college and university teachers*. Wadsworth: Cengage Learning.

Swan, B., Magleby, M., Sorensen, C., & Todd, R. (1994). A preliminary analysis of factors affecting engineering design team performance. In *Proceedings of the 1994 American Society for Engineering Education Annual Conference* (pp. 2572–2589). ASEE.

Sweedyk, E., & Keller, R. M. (2005). Fun and games: A new software engineering course. *ACM SIGCSE Bulletin, 37*(3), 138–142. doi:10.1145/1151954.1067485

Tabaka, J. (2006). *Collaboration explained: Facilitation skills for software project leaders*. Upper Saddle River, NJ: Pearson Education.

Tadayon, N. (2004). Software engineering based on the team software process with a real world project. *Journal of Computing Sciences in Colleges, 19*(4), 133–142.

Tai, G. X. L., & Yuen, M. C. (2007). Authentic assessment strategies in problem based learning. In *Proceedings ascilite Singapore* (pp. 983–999). Ascilite.

Tan, C. H., Tan, W. K., & Teo, H. H. (2008). Training students to be agile information systems developers: A pedagogical approach. In *Proceedings of the 2008 ACM SIGMIS CPR Conference on Computer Personnel Doctoral Consortium and Research* (pp. 88–96). ACM.

Tan, C., Tan, W., & Teo, H. (2008). Training students to be agile information systems developers: A pedagogical approach. In *Proceedings of the 2008 ACM SIGMIS CPR Conference on Computer Personnel Doctoral Consortium and Research* (pp. 88–96). ACM.

Tan, O. S., & Little, S. Y. Hee, & J. Conway, E. (2000). Problem-based learning: Educational innovation across disciplines. Singapore.

Tan, J., & Jones, M. (2008). A case study of classroom experience with client-based team projects. *Journal of Computing Sciences in Colleges, 23*(5), 150–159.

Tan, O. S. (2003). *Problem-based learning innovation*. Singapore.

Tatnall, A., & Burgess, S. (2009). Evolution of information systems curriculum in an Australian university over the last twenty-five years. In A. Tatnall, & A. Jones (Eds.), *Education and technology for a better world* (Vol. 302, pp. 238–246). Berlin: Springer. doi:10.1007/978-3-642-03115-1_25

Taylor, R., Medvidovic, N., & Dashofy, E. (2009). *Software architecture: Foundations, theory, and practice*. New York: Wiley.

Teel, S., Schweitzer, D., & Fulton, S. (2012). Teaching undergraduate software engineering using open source development tools. *Issues in Informing Science & Information Technology, 9*, 62–73.

Teles, V. M., & de Oliveira, C. E. T. (2003). Reviewing the curriculum of software engineering undergraduate courses to incorporate communication and interpersonal skills teaching. In *Proceedings of the 16th Conference on Software Engineering Education and Training* (pp. 158–165). IEEE.

Terhart, E. (2009). *Didaktik: Eine einführung*. Stuttgart: Reclam.

Tindale, R. S., Heath, L., Edwards, J., Posavac, E. J., & Bryant, F. B. Myers. J., Suarez-Balcazar, Y., & Henderson-King, E. (2002). Theory and research on small groups. Boston, MA: Springer US.

Tiwana, A., & Keil, M. (2004). The one-minute risk assessment tool. *Communications of the ACM*, *47*(11), 73–77. doi:10.1145/1029496.1029497

Tockey, S. (1997). A missing link in software engineering. *IEEE Software*, *14*(6), 31–36. doi:10.1109/52.636594

Todd, R. H., & Magleby, S. P. (2005). Elements of a successful capstone course considering the needs of stakeholders. *European Journal of Engineering Education*, *30*(2), 203–214. doi:10.1080/03043790500087332

Tomcat. (n.d.). Retrieved from http://tomcat.apache.org/

Tomey, A. M. (2003). Learning with cases. *Journal of Continuing Education in Nursing*, *34*(1). PMID:12546132

Tomlinson, M. (2012). Graduate employability: A review of conceptual and empirical themes. *Higher Education Policy*, *25*, 407–431. doi:10.1057/hep.2011.26

Topi, H., Valacich, J. S., Wright, R. T., Kaiser, K. M., Nunamaker, F. J., Sipior, J. C., & de Vreede, G. J. (2010). *IS 2010: Curriculum guidelines for undergraduate degree programs in information systems*. Association for Computing Machinery and Association for Information Systems.

Topping, K., & Ehly, S. (1998). *Peer-assisted learning*. Hoboken, NJ: Lawrence Erlbaum Associates, Inc.

Trello. (n.d.). Retrieved from http://trello.com

Trendowicz, A., Heidrich, J., & Shintani, K. (2011). Aligning software projects with business objectives. In *Proceedings of the 2011 Joint Conference of the 21st International Workshop on Software Measurement and the 6th International Conference on Software Process and Product Measurement* (pp. 142–150). IEEE.

Tsai, M. J. (2009). The model of strategic e-learning: Understanding and evaluating student e-learning from metacognitive perspectives. *Journal of Educational Technology & Society*, *12*(1), 34–48.

Tsai, M. J., & Tsai, C. C. (2003). Information searching strategies in web-based science learning: The role of internet self-efficacy. *Innovations in Education and Teaching International*, *40*(1), 43–49. doi:10.1080/1355800032000038822

Tucker, A., Morelli, R., & de Silva, C. (2011). *Software development: An open source approach*. Boca Raton, FL: CRC Press.

Tucker, M. L., Sojka, J. Z., Barone, F. J., & McCarthy, A. M. (2000). Training tomorrow's leaders: Enhancing the emotional intelligence of business graduates. *Journal of Education for Business*, *75*(6), 331–337. doi:10.1080/08832320009599036

Tuckman, B. (1965). Developmental sequence in small groups. *Psychological Bulletin*, *63*(6), 384–399. doi:10.1037/h0022100 PMID:14314073

Tuckman, B. W., & Jensen, M. A. C. (1977). Stages of small-group development revisited. *Group & Organization Management*, *2*(4), 419–427. doi:10.1177/105960117700200404

Turley, R. T. (1991). *Essential competencies of exceptional professional software engineers*. Fort Collins, CO: Colorado State University.

Turley, R. T., & Bieman, J. M. (1995). Competencies of exceptional and non-exceptional software engineers. *Journal of Systems and Software*, *28*(1), 19–38. doi:10.1016/0164-1212(94)00078-2

Turner, R., & Lowry, G. (2003). Education for a technology-based profession: Softening the information systems curriculum. In T. McGill (Ed.), *Current issues in IT education* (pp. 153–172). Hershey, PA: IRM Press.

Tyler, R. W. (1949). *Basic principles of curriculum and instruction*. Chicago: University of Chicago Press.

Udacity. (n.d.). Retrieved from https://www.udacity.com

Ulloa, B. C. R., & Adams, S. G. (2004). Attitude toward teamwork and effective teaming. *Team Performance Management*, *10*(7/8), 145–151. doi:10.1108/13527590410569869

Unity 3D . (n.d.). Retrieved from http://unity3d.com/

University of Nottingham. (2003). *PBLE (project based learning in engineering)*. Retrieved 09-07-2013, 2013, from http://www.pble.ac.uk/

USVN . (n.d.). Retrieved from http://www.usvn.info/

Vat, K. H. (2006). Integrating industrial practices in software development through scenario-based design of PBL activities: A pedagogical re-organization perspective. *Issues in Informing Science and Information Technology, 3*.

Vaughn, R. B. (2001). Teaching industrial practices in an undergraduate software engineering course. *Computer Science Education*, *11*(1), 21–32. doi:10.1076/csed.11.1.21.3844

Venkatagiri, S. (2006). Engineering the software requirements of nonprofits: A service-learning approach. In *Proceedings of the 28th IEEE International Conference on Software Engineering* (pp. 643–648). IEEE.

Venkatagiri, S. (2011). Teach project management, pack an agile punch. In *Proceedings of the 24th Software Engineering Education and Training Conference* (pp. 351–360). IEEE.

Versionone. (n.d.). *State of agile development survey results*. Retrieved from http://www.versionone.com/pdf/2012_State_of_Agile_Development_Survey_Results.pdf

Villa, E. Q., Kephart, K., Gates, A. Q., Thiry, H., & Hug, S. (2013). Affinity research groups in practice: Apprenticing students in research. *Journal of Engineering Education*, *102*(3), 444–466. doi:10.1002/jee.20016

Visser, L. (1998). *The development of motivational communication in distance education support*. (Unpublished doctoral dissertation). Educational Technology Department, The University of Twente, Twente, The Netherlands.

Visual-Paradigm. (n.d.). Retrieved from http://www.visual-paradigm.com/product/vpuml/

Voigt, C. (2008). *Educational design and media choice for collaborative, electronic case-based learning (E-Cbl)*. (PhD thesis). School of Computer and Information Science, Division of Information Technology, Engineering and the Environment, University of South Australia.

Voogt, J., Erstad, O., Dede, C., & Mishra, P. (2013). Challenges to learning and schooling in the digital networked world of the 21st century. *Journal of Computer Assisted Learning*, *29*, 403–413. doi:10.1111/jcal.12029

Voorhees, R. A. (2001). Competency-based learning models: A necessary future. *New Directions for Institutional Research*, (110): 5–13. doi:10.1002/ir.7

Vrasidas, C. (2000). Constructivism versus objectivism: Implications for interaction, course design, and evaluation in distance education. *International Journal of Educational Telecommunications*, *6*(4), 339–362.

Vygotsky, L. S. (1978). *Mind and society: The development of higher psychological processes*. Cambridge, MA: Harvard University Press.

Vygotsky, L. S. (1986). *Thought and language*. Cambridge, MA: The MIT Press.

Walther, J. B. (1997). Group and interpersonal effects in international computer-mediated collaboration. *Human Communication Research*, *23*(3), 342–369. doi:10.1111/j.1468-2958.1997.tb00400.x

Wang, A. I., & Stalhane, T. (2005). Using post mortem analysis to evaluate software architecture student projects. In *Proceedings of the 18th International Conference on Software Engineering Education & Training* (pp. 43–50). IEEE.

Wang, A. I., & Wu, B. (2011). Using game development to teach software architecture. *International Journal of Computer Games Technology*, 4.

Waraich, A. (2004). Using narrative as a motivating device to teach binary arithmetic and logic gates. In *Proceedings of the 9th Annual SIGCSE Conference on Innovation and Technology in Computer Science Education* (pp. 97–101). Leeds, UK: ACM.

Washbourn, P. (1996). Experiential learning: Is experience the best teacher? *Liberal Education*, *82*(3), 1–10.

Wasserman, A. I. (1996). Toward a discipline of software engineering. *IEEE Software*, *13*, 23–31. doi:10.1109/52.542291

Waterman, R. H., Waterman, J. A., & Collard, B. A. (1994). Toward a career resilient workforce. *Harvard Business Review*, *69*, 87–95.

Webpaproject. (n.d.). Retrieved from http://Webpaproject. lboro.ac.uk/

Wee, M., & Abrizah, A. (2011). An analysis of an assessment model for participation in online forums. *Computer Science and Information Systems, 8*(1), 121–140. doi:10.2298/CSIS100113036C

Weibelzahl, S., & Lahart, O. (2011a). Developing student induction sessions for problem-based learning. In *Proceedings of International Conference on Engaging Pedagogy.* Academic Press.

Weibelzahl, S., & Lahart, O. (2011b). PBL induction made easy: A web-based resource and a toolbox. In *Proceedings of FACILITATE Conference: Problem-Based Learning (PBL) Today and Tomorrow.* Dublin, Ireland: FACILITATE.

Weilkiens, T. (2008). *Systems engineering with SysML/UML: Modeling, analysis, design.* San Francisco: Morgan Kaufmann / The OMG Press.

Weinstein, C. E., & McCombs, B. L. (1998). *Strategic learning: The merging of skill, will and self-regulation.* Hillsdale, NJ: Lawrence Erlbaum Associates. doi:10.1037/10296-001

Welland, R. (2006). *Requirements engineering.* Simula Research Laboratory.

Wenger, E. (2013). *Communities of practice.* Retrieved August 9, 2013, from http://www.ewenger.com/theory/

Wenger, E. (1998). *Communities of practice: Learning, meaning, and identity.* New York: Cambridge University Press. doi:10.1017/CBO9780511803932

Wenger, E. (2000). Communities of practice and social learning systems. *Organization Science, 7*(2), 225–246. doi:10.1177/135050840072002

Wen, M. L., & Tsai, C. C. (2006). University students' perceptions of and attitudes toward (online) peer assessment. *Higher Education, 51*(1), 27–44. doi:10.1007/s10734-004-6375-8

Werner, L., Arcamone, D., & Ross, B. (2012). Using scrum in a quarter-length undergraduate software engineering course. *Journal of Computing Sciences in Colleges, 27*(4), 140–150.

Westfall, L. (2005). Software requirements engineering: What, why, who, when, and how. *Software Quality Professional, 7*(4), 17.

Whatsapp. (n.d.). Retrieved from http://www.whatsapp.com/

Wheeler, D. A. (2001). *More than a gigabuck: Estimating GNU/Linux's size.* Retrieved November 25, 2013, from http://www.dwheeler.com/sloc/redhat71-v1/redhat71sloc.html

Whitehead, J. (2007). Collaboration in software engineering: A roadmap. In *Proceedings of 2007 Future of Software Engineering Conference* (pp. 214–225). IEEE.

Whitehead, J. (2007). *Collaboration in software engineering: A roadmap.* Santa Cruz, CA: University of California.

Wilkins, D. E., & Lawhead, P. B. (2000). Evaluating individuals in team projects. *ACM SIGCSE Bulletin, 32*(1), 172–175. doi:10.1145/331795.331849

Willey, K., & Gardner, A. (2010). Investigating the capacity of self and peer assessment activities to engage students and promote learning. *European Journal of Engineering Education, 35*(4), 429–443. doi:10.1080/03043797.2010.490577

Williams, L., Brown, G., Meltzer, A., & Nagappan, N. (2011). Scrum + engineering practices: Experiences of three Microsoft teams. In *Proceedings of International Symposium on Empirical Software Engineering and Measurement* (pp. 463–471). IEEE Society.

Williams, L. (2010). Agile software development methodologies and practices. *Advances in Computers, 80,* 1–44. doi:10.1016/S0065-2458(10)80001-4

Wilson, K., & Fowler, J. (2005). Assessing the impact of learning environments on students' approaches to learning: Comparing conventional and action learning designs. *Assessment & Evaluation in Higher Education, 30*(1), 87–101. doi:10.1080/0260293042003251770

Wirth, M. A. (2003). E-notes: Using electronic lecture notes to support active learning in computer science. *ACM SIGCSE Bulletin, 35*(2), 57–60. doi:10.1145/782941.782981

Wohlin, C. (1997). Meeting the challenge of large-scale software development in an educational environment. In *Proceedings of the 10th Conference on Software Engineering Education and Training* (pp. 40–52). IEEE.

Wohlin, C., Runeson, P., Höst, M., Ohlsson, M. C., Regnell, B., & Wesslén, A. (2012). *Experimentation in software engineering*. Berlin: Springer. doi:10.1007/978-3-642-29044-2

Wood, D. (1999). Editorial: Representing, learning and understanding. *Computers & Education*, *33*, 83–90. doi:10.1016/S0360-1315(99)00026-3

Woods, D. R., Felder, R. M., Rugarcia, A., & Stice, J. E. (2000). The future of engineering education: III: Developing critical skills. *Chemical Engineering Education*, *34*(2), 1–20.

Wrycza, S., & Marcinkowski, B. (2005). UML 2 teaching at postgraduate studies–Prerequisites and practice. In *Proceedings of ISECON* (Vol. 22). ISECON.

Xie, K. (2013). What do the numbers say? The influence of motivation and peer feedback on students' behaviour in online discussions. *British Journal of Educational Technology*, *44*(2), 288–301. doi:10.1111/j.1467-8535.2012.01291.x

Xwiki. (n.d.). Retrieved from http://www.xwiki.org/

Ye, E., Liu, C., & Polack-Wahl, J. A. (2007). Enhancing software engineering education using teaching aids in 3-D online virtual worlds. In *Proceedings of the 37th Frontiers in Education Conference* (pp. T1E–8). IEEE.

Zaina, L. A. M., Bressan, G., Rodrigues, J. F., & Cardieri, M. A. C. A. (2011). Learning profile identification based on the analysis of the user's context of interaction. *IEEE Latin America Transactions*, *9*(5), 845–850. doi:10.1109/TLA.2011.6030999

Zave, P. (1997). Classification of research efforts in requirements engineering. *ACM Computing Surveys*, *29*(4), 315–321. doi:10.1145/267580.267581

Zdravkova, K., Ivanović, M., & Putnik, Z. (2009). Evolution of professional ethics courses from web supported learning towards e-learning 2.0. In *Learning in the synergy of multiple disciplines* (pp. 657–663). Berlin: Springer. doi:10.1007/978-3-642-04636-0_64

Zdravkova, K., Ivanović, M., & Putnik, Z. (2012). Experience of integrating web 2.0 technologies. *Educational Technology Research and Development*, *60*(2), 361–381. doi:10.1007/s11423-011-9228-z

Zhu, Q., Wang, T., & Tan, S. (2007). Adapting game technology to support software engineering process teaching: From SimSE to Mo–SEProcess. In *Proceedings* of the 3rd *International Conference on Natural Computation* (Vol. 5, pp. 777–780). IEEE.

Zowghi, D., & Coulin, C. (2005). Requirements elicitation: A survey of techniques, approaches, and tools. In *Engineering and managing software requirements* (pp. 19–46). Berlin: Springer. doi:10.1007/3-540-28244-0_2

Zuppiroli, S., Ciancarini, P., & Gabbrielli, M. (2012). A role-playing game for a software engineering lab: Developing a product line. In *Proceedings of the 25th IEEE Conference on Software Engineering Education and Training* (pp. 13–22). IEEE.

Zwieg, P., Kaiser, K. M., Beath, C. M., Gallagher, K. P., Goles, T., & Howland, J. (2006). The information technology workforce: Trends and implications 2005-2008. *MIS Quarterly Executive*, *5*(2), 101–108.

About the Contributors

Liguo Yu is an associate professor at Computer Science Department, Indiana University South Bend. He received his Ph.D. degree in computer science from Vanderbilt University in 2004. He received his M.S. degree from Institute of Metal Research, Chinese Academy of Science and his BS degree in physics from Jilin University. Before joining Indiana University South Bend, he was a visiting assistant professor at Tennessee Tech University. His research areas include software coupling, software maintenance and software evolution, empirical software engineering, open-source development, and software engineering education. He is also interested in social network analysis, knowledge management, and complex systems.

* * *

Zuhoor Abdullah Salim Al-Khanjari is an associate professor of Department of Computer Science, Sultan Qaboos University, Oman. She received her Ph.D. in software engineering from University of Liverpool, UK in 1999. She has chaired college e-learning committee. Currently, she is chairing the software engineering research and e-learning group in her department. She was the college assistant dean for postgraduate studies and research during 2003-2006. She participates in the evaluation committees for IT international awards, the editorial boards of IT international journals and conferences, and other professional bodies and societies. She has published 23 journal and 37 conference papers. Her research interests include software engineering, software testing, security, database, and e-learning.

Luís M. Alves has an MSc in computer science from the University of Porto in 2001. Currently he is an information systems and technologies Ph.D. student finalist at University of Minho and an assistant lecturer of Computing and Communications Department, Technology and Management School of the Polytechnic Institute of Bragança. He is also a research member at Algoritmi Research Center in the Software Engineering and Management Group integrated in the Information Systems and Technologies line. His research interests and publications are in the area of empirical software engineering, project management, software development, and software process improvement reference models.

Jocelyn Armarego is a senior lecturer in the School of Engineering and Information Technology at Murdoch University and has worked in the ICT industry for 10 years. She was the initiator of the design studio model deployed across all 3rd-4th year engineering degrees at Murdoch. She has experience in action research, studio and problem-based learning, structured workplace learning, and has had a long interest in the issues faced by students in 'learning to learn'. She has been involved in several national projects addressing ICT education. During 2011-2012 she was seconded to a role of project manager responsible for: development of a resource base for learning & teaching in ICT and engineering; development and facilitation of workshops for good practice based on these resources; and design and management of a Web portal for dissemination of information to the disciplines.

Verónica A. Bollati is a visiting professor at the Department of Computing Languages and Systems of the Rey Juan Carlos University sited in Madrid, Spain. She received her MSc degree in information systems from the National Technological University, Argentine in 2002, and her Ph.D. in computer science from the Rey Juan Carlos University in 2011. She is a member of Kybele Research Group where she leads the model-driven engineering research line. Her research interests include model-driven engineering, model transformations and languages, software engineering, tool development, services engineering, etc. She has co-authored several publications in national and international conferences and journals and has participated in several research projects.

Zoran Budimac holds the position of a full professor and is affiliated with the Department of Mathematics and Informatics, Faculty of Science, University of Novi Sad, Serbia. His current research interests include software engineering, mobile agent systems, and e-learning. He is a researcher and principal investigator in several international projects and author or co-author of more than 180 papers and 14 books.

Marcelo Campo received his computer engineer degree from Universidad Nacional del Centro de la Provincia de Buenos Aires (UNICEN), Tandil, Argentina in 1988, and his Ph.D degree in computer science from Instituto de Informática de la Universidad Federal de Rio Grande do Sul (UFRGS), Brazil in 1997. He is currently an associate professor at Computer Science Department and director of the ISISTAN Research Institute (CONICET - UNICEN). His research interests include intelligent aided software engineering, software architecture and frameworks, agent technology, and software visualization.

Oisín Cawley is a lecturer in computing at the Institute of Technology Tallaght, Dublin, Ireland. Prior to this, he was a lecturer in computing at the National College of Ireland. He has worked for 17 years in software development, predominantly for multinational companies, and has held many positions including global software development and application support manager. He holds a BSc in computer science from University College Dublin, Ireland, an MBA from Dublin City University, Ireland, and a Ph.D. in software engineering from the University of Limerick, Ireland. His research interests are in the area of software development processes and methodologies, particularly within regulated environments. He has specific interests in lean and agile software development methodologies within safety critical domains. His teaching interests lie in improving the learning process for third level computing students. In addition, he is interested in getting primary school children introduced to software development through fun approaches.

Paolo Ciancarini is a professor of computer science at the University of Bologna, where he lectures on software architectures and technologies in the Faculty of Sciences. His research interests include coordination languages and models, agent oriented software engineering, formal methods for software architectures, programming systems for high performance computing, and advanced Web technologies for groupware and digital documents. He has been involved in several projects funded by the European Commission. He is the director of CINI (Consorzio Interuniversitario Nazionale per l'Informatica), a consortium of 36 Italian universities for research in informatics and information technologies. He has been chair of the undergraduate school in computer science, and chair of the graduate school in information and communication technologies at the University of Bologna.

Cyrille Dejemeppe graduated from Louvain School of Engineering in University Louvain in 2012 where he began his Ph.D. research the same year. He is an active member of the constraint programming community and a member of the Constraint Programming Lab. His Ph.D. thesis subject aims at easing the management of patient appointment in medical centers. This research involves collaboration with industrial partners such as IBA (Ion Beam Applications). In addition to his Ph.D. research, he is involved in several teaching tasks and eager to help teaching and keep the courses up to date.

Yvonne Delaney, ACCA, MBA, and Ph.D. candidate (University of Limerick), holds a fellowship in APICS® the supply chain institute and an MBA from the University of Limerick, Ireland. She is also a qualified accountant and has over 15 years senior management experience in project management, material management, supply chain management, and managerial finance. With a proven track record in lecturing across a broad range of disciplines: academic institutions, multi-national manufacturing, including high-volume electronics, and the nutritional food business, her career spans a number of high profile companies. In 2002, she joined the University of Limerick and has been focusing on continuous professional development. She was short-listed for the University of Limericks 'Excellence in Teaching Award' in 2007. Her research area has focused on the development of frontline managers, and her teaching and learning stance is using problem based learning to aid this development.

Adrien Dessy received an M.S. degree in computer science engineering from the University Louvain, Belgium in 2010. In October 2011, he began his Ph.D. study under the supervision of Prof. Pierre Dupont at University Louvain. His primary research interests include machine learning, bioinformatics, and more specifically, the inference of gene regulatory networks from gene expression data.

Yves Deville graduated from the University Namur in 1983 where he also obtained his Ph.D. degree in 1987. He spent one year at Syracuse University (USA), where he got an M.S. degree in computer science. He received the IBM Belgium computer science award for his Ph.D. with his thesis on logic program construction. He is a full professor at the Louvain School of Engineering in the University Louvain where he leads the Constraint Programming Lab. He spent his sabbatical at Glasgow University (2003) and Brown University (2010). He has published more than one hundred papers. He has been elected to the executive committee of the Association for Constraint Programming (2011-2014), and serves in the editorial board of the *Journal of Constraint Programming*.

Jonas Eckhardt is a Ph.D. candidate at Technische Universität München, Faculty of Informatics. He received his master in computer science from University of Augsburg, Ludwig-Maximilians-Universität München and Technische Universität München. His work includes requirements engineering, model-based software engineering, traceability, and agile software engineering.

Henning Femmer is a Ph.D. candidate at Technische Universität München, Faculty of Informatics. He received his master in computer science from University of Augsburg, Ludwig-Maximilians-Universität München and Technische Universität München. His work includes requirements engineering quality, natural language processing, and agile software engineering.

Yujian Fu is an associate professor of Department of Electrical Engineering and Computer Science at the Alabama A&M University. She received her Ph.D. from Florida International University and joined Alabama A&M University in 2007. Her research areas include software verification, quality assurance, real-time embedded system design, cyber security, and formal methods, typically in the safety and mission critical systems. She has authored and coauthored several research and educational papers in refereed journals, conferences and workshops. She is a member of IEEE, ACM, ASEE, and NEA.

Ann Quiroz Gates is the chair of Computer Science Department at the University of Texas at El Paso and past associate vice president of Research and Sponsored Projects. Her research areas are software property elicitation and specification, and semantic-enabled technologies. She directs the NSF-funded Cyber-ShARE Center that focuses on the development of cyber-infrastructure to support interdisciplinary teams. She was a founding member of the NSF Advisory Committee for Cyberinfrastructure, and served on the Board of Governors of IEEE-Computer Society (2004-2009). She is a member of the Computer Science Accreditation Board (2011-2013). She leads the Computing Alliance for Hispanic-Serving Institutions, an NSF-funded consortium that is focused on the recruitment, retention, and advancement of Hispanics in computing and is a founding member of the National Center for Women in Information Technology, a national network to advance participation of women in IT.

Venkat N. Gudivada is a professor and interim chair of the Weisberg Division of Computer Science at Marshall University, Huntington, West Virginia. He received his Ph.D. in computer science from the University of Louisiana at Lafayette. His current research interests encompass big data analytics, verification and validation of SQL queries, HPC-driven applications, and personalized e-learning. He has published over 60 peer-reviewed technical papers. His industry experience includes tenure as a vice president of Wall Street companies in New York City for over 6 years. His previous academic appointments include work at the University of Michigan, University of Missouri, and Ohio University. He is a member of the IEEE Computer Society.

Hossein Hakimzadeh is an associate professor at Computer Science Department, Indiana University South Bend. He received his Ph.D. degree in computer science from North Dakota State University. His research interests include database management systems, operating systems, distributed computing and object oriented software engineering. Currently, he is the director of Informatics Program at IU South Bend.

Thomas B. Hilburn is a professor emeritus of software engineering at Embry-Riddle Aeronautical University and was a visiting scientist at the Software Engineering Institute, Carnegie-Mellon from 1997 to 2009. His research interests include software processes, object-oriented analysis and design, formal specification techniques, and curriculum development. He is an IEEE certified software developer, an IEEE software engineering certified instructor, and has chaired committees on the Professional Activities Board and the Educational Activities Board of the IEEE Computer Society.

Mirjana Ivanović is a full professor at the Department of Mathematics and Informatics, Faculty of Science, University of Novi Sad, Novi Sad, Serbia. She is author or co-author of 10 textbooks and more than 200 research papers on multi-agent systems, e-learning and Web-based learning, software engineering education, intelligent techniques (CBR, data, and Web mining), most of which are published in international journals and international conferences. Currently, she is the editor-in-chief of *Computer Science and Information Systems Journal*.

Lynette Johns-Boast is a lecturer in software engineering at the Australian National University (ANU) College of Engineering and Computer Science. Her research interests include curriculum theory, design and development; experiential and cooperative learning; personality and successful teams in software engineering; and engineering education. Prior to joining the ANU, she had 20 years' experience in the information technology industry in Australia and the United Kingdom, including establishing a successful small business providing bespoke software, training and consultancy services to the Australian Federal Government. She holds a bachelor of arts in modern European languages from the ANU and a graduate diploma in information systems from the University of Canberra. She is currently a Ph.D. candidate in higher education at the Centre for Higher Education, Learning & Teaching at the ANU. In 2012, Lynette received the Australian Council of Engineering Deans National Award for engineering education (high commendation).

Pankaj Kamthan has degrees in mathematics, mathematics education, and computer science. His teaching and research interests include agile methodologies, conceptual modeling, requirements engineering, software engineering education, and software quality. He has been teaching in academia and industry for the past two decades. He has taught over 20 courses in undergraduate and graduate programs, and is an avid user of information technology in his courses. He has been a technical editor, participated in standards development, and served on program committees of conferences and editorial boards of journals related to e-learning, information technology, and resource management. He has authored over 75 scholarly publications.

Živana Komlenov is a research and teaching assistant at the Department of Mathematics and Informatics, Faculty of Sciences, University of Novi Sad, Serbia. Her main research interests include learning technologies, application of intelligent techniques, software engineering, programming languages and tools. She has been involved as a researcher in 10 scientific and educational projects, mostly international. In addition, she participated in 31 conferences and authored or co-authored 19 research papers. Moreover, she is a certified instructional designer, Moodle server administrator at the Faculty of Sciences, official maintainer of Serbian language packs for Moodle, and conductor of various seminars in the domain of e-learning.

Marco Kuhrmann is a senior researcher at the Technische Universität München, Faculty of Informatics. He has a Ph.D. in computer sciences from the Technische Universität München. His research interests include software process improvement and management, software project management, and global software development.

Marc Lainez is an experienced agile coach, speaker and entrepreneur who is deeply involved in the Belgian startup community. He is one of the maintainers of the agile and lean community in Belgium (since 2011) and also a conference speaker and organizer. He has a university degree in computer science from the University Louvain in Belgium (2007). He started his career as a business analyst but quickly got back to software development, with a strong focus on agile project management and system thinking. He joined the Agilar team in 2011 and has been working on agile coaching and mentoring since then.

Dieter Landes holds a diploma in informatics from the University of Erlangen-Nuremberg, Germany, and a Ph.D. in economics from the University of Karlsruhe, Germany. After several years in industry (e.g. Daimler Research), he became a full professor of software engineering and database systems at the University of Applied Sciences, Coburg, Germany. His research interests include requirements engineering, software engineering education, and data mining. He (co)authored about 50 papers in books, journals, and conferences in these areas.

Marcos López-Sanz received his Ph.D. in computer science from the Rey Juan Carlos University with a work on service-oriented model-driven specification of software architectures. He also received the MSc in computer science from the same University. His research interests include software architecture, model-driven engineering, service-orientation, software engineering of distributed systems and grid computing, database design, etc. Currently, he belongs to the Kybele Research Group and works as a visiting teacher for the Rey Juan Carlos University. He received his BSc in ITS from the Polytechnic University of Madrid. He is the co-author of numerous journal articles and conference papers and has participated in many joint research projects.

Ricardo Machado is an associate professor at the Department of Information Systems, Minho University, where he teaches analysis and design of information systems, information systems project management, quality of software product and process, architecture and design of large-scale software systems. He is the director of Algoritmi Research Center, a research unit of the School of Engineering – University of Minho that develops R&D activity in information and communications technology and electronics. He is also the president of General Assembly of TICE.pt (Portuguese Competitiveness and Technology Centre for Information Technologies, Telecommunications and Electronics) and president of CT 128 (Portuguese Technical Commission for Software Engineering and Information Systems). His scientific interests focus on software engineering and management (modeling approaches for analysis and design, process and project management life-cycles).

Bonnie K. MacKellar is an associate professor of computer science at St John's University in Queens, NY. Her current research focuses on semantic search and social networking in the field of healthcare, as well as software engineering education. She received her Ph.D. in computer science from the University of Connecticut. She taught computer science at New Jersey Institute of Technology and Western Connecticut State University. In 1998, she left for industry but returned to academia in 2009 to work at St John's University, where she teaches both computer science program and healthcare informatics. She serves on the Northeastern Regional Board of the Consortium for Computing Sciences in Colleges, and is a member of the Non-Profit Free and Open Source Software Institute. She was also the St John's chapter coordinator in 2011 for the Humanitarian Free and Open Source Software Initiative.

Jean-Baptiste Mairy obtained his master degree from University Louvain in computer science and engineering in 2010. Since then, he has been pursuing a Ph.D. in computer sciences at University Louvain. He is working in the Constraint Programming Lab and has been a teaching assistant for various courses, which include introduction to programming, computability, artificial intelligence, and project in computer science. Those courses are given respectively to the first year bachelor students in computer science, the third year bachelor students in computer science and engineering, the first year master students in computer science and engineering and finally, the third year bachelor students in engineering and computer science.

Esperanza Marcos Martínez is a full professor at the Department of Computing Languages and Systems at the Rey Juan Carlos University sited in Madrid, Spain. She is the leader of the Kybele Research Group. Main research interests of Kybele include service engineering, software engineering, model driven engineering, information systems development, and philosophical foundations on information systems engineering. Marcos received a Ph.D. in computer sciences from Polytechnic University of Madrid (Spain) in 1997. She is the co-author of several books and she has published several book chapters and articles in journals and conferences. She has participated and managed several research projects.

Elizabeth Suescún Monsalve is enthusiastic about educational software and games. She is currently a Ph.D. candidate in computer science at the Department of Informatics, Pontifical Catholic University of Rio de Janeiro (Brazil) under the supervision of Prof. Julio Cesar Sampaio do Prado Leite. She got her master degree at the same department and under the supervision of the same professor. She is an informatics engineer graduated from PCJIC University Institution (Colombia). Her currently research areas include game-based learning, transparency software, pedagogy, and intentional modeling.

Jagadeesh Nandigam, Ph.D., is a professor of computer science in the School of Computing and Information Systems at Grand Valley State University (GVSU), Allendale, Michigan. He received his Ph.D. in computer science from the University of Louisiana at Lafayette in 1995. Since 1991, he has been teaching various undergraduate and graduate courses in object-oriented programming, software engineering, and principles of programming languages. Prior to joining GVSU, he worked in software industry for about three years where he was involved in the design and development of Web applications using Java, Microsoft.NET, and Web services technologies. His current academic and research interests include software engineering, software engineering education, open source software tools, mock-object frameworks, and concurrent/parallel programming in functional languages. He is a member of IEEE Computer Society.

Marcel Fouda Ndjodo is the head of the Department of Computer Science and Instructional Technology, the Higher Teachers' Training College of the University of Yaounde I (Cameroon). He received his Ph.D. in computer science from the University of Aix-Marseille II (France, 1992). His main scientific research fields include software product line engineering, business process modeling, and software engineering education. He has supervised many Ph.D. theses in information systems and software engineering.

Virginie Blanche Ngah is a secondary education computer science teacher. She obtained a master degree in computer science education at the Higher Teachers' training college of Yaoundé (Cameroon, 2009). She is currently a Ph.D. student. Her research focuses on the didactics of computer science and the training models of the secondary education computer science and ICT teachers.

Kasi Periyasamy is a professor and the program director of the Master of Software Engineering (MSE) program in the Department of Computer Science at the University of Wisconsin-La Crosse. He designed the curriculum for the MSE program and is teaching a majority of the courses in the program. Prior to joining UW-L, he was in the Department of Computer Science at the University of Manitoba, Winnipeg, Canada, where he was representing the university for a joint master program in software engineering by several universities and industries in western Canada. He has published numerous papers in software engineering. His primary research interests include formal specifications, object-oriented design, verification and testing, and software project management. He is the co-author of the book specification of software systems, the first edition was published in 1998 and the second edition in 2011, both by Springer-Verlag.

Zoran Putnik is a research assistant at the Faculty of Science, Department of Mathematics and Informatics, University of Novi Sad, Serbia. His research interests include e-learning and virtual learning environments, software engineering, and mobile agents. He participated in more than 17 international and national projects. He has published over 80 scientific articles in proceedings of international conferences and journals, and written several university textbooks in different fields of informatics.

Damith C. Rajapakse received his BSc in computer science and engineering from the University of Moratuwa, Sri Lanka (2001), and his Ph.D. from National University of Singapore (2006). He spent several years in the software industry as a practitioner before becoming a faculty member in the Department of Computer Science at the School of Computing of NUS. His research interests are mainly in the area of software engineering education. He has authored or co-authored several books and is the founder of the TEAMMATES project, a public domain tool for managing feedback paths in a class, used by many universities. He is also a fellow of the College of Alice & Peter Tan and a fellow of the Teaching Academy at NUS.

Lachana Ramingwong is an assistant professor at Department of Computer Engineering, Faculty of Engineering, Chiang Mai University, Thailand. She earned her Ph.D. from University of New England, Australia, in 2008. Software process analysis, software testing, and human-computer interaction are her main research interests.

Sakgasit Ramingwong received his Ph.D. from University of New England, Australia in 2009. He is currently an assistant professor at Department of Computer Engineering, Faculty of Engineering, Chiang Mai University, Thailand. His main research focuses on software project management, risk management, software process improvement, and general software engineering aspects.

Pedro Ribeiro is an assistant professor at the Department of Information Systems, Minho University, where he teaches software project management and software engineering. He is a member of IIBA (International Institute of Business Analysis) and SEMAG (Software Engineering and Management Group) and vowel of CT128 (Portuguese Technical Commission for Software Engineering and Information Systems). His scientific interests include project management, software requirements management, and software quality. He is an external consultant in software development organizations, for process improvement, software requirements, and project management.

Ita Richardson is a principal investigator with Lero – the Irish Software Engineering Research Centre, and a senior lecturer at the University of Limerick. Her research interests include software process improvement in medical, automotive, and financial services domains, within global software development environments and within small- to medium-sized enterprises. She has a Ph.D. in software engineering process quality from the University of Limerick. She has used problem-based learning in undergraduate and postgraduate classes for the past 5 years.

Guillermo Rodríguez received his computer engineer degree from Universidad Nacional del Centro de la Provincia de Buenos Aires (UNICEN), Tandil, Argentina, in 2010, and is currently pursuing his Ph.D. degree in computer science at the same university. Since 2008, he has been part of ISISTAN Research Institute (CONICET - UNICEN), where he has worked on projects related to software engineering, virtual reality, and games for education. His research interests focus on software architecture materialization. He has obtained a scholarship from CONICET to complete his doctoral studies.

Stefano Russo is a professor in computer engineering at the Federico II University of Naples, Italy, where he teaches software engineering and distributed systems, and leads the MOBILAB research group on distributed and mobile systems. From 2008 to 2013, he was chairman of the curriculum in computer engineering. From 2007 to 2012, he was department deputy head. From 2005 to 2013, he was director of the "C. Savy" National Laboratory of CINI (National Inter-Universities Consortium for Informatics) in Naples. He is an associate editor of the *IEEE Transactions on Services Computing*, and he served as the guest co-editor for *Performance Evaluation* and for *Journal of Software*. He authored about 140 scientific papers, and has been co-chair or program committee member of many IEEE conferences and workshops. His research has received public support by European Union, Italian Ministry, Regione Campania, and private sponsorships, such as Selex Sistemi Integrati, Ansaldo Breda, SESM, NEC Italia, and FIAT Elasis. Stefano obtained a degree in electronic engineering in 1988 and a Ph.D. in computer and systems engineering in 1993, both from the Federico II University of Naples, where he then was assistant professor from 1994 to 1998, and associate professor from 1998 to 2002.

Mihaela Sabin is an associate professor at University of New Hampshire Manchester. Her research interests are computing education, open content and open source software, public engagement models and assessment, and constraint satisfaction modeling and library software. She has a M.S. in computer science from the Politehnica University in Bucharest, Romania, and a M.S. for teachers in college teaching, and a Ph.D. in computer science from University of New Hampshire. Sabin is the university liaison for the New Hampshire Chapter of Computer Science Teacher Association, vice-chair of the Executive

Committee of the ACM SIG Information Technology Education, member of the Education Committee of the New Hampshire High Tech Council, coordinator of the Aspirations in Computing ME-NH-VT regional affiliate of the National Center for Women and Information Technology, and member of the Non-Profit Free and Open Source Software Institute.

Salamah Salamah is a clinical associate professor and the director of the Master's of Software Engineering Program at the University of Texas at El Paso. He also served for six years as an assistant professor and later as an associate professor in computer science and software engineering at Embry Riddle Aeronautical University. His research interests include formal methods in software engineering, software architecture, and software engineering education. He is a member of the Educational Activities Board of the IEEE Computer Society, and chairs the competitions committee of the board.

Ezequiel Scott received his computer engineer degree from Universidad Nacional del Centro de la Provincia de Buenos Aires (UNICEN), Tandil, Argentina, in 2012, and is currently pursuing his Ph.D. degree in computer science at the same university. Since 2012, he has been part of ISISTAN Research Institute (CONICET - UNICEN), where he has worked on projects related to software engineering, virtual reality and games for education. His research interests include virtual learning environments and agile methodologies. He has obtained a scholarship from CONICET to complete his doctoral studies.

Yvonne Sedelmaier studied pedagogy with a major focus on adult learning and continuing education at the University of Bamberg, Germany. After ten years of working in the educational sector and in quality management, she is now an academic researcher in the project experimental improvement of learning software engineering and investigating students and their learning processes. Her research interests are teaching and learning software engineering at universities and software engineering didactics.

Álvaro Soria received his computer engineer degree from Universidad Nacional del Centro de la Provincia de Buenos Aires (UNICEN), Tandil, Argentina, in 2001, and his Ph.D. degree in computer science at the same university in 2009. Since 2001, he has been part of ISISTAN Research Institute (CONICET - UNICEN). His research interests include software architectures, quality-driven design, object-oriented frameworks, fault localization, virtual learning environments, and agile methodologies.

David Surma is an associate professor at Computer Science Department, Indiana University South Bend. He received his Ph.D. degree in computer science and Engineering from University of Notre Dame in 1998. His research interest is computer network. Currently, he is the chair of Computer Science Department at IU South Bend.

Massood Towhidnejad is the director of NextGeneration ERAU Applied Research (NEAR) laboratory, and professor of software engineering in the Department of Electrical, Computer, Software, and Systems Engineering at Embry-Riddle Aeronautical University. His research interests include software engineering, software quality assurance and testing, autonomous systems, and air traffic management (NextGen). In addition to his university position, he has served as visiting research associate at the Federal Aviation Administration, Faculty Fellow at NASA Goddard Flight Research Center, and software quality assurance manager at Carrier Corporations. He also contributed to the software and system engineering profession by serving as an author for Graduate Software Engineering Reference Curriculum (GSwE2009), Graduate Reference Curriculum for Systems Engineering (GRCSE), Systems Engineering Body of Knowledge (SEBoK), and subject matter expert for IEEE Certified Software Development Associate (CSDA) training material. He is senior member of IEEE.

Allen B. Tucker is the Anne T. and Robert M. Bass Professor Emeritus in the Department of Computer Science at Bowdoin College. Prior to Bowdoin, he held similar positions at Colgate and Georgetown Universities. He earned a B.A. in mathematics from Wesleyan University and an M.S. and Ph.D. in computer science from Northwestern University. He is the author or coauthor of several books and articles in the areas of programming languages, natural language processing, and software engineering. He now teaches about and develops open source software for non-profit organizations. Professor Tucker co-authored the 1986 Liberal Arts Model Curriculum in Computer Science and co-chaired the 1991 ACM/IEEE-CS Joint Curriculum Task Force. He is an ACM Fellow, a Fulbright Lecturer (Ukraine), an Erskine Lecturer (New Zealand), and a member of the ACM, the IEEE Computer Society, the Humanitarian Free and Open Source Software (H-FOSS) Project, and the Non-Profit FOSS Institute.

Sascha Van Cauwelaert graduated from the University Louvain, Belgium in 2011. He is currently a Ph.D. student and a teaching assistant at the University of Louvain, including for a course promoting the agile development. His research focuses on constraint programming and machine learning, targeting scheduling problems solving and optimization. He also did some research in the computer music domain.

Juan M. Vara Mesa is an associate professor of Department of Computing Languages and Systems II at the University Rey Juan Carlos of Madrid where he is a member of the Kybele R&D group and he received his Ph.D. in 2009. His current research interests focus on model-driven engineering, human aspects of software engineering, and software quality. He has been a doctoral researcher at the University of Nantes and a post-doctoral researcher at the European Research Institute in Service Science at the University of Tilburg.

Elsa Q. Villa is a research assistant professor and co-director of the Center for Research in Engineering and Technology Education (CREaTE) at The University of Texas at El Paso (UTEP). She received her Ph.D. in curriculum and instruction from New Mexico State University, and her M.S. in computer science and M.A. in education from UTEP. She has led and co-led numerous grants from corporate foundations and state and federal agencies, and has numerous publications in refereed journals and edited books. Her research interests include communities of practice, gender, transformative learning, and identity.

Stephan Weibelzahl is a professor of business psychology at Private University of Applied Sciences Göttingen (PFH), Germany. He received a diploma in psychology and a Ph.D. from University of Trier, Germany. After heading a research group at the Fraunhofer Institute of Experimental Software Engineering (IESE), Kaiserslautern, Germany, and holding a lecturer position at National College of Ireland, he joined PFH in 2013. Parts of the work reported in this chapter is based on his teaching at NCI where he was in charge of coordinating PBL activities in the School of Computing and adopted the PBL methodology in several modules. With his background in psychology and computer science, he has long-standing research expertise in developing and evaluating adaptive e-learning systems. His research interests include user modelling, learning technologies, evaluation, and user experience.

Vera Maria Benjamim Werneck is an associate professor in Department of Informatics and Computer Science, State University of Rio de Janeiro. She received her Ph.D. in computer and systems engineering from the Federal University of Rio de Janeiro (1995), graduate at mathematics emphasis in informatics from Federal University of Rio de Janeiro (1982), and master at computer science from Federal University of Rio de Janeiro (1990). She has experience in computer science, focusing on software engineering teaching since 1988; she also has industry experience on software engineering, project management and consulting. She has been acting on the following subjects: software engineering, software quality, expert systems, medical systems, goal oriented and agent-oriented software engineering.

Allan Ximenes Pereira holds a degree in computer science from the State University of Rio de Janeiro (2010). He is currently enrolled in the master program of computer science at the University of the State of Rio de Janeiro - UERJ. He has experience in the area of computer science, with an emphasis on software engineering, working mainly in the following lines: requirements engineering, agile process, software development, and programming language.

Index